INTRODUCTION TO TIME SERIES ANALYSIS AND FORECASTING

INTRODUCTION TO TIME SERIES ANALYSIS AND FORECASTING

Second Edition

DOUGLAS C. MONTGOMERY
Arizona State University
Tempe, Arizona, USA

CHERYL L. JENNINGS
Arizona State University
Tempe, Arizona, USA

MURAT KULAHCI
Technical University of Denmark
Lyngby, Denmark
and
Luleå University of Technology
Luleå, Sweden

Published by John Wiley & Sons, Inc., Hoboken, New Jersey.
Published simultaneously in Canada.

No part of this publication may be reproduced, stored in a retrieval system, or transmitted in any form
or by any means, electronic, mechanical, photocopying, recording, scanning, or otherwise, except as
permitted under Section 107 or 108 of the 1976 United States Copyright Act, without either the prior
written permission of the Publisher, or authorization through payment of the appropriate per-copy fee
to the Copyright Clearance Center, Inc., 222 Rosewood Drive, Danvers, MA 01923, (978) 750-8400,
fax (978) 750-4470, or on the web at www.copyright.com. Requests to the Publisher for permission
should be addressed to the Permissions Department, John Wiley & Sons, Inc., 111 River Street,
Hoboken, NJ 07030, (201) 748-6011, fax (201) 748-6008, or online at
http://www.wiley.com/go/permission.

Limit of Liability/Disclaimer of Warranty: While the publisher and author have used their best efforts
in preparing this book, they make no representations or warranties with respect to the accuracy or
completeness of the contents of this book and specifically disclaim any implied warranties of
merchantability or fitness for a particular purpose. No warranty may be created or extended by sales
representatives or written sales materials. The advice and strategies contained herein may not be
suitable for your situation. You should consult with a professional where appropriate. Neither the
publisher nor author shall be liable for any loss of profit or any other commercial damages, including
but not limited to special, incidental, consequential, or other damages.

For general information on our other products and services or for technical support, please contact
our Customer Care Department within the United States at (800) 762-2974, outside the United States
at (317) 572-3993 or fax (317) 572-4002.

Wiley also publishes its books in a variety of electronic formats. Some content that appears in print
may not be available in electronic formats. For more information about Wiley products, visit our web
site at www.wiley.com.

Library of Congress Cataloging-in-Publication Data applied for.

Printed in the United States of America

10 9 8 7 6 5 4 3 2

CONTENTS

PREFACE

Analyzing time-oriented data and forecasting future values of a time series are among the most important problems that analysts face in many fields, ranging from finance and economics to managing production operations, to the analysis of political and social policy sessions, to investigating the impact of humans and the policy decisions that they make on the environment. Consequently, there is a large group of people in a variety of fields, including finance, economics, science, engineering, statistics, and public policy who need to understand some basic concepts of time series analysis and forecasting. Unfortunately, most basic statistics and operations management books give little if any attention to time-oriented data and little guidance on forecasting. There are some very good high level books on time series analysis. These books are mostly written for technical specialists who are taking a doctoral-level course or doing research in the field. They tend to be very theoretical and often focus on a few specific topics or techniques. We have written this book to fill the gap between these two extremes.

We have made a number of changes in this revision of the book. New material has been added on data preparation for forecasting, including dealing with outliers and missing values, use of the variogram and sections on the spectrum, and an introduction to Bayesian methods in forecasting. We have added many new exercises and examples, including new data sets in Appendix B, and edited many sections of the text to improve the clarity of the presentation.

Like the first edition, this book is intended for practitioners who make real-world forecasts. We have attempted to keep the mathematical level modest to encourage a variety of users for the book. Our focus is on short- to medium-term forecasting where statistical methods are useful. Since many organizations can improve their effectiveness and business results by making better short- to medium-term forecasts, this book should be useful to a wide variety of professionals. The book can also be used as a textbook for an applied forecasting and time series analysis course at the advanced undergraduate or first-year graduate level. Students in this course could come from engineering, business, statistics, operations research, mathematics, computer science, and any area of application where making forecasts is important. Readers need a background in basic statistics (previous exposure to linear regression would be helpful, but not essential), and some knowledge of matrix algebra, although matrices appear mostly in the chapter on regression, and if one is interested mainly in the results, the details involving matrix manipulation can be skipped. Integrals and derivatives appear in a few places in the book, but no detailed working knowledge of calculus is required.

Successful time series analysis and forecasting requires that the analyst interact with computer software. The techniques and algorithms are just not suitable to manual calculations. We have chosen to demonstrate the techniques presented using three packages: Minitab®, JMP®, and R, and occasionally SAS®. We have selected these packages because they are widely used in practice and because they have generally good capability for analyzing time series data and generating forecasts. Because R is increasingly popular in statistics courses, we have included a section in each chapter showing the R code necessary for working some of the examples in the chapter. We have also added a brief appendix on the use of R. The basic principles that underlie most of our presentation are not specific to any particular software package. Readers can use any software that they like or have available that has basic statistical forecasting capability. While the text examples do utilize these particular software packages and illustrate some of their features and capability, these features or similar ones are found in many other software packages.

There are three basic approaches to generating forecasts: regression-based methods, heuristic smoothing methods, and general time series models. Because all three of these basic approaches are useful, we give an introduction to all of them. Chapter 1 introduces the basic forecasting problem, defines terminology, and illustrates many of the common features of time series data. Chapter 2 contains many of the basic statistical tools used in analyzing time series data. Topics include plots, numerical summaries of time series data including the autocovariance and autocorrelation functions, transformations, differencing, and decomposing a time series into trend and seasonal components. We also introduce metrics for evaluating forecast errors and methods for evaluating and tracking forecasting performance over time. Chapter 3 discusses regression analysis and its use in forecasting. We discuss both crosssection and time series regression data, least squares and maximum likelihood model fitting, model adequacy checking, prediction intervals, and weighted and generalized least squares. The first part of this chapter covers many of the topics typically seen in an introductory treatment of regression, either in a stand-alone course or as part of another applied statistics course. It should be a reasonable review for many readers. Chapter 4 presents exponential smoothing techniques, both for time series with polynomial components and for seasonal data. We discuss and illustrate methods for selecting the smoothing constant(s), forecasting, and constructing prediction intervals. The explicit time series modeling approach to forecasting that we have chosen to emphasize is the autoregressive integrated moving average (ARIMA) model approach. Chapter 5 introduces ARIMA models and illustrates how to identify and fit these models for both nonseasonal and seasonal time series. Forecasting and prediction interval construction are also discussed and illustrated. Chapter 6 extends this discussion into transfer function models and intervention modeling and analysis. Chapter 7 surveys several other useful topics from time series analysis and forecasting, including multivariate time series problems, ARCH and GARCH models, and combinations of forecasts. We also give some practical advice for using statistical approaches to forecasting and provide some information about realistic expectations. The last two chapters of the book are somewhat higher in level than the first five.

Each chapter has a set of exercises. Some of these exercises involve analyzing the data sets given in Appendix B. These data sets represent an interesting cross section of real time series data, typical of those encountered in practical forecasting problems. Most of these data sets are used in exercises in two or more chapters, an indication that there are usually several approaches to analyzing, modeling, and forecasting a time series. There are other good sources of data for practicing the techniques given in this book. Some of the ones that we have found very interesting and useful include the U.S. Department of Labor—Bureau of Labor Statistics (http://www.bls.gov/data/home.htm), the U.S. Department of Agriculture—National Agricultural Statistics Service, Quick Stats Agricultural Statistics Data (http://www.nass.usda.gov/Data_and_Statistics/Quick_Stats/index.asp), the U.S. Census Bureau (http://www.census.gov), and the U.S.

Department of the Treasury (http://www.treas.gov/offices/domestic-finance/debt-management/interest-rate/). The time series data library created by Rob Hyndman at Monash University (http://www-personal.buseco.monash.edu.au/~hyndman/TSDL/index.htm) and the time series data library at the Mathematics Department of the University of York (http://www.york.ac.uk/depts/maths/data/ts/) also contain many excellent data sets. Some of these sources provide links to other data. Data sets and other materials related to this book can be found at ftp://ftp.wiley.com/public/scitechmed/ timeseries.

We would like to thank the many individuals who provided feedback and suggestions for improvement to the first edition. We found these suggestions most helpful. We are indebted to Clifford Long who generously provided the R codes he used with his students when he taught from the book. We found his codes very helpful in putting the end-of-chapter R code sections together. We also have placed a premium in the book on bridging the gap between theory and practice. We have not emphasized proofs or technical details and have tried to give intuitive explanations of the material whenever possible. The result is a book that can be used with a wide variety of audiences, with different interests and technical backgrounds, whose common interests are understanding how to analyze time-oriented data and constructing good short-term statistically based forecasts.

We express our appreciation to the individuals and organizations who have given their permission to use copyrighted material. These materials are noted in the text. Portions of the output contained in this book are printed with permission of Minitab Inc. All material remains the exclusive property and copyright of Minitab Inc. All rights reserved.

DOUGLAS C. MONTGOMERY
CHERYL L. JENNINGS
MURAT KULAHCI

CHAPTER 1

INTRODUCTION TO FORECASTING

It is difficult to make predictions, especially about the future
NEILS BOHR, *Danish physicist*

1.1 THE NATURE AND USES OF FORECASTS

A **forecast** is a prediction of some future event or events. As suggested by Neils Bohr, making good predictions is not always easy. Famously "bad" forecasts include the following from the book *Bad Predictions*:

- "The population is constant in size and will remain so right up to the end of mankind." *L'Encyclopedie*, 1756.
- "1930 will be a splendid employment year." U.S. Department of Labor, *New Year's Forecast* in 1929, just before the market crash on October 29.
- "Computers are multiplying at a rapid rate. By the turn of the century there will be 220,000 in the U.S." *Wall Street Journal*, 1966.

Introduction to Time Series Analysis and Forecasting, Second Edition.
Douglas C. Montgomery, Cheryl L. Jennings and Murat Kulahci.
© 2015 John Wiley & Sons, Inc. Published 2015 by John Wiley & Sons, Inc.

Forecasting is an important problem that spans many fields including business and industry, government, economics, environmental sciences, medicine, social science, politics, and finance. Forecasting problems are often classified as short-term, medium-term, and long-term. Short-term forecasting problems involve predicting events only a few time periods (days, weeks, and months) into the future. Medium-term forecasts extend from 1 to 2 years into the future, and long-term forecasting problems can extend beyond that by many years. Short- and medium-term forecasts are required for activities that range from operations management to budgeting and selecting new research and development projects. Long-term forecasts impact issues such as strategic planning. Short- and medium-term forecasting is typically based on identifying, modeling, and extrapolating the patterns found in historical data. Because these historical data usually exhibit inertia and do not change dramatically very quickly, statistical methods are very useful for short- and medium-term forecasting. This book is about the use of these statistical methods.

Most forecasting problems involve the use of time series data. A **time series** is a time-oriented or chronological sequence of observations on a variable of interest. For example, Figure 1.1 shows the market yield on US Treasury Securities at 10-year constant maturity from April 1953 through December 2006 (data in Appendix B, Table B.1). This graph is called a **time**

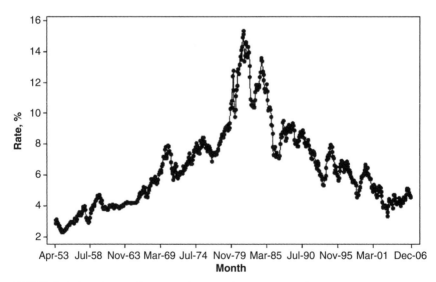

FIGURE 1.1   Time series plot of the market yield on US Treasury Securities at 10-year constant maturity. *Source:* US Treasury.

series plot. The rate variable is collected at equally spaced time periods, as is typical in most time series and forecasting applications. Many business applications of forecasting utilize daily, weekly, monthly, quarterly, or annual data, but any reporting interval may be used. Furthermore, the data may be instantaneous, such as the viscosity of a chemical product at the point in time where it is measured; it may be cumulative, such as the total sales of a product during the month; or it may be a statistic that in some way reflects the activity of the variable during the time period, such as the daily closing price of a specific stock on the New York Stock Exchange.

The reason that forecasting is so important is that prediction of future events is a critical input into many types of planning and decision-making processes, with application to areas such as the following:

1. *Operations Management*. Business organizations routinely use forecasts of product sales or demand for services in order to schedule production, control inventories, manage the supply chain, determine staffing requirements, and plan capacity. Forecasts may also be used to determine the mix of products or services to be offered and the locations at which products are to be produced.
2. *Marketing*. Forecasting is important in many marketing decisions. Forecasts of sales response to advertising expenditures, new promotions, or changes in pricing polices enable businesses to evaluate their effectiveness, determine whether goals are being met, and make adjustments.
3. *Finance and Risk Management*. Investors in financial assets are interested in forecasting the returns from their investments. These assets include but are not limited to stocks, bonds, and commodities; other investment decisions can be made relative to forecasts of interest rates, options, and currency exchange rates. Financial risk management requires forecasts of the volatility of asset returns so that the risks associated with investment portfolios can be evaluated and insured, and so that financial derivatives can be properly priced.
4. *Economics*. Governments, financial institutions, and policy organizations require forecasts of major economic variables, such as gross domestic product, population growth, unemployment, interest rates, inflation, job growth, production, and consumption. These forecasts are an integral part of the guidance behind monetary and fiscal policy, and budgeting plans and decisions made by governments. They are also instrumental in the strategic planning decisions made by business organizations and financial institutions.

5. *Industrial Process Control.* Forecasts of the future values of critical quality characteristics of a production process can help determine when important controllable variables in the process should be changed, or if the process should be shut down and overhauled. Feedback and feedforward control schemes are widely used in monitoring and adjustment of industrial processes, and predictions of the process output are an integral part of these schemes.

6. *Demography.* Forecasts of population by country and regions are made routinely, often stratified by variables such as gender, age, and race. Demographers also forecast births, deaths, and migration patterns of populations. Governments use these forecasts for planning policy and social service actions, such as spending on health care, retirement programs, and antipoverty programs. Many businesses use forecasts of populations by age groups to make strategic plans regarding developing new product lines or the types of services that will be offered.

These are only a few of the many different situations where forecasts are required in order to make good decisions. Despite the wide range of problem situations that require forecasts, there are only two broad types of forecasting techniques—qualitative methods and quantitative methods.

Qualitative forecasting techniques are often subjective in nature and require judgment on the part of experts. Qualitative forecasts are often used in situations where there is little or no historical data on which to base the forecast. An example would be the introduction of a new product, for which there is no relevant history. In this situation, the company might use the expert opinion of sales and marketing personnel to subjectively estimate product sales during the new product introduction phase of its life cycle. Sometimes qualitative forecasting methods make use of marketing tests, surveys of potential customers, and experience with the sales performance of other products (both their own and those of competitors). However, although some data analysis may be performed, the basis of the forecast is subjective judgment.

Perhaps the most formal and widely known qualitative forecasting technique is the **Delphi Method**. This technique was developed by the RAND Corporation (see Dalkey [1967]). It employs a panel of experts who are assumed to be knowledgeable about the problem. The panel members are physically separated to avoid their deliberations being impacted either by social pressures or by a single dominant individual. Each panel member responds to a questionnaire containing a series of questions and returns the information to a coordinator. Following the first questionnaire, subsequent questions are submitted to the panelists along with information about the opinions of the panel as a group. This allows panelists to review their predictions relative to the opinions of the entire group. After several rounds, it is hoped that the opinions of the panelists converge to a consensus, although achieving a consensus is not required and justified differences of opinion can be included in the outcome. Qualitative forecasting methods are not emphasized in this book.

Quantitative forecasting techniques make formal use of historical data and a **forecasting model**. The model formally summarizes patterns in the data and expresses a statistical relationship between previous and current values of the variable. Then the model is used to project the patterns in the data into the future. In other words, the forecasting model is used to extrapolate past and current behavior into the future. There are several types of forecasting models in general use. The three most widely used are regression models, smoothing models, and general time series models. Regression models make use of relationships between the variable of interest and one or more related predictor variables. Sometimes regression models are called **causal forecasting models,** because the predictor variables are assumed to describe the forces that cause or drive the observed values of the variable of interest. An example would be using data on house purchases as a predictor variable to forecast furniture sales. The method of least squares is the formal basis of most regression models. **Smoothing models** typically employ a simple function of previous observations to provide a forecast of the variable of interest. These methods may have a formal statistical basis, but they are often used and justified heuristically on the basis that they are easy to use and produce satisfactory results. General **time series models** employ the statistical properties of the historical data to specify a formal model and then estimate the unknown parameters of this model (usually) by least squares. In subsequent chapters, we will discuss all three types of quantitative forecasting models.

The form of the forecast can be important. We typically think of a forecast as a single number that represents our best estimate of the future value of the variable of interest. Statisticians would call this a **point estimate** or **point forecast.** Now these forecasts are almost always wrong; that is, we experience **forecast error**. Consequently, it is usually a good practice to accompany a forecast with an estimate of how large a forecast error might be experienced. One way to do this is to provide a **prediction interval** (PI) to accompany the point forecast. The PI is a range of values for the future observation, and it is likely to prove far more useful in decision-making than a single number. We will show how to obtain PIs for most of the forecasting methods discussed in the book.

Other important features of the forecasting problem are the **forecast horizon** and the **forecast interval.** The forecast horizon is the number of future periods for which forecasts must be produced. The horizon is often dictated by the nature of the problem. For example, in production planning, forecasts of product demand may be made on a monthly basis. Because of the time required to change or modify a production schedule, ensure that sufficient raw material and component parts are available from the supply chain, and plan the delivery of completed goods to customers or inventory facilities, it would be necessary to forecast up to 3 months ahead. The forecast horizon is also often called the forecast **lead time.** The **forecast interval** is the frequency with which new forecasts are prepared. For example, in production planning, we might forecast demand on a monthly basis, for up to 3 months in the future (the lead time or horizon), and prepare a new forecast each month. Thus the forecast interval is 1 month, the same as the basic period of time for which each forecast is made. If the forecast lead time is always the same length, say, T periods, and the forecast is revised each time period, then we are employing a **rolling** or **moving horizon** forecasting approach. This system updates or revises the forecasts for $T-1$ of the periods in the horizon and computes a forecast for the newest period T. This rolling horizon approach to forecasting is widely used when the lead time is several periods long.

1.2 SOME EXAMPLES OF TIME SERIES

Time series plots can reveal **patterns** such as random, trends, level shifts, periods or cycles, unusual observations, or a combination of patterns. Patterns commonly found in time series data are discussed next with examples of situations that drive the patterns.

The sales of a mature pharmaceutical product may remain relatively flat in the absence of unchanged marketing or manufacturing strategies. Weekly sales of a generic pharmaceutical product shown in Figure 1.2 appear to be constant over time, at about $10,400 \times 10^3$ units, in a random sequence with no obvious patterns (data in Appendix B, Table B.2).

To assure conformance with customer requirements and product specifications, the production of chemicals is monitored by many characteristics. These may be input variables such as temperature and flow rate, and output properties such as viscosity and purity.

Due to the continuous nature of chemical manufacturing processes, output properties often are **positively autocorrelated;** that is, a value above the long-run average tends to be followed by other values above the

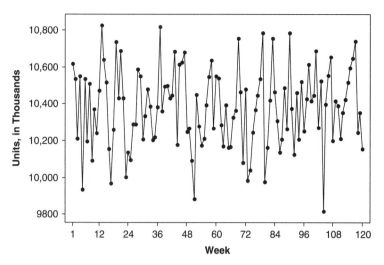

FIGURE 1.2   Pharmaceutical product sales.

average, while a value below the average tends to be followed by other values below the average.

The viscosity readings plotted in Figure 1.3 exhibit autocorrelated behavior, tending to a long-run average of about 85 centipoises (cP), but with a structured, not completely random, appearance (data in Appendix B, Table B.3). Some methods for describing and analyzing autocorrelated data will be described in Chapter 2.

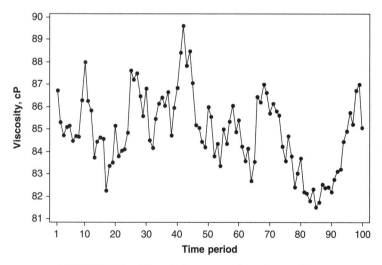

FIGURE 1.3   Chemical process viscosity readings.

The USDA National Agricultural Statistics Service publishes agricultural statistics for many commodities, including the annual production of dairy products such as butter, cheese, ice cream, milk, yogurt, and whey. These statistics are used for market analysis and intelligence, economic indicators, and identification of emerging issues.

Blue and gorgonzola cheese is one of 32 categories of cheese for which data are published. The annual US production of blue and gorgonzola cheeses (in 10^3 lb) is shown in Figure 1.4 (data in Appendix B, Table B.4). Production quadrupled from 1950 to 1997, and the **linear trend** has a constant positive slope with random, year-to-year variation.

The US Census Bureau publishes historic statistics on manufacturers' shipments, inventories, and orders. The statistics are based on North American Industry Classification System (NAICS) code and are utilized for purposes such as measuring productivity and analyzing relationships between employment and manufacturing output.

The manufacture of beverage and tobacco products is reported as part of the nondurable subsector. The plot of monthly beverage product shipments (Figure 1.5) reveals an overall increasing trend, with a distinct **cyclic pattern** that is repeated within each year. January shipments appear to be the lowest, with highs in May and June (data in Appendix B, Table B.5). This monthly, or **seasonal,** variation may be attributable to some cause

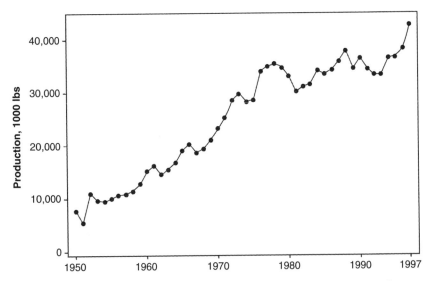

FIGURE 1.4   The US annual production of blue and gorgonzola cheeses. *Source:* USDA–NASS.

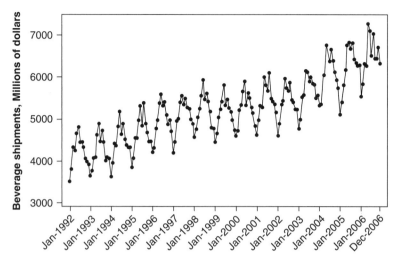

FIGURE 1.5   The US beverage manufacturer monthly product shipments, unadjusted. *Source:* US Census Bureau.

such as the impact of weather on the demand for beverages. Techniques for making seasonal adjustments to data in order to better understand general trends will be discussed in Chapter 2.

To determine whether the Earth is warming or cooling, scientists look at annual mean temperatures. At a single station, the warmest and the coolest temperatures in a day are averaged. Averages are then calculated at stations all over the Earth, over an entire year. The change in global annual mean surface air temperature is calculated from a base established from 1951 to 1980, and the result is reported as an "anomaly."

The plot of the annual mean anomaly in global surface air temperature (Figure 1.6) shows an increasing trend since 1880; however, the slope, or rate of change, varies with time periods (data in Appendix B, Table B.6). While the slope in earlier time periods appears to be constant, slightly increasing, or slightly decreasing, the slope from about 1975 to the present appears much steeper than the rest of the plot.

Business data such as stock prices and interest rates often exhibit **nonstationary** behavior; that is, the time series has no natural mean. The daily closing price adjusted for stock splits of Whole Foods Market (WFMI) stock in 2001 (Figure 1.7) exhibits a combination of patterns for both mean level and slope (data in Appendix B, Table B.7).

While the price is constant in some short time periods, there is no consistent mean level over time. In other time periods, the price changes

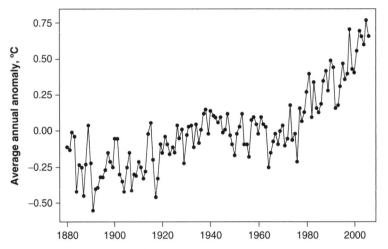

FIGURE 1.6 Global mean surface air temperature annual anomaly. *Source:* NASA-GISS.

at different rates, including occasional abrupt shifts in level. This is an example of nonstationary behavior, which will be discussed in Chapter 2.

The Current Population Survey (CPS) or "household survey" prepared by the US Department of Labor, Bureau of Labor Statistics, contains national data on employment, unemployment, earnings, and other labor market topics by demographic characteristics. The data are used to report

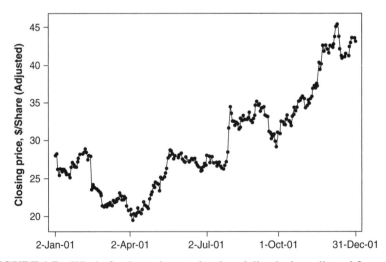

FIGURE 1.7 Whole foods market stock price, daily closing adjusted for splits.

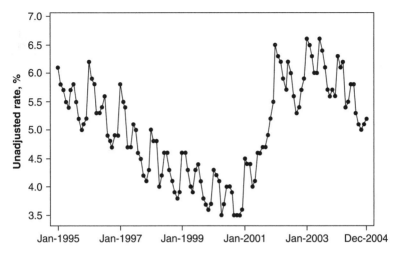

FIGURE 1.8   Monthly unemployment rate—full-time labor force, unadjusted.
Source: US Department of Labor-BLS.

on the employment situation, for projections with impact on hiring and training, and for a multitude of other business planning activities. The data are reported unadjusted and with seasonal adjustment to remove the effect of regular patterns that occur each year.

The plot of monthly unadjusted unemployment rates (Figure 1.8) exhibits a mixture of patterns, similar to Figure 1.5 (data in Appendix B, Table B.8). There is a distinct cyclic pattern within a year; January, February, and March generally have the highest unemployment rates. The overall level is also changing, from a gradual decrease, to a steep increase, followed by a gradual decrease. The use of seasonal adjustments as described in Chapter 2 makes it easier to observe the nonseasonal movements in time series data.

Solar activity has long been recognized as a significant source of noise impacting consumer and military communications, including satellites, cell phone towers, and electric power grids. The ability to accurately forecast solar activity is critical to a variety of fields. The International Sunspot Number R is the oldest solar activity index. The number incorporates both the number of observed sunspots and the number of observed sunspot groups. In Figure 1.9, the plot of annual sunspot numbers reveals cyclic patterns of varying magnitudes (data in Appendix B, Table B.9).

In addition to assisting in the identification of steady-state patterns, time series plots may also draw attention to the occurrence of **atypical events.** Weekly sales of a generic pharmaceutical product dropped due to limited

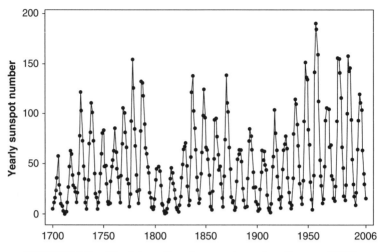

FIGURE 1.9   The international sunspot number. *Source:* SIDC.

availability resulting from a fire at one of the four production facilities. The 5-week reduction is apparent in the time series plot of weekly sales shown in Figure 1.10.

Another type of unusual event may be the failure of the data measurement or collection system. After recording a vastly different viscosity reading at time period 70 (Figure 1.11), the measurement system was

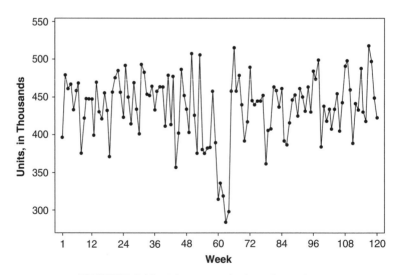

FIGURE 1.10   Pharmaceutical product sales.

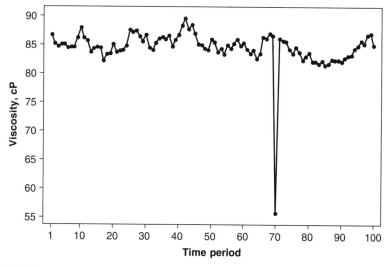

FIGURE 1.11   Chemical process viscosity readings, with sensor malfunction.

checked with a standard and determined to be out of calibration. The cause was determined to be a malfunctioning sensor.

1.3 THE FORECASTING PROCESS

A process is a series of connected activities that transform one or more inputs into one or more outputs. All work activities are performed in processes, and forecasting is no exception. The activities in the forecasting process are:

1. Problem definition
2. Data collection
3. Data analysis
4. Model selection and fitting
5. Model validation
6. Forecasting model deployment
7. Monitoring forecasting model performance

These activities are shown in Figure 1.12.

Problem definition involves developing understanding of how the forecast will be used along with the expectations of the "customer" (the user of

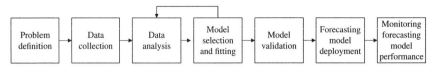

FIGURE 1.12   The forecasting process.

the forecast). Questions that must be addressed during this phase include the desired form of the forecast (e.g., are monthly forecasts required), the forecast horizon or lead time, how often the forecasts need to be revised (the forecast interval), and what level of forecast accuracy is required in order to make good business decisions. This is also an opportunity to intro-duce the decision makers to the use of prediction intervals as a measure of the risk associated with forecasts, if they are unfamiliar with this approach. Often it is necessary to go deeply into many aspects of the business system that requires the forecast to properly define the forecasting component of the entire problem. For example, in designing a forecasting system for inventory control, information may be required on issues such as product shelf life or other aging considerations, the time required to manufacture or otherwise obtain the products (production lead time), and the economic consequences of having too many or too few units of product available to meet customer demand. When multiple products are involved, the level of aggregation of the forecast (e.g., do we forecast individual products or families consisting of several similar products) can be an important consid-eration. Much of the ultimate success of the forecasting model in meeting the customer expectations is determined in the problem definition phase.

Data collection consists of obtaining the relevant history for the vari-able(s) that are to be forecast, including historical information on potential predictor variables.

The key here is "relevant"; often information collection and storage methods and systems change over time and not all historical data are useful for the current problem. Often it is necessary to deal with missing values of some variables, potential outliers, or other data-related problems that have occurred in the past. During this phase, it is also useful to begin planning how the data collection and storage issues in the future will be handled so that the reliability and integrity of the data will be preserved.

Data analysis is an important preliminary step to the selection of the forecasting model to be used. Time series plots of the data should be con-structed and visually inspected for recognizable patterns, such as trends and seasonal or other cyclical components. A trend is evolutionary move-ment, either upward or downward, in the value of the variable. Trends may be long-term or more dynamic and of relatively short duration. Seasonality is the component of time series behavior that repeats on a regular basis, such as each year. Sometimes we will smooth the data to make identification of the patterns more obvious (data smoothing will be discussed in Chapter 2). Numerical summaries of the data, such as the sample mean, standard deviation, percentiles, and autocorrelations, should also be computed and evaluated. Chapter 2 will provide the necessary background to do this. If potential predictor variables are available, scatter plots of each pair of variables should be examined. Unusual data points or potential **outliers** should be identified and flagged for possible further study. The purpose of this preliminary data analysis is to obtain some "feel" for the data, and a sense of how strong the underlying patterns such as trend and seasonality are. This information will usually suggest the initial types of quantitative forecasting methods and models to explore.

Model selection and fitting consists of choosing one or more forecasting models and fitting the model to the data. **By fitting**, we mean estimating the unknown model parameters, usually by the method of least squares. In subsequent chapters, we will present several types of time series models and discuss the procedures of model fitting. We will also discuss methods for evaluating the quality of the model fit, and determining if any of the underlying assumptions have been violated. This will be useful in discriminating between different candidate models.

Model validation consists of an evaluation of the forecasting model to determine how it is likely to perform in the intended application. This must go beyond just evaluating the "fit" of the model to the historical data and must examine what magnitude of forecast errors will be experienced when the model is used to forecast "fresh" or new data. The fitting errors will always be smaller than the forecast errors, and this is an important concept that we will emphasize in this book. A widely used method for validating a forecasting model before it is turned over to the customer is to employ some form of **data splitting,** where the data are divided into two segments—a fitting segment and a forecasting segment. The model is fit to only the fitting data segment, and then forecasts from that model are simulated for the observations in the forecasting segment. This can provide useful guidance on how the forecasting model will perform when exposed to new data and can be a valuable approach for discriminating between competing forecasting models.

Forecasting model deployment involves getting the model and the resulting forecasts in use by the customer. It is important to ensure that the customer understands how to use the model and that generating timely forecasts from the model becomes as routine as possible. Model maintenance, including making sure that data sources and other required information will continue to be available to the customer is also an important issue that impacts the timeliness and ultimate usefulness of forecasts.

Monitoring forecasting model performance should be an ongoing activity after the model has been deployed to ensure that it is still performing satisfactorily. It is the nature of forecasting that conditions change over time, and a model that performed well in the past may deteriorate in performance. Usually performance deterioration will result in larger or more systematic forecast errors. Therefore monitoring of forecast errors is an essential part of good forecasting system design. **Control charts** of forecast errors are a simple but effective way to routinely monitor the performance of a forecasting model. We will illustrate approaches to monitoring forecast errors in subsequent chapters.

1.4 DATA FOR FORECASTING

1.4.1 The Data Warehouse

Developing time series models and using them for forecasting requires data on the variables of interest to decision-makers. The data are the raw materials for the modeling and forecasting process. The terms **data** and **information** are often used interchangeably, but we prefer to use the term data as that seems to reflect a more raw or original form, whereas we think of information as something that is extracted or synthesized from data. The output of a forecasting system could be thought of as information, and that output uses data as an input.

In most modern organizations data regarding sales, transactions, company financial and business performance, supplier performance, and customer activity and relations are stored in a repository known as a **data warehouse**. Sometimes this is a single data storage system; but as the volume of data handled by modern organizations grows rapidly, the data warehouse has become an integrated system comprised of components that are physically and often geographically distributed, such as cloud data storage. The data warehouse must be able to organize, manipulate, and integrate data from multiple sources and different organizational information systems. The basic functionality required includes data extraction, data transformation, and data loading. Data extraction refers to obtaining data from internal sources and from external sources such as third party vendors or government entities and financial service organizations. Once the data are extracted, the transformation stage involves applying rules to prevent duplication of records and dealing with problems such as missing information. Sometimes we refer to the transformation activities as **data**

cleaning. We will discuss some of the important data cleaning operations subsequently. Finally, the data are loaded into the data warehouse where they are available for modeling and analysis.

Data quality has several dimensions. Five important ones that have been described in the literature are accuracy, timeliness, completeness, representativeness, and consistency. Accuracy is probably the oldest dimension of data quality and refers to how close that data conform to its "real" values. Real values are alternative sources that can be used for verification purposes. For example, do sales records match payments to accounts receivable records (although the financial records may occur in later time periods because of payment terms and conditions, discounts, etc.)? Timeliness means that the data are as current as possible. Infrequent updating of data can seriously impact developing a time series model that is going to be used for relatively short-term forecasting. In many time series model applications the time between the occurrence of the real-world event and its entry into the data warehouse must be as short as possible to facilitate model development and use. Completeness means that the data content is complete, with no missing data and no outliers. As an example of representativeness, suppose that the end use of the time series model is to forecast customer demand for a product or service, but the organization only records booked orders and the date of fulfillment. This may not accurately reflect demand, because the orders can be booked before the desired delivery period and the date of fulfillment can take place in a different period than the one required by the customer. Furthermore, orders that are lost because of product unavailability or unsatisfactory delivery performance are not recorded. In these situations demand can differ dramatically from sales. Data cleaning methods can often be used to deal with some problems of completeness. Consistency refers to how closely data records agree over time in format, content, meaning, and structure. In many organizations how data are collected and stored evolves over time; definitions change and even the types of data that are collected change. For example, consider monthly data. Some organizations define "months" that coincide with the traditional calendar definition. But because months have different numbers of days that can induce patterns in monthly data, some organizations prefer to define a year as consisting of 13 "months" each consisting of 4 weeks.

It has been suggested that the output data that reside in the data warehouse are similar to the output of a manufacturing process, where the raw data are the input. Just as in manufacturing and other service processes, the data production process can benefit by the application of quality management and control tools. Jones-Farmer et al. (2014) describe how statistical quality control methods, specifically control charts, can be used to enhance data quality in the data production process.

1.4.2 Data Cleaning

Data cleaning is the process of examining data to detect potential errors, missing data, outliers or unusual values, or other inconsistencies and then correcting the errors or problems that are found. Sometimes errors are the result of recording or transmission problems, and can be corrected by working with the original data source to correct the problem. Effective data cleaning can greatly improve the forecasting process.

Before data are used to develop a time series model, it should be subjected to several different kinds of checks, including but not necessarily limited to the following:

1. Is there missing data?
2. Does the data fall within an expected range?
3. Are there potential outliers or other unusual values?

These types of checks can be automated fairly easily. If this aspect of data cleaning is automated, the rules employed should be periodically evaluated to ensure that they are still appropriate and that changes in the data have not made some of the procedures less effective. However, it is also extremely useful to use graphical displays to assist in identifying unusual data. Techniques such as time series plots, histograms, and scatter diagrams are extremely useful. These and other graphical methods will be described in Chapter 2.

1.4.3 Imputation

Data **imputation** is the process of correcting missing data or replacing outliers with an estimation process. Imputation replaces missing or erroneous values with a "likely" value based on other available information. This enables the analysis to work with statistical techniques which are designed to handle the complete data sets.

Mean value imputation consists of replacing a missing value with the sample average calculated from the nonmissing observations. The big advantage of this method is that it is easy, and if the data does not have any specific trend or seasonal pattern, it leaves the sample mean of the complete data set unchanged. However, one must be careful if there are trends or seasonal patterns, because the sample mean of all of the data may not reflect these patterns. A variation of this is **stochastic mean value imputation**, in which a random variable is added to the mean value to capture some of the noise or variability in the data. The random variable could be assumed to follow a normal distribution with mean zero and standard deviation equal to the standard deviation of the actual observed data. A variation of mean value imputation is to use a subset of the available historical data that reflects any trend or seasonal patterns in the data. For example, consider the time series $y_1, y_2, \ldots, y_T$ and suppose that one observation y_j is missing. We can impute the missing value as

$$y_j^* = \frac{1}{2k} \left(\sum_{t=j-k}^{j-1} y_t + \sum_{t-j+1}^{j+k} y_t \right),$$

where k would be based on the seasonal variability in the data. It is usually chosen as some multiple of the smallest seasonal cycle in the data. So, if the data are monthly and exhibit a monthly cycle, k would be a multiple of 12. **Regression imputation** is a variation of mean value imputation where the imputed value is computed from a model used to predict the missing value. The prediction model does not have to be a linear regression model. For example, it could be a time series model.

Hot deck imputation is an old technique that is also known as the last value carried forward method. The term "hot deck" comes from the use of computer punch cards. The deck of cards was "hot" because it was currently in use. **Cold deck imputation** uses information from a deck of cards not currently in use. In hot deck imputation, the missing values are imputed by using values from similar complete observations. If there are several variables, sort the data by the variables that are most related to the missing observation and then, starting at the top, replace the missing values with the value of the immediately preceding variable. There are many variants of this procedure.

1.5 RESOURCES FOR FORECASTING

There are a variety of good resources that can be helpful to technical professionals involved in developing forecasting models and preparing forecasts. There are three professional journals devoted to forecasting:

- *Journal of Forecasting*
- *International Journal of Forecasting*
- *Journal of Business Forecasting Methods and Systems*

These journals publish a mixture of new methodology, studies devoted to the evaluation of current methods for forecasting, and case studies and applications. In addition to these specialized forecasting journals, there are several other mainstream statistics and operations research/management science journals that publish papers on forecasting, including:

- *Journal of Business and Economic Statistics*
- *Management Science*
- *Naval Research Logistics*
- *Operations Research*
- *International Journal of Production Research*
- *Journal of Applied Statistics*

This is by no means a comprehensive list. Research on forecasting tends to be published in a variety of outlets.

There are several books that are good complements to this one. We recommend Box, Jenkins, and Reinsel (1994); Chatfield (1996); Fuller (1995); Abraham and Ledolter (1983); Montgomery, Johnson, and Gardiner (1990); Wei (2006); and Brockwell and Davis (1991, 2002). Some of these books are more specialized than this one, in that they focus on a specific type of forecasting model such as the autoregressive integrated moving average [ARIMA] model, and some also require more background in statistics and mathematics.

Many statistics software packages have very good capability for fitting a variety of forecasting models. Minitab® Statistical Software, JMP®, the Statistical Analysis System (SAS) and R are the packages that we utilize and illustrate in this book. At the end of most chapters we provide R code for working some of the examples in the chapter. Matlab and S-Plus are also two packages that have excellent capability for solving forecasting problems.

EXERCISES

1.1 Why is forecasting an essential part of the operation of any organization or business?

1.2 What is a time series? Explain the meaning of trend effects, seasonal variations, and random error.

1.3 Explain the difference between a point forecast and an interval forecast.

1.4 What do we mean by a causal forecasting technique?

1.5 Everyone makes forecasts in their daily lives. Identify and discuss a situation where you employ forecasts.

a. What decisions are impacted by your forecasts?

b. How do you evaluate the quality of your forecasts?

c. What is the value to you of a good forecast?

d. What is the harm or penalty associated with a bad forecast?

1.6 What is meant by a rolling horizon forecast?

1.7 Explain the difference between forecast horizon and forecast interval.

1.8 Suppose that you are in charge of capacity planning for a large electric utility. A major part of your job is ensuring that the utility has sufficient generating capacity to meet current and future customer needs. If you do not have enough capacity, you run the risks of brownouts and service interruption. If you have too much capacity, it may cost more to generate electricity.

a. What forecasts do you need to do your job effectively?

b. Are these short-range or long-range forecasts?

c. What data do you need to be able to generate these forecasts?

1.9 Your company designs and manufactures apparel for the North American market. Clothing and apparel is a style good, with a relatively limited life. Items not sold at the end of the season are usually sold through off-season outlet and discount retailers. Items not sold through discounting and off-season merchants are often given to charity or sold abroad.

a. What forecasts do you need in this business to be successful?

b. Are these short-range or long-range forecasts?

c. What data do you need to be able to generate these forecasts?

d. What are the implications of forecast errors?

1.10 Suppose that you are in charge of production scheduling at a semiconductor manufacturing plant. The plant manufactures about 20 different types of devices, all on 8-inch silicon wafers. Demand for these products varies randomly. When a lot or batch of wafers is started into production, it can take from 4 to 6 weeks before the batch is finished, depending on the type of product. The routing of each batch of wafers through the production tools can be different depending on the type of product.

a. What forecasts do you need in this business to be successful?

b. Are these short-range or long-range forecasts?

c. What data do you need to be able to generate these forecasts?

d. Discuss the impact that forecast errors can potentially have on the efficiency with which your factory operates, including work-in-process inventory, meeting customer delivery schedules, and the cycle time to manufacture product.

1.11 You are the administrator of a large metropolitan hospital that operates the only 24-hour emergency room in the area. You must schedule attending physicians, resident physicians, nurses, laboratory, and support personnel to operate this facility effectively.

a. What measures of effectiveness do you think patients use to evaluate the services that you provide?

b. How are forecasts useful to you in planning services that will maximize these measures of effectiveness?

c. What planning horizon do you need to use? Does this lead to short-range or long-range forecasts?

1.12 Consider an airline that operates a network of flights that serves 200 cities in the continental United States. What long-range forecasts do the operators of the airline need to be successful? What forecasting problems does this business face on a daily basis? What are the consequences of forecast errors for the airline?

1.13 Discuss the potential difficulties of forecasting the daily closing price of a specific stock on the New York Stock Exchange. Would the problem be different (harder, easier) if you were asked to forecast the closing price of a group of stocks, all in the same industry (say, the pharmaceutical industry)?

1.14 Explain how large forecast errors can lead to high inventory levels at a retailer; at a manufacturing plant.

1.15 Your company manufactures and distributes soft drink beverages, sold in bottles and cans at retail outlets such as grocery stores, restaurants and other eating/drinking establishments, and vending machines in offices, schools, stores, and other outlets. Your product line includes about 25 different products, and many of these are produced in different package sizes.

a. What forecasts do you need in this business to be successful?

b. Is the demand for your product likely to be seasonal? Explain why or why not?

c. Does the shelf life of your product impact the forecasting problem?

d. What data do you think that you would need to be able to produce successful forecasts?

CHAPTER 2

STATISTICS BACKGROUND FOR FORECASTING

The future ain't what it used to be.

YOGI BERRA, *New York Yankees catcher*

2.1 INTRODUCTION

This chapter presents some basic statistical methods essential to modeling, analyzing, and forecasting time series data. Both graphical displays and numerical summaries of the properties of time series data are presented. We also discuss the use of data transformations and adjustments in forecasting and some widely used methods for characterizing and monitoring the performance of a forecasting model. Some aspects of how these performance measures can be used to select between competing forecasting techniques are also presented.

Forecasts are based on data or observations on the variable of interest. These data are usually in the form of a **time series**. Suppose that there are T periods of data available, with period T being the most recent. We will let the observation on this variable at time period t be denoted by $y_t, t = 1, 2, \ldots, T$. This variable can represent a cumulative quantity, such as the

Introduction to Time Series Analysis and Forecasting, Second Edition.
Douglas C. Montgomery, Cheryl L. Jennings and Murat Kulahci.
© 2015 John Wiley & Sons, Inc. Published 2015 by John Wiley & Sons, Inc.

total demand for a product during period t, or an instantaneous quantity, such as the daily closing price of a specific stock on the New York Stock Exchange.

Generally, we will need to distinguish between a **forecast** or **predicted value** of y_t that was made at some previous time period, say, $t - \tau$, and a **fitted value** of y_t that has resulted from estimating the parameters in a time series model to historical data. Note that τ is the forecast lead time. The forecast made at time period $t - \tau$ is denoted by $\hat{y}_t(t - \tau)$. There is a lot of interest in the **lead** − **1** forecast, which is the forecast of the observation in period t, y_t, made one period prior, $\hat{y}_t(t - 1)$. We will denote the fitted value of y_t by $\hat{y}_t$.

We will also be interested in analyzing **forecast errors**. The forecast error that results from a forecast of y_t that was made at time period $t - \tau$ is the **lead** − τ **forecast error**

$$e_t(\tau) = y_t - \hat{y}_t(t - \tau). \tag{2.1}$$

For example, the lead − 1 forecast error is

$$e_t(1) = y_t - \hat{y}_t(t - 1).$$

The difference between the observation y_t and the value obtained by fitting a time series model to the data, or a fitted value $\hat{y}_t$ defined earlier, is called a **residual**, and is denoted by

$$e_t = y_t - \hat{y}_t. \tag{2.2}$$

The reason for this careful distinction between forecast errors and residuals is that models usually fit historical data better than they forecast. That is, the residuals from a model-fitting process will almost always be smaller than the forecast errors that are experienced when that model is used to forecast future observations.

2.2 GRAPHICAL DISPLAYS

2.2.1 Time Series Plots

Developing a forecasting model should always begin with graphical display and analysis of the available data. Many of the broad general features of a time series can be seen visually. This is not to say that analytical tools are not useful, because they are, but the human eye can be a very sophisticated data analysis tool. To paraphrase the great New York Yankees catcher Yogi Berra, "You can observe a lot just by watching."

The basic graphical display for time series data is the **time series plot**, illustrated in Chapter 1. This is just a graph of y_t versus the time period, t, for $t = 1, 2, \dots, T$. Features such as trend and seasonality are usually easy to see from the time series plot. It is interesting to observe that some of the classical tools of descriptive statistics, such as the histogram and the stem-and-leaf display, are not particularly useful for time series data because they do not take time order into account.

Example 2.1   Figures 2.1 and 2.2 show time series plots for viscosity readings and beverage production shipments (originally shown in Figures 1.3 and 1.5, respectively). At the right-hand side of each time series plot is a histogram of the data. Note that while the two time series display very different characteristics, the histograms are remarkably similar. Essentially, the histogram summarizes the data across the time dimension, and in so doing, the key time-dependent features of the data are lost. Stem-and-leaf plots and boxplots would have the same issues, losing time-dependent features.

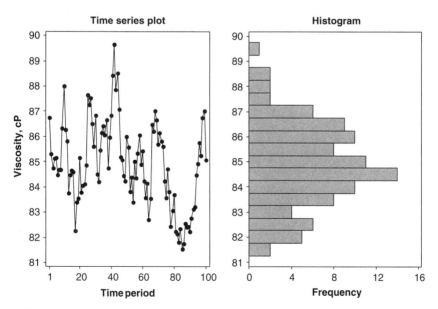

FIGURE 2.1   Time series plot and histogram of chemical process viscosity readings.

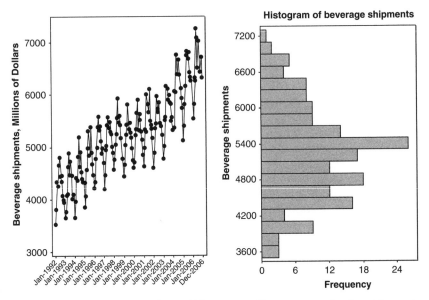

FIGURE 2.2   Time series plot and histogram of beverage production shipments.

When there are two or more variables of interest, **scatter plots** can be useful in displaying the relationship between the variables. For example, Figure 2.3 is a scatter plot of the annual global mean surface air temperature anomaly first shown in Figure 1.6 versus atmospheric CO_2 concentrations. The scatter plot clearly reveals a relationship between the two variables:

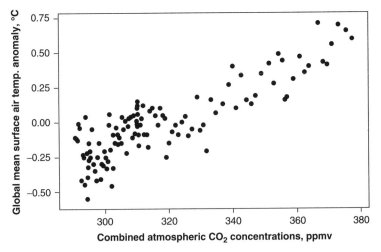

FIGURE 2.3   Scatter plot of temperature anomaly versus CO_2 concentrations. *Sources*: NASA–GISS (anomaly), DOE–DIAC (CO_2).

low concentrations of CO_2 are usually accompanied by negative anomalies, and higher concentrations of CO_2 tend to be accompanied by positive anomalies. Note that this does not imply that higher concentrations of CO_2 actually *cause* higher temperatures. The scatter plot cannot establish a causal relationship between two variables (neither can naive statistical modeling techniques, such as regression), but it is useful in displaying how the variables have varied together in the historical data set.

There are many variations of the time series plot and other graphical displays that can be constructed to show specific features of a time series. For example, Figure 2.4 displays daily price information for Whole Foods Market stock during the first quarter of 2001 (the trading days from January 2, 2001 through March 30, 2001). This chart, created in Excel®, shows the opening, closing, highest, and lowest prices experienced within a trading day for the first quarter. If the opening price was higher than the closing price, the box is filled, whereas if the closing price was higher than the opening price, the box is open. This type of plot is potentially more useful than a time series plot of just the closing (or opening) prices, because it shows the volatility of the stock within a trading day. The volatility of an asset is often of interest to investors because it is a measure of the inherent risk associated with the asset.

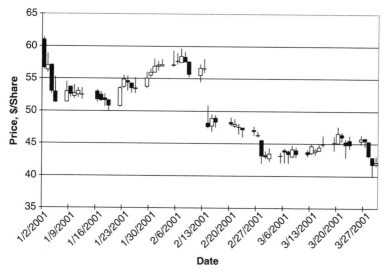

FIGURE 2.4   Open-high/close-low chart of Whole Foods Market stock price. *Source*: `finance.yahoo.com`.

2.2.2 Plotting Smoothed Data

Sometimes it is useful to overlay a **smoothed** version of the original data on the original time series plot to help reveal patterns in the original data. There are several types of data smoothers that can be employed. One of the simplest and most widely used is the ordinary or simple moving average. A simple **moving average** of span N assigns weights $1/N$ to the most recent N observations $y_T, y_{T-1}, \ldots, y_{T-N+1}$, and weight zero to all other observations. If we let M_T be the moving average, then the N-span moving average at time period T is

$$M_T = \frac{y_T + y_{T-1} + \cdots + y_{T-N+1}}{N} = \frac{1}{N} \sum_{t=T-N+1}^{T} y_t \qquad (2.3)$$

Clearly, as each new observation becomes available it is added into the sum from which the moving average is computed and the oldest observation is discarded. The moving average has less variability than the original observations; in fact, if the variance of an individual observation y_t is σ^2, then assuming that the observations are uncorrelated the variance of the moving average is

$$\mathrm{Var}(M_T) = \mathrm{Var}\left(\frac{1}{N}\sum_{t=T-N+1}^{N} y_t\right) = \frac{1}{N^2}\sum_{t=T-N+1}^{N} \mathrm{Var}(y_t) = \frac{\sigma^2}{N}$$

Sometimes a "centered" version of the moving average is used, such as in

$$M_t = \frac{1}{S+1}\sum_{i=-S}^{S} y_{t-i} \qquad (2.4)$$

where the span of the centered moving average is $N = 2S + 1$.

Example 2.2   Figure 2.5 plots the annual global mean surface air temperature anomaly data along with a five-period (a period is 1 year) moving average of the same data. Note that the moving average exhibits less variability than found in the original series. It also makes some features of the data easier to see; for example, it is now more obvious that the global air temperature decreased from about 1940 until about 1975.

Plots of moving averages are also used by analysts to evaluate stock price trends; common MA periods are 5, 10, 20, 50, 100, and 200 days. A time series plot of Whole Foods Market stock price with a 50-day moving

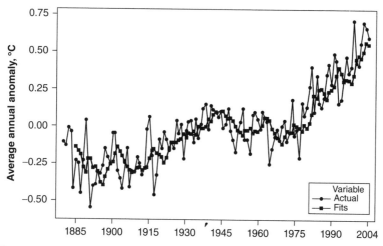

FIGURE 2.5   Time series plot of global mean surface air temperature anomaly, with five-period moving average. *Source*: NASA–GISS.

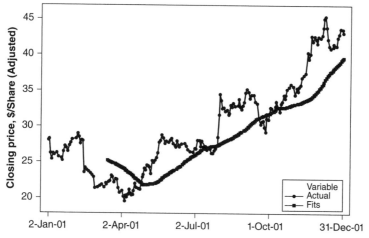

FIGURE 2.6   Time series plot of Whole Foods Market stock price, with 50-day moving average. *Source*: finance.yahoo.com.

average is shown in Figure 2.6. The moving average plot smoothes the day-to-day noise and shows a generally increasing trend.

The simple moving average is a **linear data smoother**, or a **linear filter**, because it replaces each observation y_t with a linear combination of the other data points that are near to it in time. The weights in the linear combination are equal, so the linear combination here is an average. Of course, unequal weights could be used. For example, the **Hanning filter** is a weighted, centered moving average

$$M_t^{\mathrm{H}} = 0.25y_{t+1} + 0.5y_t + 0.25y_{t-1}$$

Julius von Hann, a nineteenth century Austrian meteorologist, used this filter to smooth weather data.

An obvious disadvantage of a linear filter such as a moving average is that an unusual or erroneous data point or an outlier will dominate the moving averages that contain that observation, contaminating the moving averages for a length of time equal to the span of the filter. For example, consider the sequence of observations

$$15, 18, 13, 12, 16, 14, 16, 17, 18, 15, 18, 200, 19, 14, 21, 24, 19, 25$$

which increases reasonably steadily from 15 to 25, except for the unusual value 200. Any reasonable smoothed version of the data should also increase steadily from 15 to 25 and not emphasize the value 200. Now even if the value 200 is a legitimate observation, and not the result of a data recording or reporting error (perhaps it should be 20!), it is so unusual that it deserves special attention and should likely not be analyzed along with the rest of the data.

Odd-span **moving medians** (also called **running medians**) are an alternative to moving averages that are effective data smoothers when the time series may be contaminated with unusual values or outliers. The moving median of span N is defined as

$$m_t^{[N]} = med(y_{t-u}, \ldots, y_t, \ldots, y_{t+u}), \tag{2.5}$$

where $N = 2u + 1$. The median is the middle observation in rank order (or order of value). The moving median of span 3 is a very popular and effective data smoother, where

$$m_t^{[3]} = med(y_{t-1}, y_t, y_{t+1}).$$

This smoother would process the data three values at a time, and replace the three original observations by their median. If we apply this smoother to the data above, we obtain

$$\underline{\quad}, 15, 13, 13, 14, 16, 17, 17, 18, 18, 19, 19, 19, 21, 21, 24, \underline{\quad}.$$

This smoothed data are a reasonable representation of the original data, but they conveniently ignore the value 200. The end values are lost when using the moving median, and they are represented by "___".

In general, a moving median will pass monotone sequences of data unchanged. It will follow a step function in the data, but it will eliminate a spike or more persistent upset in the data that has duration of at most u consecutive observations. Moving medians can be applied more than once if desired to obtain an even smoother series of observations. For example, applying the moving median of span 3 to the smoothed data above results in

___, ___, 13, 13, 14, 16, 17, 17, 18, 18, 19, 19, 19, 21, 21, ___, ___.

These data are now as smooth as it can get; that is, repeated application of the moving median will not change the data, apart from the end values.

If there are a lot of observations, the information loss from the missing end values is not serious. However, if it is necessary or desirable to keep the lengths of the original and smoothed data sets the same, a simple way to do this is to "copy on" or add back the end values from the original data. This would result in the smoothed data:

15, 18, 13, 13, 14, 16, 17, 17, 18, 18, 19, 19, 19, 21, 21, 19, 25

There are also methods for smoothing the end values. Tukey (1979) is a basic reference on this subject and contains many other clever and useful techniques for data analysis.

Example 2.3   The chemical process viscosity readings shown in Figure 1.11 are an example of a time series that benefits from smoothing to evaluate patterns. The selection of a moving median over a moving average, as shown in Figure 2.7, minimizes the impact of the invalid measurements, such as the one at time period 70.

2.3 NUMERICAL DESCRIPTION OF TIME SERIES DATA

2.3.1 Stationary Time Series

A very important type of time series is a **stationary** time series. A time series is said to be **strictly stationary** if its properties are not affected

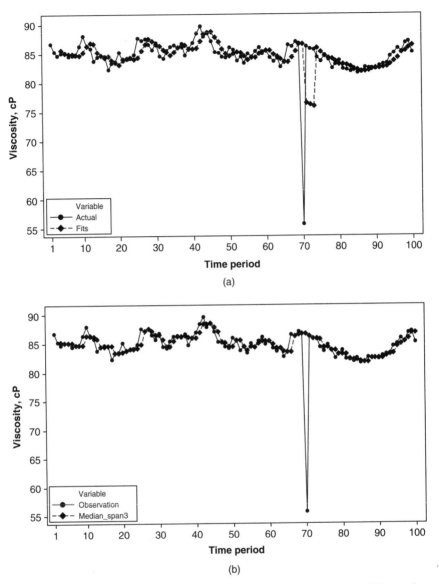

FIGURE 2.7 Viscosity readings with (a) moving average and (b) moving median.

by a change in the time origin. That is, if the joint probability distribution of the observations $y_t, y_{t+1}, \ldots, y_{t+n}$ is exactly the same as the joint probability distribution of the observations $y_{t+k}, y_{t+k+1}, \ldots, y_{t+k+n}$ then the time series is strictly stationary. When $n = 0$ the stationarity assumption means that the probability distribution of y_t is the same for all time periods

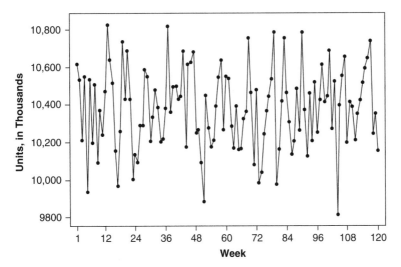

FIGURE 2.8   Pharmaceutical product sales.

and can be written as $f(y)$. The pharmaceutical product sales and chemical viscosity readings time series data originally shown in Figures 1.2 and 1.3, respectively, are examples of stationary time series. The time series plots are repeated in Figures 2.8 and 2.9 for convenience. Note that both time series seem to vary around a fixed level. Based on the earlier definition, this is a characteristic of stationary time series. On the other hand, the Whole

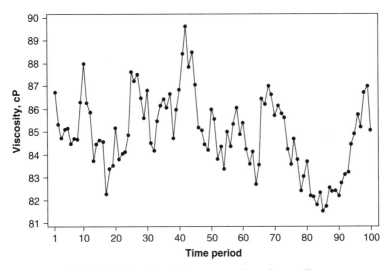

FIGURE 2.9   Chemical process viscosity readings.

Foods Market stock price data in Figure 1.7 tends to wander around or drift, with no obvious fixed level. This is behavior typical of a nonstationary time series.

Stationary implies a type of statistical **equilibrium** or **stability** in the data. Consequently, the time series has a constant mean defined in the usual way as

$$\mu_y = E(y) = \int_{-\infty}^{\infty} yf(y)dy \qquad (2.6)$$

and constant variance defined as

$$\sigma_y^2 = \text{Var}(y) = \int_{-\infty}^{\infty} (y - \mu_y)^2 f(y)dy. \qquad (2.7)$$

The sample mean and sample variance are used to estimate these parameters. If the observations in the time series are $y_1, y_2, \ldots, y_T$, then the sample mean is

$$\bar{y} = \hat{\mu}_y = \frac{1}{T} \sum_{t=1}^{T} y_t \qquad (2.8)$$

and the sample variance is

$$s^2 = \hat{\sigma}_y^2 = \frac{1}{T} \sum_{t=1}^{T} (y_t - \bar{y})^2. \qquad (2.9)$$

Note that the divisor in Eq. (2.9) is T rather than the more familiar $T - 1$. This is the common convention in many time series applications, and because T is usually not small, there will be little difference between using T instead of $T - 1$.

2.3.2 Autocovariance and Autocorrelation Functions

If a time series is stationary this means that the joint probability distribution of any two observations, say, y_t and y_{t+k}, is the same for any two time periods t and $t + k$ that are separated by the same interval k. Useful information about this joint distribution, and hence about the nature of the time series, can be obtained by plotting a scatter diagram of all of the data pairs y_t, y_{t+k} that are separated by the same interval k. The interval k is called the **lag**.

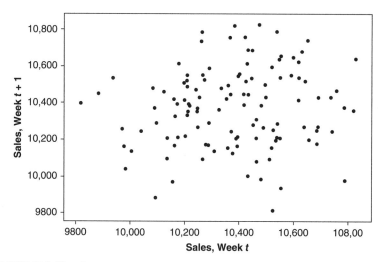

FIGURE 2.10   Scatter diagram of pharmaceutical product sales at lag $k = 1$.

Example 2.4   Figure 2.10 is a scatter diagram for the pharmaceutical product sales for lag $k = 1$ and Figure 2.11 is a scatter diagram for the chemical viscosity readings for lag $k = 1$. Both scatter diagrams were constructed by plotting y_{t+1} versus y_t. Figure 2.10 exhibits little structure; the plotted pairs of adjacent observations y_t, y_{t+1} seem to be **uncorrelated**. That is, the value of y in the current period does not provide any useful information about the value of y that will be observed in the next period. A different story is revealed in Figure 2.11, where we observe that the

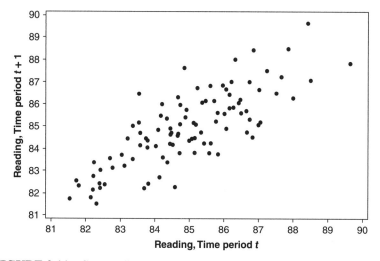

FIGURE 2.11   Scatter diagram of chemical viscosity readings at lag $k = 1$.

pairs of adjacent observations y_{t+1}, y_t are **positively correlated**. That is, a small value of y tends to be followed in the next time period by another small value of y, and a large value of y tends to be followed immediately by another large value of y. Note from inspection of Figures 2.10 and 2.11 that the behavior inferred from inspection of the scatter diagrams is reflected in the observed time series.

The covariance between y_t and its value at another time period, say, y_{t+k} is called the **autocovariance** at lag k, defined by

$$\gamma_k = \mathrm{Cov}(y_t, y_{t+k}) = E[(y_t - \mu)(y_{t+k} - \mu)]. \qquad (2.10)$$

The collection of the values of $\gamma_k, k = 0, 1, 2, \ldots$ is called the **autocovariance function**. Note that the autocovariance at lag $k = 0$ is just the variance of the time series; that is, $\gamma_0 = \sigma_y^2$, which is constant for a stationary time series. The **autocorrelation coefficient** at lag k for a stationary time series is

$$\rho_k = \frac{E[(y_t - \mu)(y_{t+k} - \mu)]}{\sqrt{E[(y_t - \mu)^2]E[(y_{t+k} - \mu)^2]}} = \frac{\mathrm{Cov}(y_t, y_{t+k})}{\mathrm{Var}(y_t)} = \frac{\gamma_k}{\gamma_0}. \qquad (2.11)$$

The collection of the values of ρ_k, $k = 0, 1, 2, \ldots$ is called the **autocorrelation function (ACF)**. Note that by definition $\rho_0 = 1$. Also, the ACF is independent of the scale of measurement of the time series, so it is a dimensionless quantity. Furthermore, $\rho_k = \rho_{-k}$; that is, the ACF is **symmetric** around zero, so it is only necessary to compute the positive (or negative) half.

If a time series has a finite mean and autocovariance function it is said to be second-order stationary (or weakly stationary of order 2). If, in addition, the joint probability distribution of the observations at all times is multivariate normal, then that would be sufficient to result in a time series that is strictly stationary.

It is necessary to estimate the autocovariance and ACFs from a time series of finite length, say, $y_1, y_2, \ldots, y_T$. The usual estimate of the autocovariance function is

$$c_k = \hat{\gamma}_k = \frac{1}{T} \sum_{t=1}^{T-k} (y_t - \bar{y})(y_{t+k} - \bar{y}), \quad k = 0, 1, 2, \ldots, K \qquad (2.12)$$

and the ACF is estimated by the **sample autocorrelation function** (or **sample ACF**)

$$r_k = \hat{\rho}_k = \frac{c_k}{c_0}, \quad k = 0, 1, \ldots, K \qquad (2.13)$$

A good general rule of thumb is that at least 50 observations are required to give a reliable estimate of the ACF, and the individual sample autocorrelations should be calculated up to lag K, where K is about $T/4$.

Often we will need to determine if the autocorrelation coefficient at a particular lag is zero. This can be done by comparing the sample autocorrelation coefficient at lag k, r_k, to its standard error. If we make the assumption that the observations are uncorrelated, that is, $\rho_k = 0$ for all k, then the variance of the sample autocorrelation coefficient is

$$\text{Var}(r_k) \cong \frac{1}{T} \tag{2.14}$$

and the standard error is

$$se(r_k) \cong \frac{1}{\sqrt{T}} \tag{2.15}$$

Example 2.5   Consider the chemical process viscosity readings plotted in Figure 2.9; the values are listed in Table 2.1.

The sample ACF at lag $k = 1$ is calculated as

$$c_0 = \frac{1}{100} \sum_{t=1}^{100-0} (y_t - \bar{y})(y_{t+0} - \bar{y})$$

$$= \frac{1}{100}[(86.7418 - 84.9153)(86.7418 - 84.9153) + \cdots$$
$$+ (85.0572 - 84.9153)(85.0572 - 84.9153)]$$
$$= 280.9332$$

$$c_1 = \frac{1}{100} \sum_{t=1}^{100-1} (y_t - \bar{y})(y_{t+1} - \bar{y})$$

$$= \frac{1}{100}[(86.7418 - 84.9153)(85.3195 - 84.9153) + \cdots$$
$$+ (87.0048 - 84.9153)(85.0572 - 84.9153)]$$
$$= 220.3137$$

$$r_1 = \frac{c_1}{c_0} = \frac{220.3137}{280.9332} = 0.7842$$

A plot and listing of the sample ACFs generated by Minitab for the first 25 lags are displayed in Figures 2.12 and 2.13, respectively.

TABLE 2.1   Chemical Process Viscosity Readings

Time Period	Reading	Time Period	Reading	Time Period	Reading	Time Period	Reading
1	86.7418	26	87.2397	51	85.5722	76	84.7052
2	85.3195	27	87.5219	52	83.7935	77	83.8168
3	84.7355	28	86.4992	53	84.3706	78	82.4171
4	85.1113	29	85.6050	54	83.3762	79	83.0420
5	85.1487	30	86.8293	55	84.9975	80	83.6993
6	84.4775	31	84.5004	56	84.3495	81	82.2033
7	84.6827	32	84.1844	57	85.3395	82	82.1413
8	84.6757	33	85.4563	58	86.0503	83	81.7961
9	86.3169	34	86.1511	59	84.8839	84	82.3241
10	88.0006	35	86.4142	60	85.4176	85	81.5316
11	86.2597	36	86.0498	61	84.2309	86	81.7280
12	85.8286	37	86.6642	62	83.5761	87	82.5375
13	83.7500	38	84.7289	63	84.1343	88	82.3877
14	84.4628	39	85.9523	64	82.6974	89	82.4159
15	84.6476	40	86.8473	65	83.5454	90	82.2102
16	84.5751	41	88.4250	66	86.4714	91	82.7673
17	82.2473	42	89.6481	67	86.2143	92	83.1234
18	83.3774	43	87.8566	68	87.0215	93	83.2203
19	83.5385	44	88.4997	69	86.6504	94	84.4510
20	85.1620	45	87.0622	70	85.7082	95	84.9145
21	83.7881	46	85.1973	71	86.1504	96	85.7609
22	84.0421	47	85.0767	72	85.8032	97	85.2302
23	84.1023	48	84.4362	73	85.6197	98	86.7312
24	84.8495	49	84.2112	74	84.2339	99	87.0048
25	87.6416	50	85.9952	75	83.5737	100	85.0572

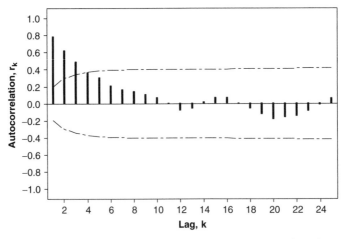

FIGURE 2.12   Sample autocorrelation function for chemical viscosity readings, with 5% significance limits.

Autocorrelation function: reading

Lag	ACF	T	LBQ
1	0.784221	7.84	63.36
2	0.628050	4.21	104.42
3	0.491587	2.83	129.83
4	0.362880	1.94	143.82
5	0.304554	1.57	153.78
6	0.208979	1.05	158.52
7	0.164320	0.82	161.48
8	0.144789	0.72	163.80
9	0.103625	0.51	165.01
10	0.066559	0.33	165.51
11	0.003949	0.02	165.51
12	−0.077226	−0.38	166.20
13	−0.051953	−0.25	166.52
14	0.020525	0.10	166.57
15	0.072784	0.36	167.21
16	0.070753	0.35	167.81
17	0.001334	0.01	167.81
18	−0.057435	−0.28	168.22
19	−0.123122	−0.60	170.13
20	−0.180546	−0.88	174.29
21	−0.162466	−0.78	177.70
22	−0.145979	−0.70	180.48
23	−0.087420	−0.42	181.50
24	−0.011579	−0.06	181.51
25	0.063170	0.30	182.06

FIGURE 2.13 Listing of sample autocorrelation functions for first 25 lags of chemical viscosity readings, Minitab session window output (the definition of T and LBQ will be given later).

Note the rate of decrease or decay in ACF values in Figure 2.12 from 0.78 to 0, followed by a sinusoidal pattern about 0. This ACF pattern is typical of stationary time series. The importance of ACF estimates exceeding the 5% significance limits will be discussed in Chapter 5. In contrast, the plot of sample ACFs for a time series of random values with constant mean has a much different appearance. The sample ACFs for pharmaceutical product sales plotted in Figure 2.14 appear randomly positive or negative, with values near zero.

While the ACF is strictly speaking defined only for a stationary time series, the sample ACF can be computed for *any* time series, so a logical question is: What does the sample ACF of a nonstationary time series look like? Consider the daily closing price for Whole Foods Market stock in Figure 1.7. The sample ACF of this time series is shown in Figure 2.15. Note that this sample ACF plot behaves quite differently than the ACF plots in Figures 2.12 and 2.14. Instead of cutting off or tailing off near zero after a few lags, this sample ACF is very **persistent**; that is, it decays very slowly and exhibits sample autocorrelations that are still rather large even at long lags. This behavior is characteristic of a nonstationary time series. Generally, if the sample ACF does not dampen out within about 15 to 20 lags, the time series is nonstationary.

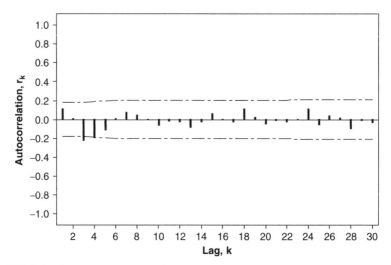

FIGURE 2.14   Autocorrelation function for pharmaceutical product sales, with 5% significance limits.

2.3.3 The Variogram

We have discussed two techniques for determining if a time series is nonstationary, plotting a reasonable long series of the data to see if it drifts or wanders away from its mean for long periods of time, and computing the sample ACF. However, often in practice there is no clear demarcation

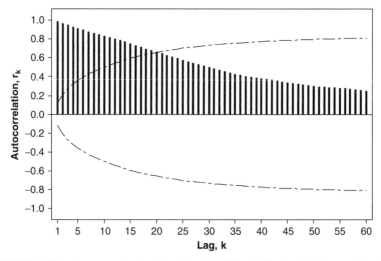

FIGURE 2.15   Autocorrelation function for Whole Foods Market stock price, with 5% significance limits.

between a stationary and a nonstationary process for many real-world time series. An additional diagnostic tool that is very useful is the **variogram**.

Suppose that the time series observations are represented by y_t. The variogram G_k measures variances of the differences between observations that are k lags apart, relative to the variance of the differences that are one time unit apart (or at lag 1). The variogram is defined mathematically as

$$G_k = \frac{\text{Var}(y_{t+k} - y_t)}{\text{Var}(y_{t+1} - y_t)} \quad k = 1, 2, \dots \tag{2.16}$$

and the values of G_k are plotted as a function of the lag k. If the time series is stationary, it turns out that

$$G_k = \frac{1 - \rho_k}{1 - \rho_1},$$

but for a stationary time series $\rho_k \to 0$ as k increases, so when the variogram is plotted against lag k, G_k will reach an asymptote $1/(1 - \rho_1)$. However, if the time series is nonstationary, G_k will increase monotonically.

Estimating the variogram is accomplished by simply applying the usual sample variance to the differences, taking care to account for the changing sample sizes when the differences are taken (see Haslett (1997)). Let

$$d_t^k = y_{t+k} - y_t$$
$$\bar{d}^k = \frac{1}{T-k} \sum d_t^k.$$

Then an estimate of $\text{Var}(y_{t+k} - y_t)$ is

$$s_k^2 = \frac{\sum_{t=1}^{T-k} (d_t^k - \bar{d}^k)^2}{T - k - 1}.$$

Therefore the sample variogram is given by

$$\hat{G}_k = \frac{s_k^2}{s_1^2} \quad k = 1, 2, \dots \tag{2.17}$$

To illustrate the use of the variogram, consider the chemical process viscosity data plotted in Figure 2.9. Both the data plot and the sample ACF in

Lag	Variogram	Plot Variogram
1	1.0000	
2	1.7238	
3	2.3562	
4	2.9527	
5	3.2230	
6	3.6659	
7	3.8729	
8	3.9634	
9	4.1541	
10	4.3259	
11	4.6161	
12	4.9923	
13	4.8752	
14	4.5393	
15	4.2971	
16	4.3065	
17	4.6282	
18	4.9006	
19	5.2050	
20	5.4711	
21	5.3873	
22	5.3109	
23	5.0395	
24	4.6880	
25	4.3416	

FIGURE 2.16   JMP output for the sample variogram of the chemical process viscosity data from Figure 2.19.

Figures 2.12 and 2.13 suggest that the time series is stationary. Figure 2.16 is the variogram. Many software packages do not offer the variogram as a standard pull-down menu selection, but the JMP package does. Without software, it is still fairly easy to compute.

Start by computing the successive differences of the time series for a number of lags and then find their sample variances. The ratios of these sample variances to the sample variance of the first differences will produce the sample variogram. The JMP calculations of the sample variogram are shown in Figure 2.16 and a plot is given in Figure 2.17. Notice that the sample variogram generally converges to a stable level and then fluctuates around it. This is consistent with a stationary time series, and it provides additional evidence that the chemical process viscosity data are stationary.

Now let us see what the sample variogram looks like for a nonstationary time series. The Whole Foods Market stock price data from Appendix Table B.7 originally shown in Figure 1.7 are apparently nonstationary, as it wanders about with no obvious fixed level. The sample ACF in Figure 2.15 decays very slowly and as noted previously, gives the impression that the time series is nonstationary. The calculations for the variogram from JMP are shown in Figure 2.18 and the variogram is plotted in Figure 2.19.

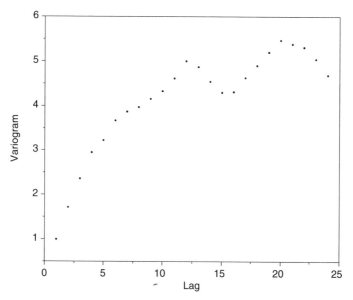

FIGURE 2.17   JMP sample variogram of the chemical process viscosity data from Figure 2.9.

Lag	Variogram	Plot Variogram
1	1.0000	
2	2.0994	
3	3.2106	
4	4.3960	
5	5.4982	
6	6.5810	
7	7.5690	
8	8.5332	
9	9.4704	
10	10.4419	
11	11.4154	
12	12.3452	
13	13.3759	
14	14.4411	
15	15.6184	
16	16.9601	
17	18.2442	
18	19.3782	
19	20.3934	
20	21.3618	
21	22.4010	
22	23.4788	
23	24.5450	
24	25.5906	
25	26.6620	

FIGURE 2.18   JMP output for the sample variogram of the Whole Foods Market stock price data from Figure 1.7 and Appendix Table B.7.

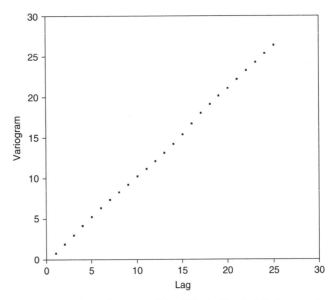

FIGURE 2.19 Sample variogram of the Whole Foods Market stock price data from Figure 1.7 and Appendix Table B.7.

Notice that the sample variogram in Figure 2.19 increases monotonically for all 25 lags. This is a strong indication that the time series is nonstationary.

2.4 USE OF DATA TRANSFORMATIONS AND ADJUSTMENTS

2.4.1 Transformations

Data transformations are useful in many aspects of statistical work, often for stabilizing the variance of the data. Nonconstant variance is quite common in time series data. For example, the International Sunspot Numbers plotted in Figure 2.20a show cyclic patterns of varying magnitudes. The variability from about 1800 to 1830 is smaller than that from about 1830 to 1880; other small periods of constant, but different, variances can also be identified.

A very popular type of data transformation to deal with nonconstant variance is the **power family** of transformations, given by

$$y^{(\lambda)} = \begin{cases} \dfrac{y^{\lambda} - 1}{\lambda \dot{y}^{\lambda-1}}, & \lambda \neq 0 \\ \dot{y} \ln y, & \lambda = 0 \end{cases}, \tag{2.18}$$

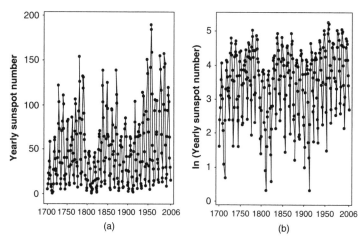

FIGURE 2.20   Yearly International Sunspot Number, (a) untransformed and (b) natural logarithm transformation. *Source*: SIDC.

where $\dot{y} = \exp[(1/T)\sum_{t=1}^{T} \ln y_t]$ is the geometric mean of the observations. If $\lambda = 1$, there is no transformation. Typical values of λ used with time series data are $\lambda = 0.5$ (a square root transformation), $\lambda = 0$ (the log transformation), $\lambda = -0.5$ (reciprocal square root transformation), and $\lambda = -1$ (inverse transformation). The divisor $\dot{y}^{\lambda-1}$ is simply a scale factor that ensures that when different models are fit to investigate the utility of different transformations (values of λ), the residual sum of squares for these models can be meaningfully compared. The reason that $\lambda = 0$ implies a log transformation is that $(y^{\lambda} - 1)/\lambda$ approaches the log of y as λ approaches zero. Often an appropriate value of λ is chosen empirically by fitting a model to $y^{(\lambda)}$ for various values of λ and then selecting the transformation that produces the minimum residual sum of squares.

The log transformation is used frequently in situations where the variability in the original time series increases with the average level of the series. When the standard deviation of the original series increases linearly with the mean, the log transformation is in fact an optimal variance-stabilizing transformation. The log transformation also has a very nice physical interpretation as percentage change. To illustrate this, let the time series be $y_1, y_2, \ldots, y_T$ and suppose that we are interested in the percentage change in y_t, say,

$$x_t = \frac{100(y_t - y_{t-1})}{y_{t-1}},$$

The approximate percentage change in y_t can be calculated from the differences of the log-transformed time series $x_t \cong 100[\ln(y_t) - \ln(y_{t-1})]$ because

$$100[\ln(y_t) - \ln(y_{t-1})] = 100 \ln \left(\frac{y_t}{y_{t-1}} \right) = 100 \ln \left(\frac{y_{t-1} + (y_t - y_{t-1})}{y_{t-1}} \right)$$

$$= 100 \ln \left(1 + \frac{x_t}{100} \right) \cong x_t$$

since $\ln(1 + z) \cong z$ when z is small.

The application of a natural logarithm transformation to the International Sunspot Number, as shown in Figure 2.20b, tends to stabilize the variance and leaves just a few unusual values.

2.4.2 Trend and Seasonal Adjustments

In addition to transformations, there are also several types of adjustments that are useful in time series modeling and forecasting. Two of the most widely used are **trend adjustments** and **seasonal adjustments**. Sometimes these procedures are called trend and seasonal decomposition.

A time series that exhibits a trend is a **nonstationary** time series. Modeling and forecasting of such a time series is greatly simplified if we can eliminate the trend. One way to do this is to fit a **regression model** describing the trend component to the data and then subtracting it out of the original observations, leaving a set of residuals that are free of trend. The trend models that are usually considered are the linear trend, in which the mean of y_t is expected to change linearly with time as in

$$E(y_t) = \beta_0 + \beta_1 t \tag{2.19}$$

or as a quadratic function of time

$$E(y_t) = \beta_0 + \beta_1 t + \beta_2 t^2 \tag{2.20}$$

or even possibly as an exponential function of time such as

$$E(y_t) = \beta_0 e^{\beta_1 t}. \tag{2.21}$$

The models in Eqs. (2.19)–(2.21) are usually fit to the data by using ordinary least squares.

Example 2.6   We will show how least squares can be used to fit regression models in Chapter 3. However, it would be useful at this point to illustrate how trend adjustment works. Minitab can be used to perform trend adjustment. Consider the annual US production of blue and gorgonzola cheeses

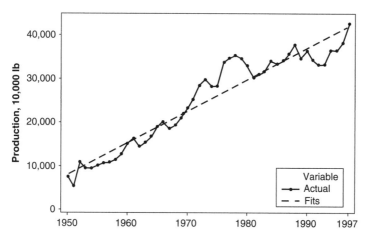

FIGURE 2.21   Blue and gorgonzola cheese production, with fitted regression line. *Source*: USDA–NASS.

shown in Figure 1.4. There is clearly a positive, nearly linear trend. The trend analysis plot in Figure 2.21 shows the original time series with the fitted line.

Plots of the residuals from this model indicate that, in addition to an underlying trend, there is additional structure. The normal probability plot (Figure 2.22a) and histogram (Figure 2.22c) indicate the residuals are

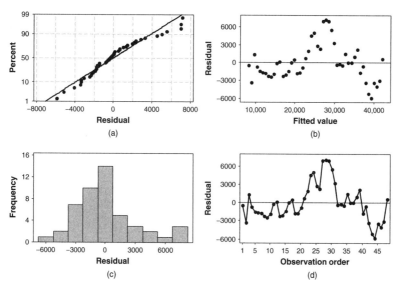

FIGURE 2.22   Residual plots for simple linear regression model of blue and gorgonzola cheese production.

approximately normally distributed. However, the plots of residuals versus fitted values (Figure 2.22b) and versus observation order (Figure 2.22d) indicate nonconstant variance in the last half of the time series. Analysis of model residuals is discussed more fully in Chapter 3.

Another approach to removing trend is by **differencing** the data; that is, applying the difference operator to the original time series to obtain a new time series, say,

$$x_t = y_t - y_{t-1} = \nabla y_t, \tag{2.22}$$

where ∇ is the (backward) difference operator. Another way to write the differencing operation is in terms of a **backshift operator** B, defined as $By_t = y_{t-1}$, so

$$x_t = (1 - B)y_t = \nabla y_t = y_t - y_{t-1} \tag{2.23}$$

with $\nabla = (1 - B)$. Differencing can be performed successively if necessary until the trend is removed; for example, the second difference is

$$x_t = \nabla^2 y_t = \nabla(\nabla y_t) = (1 - B)^2 y_t = (1 - 2B + B^2) = y_t - 2y_{t-1} + y_{t-2} \tag{2.24}$$

In general, powers of the backshift operator and the backward difference operator are defined as

$$B^d y_t = y_{t-d}$$
$$\nabla^d = (1 - B)^d \tag{2.25}$$

Differencing has two advantages relative to fitting a trend model to the data. First, it does not require estimation of any parameters, so it is a more **parsimonious** (i.e., simpler) approach; and second, model fitting assumes that the trend is fixed throughout the time series history and will remain so in the (at least immediate) future. In other words, the trend component, once estimated, is assumed to be **deterministic**. Differencing can allow the trend component to change through time. The first difference accounts for a trend that impacts the change in the mean of the time series, the second difference accounts for changes in the slope of the time series, and so forth. Usually, one or two differences are all that is required in practice to remove an underlying trend in the data.

Example 2.7   Reconsider the blue and gorgonzola cheese production data. A difference of one applied to this time series removes the increasing trend (Figure 2.23) and also improves the appearance of the residuals plotted versus fitted value and observation order when a linear model is fitted to the detrended time series (Figure 2.24). This illustrates that differencing may be a very good alternative to detrending a time series by using a regression model.

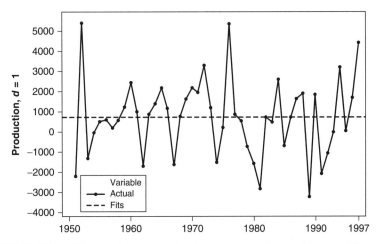

FIGURE 2.23   Blue and gorgonzola cheese production, with one difference. *Source*: USDA–NASS.

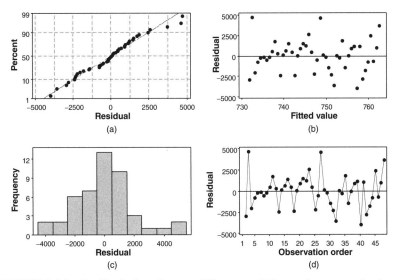

FIGURE 2.24   Residual plots for one difference of blue and gorgonzola cheese production.

Seasonal, or both trend *and* seasonal, components are present in many time series. Differencing can also be used to eliminate seasonality. Define a lag—*d* **seasonal difference** operator as

$$\nabla_d y_t = (1 - B^d) = y_t - y_{t-d}. \tag{2.26}$$

For example, if we had monthly data with an annual season (a very common situation), we would likely use $d = 12$, so the seasonally differenced data would be

$$\nabla_{12} y_t = (1 - B^{12}) y_t = y_t - y_{t-12}.$$

When both trend *and* seasonal components are simultaneously present, we can sequentially difference to eliminate these effects. That is, first seasonally difference to remove the seasonal component and then difference one or more times using the regular difference operator to remove the trend.

Example 2.8   The beverage shipment data shown in Figure 2.2 appear to have a strong monthly pattern—January consistently has the lowest shipments in a year while the peak shipments are in May and June. There is also an overall increasing trend from year to year that appears to be the same regardless of month.

A seasonal difference of twelve followed by a trend difference of one was applied to the beverage shipments, and the results are shown in Figure 2.25. The seasonal differencing removes the monthly pattern (Figure 2.25a), and the second difference of one removes the overall increasing trend (Figure 2.25b). The fitted linear trend line in Figure 2.25b has a slope of virtually zero. Examination of the residual plots in Figure 2.26 does not reveal any problems with the linear trend model fit to the differenced data.

Regression models can also be used to eliminate seasonal (or trend and seasonal components) from time series data. A simple but useful model is

$$E(y_t) = \beta_0 + \beta_1 \sin \frac{2\pi}{d} t + \beta_2 \cos \frac{2\pi}{d} t, \tag{2.27}$$

where *d* is the period (or length) of the season and $2\pi/d$ is expressed in radians. For example, if we had monthly data and an annual season, then $d = 12$. This model describes a simple, symmetric seasonal pattern that

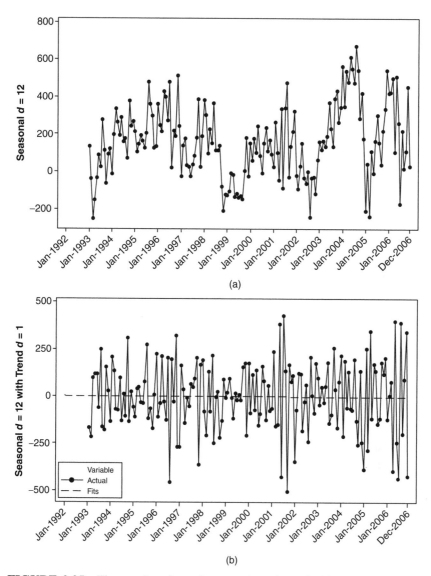

FIGURE 2.25 Time series plots of seasonal- and trend-differenced beverage data.

repeats every 12 periods. The model is actually a sine wave. To see this, recall that a sine wave with amplitude β, phase angle or origin θ, and period or cycle length ω can be written as

$$E(y_t) = \beta \sin \omega(t + \theta). \tag{2.28}$$

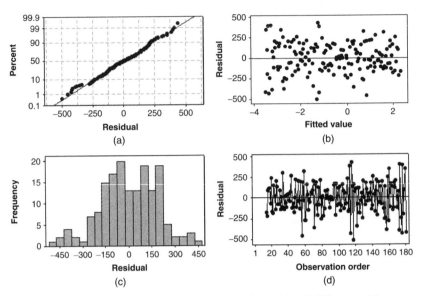

FIGURE 2.26   Residual plots for linear trend model of differenced beverage shipments.

Equation (2.27) was obtained by writing Eq. (2.28) as a sine–cosine pair using the trigonometric identity $\sin(u + v) = \cos u \sin v + \sin u \cos v$ and adding an intercept term β_0:

$$E(y_t) = \beta \sin \omega(t + \theta)$$
$$= \beta \cos \omega\theta \sin \omega t + \beta \sin \omega\theta \cos \omega t$$
$$= \beta_1 \sin \omega t + \beta_2 \cos \omega t$$

where $\beta_1 = \beta \cos \omega\theta$ and $\beta_2 = \beta \sin \omega\theta$. Setting $\omega = 2\pi/12$ and adding the intercept term β_0 produces Eq. (2.27). This model is very flexible; for example, if we set $\omega = 2\pi/52$ we can model a yearly seasonal pattern that is observed weekly, if we set $\omega = 2\pi/4$ we can model a yearly seasonal pattern observed quarterly, and if we set $\omega = 2\pi/13$ we can model an annual seasonal pattern observed in 13 four-week periods instead of the usual months.

Equation (2.27) incorporates a single sine wave at the **fundamental frequency** $\omega = 2\pi/12$. In general, we could add **harmonics** of the fundamental frequency to the model in order to model more complex seasonal patterns. For example, a very general model for monthly data and an annual season that uses the fundamental frequency and the first three harmonics is

$$E(y_t) = \beta_0 + \sum_{j=1}^{4} \left(\beta_j \sin \frac{2\pi j}{12} t + \beta_{4+j} \cos \frac{2\pi j}{12} t \right). \qquad (2.29)$$

If the data are observed in 13 four-week periods, the model would be

$$E(y_t) = \beta_0 + \sum_{j=1}^{4} \left(\beta_j \sin \frac{2\pi j}{13} t + \beta_{4+j} \cos \frac{2\pi j}{13} t \right). \qquad (2.30)$$

There is also a "classical" approach to decomposition of a time series into trend and seasonal components (actually, there are a lot of different decomposition algorithms; here we explain a very simple but useful approach). The general mathematical model for this decomposition is

$$y_t = f(S_t, T_t, \varepsilon_t),$$

where S_t is the seasonal component, T_t is the trend effect (sometimes called the trend-cycle effect), and ε_t is the random error component. There are usually two forms for the function f; an additive model

$$y_t = S_t + T_t + \varepsilon_t$$

and a multiplicative model

$$y_t = S_t T_t \varepsilon_t.$$

The additive model is appropriate if the magnitude (amplitude) of the seasonal variation does not vary with the level of the series, while the multiplicative version is more appropriate if the amplitude of the seasonal fluctuations increases or decreases with the average level of the time series.

Decomposition is useful for breaking a time series down into these component parts. For the additive model, it is relatively easy. First, we would model and remove the trend. A simple linear model could be used to do this, say, $T_t = \beta_0 + \beta_1 t$. Other methods could also be used. Moving averages can be used to isolate a trend and remove it from the original data, as could more sophisticated regression methods. These techniques might be appropriate when the trend is not a straight line over the history of the time series. Differencing could also be used, although it is not typically in the classical decomposition approach.

Once the trend or trend-cycle component is estimated, the series is detrended:

$$y_t - T_t = S_t + \varepsilon_t.$$

Now a seasonal factor can be calculated for each period in the season. For example, if the data are monthly and an annual season is anticipated, we would calculate a season effect for each month in the data set. Then the seasonal indices are computed by taking the average of all of the seasonal factors for each period in the season. In this example, all of the January seasonal factors are averaged to produce a January season index; all of the February seasonal factors are averaged to produce a February season index; and so on. Sometimes medians are used instead of averages. In multiplicative decomposition, ratios are used, so that the data are detrended by

$$\frac{y_t}{T_t} = S_t \varepsilon_t.$$

The seasonal indices are estimated by taking the averages over all of the detrended values for each period in the season.

Example 2.9   The decomposition approach can be applied to the beverage shipment data. Examining the time series plot in Figure 2.2, there is both a strong positive trend as well as month-to-month variation, so the model should include both a trend and a seasonal component. It also appears that the magnitude of the seasonal variation does not vary with the level of the series, so an additive model is appropriate.

Results of a time series decomposition analysis from Minitab of the beverage shipments are in Figure 2.27, showing the original data (labeled "Actual"); along with the fitted trend line ("Trend") and the predicted values ("Fits") from the additive model with both the trend and seasonal components.

Details of the seasonal analysis are shown in Figure 2.28. Estimates of the monthly variation from the trend line for each season (seasonal indices) are in Figure 2.28a with boxplots of the actual differences in Figure 2.28b. The percent of variation by seasonal period is in Figure 2.28c, and model residuals by seasonal period are in Figure 2.28d.

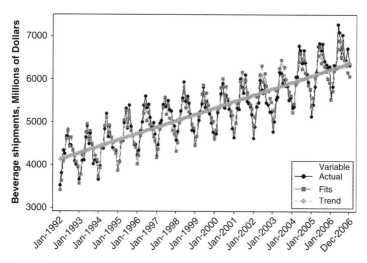

FIGURE 2.27   Time series plot of decomposition model for beverage shipments.

Additional details of the component analysis are shown in Figure 2.29. Figure 2.29a is the original time series, Figure 2.29b is a plot of the time series with the trend removed, Figure 2.29c is a plot of the time series with the seasonality removed, and Figure 2.29d is essentially a residual plot of the detrended and seasonally adjusted data. The wave-like pattern in Figure 2.29d suggests a potential issue with the assumption of constant variance over time.

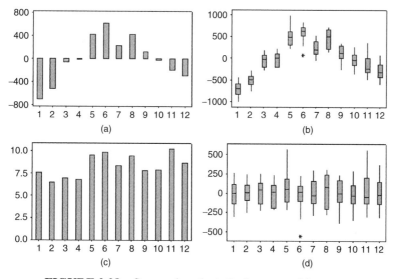

FIGURE 2.28   Seasonal analysis for beverage shipments.

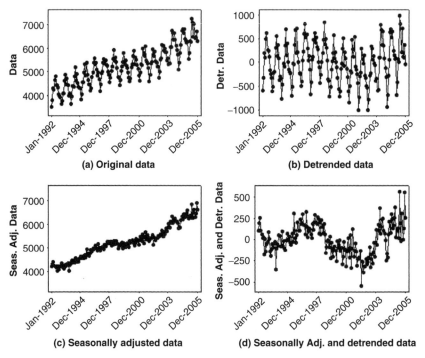

FIGURE 2.29   Component analysis of beverage shipments.

Looking at the normal probability plot and histogram of residuals (Figure 2.30a,c), there does not appear to be an issue with the normality assumption. Figure 2.30d is the same plot as Figure 2.29d. However, variance does seem to increase as the predicted value increases; there is a funnel shape to the residuals plotted in Figure 2.30b. A natural logarithm transformation of the data may stabilize the variance and allow a useful decomposition model to be fit.

Results from the decomposition analysis of the natural log-transformed beverage shipment data are plotted in Figure 2.31, with the transformed data, fitted trend line, and predicted values. Figure 2.32a shows the transformed data, Figure 2.32b the transformed data with the trend removed, Figure 2.32c the transformed data with seasonality removed, and Figure 2.32d the residuals plot of the detrended and seasonally adjusted transformed data. The residual plots in Figure 2.33 indicate that the variance over the range of the predicted values is now stable (Figure 2.33b), and there are no issues with the normality assumption (Figures 2.33a,c). However, there is still a wave-like pattern in the plot of residuals versus time,

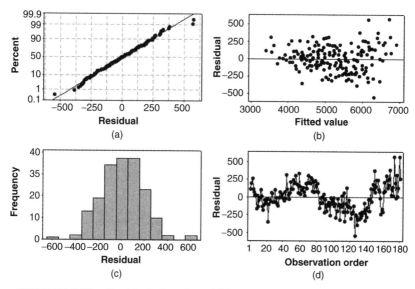

FIGURE 2.30   Residual plots for additive model of beverage shipments.

both Figures 2.32d and 2.33d, indicating that some other structure in the transformed data over time is not captured by the decomposition model. This was not an issue with the model based on seasonal and trend differencing (Figures 2.25 and 2.26), which may be a more appropriate model for monthly beverage shipments.

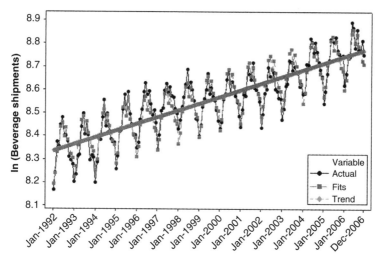

FIGURE 2.31   Time series plot of decomposition model for transformed beverage data.

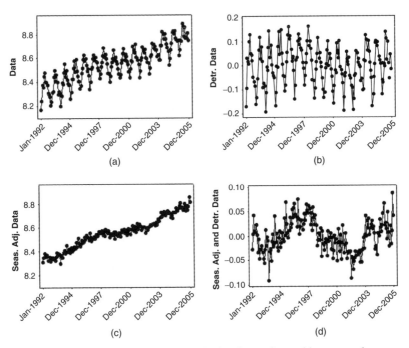

FIGURE 2.32 Component analysis of transformed beverage data.

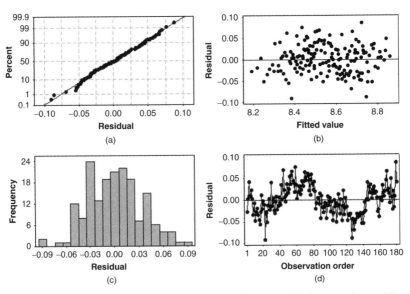

FIGURE 2.33 Residual plots from decomposition model for transformed beverage data.

Another technique for seasonal adjustment that is widely used in modeling and analyzing economic data is the *X*-**11 method**. Much of the development work on this method was done by Julian Shiskin and others at the US Bureau of the Census beginning in the mid-1950s and culminating into the *X*-11 Variant of the Census Method II Seasonal Adjustment Program. References for this work during this period include Shiskin (1958), and Marris (1961). Authoritative documentation for the *X*-11 procedure is in Shiskin, Young, and Musgrave (1967). The *X*-11 method uses symmetric moving averages in an iterative approach to estimating the trend and seasonal components. At the end of the series, however, these symmetric weights cannot be applied. Asymmetric weights have to be used.

JMP (V12 and higher) provides the *X*-11 technique. Figure 2.34 shows the JMP *X*-11 output for the beverage shipment data from Figure 2.2. The upper part of the output contains a plot of the original time series, followed by the sample ACF and PACF. Then Display D10 in the figure shows the final estimates of the seasonal factors by month followed in Display D13 by the irregular or deseasonalized series. The final display is a plot of the original and adjusted time series.

While different variants of the *X*-11 technique have been proposed, the most important method to date has been the *X-11-ARIMA* method developed at Statistics Canada. This method uses Box–Jenkins autoregressive integrated moving average models (which are discussed in Chapter 5) to extend the series. The use of ARIMA models will result in differences in the final component estimates. Details of this method are in Dagum (1980, 1983, 1988).

2.5 GENERAL APPROACH TO TIME SERIES MODELING AND FORECASTING

The techniques that we have been describing form the basis of a general approach to modeling and forecasting time series data. We now give a broad overview of the approach. This should give readers a general understanding of the connections between the ideas we have presented in this chapter and guidance in understanding how the topics in subsequent chapters form a collection of useful techniques for modeling and forecasting time series.

The basic steps in modeling and forecasting a time series are as follows:

1. Plot the time series and determine its basic features, such as whether trends or seasonal behavior or both are present. Look for possible outliers or any indication that the time series has changed with respect

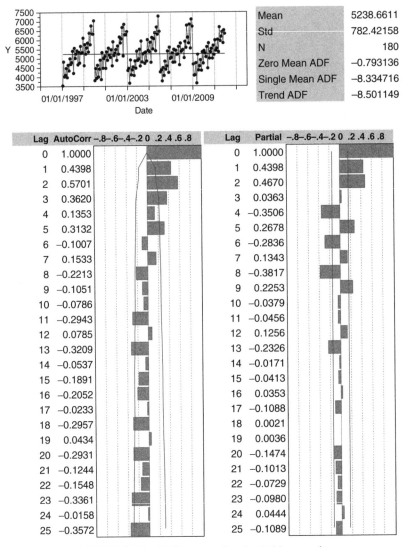

FIGURE 2.34 JMP output for the X-11 procedure.

to its basic features (such as trends or seasonality) over the time period history.

2. Eliminate any trend or seasonal components, either by differencing or by fitting an appropriate model to the data. Also consider using data transformations, particularly if the variability in the time series seems to be proportional to the average level of the series. The objective of these operations is to produce a set of stationary residuals.

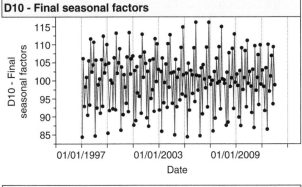

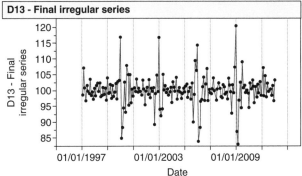

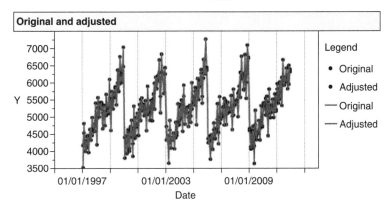

FIGURE 2.34 (*Continued*)

3. Develop a forecasting model for the residuals. It is not unusual to find that there are several plausible models, and additional analysis will have to be performed to determine the best one to deploy. Sometimes potential models can be eliminated on the basis of their fit to the historical data. It is unlikely that a model that fits poorly will produce good forecasts.

4. Validate the performance of the model (or models) from the previous step. This will probably involve some type of split-sample or cross-validation procedure. The objective of this step is to select a model to use in forecasting. We will discuss this more in the next section and illustrate these techniques throughout the book.

5. Also of interest are the differences between the original time series y_t and the values that would be forecast by the model on the original scale. To forecast values on the scale of the original time series y_t, reverse the transformations and any differencing adjustments made to remove trends or seasonal effects.

6. For forecasts of future values in period $T + \tau$ on the original scale, if a transformation was used, say, $x_t = \ln y_t$, then the forecast made at the end of period T for $T + \tau$ would be obtained by reversing the transformation. For the natural log this would be

$$\hat{y}_{T+\tau}(T) = \exp[\hat{x}_{T+\tau}(T)].$$

7. If prediction intervals are desired for the forecast (and we recommend doing this), construct prediction intervals for the residuals and then reverse the transformations made to produce the residuals as described earlier. We will discuss methods for finding prediction intervals for most of the forecasting methods presented in this book.

8. Develop and implement a procedure for monitoring the forecast to ensure that deterioration in performance will be detected reasonably quickly. Forecast monitoring is usually done by evaluating the stream of forecast errors that are experienced. We will present methods for monitoring forecast errors with the objective of detecting changes in performance of the forecasting model.

2.6 EVALUATING AND MONITORING FORECASTING MODEL PERFORMANCE

2.6.1 Forecasting Model Evaluation

We now consider how to evaluate the performance of a forecasting technique for a particular time series or application. It is important to carefully define the meaning of performance. It is tempting to evaluate performance on the basis of the fit of the forecasting or time series model to historical data. There are many statistical measures that describe how well a model fits a given sample of data, and several of these will be described in subsequent chapters. This goodness-of-fit approach often uses the residuals and does not really reflect the capability of the forecasting technique to successfully predict future observations. The user of the forecasts is very concerned about the accuracy of future forecasts, not model goodness of fit, so it is important to evaluate this aspect of any recommended technique. Sometimes forecast accuracy is called "out-of-sample" forecast error, to distinguish it from the residuals that arise from a model-fitting process.

Measure of forecast accuracy should always be evaluated as part of a model validation effort (see step 4 in the general approach to forecasting in the previous section). When more than one forecasting technique seems reasonable for a particular application, these forecast accuracy measures can also be used to discriminate between competing models. We will discuss this more in Section 2.6.2.

It is customary to evaluate forecasting model performance using the one-step-ahead forecast errors

$$e_t(1) = y_t - \hat{y}_t(t-1), \tag{2.31}$$

where $\hat{y}_t(t-1)$ is the forecast of y_t that was made one period prior. Forecast errors at other lags, or at several different lags, could be used if interest focused on those particular forecasts. Suppose that there are n observations for which forecasts have been made and n one-step-ahead forecast errors, $e_t(1), t = 1, 2, \ldots, n$. Standard measures of forecast accuracy are the **average error** or **mean error**

$$\text{ME} = \frac{1}{n} \sum_{t=1}^{n} e_t(1), \tag{2.32}$$

the **mean absolute deviation** (or mean absolute error)

$$\text{MAD} = \frac{1}{n} \sum_{t=1}^{n} |e_t(1)|, \tag{2.33}$$

and the **mean squared error**

$$\text{MSE} = \frac{1}{n} \sum_{t=1}^{n} [e_t(1)]^2. \tag{2.34}$$

The mean forecast error in Eq. (2.32) is an estimate of the expected value of forecast error, which we would hope to be zero; that is, the forecasting technique produces **unbiased** forecasts. If the mean forecast error differs appreciably from zero, bias in the forecast is indicated. If the mean forecast error drifts away from zero when the forecasting technique is in use, this can be an indication that the underlying time series has changed in some fashion, the forecasting technique has not tracked this change, and now biased forecasts are being generated.

Both the mean absolute deviation (MAD) in Eq. (2.33) and the mean squared error (MSE) in Eq. (2.34) measure the **variability** in forecast errors. Obviously, we want the variability in forecast errors to be small. The MSE is a direct estimator of the variance of the one-step-ahead forecast errors:

$$\hat{\sigma}^2_{e(1)} = \text{MSE} = \frac{1}{n} \sum_{t=1}^{n} [e_t(1)]^2. \tag{2.35}$$

If the forecast errors are normally distributed (this is usually not a bad assumption, and one that is easily checked), the MAD is related to the standard deviation of forecast errors by

$$\hat{\sigma}_{e(1)} = \sqrt{\frac{\pi}{2}} \text{MAD} \cong 1.25 \, \text{MAD} \tag{2.36}$$

The one-step-ahead forecast error and its summary measures, the ME, MAD, and MSE, are all scale-dependent measures of forecast accuracy; that is, their values are expressed in terms of the original units of measurement (or in the case of MSE, the square of the original units). So, for example, if we were forecasting demand for electricity in Phoenix during the summer, the units would be megawatts (MW). If the MAD for the forecast error during summer months was 5 MW, we might not know whether this was a large forecast error or a relatively small one. Furthermore, accuracy measures that are scale dependent do not facilitate comparisons of a single forecasting technique across different time series, or comparisons across different time periods. To accomplish this, we need a measure of relative forecast error.

Define the **relative forecast error** (in percent) as

$$re_t(1) = \left(\frac{y_t - \hat{y}_t(t-1)}{y_t} \right) 100 = \left(\frac{e_t(1)}{y_t} \right) 100. \tag{2.37}$$

This is customarily called the **percent forecast error**. The mean percent forecast error (MPE) is

$$\text{MPE} = \frac{1}{n} \sum_{t=1}^{n} re_t(1) \qquad (2.38)$$

and the mean absolute percent forecast error (MAPE) is

$$\text{MAPE} = \frac{1}{n} \sum_{t=1}^{n} |re_t(1)|. \qquad (2.39)$$

Knowing that the relative or percent forecast error or the MAPE is 3% (say) can be much more meaningful than knowing that the MAD is 5 MW. Note that the relative or percent forecast error only makes sense if the time series y_t does not contain zero values.

Example 2.10   Table 2.2 illustrates the calculation of the one-step-ahead forecast error, the absolute errors, the squared errors, the relative (percent) error, and the absolute percent error from a forecasting model for 20 time periods. The last row of columns (3) through (7) display the sums required to calculate the ME, MAD, MSE, MPE, and MAPE.

From Eq. (2.32), the mean (or average) forecast error is

$$\text{ME} = \frac{1}{n} \sum_{t=1}^{n} e_t(1) = \frac{1}{20}(-11.6) = -0.58,$$

the MAD is computed from Eq. (2.33) as

$$\text{MAD} = \frac{1}{n} \sum_{t=1}^{n} |e_t(1)| = \frac{1}{20}(86.6) = 4.33,$$

and the MSE is computed from Eq. (2.34) as

$$\text{MSE} = \frac{1}{n} \sum_{t=1}^{n} [e_t(1)]^2 = \frac{1}{20}(471.8) = 23.59.$$

TABLE 2.2 Calculation of Forecast Accuracy Measures

Time Period	(1) Observed Value y_t	(2) Forecast $\hat{y}_t(t-1)$	(3) Forecast Error $e_t(1)$	(4) Absolute Error $|e_t(1)|$	(5) Squared Error $[e_t(1)]^2$	(6) Relative (%) Error $(e_t(1)/y_t)\,100$	(6) Absolute (%) Error $|(e_t(1)/y_t)\,100|$
1	47	51.1	−4.1	4.1	16.81	−8.7234	8.723404
2	46	52.9	−6.9	6.9	47.61	−15	15
3	51	48.8	2.2	2.2	4.84	4.313725	4.313725
4	44	48.1	−4.1	4.1	16.81	−9.31818	9.318182
5	54	49.7	4.3	4.3	18.49	7.962963	7.962963
6	47	47.5	−0.5	0.5	0.25	−1.06383	1.06383
7	52	51.2	0.8	0.8	0.64	1.538462	1.538462
8	45	53.1	−8.1	8.1	65.61	−18	18
9	50	54.4	−4.4	4.4	19.36	−8.8	8.8
10	51	51.2	−0.2	0.2	0.04	−0.39216	0.392157
11	49	53.3	−4.3	4.3	18.49	−8.77551	8.77551
12	41	46.5	−5.5	5.5	30.25	−13.4146	13.41463
13	48	53.1	−5.1	5.1	26.01	−10.625	10.625
14	50	52.1	−2.1	2.1	4.41	−4.2	4.2
15	51	46.8	4.2	4.2	17.64	8.235294	8.235294
16	55	47.7	7.3	7.3	53.29	13.27273	13.27273
17	52	45.4	6.6	6.6	43.56	12.69231	12.69231
18	53	47.1	5.9	5.9	34.81	11.13208	11.13208
19	48	51.8	−3.8	3.8	14.44	−7.91667	7.916667
20	52	45.8	6.2	6.2	38.44	11.92308	11.92308
Totals			−11.6	86.6	471.8	−35.1588	177.3

Because the MSE estimates the variance of the one-step-ahead forecast errors, we have

$$\hat{\sigma}^2_{e(1)} = \text{MSE} = 23.59$$

and an estimate of the standard deviation of forecast errors is the square root of this quantity, or $\hat{\sigma}_{e(1)} = \sqrt{\text{MSE}} = 4.86$. We can also obtain an estimate of the standard deviation of forecasts errors from the MAD using Eq. (2.36)

$$\hat{\sigma}_{e(1)} \cong 1.25\,\text{MAD} = 1.25(4.33) = 5.41.$$

These two estimates are reasonably similar. The mean percent forecast error, MPE, is computed from Eq. (2.38) as

$$\text{MPE} = \frac{1}{n} \sum_{t=1}^{n} re_t(1) = \frac{1}{20}(-35.1588) = -1.76\%$$

and the mean absolute percent error is computed from Eq. (2.39) as

$$\text{MAPE} = \frac{1}{n} \sum_{t=1}^{n} |re_t(1)| = \frac{1}{20}(177.3) = 8.87\%.$$

There is much empirical evidence (and even some theoretical justification) that the distribution of forecast errors can be well approximated by a **normal** distribution. This can easily be checked by constructing a **normal probability plot** of the forecast errors in Table 2.2, as shown in Figure 2.35. The forecast errors deviate somewhat from the straight line, indicating that the normal distribution is not a perfect model for the distribution of forecast errors, but it is not unreasonable. Minitab calculates the Anderson–Darling statistic, a widely used test statistic for normality. The P-value is 0.088, so the hypothesis of normality of the forecast errors would not be rejected at the 0.05 level. This test assumes that the observations (in this case the forecast errors) are uncorrelated. Minitab also reports the standard deviation of the forecast errors to be 4.947, a slightly larger value than we computed from the MSE, because Minitab uses the standard method for calculating sample standard deviations.

Note that Eq. (2.31) could have been written as

$$\text{Error} = \text{Observation} - \text{Forecast}.$$

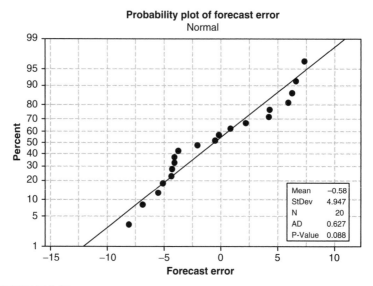

FIGURE 2.35 Normal probability plot of forecast errors from Table 2.2.

Hopefully, the forecasts do a good job of describing the structure in the observations. In an ideal situation, the forecasts would adequately model all of the structure in the data, and the sequence of forecast errors would be structureless. If they are, the sample ACF of the forecast error should look like the ACF of random data; that is, there should not be any large "spikes" on the sample ACF at low lag. Any systematic or nonrandom pattern in the forecast errors will tend to show up as significant spikes on the sample ACF. If the sample ACF suggests that the forecast errors are not random, then this is evidence that the forecasts can be improved by **refining** the forecasting model. Essentially, this would consist of taking the structure out of the forecast errors and putting it into the forecasts, resulting in forecasts that are better prediction of the data.

Example 2.11   Table 2.3 presents a set of 50 one-step-ahead errors from a forecasting model, and Table 2.4 shows the sample ACF of these forecast errors. The sample ACF is plotted in Figure 2.36. This sample ACF was obtained from Minitab. Note that sample autocorrelations for the first 13 lags are computed. This is consistent with our guideline indicating that for T observations only the first $T/4$ autocorrelations should be computed. The sample ACF does not provide any strong evidence to support a claim that there is a pattern in the forecast errors.

TABLE 2.3   One-Step-Ahead Forecast Errors

Period, t	$e_t(1)$	Period, t	$e_t(1)$	Period, t	$e_t(1)$	Period, t	$e_t(1)$	Period, t	$e_t(1)$
1	−0.62	11	−0.49	21	2.90	31	−1.88	41	−3.98
2	−2.99	12	4.13	22	0.86	32	−4.46	42	−4.28
3	0.65	13	−3.39	23	5.80	33	−1.93	43	1.06
4	0.81	14	2.81	24	4.66	34	−2.86	44	0.18
5	−2.25	15	−1.59	25	3.99	35	0.23	45	3.56
6	−2.63	16	−2.69	26	−1.76	36	−1.82	46	−0.24
7	3.57	17	3.41	27	2.31	37	0.64	47	−2.98
8	0.11	18	4.35	28	−2.24	38	−1.55	48	2.47
9	0.59	19	−4.37	29	2.95	39	0.78	49	0.66
10	−0.63	20	2.79	30	6.30	40	2.84	50	0.32

TABLE 2.4   Sample ACF of the One-Step-Ahead Forecast Errors in Table 2.3

Lag	Sample ACF, r_k	Z-Statistic	Ljung–Box Statistic, Q_{LB}
1	0.004656	0.03292	0.0012
2	−0.102647	−0.72581	0.5719
3	0.136810	0.95734	1.6073
4	−0.033988	−0.23359	1.6726
5	0.118876	0.81611	2.4891
6	0.181508	1.22982	4.4358
7	−0.039223	−0.25807	4.5288
8	−0.118989	−0.78185	5.4053
9	0.003400	0.02207	5.4061
10	0.034631	0.22482	5.4840
11	−0.151935	−0.98533	7.0230
12	−0.207710	−1.32163	9.9749
13	0.089387	0.54987	10.5363

If a time series consists of uncorrelated observations and has constant variance, we say that it is **white noise**. If, in addition, the observations in this time series are normally distributed, the time series is **Gaussian white noise**. Ideally, forecast errors are Gaussian white noise. The normal probability plot of the one-step-ahead forecast errors from Table 2.3 are shown in Figure 2.37. This plot does not indicate any serious problem, with the normality assumption, so the forecast errors in Table 2.3 are Gaussian white noise.

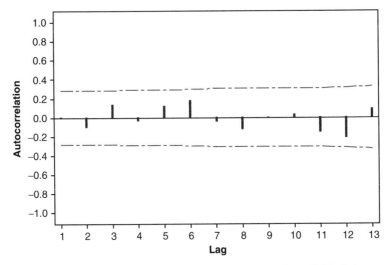

FIGURE 2.36   Sample ACF of forecast errors from Table 2.4.

If a time series is white noise, the distribution of the sample autocorrelation coefficient at lag k in large samples is approximately normal with mean zero and variance $1/T$; that is,

$$r_k \sim N\left(0, \frac{1}{T}\right).$$

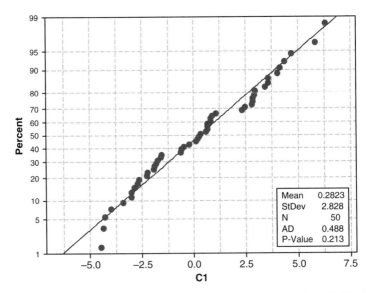

FIGURE 2.37   Normal probability plot of forecast errors from Table 2.3.

Therefore we could test the hypothesis $H_0 : \rho_k = 0$ using the test statistic

$$Z_0 = \frac{r_k}{\sqrt{\frac{1}{T}}} = r_k \sqrt{T}. \tag{2.40}$$

Minitab calculates this Z-statistic (calling it a t-statistic), and it is reported in Table 2.4 for the one-step-ahead forecast errors of Table 2.3 (this is the t-statistic reported in Figure 2.13 for the ACF of the chemical viscosity readings). Large values of this statistic (say, $|Z_0| > Z_{\alpha/2}$, where $Z_{\alpha/2}$ is the upper $\alpha/2$ percentage point of the standard normal distribution) would indicate that the corresponding autocorrelation coefficient does not equal zero. Alternatively, we could calculate a P-value for this test statistic. Since none of the absolute values of the Z-statistics in Table 2.4 exceeds $Z_{\alpha/2} = Z_{0.025} = 1.96$, we cannot conclude at significance level $\alpha = 0.05$ that any individual autocorrelation coefficient differs from zero.

   This procedure is a one-at-a-time test; that is, the significance level applies to the autocorrelations considered individually. We are often interested in evaluating a *set* of autocorrelations jointly to determine if they indicate that the time series is white noise. Box and Pierce (1970) have suggested such a procedure. Consider the square of the test statistic Z_0 in Eq. (2.40). The distribution of $Z_0^2 = r_k^2 T$ is approximately chi-square with one degree of freedom. The Box–Pierce statistic

$$Q_{\text{BP}} = T \sum_{k=1}^{K} r_k^2 \tag{2.41}$$

is distributed approximately as chi-square with K degrees of freedom under the null hypothesis that the time series is white noise. Therefore, if $Q_{\text{BP}} > \chi_{\alpha,K}^2$ we would reject the null hypothesis and conclude that the time series is not white noise because some of the autocorrelations are not zero. A P-value approach could also be used. When this test statistic is applied to a set of **residual autocorrelations** the statistic $Q_{\text{BP}} \sim \chi_{\alpha,K-p}^2$, where p is the number of parameters in the model, so the number of degrees of freedom in the chi-square distribution becomes $K - p$. Box and Pierce call this procedure a "Portmanteau" or general **goodness-of-fit statistic** (it is testing the goodness of fit of the ACF to the ACF of white noise). A modification of this test that works better for small samples was devised by Ljung and Box (1978). The Ljung–Box goodness-of-fit statistic is

$$Q_{\text{LB}} = T(T + 2) \sum_{k=1}^{K} \left(\frac{1}{T - k} \right) r_k^2. \tag{2.42}$$

Note that the Ljung–Box goodness-of-fit statistic is very similar to the original Box–Pierce statistic, the difference being that the squared sample autocorrelation at lag k is weighted by $(T + 2)/(T - k)$. For large values of T, these weights will be approximately unity, and so the Q_{LB} and Q_{BP} statistics will be very similar.

Minitab calculates the Ljung–Box goodness-of-fit statistic Q_{LB}, and the values for the first 13 sample autocorrelations of the one-step-ahead forecast errors of Table 2.3 are shown in the last column of Table 2.4. At lag 13, the value $Q_{LB} = 10.5363$, and since $\chi^2_{0.05,13} = 22.36$, there is no strong evidence to indicate that the first 13 autocorrelations of the forecast errors considered jointly differ from zero. If we calculate the P-value for this test statistic, we find that $P = 0.65$. This is a good indication that the forecast errors are white noise. Note that Figure 2.13 also gave values for the Ljung–Box statistic.

2.6.2 Choosing Between Competing Models

There are often several competing models that can be used for forecasting a particular time series. For example, there are several ways to model and forecast trends. Consequently, selecting an appropriate forecasting model is of considerable practical importance. In this section we discuss some general principles of model selection. In subsequent chapters, we will illustrate how these principles are applied in specific situations.

Selecting the model that provides the best fit to historical data generally does not result in a forecasting method that produces the best forecasts of new data. Concentrating too much on the model that produces the best historical fit often results in **overfitting**, or including too many parameters or terms in the model just because these additional terms improve the model fit. In general, the best approach is to select the model that results in the smallest standard deviation (or mean squared error) of the one-step-ahead forecast errors when the model is applied to data that were not used in the fitting process. Some authors refer to this as an **out-of-sample** forecast error standard deviation (or mean squared error). A standard way to measure this out-of-sample performance is by utilizing some form of **data splitting**; that is, divide the time series data into two segments—one for model fitting and the other for performance testing. Sometimes data splitting is called **cross-validation**. It is somewhat arbitrary as to how the data splitting is accomplished. However, a good rule of thumb is to have at least 20 or 25 observations in the performance testing data set.

When evaluating the fit of the model to historical data, there are several criteria that may be of value. The **mean squared error** of the residuals is

$$s^2 = \frac{\sum_{t=1}^{T} e_t^2}{T - p} \tag{2.43}$$

where T periods of data have been used to fit a model with p parameters and e_t is the residual from the model-fitting process in period t. The mean squared error s^2 is just the sample variance of the residuals and it is an estimator of the variance of the model errors.

Another criterion is the R-squared statistic

$$R^2 = 1 - \frac{\sum_{t=1}^{T} e_t^2}{\sum_{t=1}^{T} (y_t - \bar{y})^2}. \tag{2.44}$$

The denominator of Eq. (2.44) is just the total sum of squares of the observations, which is constant (not model dependent), and the numerator is just the residual sum of squares. Therefore, selecting the model that maximizes R^2 is equivalent to selecting the model that minimizes the sum of the squared residuals. Large values of R^2 suggest a good fit to the historical data. Because the residual sum of squares always decreases when parameters are added to a model, relying on R^2 to select a forecasting model encourages overfitting or putting in more parameters than are really necessary to obtain good forecasts. A large value of R^2 does not ensure that the out-of-sample one-step-ahead forecast errors will be small.

A better criterion is the "adjusted" R^2 statistic, defined as

$$R_{\text{Adj}}^2 = 1 - \frac{\sum_{t=1}^{T} e_t^2 /(T - p)}{\sum_{t=1}^{T} (y_t - \bar{y})^2 /(T - 1)} = 1 - \frac{s^2}{\sum_{t=1}^{T} (y_t - \bar{y})^2 /(T - 1)}. \tag{2.45}$$

The adjustment is a "size" adjustment—that is, adjust for the number of parameters in the model. Note that a model that maximizes the adjusted R^2 statistic is also the model that minimizes the residual mean square.

Two other important criteria are the **Akaike Information Criterion** (**AIC**) (see Akaike (1974)) and the **Schwarz Bayesian Information Criterion (abbreviated as BIC or SIC by various authors)** (see Schwarz (1978)):

$$\text{AIC} = \ln\left(\frac{\sum_{t=1}^{T} e_t^2}{T}\right) + \frac{2p}{T} \qquad (2.46)$$

and

$$\text{BIC} = \ln\left(\frac{\sum_{t=1}^{T} e_t^2}{T}\right) + \frac{p\ln(T)}{T}. \qquad (2.47)$$

These two criteria penalize the sum of squared residuals for including additional parameters in the model. Models that have small values of the AIC or BIC are considered good models.

One way to evaluate model selection criteria is in terms of **consistency**. A model selection criterion is consistent if it selects the true model when the true model is among those considered with probability approaching unity as the sample size becomes large, and if the true model is not among those considered, it selects the best approximation with probability approaching unity as the sample size becomes large. It turns out that s^2, the adjusted R^2, and the AIC are all inconsistent, because they do not penalize for adding parameters heavily enough. Relying on these criteria tends to result in overfitting. The BIC, which caries a heavier "size adjustment" penalty, is consistent.

Consistency, however, does not tell the complete story. It may turn out that the true model and any reasonable approximation to it are very complex. An **asymptotically efficient** model selection criterion chooses a sequence of models as T(the amount of data available) gets large for which the one-step-ahead forecast error variances approach the one-step-ahead forecast error variance for the true model at least as fast as any other criterion. The AIC is asymptotically efficient but the BIC is not.

There are a number of variations and extensions of these criteria. The AIC is a biased estimator of the discrepancy between all candidate models and the true model. This has led to developing a "corrected" version of AIC:

$$\text{AICc} = \ln \left(\frac{\sum\limits_{t=1}^{T} e_t^2}{T} \right) + \frac{2T(p+1)}{T-p-2}. \tag{2.48}$$

Sometimes we see the first term in the AIC, AICc, or BIC written as $-2 \ln L(\beta, \sigma^2)$, where $L(\beta, \sigma^2)$ is the **likelihood function** for the fitted model evaluated at the maximum likelihood estimates of the unknown parameters β and σ^2. In this context, AIC, AICc, and SIC are called penalized likelihood criteria.

Many software packages evaluate and print model selection criteria, such as those discussed here. When both AIC and SIC are available, we prefer using SIC. It generally results in smaller, and hence simpler, models, and so its use is consistent with the time-honored model-building principle of **parsimony** (all other things being equal, simple models are preferred to complex ones). We will discuss and illustrate model selection criteria again in subsequent chapters. However, remember that the best way to evaluate a candidate model's potential predictive performance is to use data splitting. This will provide a direct estimate of the one-step-ahead forecast error variance, and this method should always be used, if possible, along with the other criteria that we have discussed here.

2.6.3 Monitoring a Forecasting Model

Developing and implementing procedures to monitor the performance of the forecasting model is an essential component of good forecasting system design. No matter how much effort has been expended in developing the forecasting model, and regardless of how well the model works initially, over time it is likely that its performance will deteriorate. The underlying pattern of the time series may change, either because the internal inertial forces that drive the process may evolve through time, or because of external events such as new customers entering the market. For example, a level change or a slope change could occur in the variable that is being forecasted. It is also possible for the inherent variability in the data to increase. Consequently, performance monitoring is important.

The one-step-ahead forecast errors $e_t(1)$ are typically used for forecast monitoring. The reason for this is that changes in the underlying time series will also typically be reflected in the forecast errors. For example, if a level change occurs in the time series, the sequence of forecast errors will no longer fluctuate around zero; that is, a positive or negative bias will be introduced.

There are several ways to monitor forecasting model performance. The simplest way is to apply **Shewhart control charts** to the forecast errors. A Shewhart control chart is a plot of the forecast errors versus time containing a center line that represents the average (or the target value) of the forecast errors and a set of **control limits** that are designed to provide an indication that the forecasting model performance has changed. The center line is usually taken as either zero (which is the anticipated forecast error for an unbiased forecast) or the average forecast error (ME from Eq. (2.32)), and the control limits are typically placed at three standard deviations of the forecast errors above and below the center line. If the forecast errors plot within the control limits, we assume that the forecasting model performance is satisfactory (or in control), but if one or more forecast errors exceed the control limits, that is a signal that something has happened and the forecast errors are no longer fluctuating around zero. In control chart terminology, we would say that the forecasting process is out of control and some analysis is required to determine what has happened.

The most familiar Shewhart control charts are those applied to data that have been collected in subgroups or samples. The one-step-ahead forecast errors $e_t(1)$ are individual observations. Therefore the Shewhart control chart for individuals would be used for forecast monitoring. On this control chart it is fairly standard practice to estimate the standard deviation of the individual observations using a moving range method. The moving range is defined as the absolute value of the difference between any two successive one-step-ahead forecast errors, say, $|e_t(1) - e_{t-1}(1)|$, and the moving range based on n observations is

$$MR = \sum_{t=2}^{n} |e_t(1) - e_{t-1}(1)|. \tag{2.49}$$

The estimate of the standard deviation of the one-step-ahead forecast errors is based on the average of the moving ranges

$$\hat{\sigma}_{e(1)} = \frac{0.8865 MR}{n - 1} = \frac{0.8865 \sum_{t=2}^{n} |e_t(1) - e_{t-1}(1)|}{n - 1} = 0.8865 \overline{MR}, \tag{2.50}$$

where $\overline{MR}$ is the average of the moving ranges. This estimate of the standard deviation would be used to construct the control limits on the control chart for forecast errors. For more details on constructing and interpreting control charts, see Montgomery (2013).

Example 2.12 Minitab can be used to construct Shewhart control charts for individuals. Figure 2.38 shows the Minitab control charts for the one-step-ahead forecast errors in Table 2.3. Note that both an individuals control chart of the one-step-ahead forecast errors and a control chart of the moving ranges of these forecast errors are provided. On the individuals control chart the center line is taken to be the average of the forecast errors ME defined in Eq. (2.30) (denoted $\overline{X}$ in Figure 2.38) and the upper and lower three-sigma control limits are abbreviated as UCL and LCL, respectively. The center line on the moving average control chart is at the average of the moving ranges $\overline{MR} = MR/(n-1)$, the three-sigma upper control limit UCL is at $3.267MR/(n-1)$, and the lower control limit is at zero (for details on how the control limits are derived, see Montgomery (2013)). All of the one-step-ahead forecast errors plot within the control limits (and the moving range also plot within their control limits). Thus there is no reason to suspect that the forecasting model is performing inadequately, at least from the statistical stability viewpoint. Forecast errors that plot outside the control limits would indicate model inadequacy, or possibly the presence of unusual observations such as outliers in the data. An investigation would be required to determine why these forecast errors exceed the control limits.

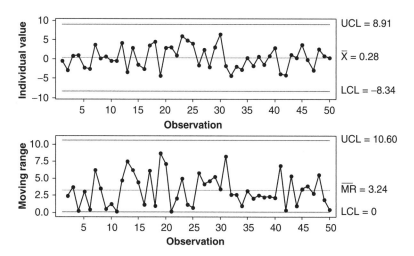

FIGURE 2.38 Individuals and moving range control charts of the one-step-ahead forecast errors in Table 2.3.

Because the control charts in Figure 2.38 exhibit statistical control, we would conclude that there is no strong evidence of statistical inadequacy in the forecasting model. Therefore, these control limits would be retained and used to judge the performance of future forecasts (in other words, we do not recalculate the control limits with each new forecast). However, the stable control chart does not imply that the forecasting performance is satisfactory in the sense that the model results in small forecast errors. In the quality control literature, these two aspects of process performance are referred to as control and capability, respectively. It is possible for the forecasting process to be stable or in statistical control but not capable— that is, produce forecast errors that are unacceptably large.

Two other types of control charts, the cumulative sum (or CUSUM) control chart and the exponentially weighted moving average (or EWMA) control chart, can also be useful for monitoring the performance of a forecasting model. These charts are more effective at detecting smaller changes or disturbances in the forecasting model performance than the individuals control chart. The CUSUM is very effective in detecting level changes in the monitored variable. It works by accumulating deviations of the forecast errors that are above the desired target value T_0 (usually either zero or the average forecast error) with one statistic C^+ and deviations that are below the target with another statistic C^-. The statistics C^+ and C^- are called the upper and lower CUSUMs, respectively. They are computed as follows:

$$C_t^+ = \max[0, e_t(1) - (T_0 + K) + C_{t-1}^+]$$
$$C_t^- = \min[0, e_t(1) - (T_0 - K) + C_{t-1}^-]$$

(2.51)

where the constant K, usually called the reference value, is usually chosen as $K = 0.5\sigma_{e(1)}$ and $\sigma_{e(1)}$ is the standard deviation of the one-step-ahead forecast errors. The logic is that if the forecast errors begin to systematically fall on one side of the target value (or zero), one of the CUSUMs in Eq. (2.51) will increase in magnitude. When this increase becomes large enough, an out-of-control signal is generated. The decision rule is to signal if the statistic C^+ exceeds a decision interval $H = 5\sigma_{e(1)}$ or if C^- exceeds $-H$. The signal indicates that the forecasting model is not performing satisfactorily (Montgomery (2013) discusses the choice of H and K in detail).

Example 2.13   The CUSUM control chart for the forecast errors shown in Table 2.3 is shown in Figure 2.39. This CUSUM chart was constructed

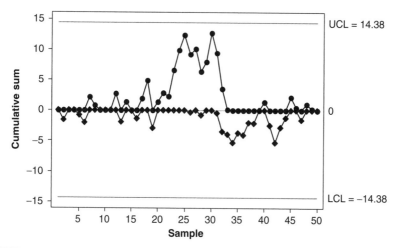

FIGURE 2.39   CUSUM control chart of the one-step-ahead forecast errors in Table 2.3.

using Minitab with a target value of $T = 0$ and $\sigma_{e(1)}$ was estimated using the moving range method described previously, resulting in $H = 5\hat{\sigma}_{e(1)} = 5(0.8865)MR/(T - 1) = 5(0.8865)3.24 = 14.36$. Minitab labels H and $-H$ as UCL and LCL, respectively. The CUSUM control chart reveals no obvious forecasting model inadequacies.

A control chart based on the EWMA is also useful for monitoring forecast errors. The EWMA applied to the one-step-ahead forecast errors is

$$\bar{e}_t(1) = \lambda e_t(1) + (1 - \lambda)\bar{e}_{t-1}(1), \tag{2.52}$$

where $0 < \lambda < 1$ is a constant (usually called the smoothing constant) and the starting value of the EWMA (required at the first observation) is either $\bar{e}_0(1) = 0$ or the average of the forecast errors. Typical values of the smoothing constant for an EWMA control chart are $0.05 < \lambda < 0.2$.

The EWMA is a weighted average of all current and previous forecast errors, and the weights decrease geometrically with the "age" of the forecast error. To see this, simply substitute recursively for $\bar{e}_{t-1}(1)$, then $\bar{e}_{t-2}(1)$, then $\bar{e}_{t-j}(1)_j$ for $j = 3, 4, \ldots$, until we obtain

$$\bar{e}_n(1) = \lambda \sum_{j=0}^{n-1} (1 - \lambda)^j e_{T-j}(1) + (1 - \lambda)^n \bar{e}_0(1)$$

and note that the weights sum to unity because

$$\lambda \sum_{j=0}^{n-1} (1 - \lambda)^j = 1 - (1 - \lambda)^n.$$

The standard deviation of the EWMA is

$$\sigma_{\bar{e}_t(1)} = \sigma_{e(1)} \sqrt{\frac{\lambda}{2 - \lambda}[1 - (1 - \lambda)^{2t}]}.$$

So an EWMA control chart for the one-step-ahead forecast errors with a center line of T (the target for the forecast errors) is defined as follows:

$$\text{UCL} = T + 3\sigma_{e(1)} \sqrt{\frac{\lambda}{2 - \lambda}[1 - (1 - \lambda)^{2t}]}$$

Center line $= T$ (2.53)

$$\text{LCL} = T - 3\sigma_{e(1)} \sqrt{\frac{\lambda}{2 - \lambda}[1 - (1 - \lambda)^{2t}]}$$

Example 2.14   Minitab can be used to construct EWMA control charts. Figure 2.40 is the EWMA control chart of the forecast errors in Table 2.3. This chart uses the mean forecast error as the center line, $\sigma_{e(1)}$ was estimated using the moving range method, and we chose $\lambda = 0.1$. None of the forecast

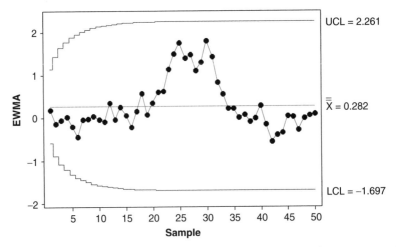

FIGURE 2.40   EWMA control chart of the one-step-ahead forecast errors in Table 2.3.

errors exceeds the control limits so there is no indication of a problem with the forecasting model.

Note from Eq. (2.51) and Figure 2.40 that the control limits on the EWMA control chart increase in width for the first few observations and then stabilize at a constant value because the term $[1 - (1 - \lambda)^{2t}]$ approaches unity as t increases. Therefore steady-state limits for the EWMA control chart are

$$
\begin{aligned}
\text{UCL} &= T_0 + 3\sigma_{e(1)}\sqrt{\frac{\lambda}{2 - \lambda}} \\
\text{Center line} &= T \\
\text{LCL} &= T_0 - 3\sigma_{e(1)}\sqrt{\frac{\lambda}{2 - \lambda}}.
\end{aligned}
\tag{2.54}
$$

In addition to control charts, other statistics have been suggested for monitoring the performance of a forecasting model. The most common of these are **tracking signals**. The cumulative error tracking signal (CETS) is based on the cumulative sum of all current and previous forecast errors, say,

$$
Y(n) = \sum_{t=1}^{n} e_t(1) = Y(n - 1) + e_n(1).
$$

If the forecasts are unbiased, we would expect $Y(n)$ to fluctuate around zero. If it differs from zero by very much, it could be an indication that the forecasts are biased. The standard deviation of $Y(n)$, say, $\sigma_{Y(n)}$, will provide a measure of how far $Y(n)$ can deviate from zero due entirely to random variation. Therefore, we would conclude that the forecast is biased if $|Y(n)|$ exceeds some multiple of its standard deviation. To operationalize this, suppose that we have an estimate $\hat{\sigma}_{Y(n)}$ of $\sigma_{Y(n)}$ and form the **cumulative error tracking signal**

$$
\text{CETS} = \left| \frac{Y(n)}{\hat{\sigma}_{Y(n)}} \right|.
\tag{2.55}
$$

If the CETS exceeds a constant, say, K_1, we would conclude that the forecasts are biased and that the forecasting model may be inadequate.

It is also possible to devise a **smoothed error tracking signal** based on the smoothed one-step-ahead forecast errors in Eq. (2.52). This would lead to a ratio

$$
\text{SETS} = \left| \frac{\bar{e}_n(1)}{\hat{\sigma}_{\bar{e}_n(1)}} \right|.
\tag{2.56}
$$

If the SETS exceeds a constant, say, K_2, this is an indication that the fore-casts are biased and that there are potentially problems with the forecasting model.

Note that the CETS is very similar to the CUSUM control chart and that the SETS is essentially equivalent to the EWMA control chart. Furthermore, the CUSUM and EWMA are available in standard statistics software (such as Minitab) and the tracking signal procedures are not. So, while tracking signals have been discussed extensively and recommended by some authors, we are not going to encourage their use. Plotting and periodically visually examining a control chart of forecast errors is also very informative, something that is not typically done with tracking signals.

2.7 R COMMANDS FOR CHAPTER 2

Example 2.15 The data are in the second column of the array called gms.data in which the first column is the year. For moving averages, we use functions from package "zoo."

```
plot(gms.data,type="l",xlab='Year',ylab='Average Amount of
Anomaly, °C')
points(gms.data,pch=16,cex=.5)
lines(gms.data[5:125,1],rollmean(gms.data[,2],5),col="red")
points(gms.data[5:125,1],rollmean(gms.data[,2],5),col="red",pch=15,
cex=.5)
legend(1980,-.3,c("Actual","Fits"), pch=c(16,15),lwd=c(.5,.5),
cex=.55,col=c("black","red"))
```

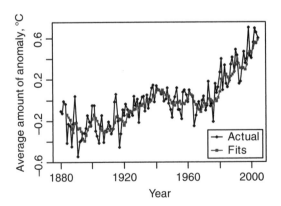

Example 2.16   The data are in the second column of the array called vis.data in which the first column is the time period (or index).

```
# Moving Average
plot(vis.data,type="l",xlab='Time Period',ylab='Viscosity, cP')
points(vis.data,pch=16,cex=.5)
lines(vis.data[5:100,1], rollmean(vis.data[,2],5),col="red")
points(vis.data[5:100,1], rollmean(vis.data[,2],5),col="red",
pch=15,cex=.5)
legend(1,61,c("Actual","Fits"), pch=c(16,15),lwd=c(.5,.5),cex=.55,
col=c("black","red"))
```

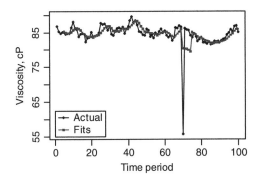

```
# Moving Median
plot(vis.data,type="l",xlab='Time Period',ylab='Viscosity, cP')
points(vis.data,pch=16,cex=.5)
lines(vis.data[5:100,1], rollmedian(vis.data[,2],5),col="red")
points(vis.data[5:100,1], rollmedian(vis.data[,2],5),col="red",
pch=15,cex=.5)
legend(1,61,c("Actual","Fits"), pch=c(16,15),lwd=c(.5,.5),cex=.55,
col=c("black","red"))
```

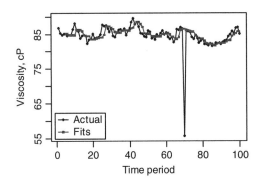

Example 2.17   The pharmaceutical sales data are in the second column of the array called pharma.data in which the first column is the week.

The viscosity data are in the second column of the array called vis.data in which the first column is the year (Note that the 70th observation is corrected).

```
nrp<-dim(pharma.data)[1]

nrv<-dim(vis.data)[1]

plot(pharma.data[1:(nrp-1),2], pharma.data[2:nrp,2],type="p",
xlab='Sales, Week t',ylab=' Sales, Week t+1',pch=20,cex=1)

plot(vis.data[1:(nrv-1),2], vis.data[2:nrv,2],type="p", xlab=
'Reading, Time Period t',ylab=' Reading, Time Period t+1',pch=20,
cex=1)
```

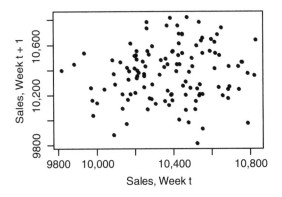

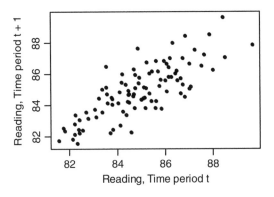

Example 2.18   The viscosity data are in the second column of the array
called vis.data in which the first column is the year (Note that the 70th
observation is corrected).

```
acf(vis.data[,2], lag.max=25,type="correlation",main="ACF of
viscosity readings")
```

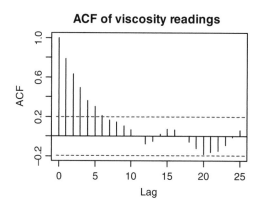

Example 2.19   The cheese production data are in the second column of
the array called cheese.data in which the first column is the year.

```
fit.cheese<-lm(cheese.data[,2]~cheese.data[,1])
plot(cheese.data,type="l",xlab='Year',ylab='Production, 10000lb')
points(cheese.data,pch=16,cex=.5)
lines(cheese.data[,1], fit.cheese$fit,col="red",lty=2)
legend(1990,12000,c("Actual","Fits"),
pch=c(16,NA),lwd=c(.5,.5),lty=c(1,2),cex=.55,col=c("black","red"))
```

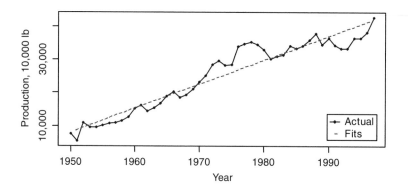

```
par(mfrow=c(2,2),oma=c(0,0,0,0))
qqnorm(fit.cheese$res,datax=TRUE,pch=16,xlab='Residual',main='')
qqline(fit.cheese$res,datax=TRUE)
plot(fit.cheese$fit,fit.cheese$res,pch=16, xlab='Fitted Value',
ylab='Residual')
abline(h=0)
hist(fit.cheese$res,col="gray",xlab='Residual',main='')
plot(fit.cheese$res,type="l",xlab='Observation Order',
ylab='Residual')
points(fit.cheese$res,pch=16,cex=.5)
abline(h=0)
```

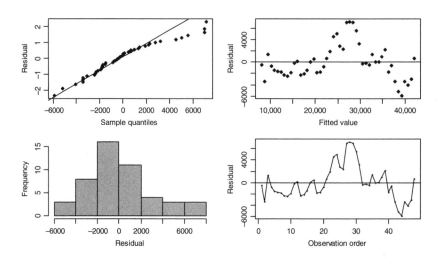

Example 2.20   The cheese production data are in the second column of the array called cheese.data in which the first column is the year.

```
nrc<-dim(cheese.data)[1]
dcheese.data<-cbind(cheese.data[2:nrc,1],diff(cheese.data[,2]))
fit.dcheese<-lm(dcheese.data[,2]~dcheese.data[,1])
plot(dcheese.data,type="l",xlab='',ylab='Production, d=1')
points(dcheese.data,pch=16,cex=.5)
lines(dcheese.data[,1], fit.dcheese$fit,col="red",lty=2)
legend(1952,-2200,c("Actual","Fits"),
pch=c(16,NA),lwd=c(.5,.5),lty=c(1,2),
cex=.75,col=c("black","red"))
```

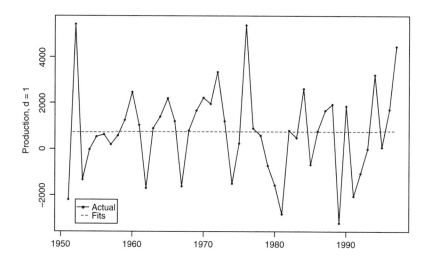

```
par(mfrow=c(2,2),oma=c(0,0,0,0))
qqnorm(fit.dcheese$res,datax=TRUE,pch=16,xlab='Residual',main='')
qqline(fit.dcheese$res,datax=TRUE)
plot(fit.dcheese$fit,fit.dcheese$res,pch=16, xlab='Fitted Value',
ylab='Residual')
abline(h=0)
hist(fit.dcheese$res,col="gray",xlab='Residual',main='')
plot(fit.dcheese$res,type="l",xlab='Observation Order',
ylab='Residual')
points(fit.dcheese$res,pch=16,cex=.5)
abline(h=0)
```

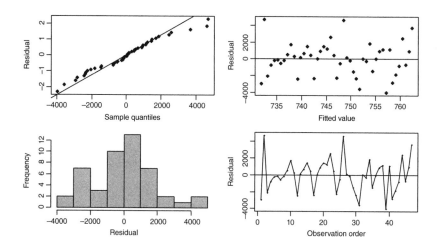

Example 2.21   The beverage sales data are in the second column of the array called bev.data in which the first column is the month of the year.

```
nrb<-dim(bev.data)[1]
tt<-1:nrb
dsbev.data<-bev.data
dsbev.data[,2]<- c(array(NA,dim=c(12,1)),diff(bev.data[,2],12))

plot(tt,dsbev.data[,2],type="l",xlab='',ylab='Seasonal d=12',
xaxt='n')axis(1,seq(1,nrb,24),labels=dsbev.data[seq(1,nrb,24),1])
points(tt,dsbev.data[,2],pch=16,cex=.5)
```

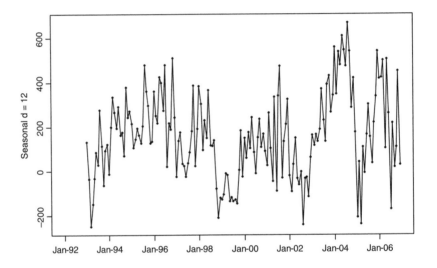

```
dstbev.data<-dsbev.data
dstbev.data[,2]<- c(NA,diff(dstbev.data[,2],1))
fit.dstbev<-lm(dstbev.data[,2]~tt)
plot(tt,dstbev.data[,2],type="l",xlab='',ylab='Seasonal d=12 with
Trend d=1',xaxt='n')
axis(1,seq(1,nrb,24),labels=dsbev.data[seq(1,nrb,24),1])
points(tt,dstbev.data[,2],pch=16,cex=.5)
lines(c(array(NA,dim=c(12,1)),fit.dstbev$fit),col="red",lty=2)
legend(2,-300,c("Actual","Fits"),
pch=c(16,NA),lwd=c(.5,.5),lty=c(1,2),cex=.75,col=c("black","red"))
```

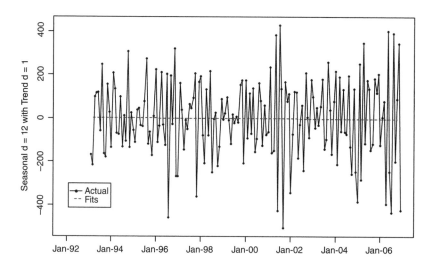

```
par(mfrow=c(2,2),oma=c(0,0,0,0))
qqnorm(fit.dstbev$res,datax=TRUE,pch=16,xlab='Residual',main='')
qqline(fit.dstbev$res,datax=TRUE)
plot(fit.dstbev$fit,fit.dstbev$res,pch=16, xlab='Fitted Value',
ylab='Residual')
abline(h=0)
hist(fit.dstbev$res,col="gray",xlab='Residual',main='')
plot(fit.dstbev$res,type="l",xlab='Observation Order',
ylab='Residual')
points(fit.dstbev$res,pch=16,cex=.5)
abline(h=0)
```

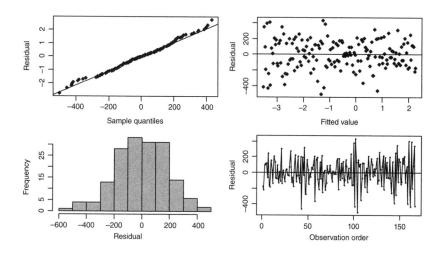

Example 2.22   The beverage sales data are in the second column of the array called bev.data in which the first column is the month of the year.

Software packages use different methods for decomposing a time series. Below we provide the code of doing it in R without using these functions. Note that we use the additive model.

```
nrb<-dim(bev.data)[1]

# De-trend the data
tt<-1:nrb
fit.tbev<-lm(bev.data[,2]~tt)
bev.data.dt<-fit.tbev$res

# Obtain seasonal medians for each month, seasonal period is sp=12
sp<-12
smed<-apply(matrix(bev.data.dt,nrow=sp),1,median)

# Adjust the medians so that their sum is zero
smed<-smed-mean(smed)

# Data without the trend and seasonal components
bev.data.dts<-bev.data.dt-rep(smed,nrb/sp)

# Note that we can also reverse the order, i.e. first take the
  seasonality out
smed2<-apply(matrix(bev.data[,2],nrow=sp),1,median)
smed2<-smed2-mean(smed2)
bev.data.ds<-bev.data[,2]-rep(smed2,nrb/sp)

# To reproduce Figure 2.25

par(mfrow=c(2,2),oma=c(0,0,0,0))
plot(tt,bev.data[,2],type="l",xlab='(a) Original Data',ylab=
'Data',xaxt='n')
axis(1,seq(1,nrb,24),labels=bev.data[seq(1,nrb,24),1])
points(tt,bev.data[,2],pch=16,cex=.75)

plot(tt, bev.data.dt,type="l",xlab='(b) Detrended Data',ylab='Detr.
Data',xaxt='n')
axis(1,seq(1,nrb,24),labels=bev.data[seq(1,nrb,24),1])

points(tt, bev.data.dt,pch=16,cex=.75)
plot(tt, bev.data.ds,type="l",xlab='(c) Seasonally Adjusted Data',
ylab='Seas.
Adj. Data',xaxt='n')
axis(1,seq(1,nrb,24),labels=bev.data[seq(1,nrb,24),1])

points(tt, bev.data.ds,pch=16,cex=.75)
```

```
plot(tt, bev.data.dts,type="l",xlab='(c) Seasonally Adj. and
Detrended Data',ylab='Seas. Adj. and Detr. Data',xaxt='n')
axis(1,seq(1,nrb,24),labels=bev.data[seq(1,nrb,24),1]) points(tt,
bev.data.dts, pch=16,cex=.75)
```

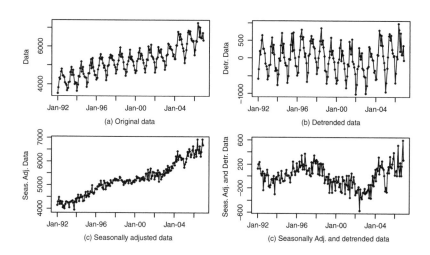

(a) Original data

(b) Detrended data

(c) Seasonally adjusted data

(c) Seasonally Adj. and detrended data

Example 2.23   Functions used to fit a time series model often also provide summary statistics. However, in this example we provide some calculations for a given set of forecast errors as provided in the text.

```
# original data and forecast errors
yt<-c(47,46,51,44,54,47,52,45,50,51,49,41,48,50,51,55,52,53,48,52)
fe<-c(-4.1,-6.9,2.2,-4.1,4.3,-.5,.8,-8.1,-4.4,-.2,-4.3,-5.5,-5.1,
-2.1,4.2,7.3,6.6,5.9,-3.8,6.2)

ME<-mean(fe)
MAD<-mean(abs(fe))
MSE<-mean(fe^2)
ret1<-(fe/yt)*100
MPE<-mean(ret1)
MAPE<-mean(abs(ret1))

> ME
[1] -0.58
> MAD
[1] 4.33
> MSE
[1] 23.59
```

```
> MPE
[1] -1.757938
> MAPE
[1] 8.865001
```

Example 2.24   The forecast error data are in the second column of the array called fe2.data in which the first column is the period.

```
acf.fe2<-acf(fe2.data[,2],main='ACF of Forecast Error (Ex 2.11)')
```

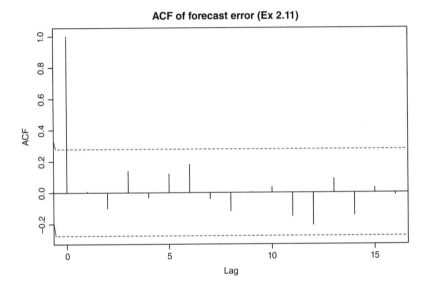

ACF of forecast error (Ex 2.11)

```
# To get the Q_LB statistic, we first define the lag K

K<-13
T<-dim(fe2.data)[1]
QLB<-T*(T+2)*sum((1/(T-1:K))*(acf.fe2$acf[2:(K+1)]^2))

# Upper 5% of χ² distribution with K degrees of freedom
qchisq(.95,K)
```

Example 2.25   The forecast error data are in the second column of the array called fe2.data in which the first column is the period.

```
# The following function can be found in qcc package
# Generating the chart for individuals
qcc(fe2.data[,2],type="xbar.one",title="Individuals Chart for the
Forecast Error")
```

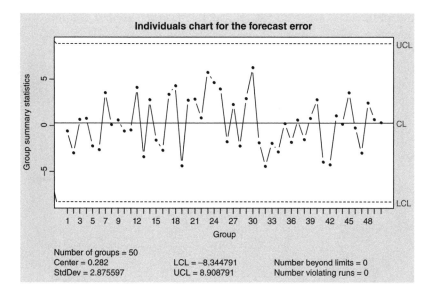

Example 2.26   The forecast error data are in the second column of the array called fe2.data in which the first column is the period.

```
# The following function can be found in qcc package
# Generating the cusum chart

cusum(fe2.data[,2], title='Cusum Chart for the Forecast
Error', sizes=1)
```

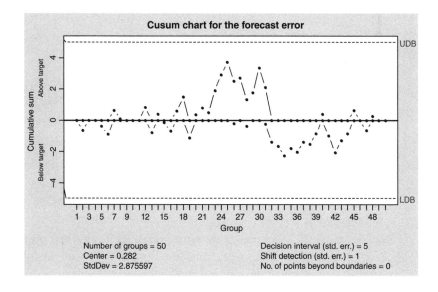

Example 2.27 The forecast error data are in the second column of the array called fe2.data in which the first column is the period.

```
# The following function can be found in qcc package
# Generating the EWMA chart
ewma(fe2.data[,2], title='EWMA Chart for the Forecast Error',
lambda=.1,sizes=1)
```

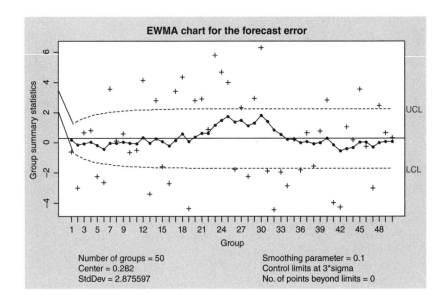

EXERCISES

2.1 Consider the US Treasury Securities rate data in Table B.1 (Appendix B). Find the sample autocorrelation function and the variogram for these data. Is the time series stationary or nonstationary?

2.2 Consider the data on US production of blue and gorgonzola cheeses in Table B.4.

  a. Find the sample autocorrelation function and the variogram for these data. Is the time series stationary or nonstationary?

  b. Take the first difference of the time series, then find the sample autocorrelation function and the variogram. What conclusions can you draw about the structure and behavior of the time series?

2.3 Table B.5 contains the US beverage product shipments data. Find the sample autocorrelation function and the variogram for these data. Is the time series stationary or nonstationary?

2.4 Table B.6 contains two time series: the global mean surface air temperature anomaly and the global CO_2 concentration. Find the sample autocorrelation function and the variogram for both of these time series. Is either one of the time series stationary?

2.5 Reconsider the global mean surface air temperature anomaly and the global CO_2 concentration time series from Exercise 2.4. Take the first difference of both time series. Find the sample autocorrelation function and variogram of these new time series. Is either one of these differenced time series stationary?

2.6 Find the closing stock price for a stock that interests you for the last 200 trading days. Find the sample autocorrelation function and the variogram for this time series. Is the time series stationary?

2.7 Reconsider the Whole Foods Market stock price data from Exercise 2.6. Take the first difference of the data. Find the sample autocorrelation function and the variogram of this new time series. Is this differenced time series stationary?

2.8 Consider the unemployment rate data in Table B.8. Find the sample autocorrelation function and the variogram for this time series. Is the time series stationary or nonstationary? What conclusions can you draw about the structure and behavior of the time series?

2.9 Table B.9 contains the annual International Sunspot Numbers. Find the sample autocorrelation function and the variogram for this time series. Is the time series stationary or nonstationary?

2.10 Table B.10 contains data on the number of airline miles flown in the United Kingdom. This is strongly seasonal data. Find the sample autocorrelation function for this time series.
   a. Is the seasonality apparent in the sample autocorrelation function?
   b. Is the time series stationary or nonstationary?

2.11 Reconsider the data on the number of airline miles flown in the United Kingdom from Exercise 2.10. Take the natural logarithm of the data and plot this new time series.
   a. What impact has the log transformation had on the time series?

b. Find the autocorrelation function for this time series.

c. Interpret the sample autocorrelation function.

2.12 Reconsider the data on the number of airline miles flown in the United Kingdom from Exercises 2.10 and 2.11. Take the first difference of the natural logarithm of the data and plot this new time series.

a. What impact has the log transformation had on the time series?

b. Find the autocorrelation function for this time series.

c. Interpret the sample autocorrelation function.

2.13 The data on the number of airline miles flown in the United Kingdom in Table B.10 are seasonal. Difference the data at a season lag of 12 months and also apply a first difference to the data. Plot the differenced series. What effect has the differencing had on the time series? Find the sample autocorrelation function and the variogram. What does the sample autocorrelation function tell you about the behavior of the differenced series?

2.14 Table B.11 contains data on the monthly champagne sales in France. This is strongly seasonal data. Find the sample autocorrelation function and variogram for this time series.

a. Is the seasonality apparent in the sample autocorrelation function?

b. Is the time series stationary or nonstationary?

2.15 Reconsider the champagne sales data from Exercise 2.14. Take the natural logarithm of the data and plot this new time series.

a. What impact has the log transformation had on the time series?

b. Find the autocorrelation function and variogram for this time series.

c. Interpret the sample autocorrelation function and variogram.

2.16 Table B.13 contains data on ice cream and frozen yogurt production. Plot the data and calculate both the sample autocorrelation function and variogram. Is there an indication of nonstationary behavior in the time series? Now plot the first difference of the time series and compute the sample autocorrelation function and variogram of the first differences. What impact has differencing had on the time series?

2.17 Table B.14 presents data on CO_2 readings from the Mauna Loa Observatory. Plot the data, then calculate the sample autocorrelation function and variogram. Is there an indication of nonstationary behavior in the time series? Now plot the first difference of the time series and compute the sample autocorrelation function and the variogram of the first differences. What impact has differencing had on the time series?

2.18 Data on violent crime rates are given in Table B.15. Plot the data and calculate the sample autocorrelation function and variogram. Is there an indication of nonstationary behavior in the time series? Now plot the first difference of the time series and compute the sample autocorrelation function and variogram of the first differences. What impact has differencing had on the time series?

2.19 Table B.16 presents data on the US Gross Domestic Product (GDP). Plot the GDP data and calculate the sample autocorrelation function and variogram. Is there an indication of nonstationary behavior in the time series? Now plot the first difference of the GDP time series and compute the sample autocorrelation function and variogram of the first differences. What impact has differencing had on the time series?

2.20 Table B.17 contains information on total annual energy consumption. Plot the energy consumption data and calculate the sample autocorrelation function and variogram. Is there an indication of nonstationary behavior in the time series? Now plot the first difference of the time series and compute the sample autocorrelation function and variogram of the first differences. What impact has differencing had on the time series?

2.21 Data on US coal production are given in Table B.18. Plot the coal production data and calculate the sample autocorrelation function and variogram. Is there an indication of nonstationary behavior in the time series? Now plot the first difference of the time series and compute the sample autocorrelation function and variogram of the first differences. What impact has differencing had on the time series?

2.22 Consider the CO_2 readings from Mauna Loa in Table B.14. Use a six-period moving average to smooth the data. Plot both the smoothed data and the original CO_2 readings on the same axes. What has the moving average done? Repeat the procedure with a three-period moving average. What is the effect of changing the span of the moving average?

2.23 Consider the violent crime rate data in Table B.15. Use a ten-period moving average to smooth the data. Plot both the smoothed data and the original CO_2 readings on the same axes. What has the moving average done? Repeat the procedure with a four-period moving average. What is the effect of changing the span of the moving average?

2.24 Table B.21 contains data from the US Energy Information Administration on monthly average price of electricity for the residential sector in Arizona. Plot the data and comment on any features that you observe from the graph. Calculate and plot the sample ACF and variogram. Interpret these graphs.

2.25 Reconsider the residential electricity price data from Exercise 2.24.
   a. Plot the first difference of the data and comment on any features that you observe from the graph. Calculate and plot the sample ACF and variogram for the differenced data. Interpret these graphs. What impact did differencing have?
   b. Now difference the data again at a seasonal lag of 12. Plot the differenced data and comment on any features that you observe from the graph. Calculate and plot the sample ACF and variogram for the differenced data. Interpret these graphs. What impact did regular differencing combined with seasonal differencing have?

2.26 Table B.22 contains data from the Danish Energy Agency on Danish crude oil production. Plot the data and comment on any features that you observe from the graph. Calculate and plot the sample ACF and variogram. Interpret these graphs.

2.27 Reconsider the Danish crude oil production data from Exercise 2.26. Plot the first difference of the data and comment on any features that you observe from the graph. Calculate and plot the sample ACF and variogram for the differenced data. Interpret these graphs. What impact did differencing have?

2.28 Use a six-period moving average to smooth the first difference of the Danish crude oil production data that you computed in Exercise 2.27. Plot both the smoothed data and the original data on the same axes. What has the moving average done? Does the moving average look like a reasonable forecasting technique for the differenced data?

2.29 Weekly data on positive laboratory test results for influenza are shown in Table B.23. Notice that these data have a number of missing values. Construct a time series plot of the data and comment on any relevant features that you observe.

a. What is the impact of the missing observations on your ability to model and analyze these data?

b. Develop and implement a scheme to estimate the missing values

2.30 Climate data collected from Remote Automated Weather Stations (RAWS) are used to monitor the weather and to assist land management agencies with projects such as monitoring air quality, rating fire danger, and other research purposes. Data from the Western Regional Climate Center for the mean daily solar radiation (in Langleys) at the Zion Canyon, Utah, station are shown in Table B.24.

a. Plot the data and comment on any features that you observe.

b. Calculate and plot the sample ACF and variogram. Comment on the plots.

c. Apply seasonal differencing to the data, plot the data, and construct the sample ACF and variogram. What was the impact of seasonal differencing?

2.31 Table B.2 contains annual US motor vehicle traffic fatalities along with other information. Plot the data and comment on any features that you observe from the graph. Calculate and plot the sample ACF and variogram. Interpret these graphs.

2.32 Reconsider the motor vehicle fatality data from Exercise 2.31.

a. Plot the first difference of the data and comment on any features that you observe from the graph. Calculate and plot the sample ACF and variogram for the differenced data. Interpret these graphs. What impact did differencing have?

b. Compute a six-period moving average for the differenced data. Plot the moving average and the original data on the same axes. Does it seem that the six-period moving average would be a good forecasting technique for the differenced data?

2.33 Apply the X-11 seasonal decomposition method (or any other seasonal adjustment technique for which you have software) to the mean daily solar radiation in Table B.24.

2.34 Consider the N-span moving average applied to data that are uncorrelated with mean μ and variance σ^2.

a. Show that the variance of the moving average is $\text{Var}(M_t) = \sigma^2/N$.

b. Show that $\text{Cov}(M_t, M_{t+k}) = \sigma^2 \sum_{j=1}^{N-k} (1/N)^2$, for $k < N$.

c. Show that the autocorrelation function is

$$\rho_k = \begin{cases} 1 - \dfrac{|k|}{N}, & k = 1, 2, \dots, N-1 \\ 0, & k \geq N \end{cases}$$

2.35 Consider an N-span moving average where each observation is weighted by a constant, say, $a_j \geq 0$. Therefore the weighted moving average at the end of period T is

$$M_T^w = \sum_{t=T-N+1}^{T} a_{T+1-t} y_t.$$

a. Why would you consider using a weighted moving average?

b. Show that the variance of the weighted moving average is $\text{Var}(M_T^w) = \sigma^2 \sum_{j=i}^{N} a_j^2$.

c. Show that $\text{Cov}(M_T^w, M_{T+k}^w) = \sigma^2 \sum_{j=1}^{N-k} a_j a_{j+k}$, $|k| < N$.

d. Show that the autocorrelation function is

$$\rho_k = \begin{cases} \left(\displaystyle\sum_{j=1}^{N-k} a_j a_{j+k} \right) \Big/ \left(\displaystyle\sum_{j=1}^{N} a_j^2 \right), & k = 1, 2, \dots, N-1 \\ 0, & k \geq N \end{cases}$$

2.36 Consider the Hanning filter. This is a weighted moving average.

a. Find the variance of the weighted moving average for the Hanning filter. Is this variance smaller than the variance of a simple span-3 moving average with equal weights?

b. Find the autocorrelation function for the Hanning filter. Compare this with the autocorrelation function for a simple span-3 moving average with equal weights.

2.37 Suppose that a simple moving average of span N is used to forecast a time series that varies randomly around a constant, that is, $y_t = \mu + \varepsilon_t$, where the variance of the error term is σ^2. The forecast error at lead one is $e_{T+1}(1) = y_{T+1} - M_T$. What is the variance of this lead-one forecast error?

2.38 Suppose that a simple moving average of span N is used to forecast a time series that varies randomly around a constant, that is,

$y_t = \mu + \varepsilon_t$, where the variance of the error term is σ^2. You are interested in forecasting the cumulative value of y over a lead time of L periods, say, $y_{T+1} + y_{T+2} + \cdots + y_{T+L}$.

a. The forecast of this cumulative demand is LM_T. Why?

b. What is the variance of the cumulative forecast error?

2.39 Suppose that a simple moving average of span N is used to forecast a time series that varies randomly around a constant mean, that is, $y_t = \mu + \varepsilon_t$. At the start of period t_1 the process shifts to a new mean level, say, $\mu + \delta$. Show that the expected value of the moving average is

$$E(M_T) = \begin{cases} \mu, & T \leq t_1 - 1 \\ \mu + \dfrac{T - t_1 + 1}{N}\delta, & t_1 \leq T \leq t_1 + N - 2 \cdot \\ \mu + \delta, & T \geq t_1 + N - 1 \end{cases}$$

2.40 Suppose that a simple moving average of span N is used to forecast a time series that varies randomly around a constant mean, that is, $y_t = \mu + \varepsilon_t$. At the start of period t_1 the process experiences a transient; that is, it shifts to a new mean level, say, $\mu + \delta$, but it reverts to its original level μ at the start of period $t_1 + 1$. Show that the expected value of the moving average is

$$E(M_T) = \begin{cases} \mu, & T \leq t_1 - 1 \\ \mu + \dfrac{\delta}{N}, & t_1 \leq T \leq t_1 + N - 1 \cdot \\ \mu, & T \geq t_1 + N \end{cases}$$

2.41 If a simple N–span moving average is applied to a time series that has a linear trend, say, $y_t = \beta_0 + \beta_1 t + \varepsilon_t$, the moving average will lag behind the observations. Assume that the observations are uncorrelated and have constant variance. Show that at time T the expected value of the moving average is

$$E(M_T) = \beta_0 + \beta_1 T - \dfrac{N-1}{2}\beta_1.$$

2.42 Use a three-period moving average to smooth the champagne sales data in Table B.11. Plot the moving average on the same axes as the original data. What impact has this smoothing procedure had on the data?

TABLE E2.1 One-Step-Ahead Forecast Errors for Exercise 2.44

Period, t	$e_t(1)$	Period, t	$e_t(1)$	Period, t	$e_t(1)$	Period, t	$e_t(1)$
1	1.83	11	-2.30	21	3.30	31	-0.07
2	-1.80	12	0.65	22	1.036	32	0.57
3	0.09	13	-0.01	23	2.042	33	2.92
4	-1.53	14	-1.11	24	1.04	34	1.99
5	-0.58	15	0.13	25	-0.87	35	1.74
6	0.21	16	-1.07	26	-0.39	36	-0.76
7	1.25	17	0.80	27	-0.29	37	2.35
8	-1.22	18	-1.98	28	2.08	38	-1.91
9	1.32	19	0.02	29	3.36	39	2.22
10	3.63	20	0.25	30	-0.53	40	2.57

2.43 Use a 12-period moving average to smooth the champagne sales data in Table B.11. Plot the moving average on the same axes as the original data. What impact has this smoothing procedure had on the data?

2.44 Table E2.1 contains 40 one-step-ahead forecast errors from a forecasting model.

a. Find the sample ACF of the forecast errors. Interpret the results.

b. Construct a normal probability plot of the forecast errors. Is there evidence to support a claim that the forecast errors are normally distributed?

c. Find the mean error, the mean squared error, and the mean absolute deviation. Is it likely that the forecasting technique produces unbiased forecasts?

2.45 Table E2.2 contains 40 one-step-ahead forecast errors from a forecasting model.

a. Find the sample ACF of the forecast errors. Interpret the results.

b. Construct a normal probability plot of the forecast errors. Is there evidence to support a claim that the forecast errors are normally distributed?

c. Find the mean error, the mean squared error, and the mean absolute deviation. Is it likely that the forecasting method produces unbiased forecasts?

2.46 Exercises 2.44 and 2.45 present information on forecast errors. Suppose that these two sets of forecast errors come from two different

TABLE E2.2   One-Step-Ahead Forecast Errors for Exercise 2.45

Period, t	$e_t(1)$	Period, t	$e_t(1)$	Period, t	$e_t(1)$	Period, t	$e_t(1)$
1	−4.26	11	3.62	21	−6.24	31	−6.42
2	−3.12	12	−5.08	22	−0.25	32	−8.94
3	−1.87	13	−1.35	23	−3.64	33	−1.76
4	0.98	14	3.46	24	5.49	34	−0.57
5	−5.17	15	−0.19	25	−2.01	35	−10.32
6	0.13	16	−7.48	26	−4.24	36	−5.64
7	1.85	17	−3.61	27	−4.61	37	−1.45
8	−2.83	18	−4.21	28	3.24	38	−5.67
9	0.95	19	−6.49	29	−8.66	39	−4.45
10	7.56	20	4.03	30	−1.32	40	−10.23

forecasting methods applied to the same time series. Which of these two forecasting methods would you recommend for use? Why?

2.47 Consider the forecast errors in Exercise 2.44. Construct individuals and moving range control charts for these forecast errors. Does the forecasting system exhibit stability over this time period?

2.48 Consider the forecast errors in Exercise 2.44. Construct a cumulative sum control chart for these forecast errors. Does the forecasting system exhibit stability over this time period?

2.49 Consider the forecast errors in Exercise 2.45. Construct individuals and moving range control charts for these forecast errors. Does the forecasting system exhibit stability over this time period?

2.50 Consider the forecast errors in Exercise 2.45. Construct a cumulative sum control chart for these forecast errors. Does the forecasting system exhibit stability over this time period?

2.51 Ten additional forecast errors for the forecasting model in Exercise 2.44 are as follows: 5.5358, −2.6183, 0.0130, 1.3543, 12.6980, 2.9007, 0.8985, 2.9240, 2.6663, and −1.6710. Plot these additional 10 forecast errors on the individuals and moving range control charts constructed in Exercise 2.47. Is the forecasting system still working satisfactorily?

2.52 Plot the additional 10 forecast errors from Exercise 2.51 on the cumulative sum control chart constructed in Exercise 2.38. Is the forecasting system still working satisfactorily?

CHAPTER 3

REGRESSION ANALYSIS AND FORECASTING

Weather forecast for tonight: dark

GEORGE CARLIN, *American comedian*

3.1 INTRODUCTION

Regression analysis is a statistical technique for modeling and investigating the relationships between an **outcome** or **response** variable and one or more **predictor** or **regressor** variables. The end result of a regression analysis study is often to generate a model that can be used to forecast or predict future values of the response variable, given specified values of the predictor variables.

The **simple linear regression model** involves a single predictor variable and is written as

$$y = \beta_0 + \beta_1 x + \varepsilon, \tag{3.1}$$

where y is the response, x is the predictor variable, β_0 and β_1 are unknown parameters, and ε is an error term. The model parameters or **regression**

Introduction to Time Series Analysis and Forecasting, Second Edition.
Douglas C. Montgomery, Cheryl L. Jennings and Murat Kulahci.
© 2015 John Wiley & Sons, Inc. Published 2015 by John Wiley & Sons, Inc.

coefficients β_0 and β_1 have a physical interpretation as the intercept and slope of a straight line, respectively. The slope β_1 measures the change in the mean of the response variable y for a unit change in the predictor variable x. These parameters are typically unknown and must be estimated from a sample of data. The error term ε accounts for deviations of the actual data from the straight line specified by the model equation. We usually think of ε as a statistical error, so we define it as a random variable and will make some assumptions about its distribution. For example, we typically assume that ε is normally distributed with mean zero and variance σ^2, abbreviated $N(0, \sigma^2)$. Note that the variance is assumed constant; that is, it does not depend on the value of the predictor variable (or any other variable).

Regression models often include more than one predictor or regressor variable. If there are k predictors, the **multiple linear regression model** is

$$y = \beta_0 + \beta_1 x_1 + \beta_2 x_2 + \cdots + \beta_k x_k + \varepsilon. \tag{3.2}$$

The parameters $\beta_0, \beta_1, \ldots, \beta_k$ in this model are often called partial regression coefficients because they convey information about the effect on y of the predictor that they multiply, given that all of the other predictors in the model do not change.

The regression models in Eqs. (3.1) and (3.2) are **linear** regression models because they are linear in the unknown parameters (the β's), and not because they necessarily describe linear relationships between the response and the regressors. For example, the model

$$y = \beta_0 + \beta_1 x + \beta_2 x^2 + \varepsilon$$

is a linear regression model because it is linear in the unknown parameters β_0, β_1, and β_2, although it describes a quadratic relationship between y and x. As another example, consider the regression model

$$y_t = \beta_0 + \beta_1 \sin \frac{2\pi}{d} t + \beta_2 \cos \frac{2\pi}{d} t + \varepsilon_t, \tag{3.3}$$

which describes the relationship between a response variable y that varies cyclically with time (hence the subscript t) and the nature of this cyclic variation can be described as a simple sine wave. Regression models such as Eq. (3.3) can be used to remove seasonal effects from time series data (refer to Section 2.4.2 where models like this were introduced). If the period d of the cycle is specified (such as $d = 12$ for monthly data with an annual cycle), then sin $(2\pi/d)t$ and cos $(2\pi/d)t$ are just numbers for each observation on the response variable and Eq. (3.3) is a standard linear regression model.

We will discuss the use of regression models for forecasting or making predictions in two different situations. The first of these is the situation where all of the data are collected on y and the regressors in a single time period (or put another way, the data are not time oriented). For example, suppose that we wanted to develop a regression model to predict the proportion of consumers who will redeem a coupon for purchase of a particular brand of milk (y) as a function of the amount of the discount or face value of the coupon (x). These data are collected over some specified study period (such as a month) and the data do not explicitly vary with time. This type of regression data is called **cross-section data**. The regression model for cross-section data is written as

$$y_i = \beta_0 + \beta_1 x_{i1} + \beta_2 x_{i2} + \cdots + \beta_k x_{ik} + \varepsilon_i, \quad i = 1, 2, \ldots, n, \quad (3.4)$$

where the subscript i is used to denote each individual observation (or case) in the data set and n represents the number of observations. In the other situation the response and the regressors are time series, so the regression model involves **time series data**. For example, the response variable might be hourly CO_2 emissions from a chemical plant and the regressor variables might be the hourly production rate, hourly changes in the concentration of an input raw material, and ambient temperature measured each hour. All of these are time-oriented or time series data.

The regression model for time series data is written as

$$y_t = \beta_0 + \beta_1 x_{t1} + \beta_2 x_{t2} + \cdots + \beta_k x_{tk} + \varepsilon_t, \quad t = 1, 2, \ldots, T \quad (3.5)$$

In comparing Eq. (3.5) to Eq. (3.4), note that we have changed the observation or case subscript from i to t to emphasize that the response and the predictor variables are time series. Also, we have used T instead of n to denote the number of observations in keeping with our convention that, when a time series is used to build a forecasting model, T represents the most recent or last available observation. Equation (3.3) is a specific example of a time series regression model.

The unknown parameters $\beta_0, \beta_1, \ldots, \beta_k$ in a linear regression model are typically estimated using the method of **least squares**. We illustrated least squares model fitting in Chapter 2 for removing trend and seasonal effects from time series data. This is an important application of regression models in forecasting, but not the only one. Section 3.1 gives a formal description of the least squares estimation procedure. Subsequent sections deal with statistical inference about the model and its parameters, and with model adequacy checking. We will also describe and illustrate several ways in which regression models are used in forecasting.

3.2 LEAST SQUARES ESTIMATION IN LINEAR REGRESSION MODELS

We begin with the situation where the regression model is used with cross-section data. The model is given in Eq. (3.4). There are $n > k$ observations on the response variable available, say, $y_1, y_2, \ldots, y_n$. Along with each observed response y_i, we will have an observation on each regressor or predictor variable and x_{ij} denotes the ith observation or level of variable x_j. The data will appear as in Table 3.1. We assume that the error term ε in the model has expected value $E(\varepsilon) = 0$ and variance Var $(\varepsilon) = \sigma^2$, and that the errors ε_i, $i = 1, 2, \ldots, n$ are uncorrelated random variables.

The method of least squares chooses the model parameters (the β's) in Eq. (3.4) so that the sum of the squares of the errors, ε_i, is minimized. The least squares function is

$$
L = \sum_{i=1}^{n} \varepsilon_i^2 = \sum_{i=1}^{n} (y_i - \beta_0 - \beta_1 x_{i1} - \beta_2 x_{i2} - \cdots - \beta_k x_{ik})^2
$$

$$
= \sum_{i=1}^{n} \left(y_i - \beta_0 - \sum_{j=1}^{k} \beta_j x_{ij} \right)^2.
$$

(3.6)

This function is to be minimized with respect to $\beta_0, \beta_1, \ldots, \beta_k$. Therefore the least squares estimators, say, $\hat{\beta}_0, \hat{\beta}_1, \ldots, \hat{\beta}_k$, must satisfy

$$
\frac{\partial L}{\partial \beta_0}\bigg|_{\beta_0,\beta_1,\ldots,\beta_k} = -2 \sum_{i=1}^{n} \left(y_i - \hat{\beta}_0 - \sum_{j=1}^{k} \hat{\beta}_j x_{ij} \right) = 0
$$

(3.7)

TABLE 3.1   Cross-Section Data for Multiple Linear Regression

Observation	Response, y	x_1	x_2	$\cdots$	x_k
1	y_1	x_{11}	x_{12}	$\cdots$	x_{1k}
2	y_2	x_{21}	x_{22}	$\cdots$	x_{2k}
$\vdots$	$\vdots$	$\vdots$	$\vdots$	$\vdots$	$\vdots$
n	y_n	x_{n1}	x_{n2}	$\cdots$	x_{nk}

and

$$\left.\frac{\partial L}{\partial \beta_j}\right|_{\beta_0,\beta_1,\ldots,\beta_k} = -2 \sum_{i=1}^{n} \left(y_i - \hat{\beta}_0 - \sum_{j=1}^{k} \hat{\beta}_j x_{ij} \right) x_{ij} = 0, \quad j = 1, 2, \ldots, k$$

$$(3.8)$$

Simplifying Eqs. (3.7) and (3.8), we obtain

$$n\hat{\beta}_0 + \hat{\beta}_1 \sum_{i=1}^{n} x_{i1} + \hat{\beta}_2 \sum_{i=1}^{n} x_{i2} + \cdots + \hat{\beta}_k \sum_{i=1}^{n} x_{ik} = \sum_{i=1}^{n} y_i \qquad (3.9)$$

$$\hat{\beta}_0 \sum_{i=1}^{n} x_{i1} + \hat{\beta}_1 \sum_{i=1}^{n} x_{i1}^2 + \hat{\beta}_2 \sum_{i=1}^{n} x_{i2} x_{i1} + \cdots + \hat{\beta}_k \sum_{i=1}^{n} x_{ik} x_{i1} = \sum_{i=1}^{n} y_i x_{i1}$$

$$\vdots \qquad (3.10)$$

$$\hat{\beta}_0 \sum_{i=1}^{n} x_{ik} + \hat{\beta}_1 \sum_{i=1}^{n} x_{i1} x_{ik} + \hat{\beta}_2 \sum_{i=1}^{n} x_{i2} x_{ik} + \cdots + \hat{\beta}_k \sum_{i=1}^{n} x_{ik}^2 = \sum_{i=1}^{n} y_i x_{ik}$$

These equations are called the **least squares normal equations**. Note that there are $p = k + 1$ normal equations, one for each of the unknown regression coefficients. The solutions to the normal equations will be the least squares estimators of the model regression coefficients.

It is simpler to solve the normal equations if they are expressed in matrix notation. We now give a matrix development of the normal equations that parallels the development of Eq. (3.10). The multiple linear regression model may be written in matrix notation as

$$\mathbf{y} = \mathbf{X}\boldsymbol{\beta} + \boldsymbol{\varepsilon}, \qquad (3.11)$$

where

$$\mathbf{y} = \begin{bmatrix} y_1 \\ y_2 \\ \vdots \\ y_n \end{bmatrix}, \quad \mathbf{X} = \begin{bmatrix} 1 & x_{11} & x_{12} & \cdots & x_{1k} \\ 1 & x_{21} & x_{22} & \cdots & x_{2k} \\ \vdots & \vdots & \vdots & & \vdots \\ 1 & x_{n1} & x_{n2} & \cdots & x_{nk} \end{bmatrix}, \quad \boldsymbol{\beta} = \begin{bmatrix} \beta_0 \\ \beta_1 \\ \vdots \\ \beta_k \end{bmatrix}, \quad \text{and} \quad \boldsymbol{\varepsilon} = \begin{bmatrix} \varepsilon_1 \\ \varepsilon_2 \\ \vdots \\ \varepsilon_n \end{bmatrix}$$

In general, $\mathbf{y}$ is an $(n \times 1)$ vector of the observations, $\mathbf{X}$ is an $(n \times p)$ matrix of the levels of the regressor variables, $\boldsymbol{\beta}$ is a $(p \times 1)$ vector of the regression coefficients, and ε is an $(n \times 1)$ vector of random errors. $\mathbf{X}$ is usually called the **model matrix**, because it is the original data table for the problem expanded to the form of the regression model that you desire to fit.

The vector of least squares estimators minimizes

$$L = \sum_{i=1}^{n} \varepsilon_i^2 = \varepsilon'\varepsilon = (\mathbf{y} - \mathbf{X}\beta)'(\mathbf{y} - \mathbf{X}\beta)$$

We can expand the right-hand side of L and obtain

$$L = \mathbf{y}'\mathbf{y} - \beta'\mathbf{X}'\mathbf{y} - \mathbf{y}'\mathbf{X}\beta + \beta'\mathbf{X}'\mathbf{X}\beta = \mathbf{y}'\mathbf{y} - 2\beta'\mathbf{X}'\mathbf{y} + \beta'\mathbf{X}'\mathbf{X}\beta,$$

because $\beta'\mathbf{X}'\mathbf{y}$ is a (1×1) matrix, or a scalar, and its transpose $(\beta'\mathbf{X}'\mathbf{y})' = \mathbf{y}'\mathbf{X}\beta$ is the same scalar. The least squares estimators must satisfy

$$\left.\frac{\partial L}{\partial \beta}\right|_{\hat{\beta}} = -2\mathbf{X}'\mathbf{y} + 2(\mathbf{X}'\mathbf{X})\hat{\beta} = \mathbf{0},$$

which simplifies to

$$(\mathbf{X}'\mathbf{X})\hat{\beta} = \mathbf{X}'\mathbf{y} \tag{3.12}$$

In Eq. (3.12) $\mathbf{X}'\mathbf{X}$ is a $(p \times p)$ symmetric matrix and $\mathbf{X}'\mathbf{y}$ is a $(p \times 1)$ column vector. Equation (3.12) is just the matrix form of the least squares normal equations. It is identical to Eq. (3.10). To solve the normal equations, multiply both sides of Eq. (3.12) by the inverse of $\mathbf{X}'\mathbf{X}$ (we assume that this inverse exists). Thus the least squares estimator of $\hat{\beta}$ is

$$\hat{\beta} = (\mathbf{X}'\mathbf{X})^{-1}\mathbf{X}'\mathbf{y} \tag{3.13}$$

The fitted values of the response variable from the regression model are computed from

$$\hat{\mathbf{y}} = \mathbf{X}\hat{\beta} \tag{3.14}$$

or in scalar notation,

$$\hat{y}_i = \hat{\beta}_0 + \hat{\beta}_1 x_{i1} + \hat{\beta}_2 x_{i2} + \cdots + \hat{\beta}_k x_{ik}, \quad i = 1, 2, \ldots, n \tag{3.15}$$

The difference between the actual observation y_i and the corresponding fitted value is the **residual** $e_i = y_i - \hat{y}_i, i = 1, 2, \ldots, n$. The n residuals can be written as an $(n \times 1)$ vector denoted by

$$\mathbf{e} = \mathbf{y} - \hat{\mathbf{y}} = \mathbf{y} - \mathbf{X}\hat{\boldsymbol{\beta}} \tag{3.16}$$

In addition to estimating the regression coefficients $\beta_0, \beta_1, \ldots, \beta_k$, it is also necessary to estimate the variance of the model errors, σ^2. The estimator of this parameter involves the sum of squares of the residuals

$$SS_E = (\mathbf{y} - \mathbf{X}\hat{\boldsymbol{\beta}})'(\mathbf{y} - \mathbf{X}\hat{\boldsymbol{\beta}})$$

We can show that $E(SS_E) = (n - p)\sigma^2$, so the estimator of σ^2 is the **residual** or **mean square error**

$$\hat{\sigma}^2 = \frac{SS_E}{n - p} \tag{3.17}$$

The method of least squares is not the only way to estimate the parameters in a linear regression model, but it is widely used, and it results in estimates of the model parameters that have nice properties. If the model is correct (it has the right form and includes all of the relevant predictors), the least squares estimator $\hat{\boldsymbol{\beta}}$ is an unbiased estimator of the model parameters $\boldsymbol{\beta}$; that is,

$$E(\hat{\boldsymbol{\beta}}) = \boldsymbol{\beta}.$$

The variances and covariances of the estimators $\hat{\boldsymbol{\beta}}$ are contained in a $(p \times p)$ covariance matrix

$$\text{Var}(\hat{\boldsymbol{\beta}}) = \sigma^2 (\mathbf{X}'\mathbf{X})^{-1} \tag{3.18}$$

The variances of the regression coefficients are on the main diagonal of this matrix and the covariances are on the off-diagonals.

Example 3.1   A hospital is implementing a program to improve quality and productivity. As part of this program, the hospital is attempting to measure and evaluate patient satisfaction. Table 3.2 contains some of the data that have been collected for a random sample of 25 recently discharged patients. The "severity" variable is an index that measures the severity of

TABLE 3.2 Patient Satisfaction Survey Data

Observation	Age (x_1)	Severity (x_2)	Satisfaction (y)
1	55	50	68
2	46	24	77
3	30	46	96
4	35	48	80
5	59	58	43
6	61	60	44
7	74	65	26
8	38	42	88
9	27	42	75
10	51	50	57
11	53	38	56
12	41	30	88
13	37	31	88
14	24	34	102
15	42	30	88
16	50	48	70
17	58	61	52
18	60	71	43
19	62	62	46
20	68	38	56
21	70	41	59
22	79	66	26
23	63	31	52
24	39	42	83
25	49	40	75

the patient's illness, measured on an increasing scale (i.e., more severe illnesses have higher values of the index), and the response satisfaction is also measured on an increasing scale, with larger values indicating greater satisfaction.

We will fit a multiple linear regression model to the patient satisfaction data. The model is

$$y = \beta_0 + \beta_1 x_1 + \beta_2 x_2 + \varepsilon,$$

where y = patient satisfaction, x_1 = patient age, and x_2 = illness severity. To solve the least squares normal equations, we will need to set up the $\mathbf{X'X}$

matrix and the $\mathbf{X'y}$ vector. The model matrix $\mathbf{X}$ and observation vector $\mathbf{y}$ are

$$
\mathbf{X} = \begin{bmatrix}
1 & 55 & 50 \\
1 & 46 & 24 \\
1 & 30 & 46 \\
1 & 35 & 48 \\
1 & 59 & 58 \\
1 & 61 & 60 \\
1 & 74 & 65 \\
1 & 38 & 42 \\
1 & 27 & 42 \\
1 & 51 & 50 \\
1 & 53 & 38 \\
1 & 41 & 30 \\
1 & 37 & 31 \\
1 & 24 & 34 \\
1 & 42 & 30 \\
1 & 50 & 48 \\
1 & 58 & 61 \\
1 & 60 & 71 \\
1 & 62 & 62 \\
1 & 68 & 38 \\
1 & 70 & 41 \\
1 & 79 & 66 \\
1 & 63 & 31 \\
1 & 39 & 42 \\
1 & 49 & 40
\end{bmatrix}, \qquad
\mathbf{y} = \begin{bmatrix}
68 \\
77 \\
96 \\
80 \\
43 \\
44 \\
26 \\
88 \\
75 \\
57 \\
56 \\
88 \\
88 \\
102 \\
88 \\
70 \\
52 \\
43 \\
46 \\
56 \\
59 \\
26 \\
52 \\
83 \\
75
\end{bmatrix}
$$

The $\mathbf{X'X}$ matrix and the $\mathbf{X'y}$ vector are

$$
\mathbf{X'X} = \begin{bmatrix}
1 & 1 & \cdots & 1 \\
55 & 46 & \cdots & 49 \\
50 & 24 & \cdots & 40
\end{bmatrix}
\begin{bmatrix}
1 & 55 & 50 \\
1 & 46 & 24 \\
\vdots & \vdots & \vdots \\
1 & 49 & 40
\end{bmatrix}
= \begin{bmatrix}
25 & 1271 & 1148 \\
1271 & 69881 & 60814 \\
1148 & 60814 & 56790
\end{bmatrix}
$$

and

$$
\mathbf{X'y} = \begin{bmatrix}
1 & 1 & \cdots & 1 \\
55 & 46 & \cdots & 49 \\
50 & 24 & \cdots & 40
\end{bmatrix}
\begin{bmatrix}
68 \\
77 \\
\vdots \\
75
\end{bmatrix}
= \begin{bmatrix}
1638 \\
76487 \\
70426
\end{bmatrix}
$$

Using Eq. (3.13), we can find the least squares estimates of the parameters in the regression model as

$$\hat{\beta} = (\mathbf{X}'\mathbf{X})^{-1}\mathbf{X}'\mathbf{y}$$

$$= \begin{bmatrix} 25 & 1271 & 1148 \\ 1271 & 69881 & 60814 \\ 1148 & 60814 & 56790 \end{bmatrix}^{-1} \begin{bmatrix} 1638 \\ 76487 \\ 70426 \end{bmatrix}$$

$$= \begin{bmatrix} 0.699946097 & -0.006128086 & -0.007586982 \\ -0.006128086 & 0.00026383 & -0.000158646 \\ -0.007586982 & -0.000158646 & 0.000340866 \end{bmatrix} \begin{bmatrix} 1638 \\ 76487 \\ 70426 \end{bmatrix}$$

$$= \begin{bmatrix} 143.4720118 \\ -1.031053414 \\ -0.55603781 \end{bmatrix}$$

Therefore the regression model is

$$\hat{y} = 143.472 - 1.031x_1 - 0.556x_2,$$

where x_1 = patient age and x_2 = severity of illness, and we have reported the regression coefficients to three decimal places.

Table 3.3 shows the output from the JMP regression routine for the patient satisfaction data. At the top of the table JMP displays a plot of the actual satisfaction data points versus the fitted values from the regression. If the fit is "perfect" then the actual-predicted and the plotted points would lie on a straight 45° line. The points do seem to scatter closely along the 45° line, suggesting that the model is a reasonably good fit to the data. Note that, in addition to the fitted regression model, JMP provides a list of the residuals computed from Eq. (3.16) along with other output that will provide information about the quality of the regression model. This output will be explained in subsequent sections, and we will frequently refer back to Table 3.3.

Example 3.2 Trend Adjustment   One way to forecast time series data that contain a linear trend is with a trend adjustment procedure. This involves fitting a model with a linear trend term in time, subtracting the fitted values from the original observations to obtain a set of residuals that are trend-free, then forecast the residuals, and compute the forecast by adding the forecast of the residual value(s) to the estimate of trend. We

TABLE 3.3   JMP Output for the Patient Satisfaction Data in Table 3.2

Actual by Predicted Plot

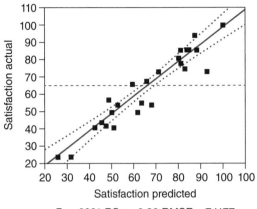

P < .0001 RSq = 0.90 RMSE = 7.1177

Summary of Fit

RSquare	0.896593
RSquare Adj	0.887192
Root mean square error	7.117667
Mean of response	65.52
Observations (or Sum Wgts)	25

Analysis of Variance

Source	DF	Sum of Squares	Mean Square	F Ratio
Model	2	9663.694	4831.85	95.3757
Error	22	1114.546	50.66	Prob > F
C. Total	24	10778.240		<.0001*

Parameter Estimates

Term	Estimate	Std Error	t Ratio	Prob> |t|
Intercept	143.47201	5.954838	24.09	<.0001*
Age	−1.031053	0.115611	−8.92	<.0001*
Severity	−0.556038	0.13141	−4.23	0.0003*

(*continued*)

TABLE 3.3 *(Continued)*

Observation	Age	Severity	Satisfaction	Residual
1	55	50	68	9.03781647
2	46	24	77	−5.6986473
3	30	46	96	9.03732988
4	35	48	80	−0.6953274
5	59	58	43	−7.3896674
6	61	60	44	−3.2154849
7	74	65	26	−5.0316015
8	38	42	88	7.06160595
9	27	42	75	−17.279982
10	51	50	57	−6.0863972
11	53	38	56	−11.696744
12	41	30	88	3.48231247
13	37	31	88	−0.0858634
14	24	34	102	2.17855567
15	42	30	88	4.51336588
16	50	48	70	4.77047378
17	58	61	52	2.24739262
18	60	71	43	0.86987755
19	62	62	46	0.92764409
20	68	38	56	3.76905713
21	70	41	59	10.4992774
22	79	66	26	0.67970337
23	63	31	52	−9.2784746
24	39	42	83	3.09265936
25	49	40	75	4.29111788

described and illustrated trend adjustment in Section 2.4.2, and the basic trend adjustment model introduced there was

$$y_t = \beta_0 + \beta_1 t + \varepsilon, \quad t = 1, 2, \ldots, T.$$

The least squares normal equations for this model are

$$T\hat{\beta}_0 + \hat{\beta}_1 \frac{T(T+1)}{2} = \sum_{t=1}^{T} y_t$$

$$\hat{\beta}_0 \frac{T(T+1)}{2} + \hat{\beta}_1 \frac{T(T+1)(2T+1)}{6} = \sum_{t=1}^{T} t y_t$$

Because there are only two parameters, it is easy to solve the normal equations directly, resulting in the least squares estimators

$$\hat{\beta}_0 = \frac{2(2T+1)}{T(T-1)} \sum_{t=1}^{T} y_t - \frac{6}{T(T-1)} \sum_{t=1}^{T} ty_t$$

$$\hat{\beta}_1 = \frac{12}{T(T^2-1)} \sum_{t=1}^{T} ty_t - \frac{6}{T(T-1)} \sum_{t=1}^{T} y_t$$

Minitab computes these parameter estimates in its trend adjustment procedure, which we illustrated in Example 2.6. The least squares estimates obtained from this trend adjustment model depend on the point in time at which they were computed, that is, T. Sometimes it may be convenient to keep track of the period of computation and denote the estimates as functions of time, say, $\hat{\beta}_0(T)$ and $\hat{\beta}_1(T)$. The model can be used to predict the next observation by predicting the point on the trend line in period $T + 1$, which is $\hat{\beta}_0(T) + \hat{\beta}_1(T)(T + 1)$, and adding to the trend a forecast of the next residual, say, $\hat{e}_{T+1}(1)$. If the residuals are structureless and have average value zero, the forecast of the next residual would be zero. Then the forecast of the next observation would be

$$\hat{y}_{T+1}(T) = \hat{\beta}_0(T) + \hat{\beta}_1(T)(T + 1)$$

When a new observation becomes available, the parameter estimates $\hat{\beta}_0(T)$ and $\hat{\beta}_1(T)$ could be updated to reflect the new information. This could be done by solving the normal equations again. In some situations it is possible to devise simple updating equations so that new estimates $\hat{\beta}_0(T + 1)$ and $\hat{\beta}_1(T + 1)$ can be computed directly from the previous ones $\hat{\beta}_0(T)$ and $\hat{\beta}_1(T)$ without having to directly solve the normal equations. We will show how to do this later.

3.3 STATISTICAL INFERENCE IN LINEAR REGRESSION

In linear regression problems, certain tests of hypotheses about the model parameters and confidence interval estimates of these parameters are helpful in measuring the usefulness of the model. In this section, we describe several important hypothesis-testing procedures and a confidence interval estimation procedure. These procedures require that the errors ε_i in the model are normally and independently distributed with mean zero and variance σ^2, abbreviated NID(0, σ^2). As a result of this assumption, the observations y_i are normally and independently distributed with mean $\beta_0 + \sum_{j=1}^{k} \beta_j x_{ij}$ and variance σ^2.

3.3.1 Test for Significance of Regression

The test for significance of regression is a test to determine whether there is a linear relationship between the response variable y and a subset of the predictor or regressor variables $x_1, x_2, \ldots, x_k$. The appropriate hypotheses are

$$H_0 : \beta_1 = \beta_2 = \cdots = \beta_k = 0$$
$$H_1 : \text{at least one } \beta_j \neq 0 \tag{3.19}$$

Rejection of the null hypothesis H_0 in Eq. (3.19) implies that at least one of the predictor variables $x_1, x_2, \ldots, x_k$ contributes significantly to the model. The test procedure involves an analysis of variance partitioning of the total sum of squares

$$SS_T = \sum_{i=1}^{n} (y_i - \bar{y})^2 \tag{3.20}$$

into a sum of squares due to the **model** (or to **regression**) and a sum of squares due to **residual** (or **error**), say,

$$SS_T = SS_R + SS_E \tag{3.21}$$

Now if the null hypothesis in Eq. (3.19) is true and the model errors are normally and independently distributed with constant variance as assumed, then the test statistic for significance of regression is

$$F_0 = \frac{SS_R/k}{SS_E/(n-p)} \tag{3.22}$$

and one rejects H_0 if the test statistic F_0 exceeds the upper tail point of the F distribution with k numerator degrees of freedom and $n - p$ denominator degrees of freedom, $F_{\alpha,k,n-p}$. Table A.4 in Appendix A contains these upper tail percentage points of the F distribution.

Alternatively, we could use the P-value approach to hypothesis testing and thus reject the null hypothesis if the P-value for the statistic F_0 is

TABLE 3.4 Analysis of Variance for Testing Significance of Regression

Source of Variation	Sum of Squares	Degrees of Freedom	Mean Square	Test Statistic, F_0
Regression	SS_R	k	$\dfrac{SS_R}{k}$	$F_0 = \dfrac{SS_R/k}{SS_E/(n-p)}$
Residual (error)	SS_E	$n - p$	$\dfrac{SS_E}{n-p}$	
Total	SS_T	$n-1$		

less than α. The quantities in the numerator and denominator of the test statistic F_0 are called **mean squares**. Recall that the mean square for error or residual estimates σ^2.

The test for significance of regression is usually summarized in an analysis of variance (ANOVA) table such as Table 3.4. Computational formulas for the sums of squares in the ANOVA are

$$SS_T = \sum_{i=1}^{n} (y_i - \bar{y})^2 = \mathbf{y}'\mathbf{y} - n\bar{y}^2$$

$$SS_R = \hat{\boldsymbol{\beta}}' \mathbf{X}'\mathbf{y} - n\bar{y}^2 \tag{3.23}$$

$$SS_E = \mathbf{y}'\mathbf{y} - \hat{\boldsymbol{\beta}}' \mathbf{X}'\mathbf{y}$$

Regression model ANOVA computations are almost always performed using a computer software package. The JMP output in Table 3.3 shows the ANOVA test for significance of regression for the regression model for the patient satisfaction data. The hypotheses in this problem are

$$H_0 : \beta_1 = \beta_2 = 0$$
$$H_1 : \text{ at least one } \beta_j \neq 0$$

The reported value of the F-statistic from Eq. (3.22) is

$$F_0 = \frac{9663.694/2}{1114.546/22} = \frac{4831.85}{50.66} = 95.38.$$

and the P-value is reported as <0.0001. The actual P-value is approximately 1.44×10^{-11}, a very small value, so there is strong evidence to reject the null hypothesis and we conclude that either patient age or severity are useful predictors for patient satisfaction.

Table 3.3 also reports the coefficient of multiple determination R^2, first introduced in Section 2.6.2 in the context of choosing between competing forecasting models. Recall that

$$R^2 = \frac{SS_{\text{R}}}{SS_{\text{T}}} = 1 - \frac{SS_{\text{E}}}{SS_{\text{T}}} \tag{3.24}$$

For the regression model for the patient satisfaction data, we have

$$R^2 = \frac{SS_{\text{R}}}{SS_{\text{T}}} = \frac{9663.694}{10778.24} = 0.8966$$

So this model explains about 89.7% of the variability in the data.

The statistic R^2 is a measure of the amount of reduction in the variability of y obtained by using the predictor variables $x_1, x_2, \ldots, x_k$ in the model. It is a measure of how well the regression model fits the data sample. However, as noted in Section 2.6.2, a large value of R^2 does not necessarily imply that the regression model is a good one. Adding a variable to the model will never cause a decrease in R^2, even in situations where the additional variable is not statistically significant. In almost all cases, when a variable is added to the regression model R^2 increases. As a result, over reliance on R^2 as a measure of model adequacy often results in **overfitting**; that is, putting too many predictors in the model. In Section 2.6.2 we introduced the adjusted R^2 statistic

$$R^2_{\text{Adj}} = 1 - \frac{SS_{\text{E}}/(n-p)}{SS_{\text{T}}/(n-1)} \tag{3.25}$$

In general, the adjusted R^2 statistic will not always increase as variables are added to the model. In fact, if unnecessary regressors are added, the value of the adjusted R^2 statistic will often decrease. Consequently, models with a large value of the adjusted R^2 statistic are usually considered good regression models. Furthermore, the regression model that maximizes the adjusted R^2 statistic is also the model that minimizes the residual mean square.

JMP reports both R^2 and R^2_{Adj} in Table 3.4. The value of $R^2 = 0.897$ (or 89.7%), and the adjusted R^2 statistic is

$$R^2_{\text{Adj}} = 1 - \frac{SS_E/(n-p)}{SS_T/(n-1)}$$

$$= 1 - \frac{1114.546/(25-3)}{10778.24/(25-1)}$$

$$= 0.887.$$

Both R^2 and R^2_{Adj} are very similar, usually a good sign that the regression model does not contain unnecessary predictor variables. It seems reasonable to conclude that the regression model involving patient age and severity accounts for between about 88% and 90% of the variability in the patient satisfaction data.

3.3.2 Tests on Individual Regression Coefficients and Groups of Coefficients

Tests on Individual Regression Coefficients   We are frequently interested in testing hypotheses on the individual regression coefficients. These tests would be useful in determining the value or contribution of each predictor variable in the regression model. For example, the model might be more effective with the inclusion of additional variables or perhaps with the deletion of one or more of the variables already in the model.

Adding a variable to the regression model always causes the sum of squares for regression to increase and the error sum of squares to decrease. We must decide whether the increase in the regression sum of squares is sufficient to warrant using the additional variable in the model. Furthermore, adding an unimportant variable to the model can actually increase the mean squared error, thereby decreasing the usefulness of the model.

The hypotheses for testing the significance of any individual regression coefficient, say, β_j, are

$$\begin{aligned} H_0 &: \beta_j = 0 \\ H_1 &: \beta_j \neq 0 \end{aligned} \qquad (3.26)$$

If the null hypothesis $H_0 : \beta_j = 0$ is not rejected, then this indicates that the predictor variable x_j can be deleted from the model.

The test statistic for this hypothesis is

$$t_0 = \frac{\hat{\beta}_j}{\sqrt{\hat{\sigma}^2 C_{jj}}}, \qquad (3.27)$$

where C_{jj} is the diagonal element of the $(\mathbf{X'X})^{-1}$ matrix corresponding to the regression coefficient $\hat{\beta}_j$ (in numbering the elements of the matrix $\mathbf{C} = (\mathbf{X'X})^{-1}$, it is necessary to number the first row and column as zero so that the first diagonal element C_{00} will correspond to the subscript number on the intercept). The null hypothesis $H_0 : \beta_j = 0$ is rejected if the absolute value of the test statistic $|t_0| > t_{\alpha/2,n-p}$, where $t_{\alpha/2,n-p}$ is the upper $\alpha/2$ percentage point of the t distribution with $n - p$ degrees of freedom. Table A.3 in Appendix A contains these upper tail points of the t distribution. A P-value approach could also be used. This t-test is really a partial or marginal test because the regression coefficient $\hat{\beta}_j$ depends on all the other regressor variables x_i $(i \neq j)$ that are in the model.

The denominator of Eq. (3.27), $\sqrt{\hat{\sigma}^2 C_{jj}}$, is usually called the **standard error** of the regression coefficient. That is,

$$se(\hat{\beta}_j) = \sqrt{\hat{\sigma}^2 C_{jj}}. \qquad (3.28)$$

Therefore an equivalent way to write the t-test statistic in Eq. (3.27) is

$$t_0 = \frac{\hat{\beta}_j}{se(\hat{\beta}_j)}. \qquad (3.29)$$

Most regression computer programs provide the t-test for each model parameter. For example, consider Table 3.3, which contains the JMP output for Example 3.1. The upper portion of this table gives the least squares estimate of each parameter, the standard error, the t statistic, and the corresponding P-value. To illustrate how these quantities are computed, suppose that we wish to test the hypothesis that x_1 = patient age contributes significantly to the model, given that x_2 = severity is included in the regression equation. Stated formally, the hypotheses are

$$H_0 : \beta_1 = 0$$
$$H_1 : \beta_1 \neq 0$$

The regression coefficient for x_1 = patient age is $\hat{\beta}_1 = -1.0311$. The standard error of this estimated regression coefficient is

$$se(\hat{\beta}_1) = \sqrt{\hat{\sigma}^2 C_{11}} = \sqrt{(50.66)(0.00026383)} = 0.1156.$$

which when rounded agrees with the JMP output. (Often manual calculations will differ slightly from those reported by the computer, because the computer carries more decimal places. For instance, in this example if the mean squared error is computed to four decimal places as $MS_E = SS_E/(n-p) = 1114.546/(25-3) = 50.6612$ instead of the two places reported in the JMP output, and this value of the MS_E is used as the estimate $\hat{\sigma}^2$ in calculating the standard error, then the standard error of $\hat{\beta}_1$ will match the JMP output.) The test statistic is computed from Eq. (3.29) as

$$t_0 = \frac{\hat{\beta}_1}{se(\hat{\beta}_1)} = \frac{-1.031053}{0.115611} = -8.92$$

This is agrees with the results reported by JMP. Because the P-value reported is small, we would conclude that patient age is statistically significant; that is, it is an important predictor variable, given that severity is also in the model. Similarly, because the t-test statistic for x_2 = severity is large, we would conclude that severity is a significant predictor, given that patient age is in the model.

Tests on Groups of Coefficients   We may also directly examine the contribution to the regression sum of squares for a particular predictor, say, x_j, or a **group** of predictors, given that other predictors x_i $(i \neq j)$ are included in the model. The procedure for doing this is the general regression significance test or, as it is more often called, the **extra sum of squares method**. This procedure can also be used to investigate the contribution of a *subset* involving several regressor or predictor variables to the model. Consider the regression model with k regressor variables

$$\mathbf{y} = \mathbf{X}\boldsymbol{\beta} + \boldsymbol{\varepsilon}, \tag{3.30}$$

where $\mathbf{y}$ is $(n \times 1)$, $\mathbf{X}$ is $(n \times p)$, $\boldsymbol{\beta}$ is $(p \times 1)$, $\boldsymbol{\varepsilon}$ is $(n \times 1)$, and $p = k + 1$. We would like to determine if a subset of the predictor variables $x_1, x_2, \ldots, x_r$

$(r < k)$ contributes significantly to the regression model. Let the vector of regression coefficients be partitioned as follows:

$$\beta = \begin{bmatrix} \beta_1 \\ \beta_2 \end{bmatrix},$$

where β_1 is $(r \times 1)$ and β_2 is $[(p - r) \times 1]$. We wish to test the hypotheses

$$\begin{aligned} H_0 &: \beta_1 = 0 \\ H_1 &: \beta_1 \neq 0 \end{aligned} \tag{3.31}$$

The model may be written as

$$y = X\beta + \varepsilon = X_1\beta_1 + X_2\beta_2 + \varepsilon, \tag{3.32}$$

where X_1 represents the columns of X (or the predictor variables) associated with β_1 and X_2 represents the columns of X (predictors) associated with β_2.

For the **full model** (including both β_1 and β_2), we know that $\hat{\beta} = (X'X)^{-1}X'y$. Also, the regression sum of squares for all predictor variables including the intercept is

$$SS_R(\beta) = \hat{\beta}'X'y \qquad (p \text{ degrees of freedom}) \tag{3.33}$$

and the estimate of σ^2 based on this full model is

$$\hat{\sigma}^2 = \frac{y'y - \hat{\beta}'X'y}{n - p} \tag{3.34}$$

$SS_R(\beta)$ is called the regression sum of squares due to β. To find the contribution of the terms in β_1 to the regression, we fit the model assuming that the null hypothesis $H_0: \beta_1 = 0$ is true. The **reduced model** is found from Eq. (3.32) with $\beta_1 = 0$:

$$y = X_2\beta_2 + \varepsilon \tag{3.35}$$

The least squares estimator of β_2 is $\hat{\beta}_2 = (X'_2X_2)^{-1}X'_2y$ and the regression sum of squares for the reduced model is

$$SS_R(\beta_2) = \hat{\beta}'_2X'_2y \;(p - r \text{ degrees of freedom}) \tag{3.36}$$

The regression sum of squares due to β_1, given that β_2 is already in the model is

$$SS_R(\beta_1|\beta_2) = SS_R(\beta) - SS_R(\beta_2) = \hat{\beta}'X'y - \hat{\beta}'_2X'_2y \qquad (3.37)$$

This sum of squares has r degrees of freedom. It is the "**extra sum of squares**" due to β_1. Note that $SS_R(\beta_1|\beta_2)$ is the increase in the regression sum of squares due to including the predictor variables $x_1, x_2, \ldots, x_r$ in the model. Now $SS_R(\beta_1|\beta_2)$ is independent of the estimate of σ^2 based on the full model from Eq. (3.34), so the null hypothesis $H_0: \beta_1 = \mathbf{0}$ may be tested by the statistic

$$F_0 = \frac{SS_R(\beta_1|\beta_2)/r}{\hat{\sigma}^2}, \qquad (3.38)$$

where $\hat{\sigma}^2$ is computed from Eq. (3.34). If $F_0 > F_{\alpha,r,n-p}$ we reject H_0, concluding that at least one of the parameters in β_1 is not zero, and, consequently, at least one of the predictor variables $x_1, x_2, \ldots, x_r$ in $\mathbf{X}_1$ contributes significantly to the regression model. A P-value approach could also be used in testing this hypothesis. Some authors call the test in Eq. (3.38) a **partial F test**.

The partial F test is very useful. We can use it to evaluate the contribution of an individual predictor or regressor x_j as if it were the last variable added to the model by computing

$$SS_R(\beta_j|\beta_i; i \neq j)$$

This is the increase in the regression sum of squares due to adding x_j to a model that already includes $x_1, \ldots, x_{j-1}, x_{j+1}, \ldots, x_k$. The partial F test on a single variable x_j is equivalent to the t-test in Equation (3.27). The computed value of F_0 will be exactly equal to the square of the t-test statistic t_0. However, the partial F test is a more general procedure in that we can evaluate simultaneously the contribution of more than one predictor variable to the model.

Example 3.3   To illustrate this procedure, consider again the patient satisfaction data from Table 3.2. Suppose that we wish to consider fitting a more elaborate model to this data; specifically, consider the second-order polynomial

$$y = \beta_0 + \beta_1 x_1 + \beta_2 x_2 + \beta_{12} x_1 x_2 + \beta_{11} x_1^2 + \beta_{22} x_2^2 + \varepsilon$$

TABLE 3.5 JMP Output for the Second-Order Model for the Patient Satisfaction Data

Summary of Fit

RSquare	0.900772
RSquare Adj	0.874659
Root mean square error	7.502639
Mean of response	65.52
Observations (or Sum Wgts)	25

Analysis of Variance

Source	DF	Sum of Squares	Mean Square	F Ratio
Model	5	9708.738	1941.75	34.4957
Error	19	1069.502	56.29	Prob > F
C. Total	24	10,778.240		<.0001*

Parameter Estimates

Term	Estimate	Std Error	t Ratio	Prob> |t|
Intercept	143.74009	6.774622	21.22	<0.0001*
Age	−0.986524	0.135366	−7.29	<0.0001*
Severity	−0.571637	0.158928	−3.60	0.0019*
(Severity-45.92)*(Age-50.84)	0.0064566	0.016546	0.39	0.7007
(Age-50.84)*(Age-50.84)	−0.00283	0.008588	−0.33	0.7453
(Severity-45.92)* (Severity-45.92)	−0.011368	0.013533	−0.84	0.4113

where x_1 = patient age and x_2 = severity. To fit the model, the model matrix would need to be expanded to include columns for the second-order terms $x_1 x_2, x_1^2$, and x_2^2. The results of fitting this model using JMP are shown in Table 3.5.

Suppose that we want to test the significance of the additional second-order terms. That is, the hypotheses are

$$H_0 : \beta_{12} = \beta_{11} = \beta_{22} = 0$$
$$H_1 : \text{at least one of the parameters } \beta_{12}, \beta_{11}, \text{ or } \beta_{22} \neq 0$$

In the notation used in this section, these second-order terms are the parameters in the vector β_1. Since the quadratic model is the full model, we can find $SS_R(\beta)$ directly from the JMP output in Table 3.5 as

$$SS_R(\beta) = 9708.738$$

with 5 degrees of freedom (because there are five predictors in this model). The reduced model is the model with all of the predictors in the vector β_1 equal to zero. This reduced model is the original regression model that we fit to the data in Table 3.3. From Table 3.3, we can find the regression sum of squares for the reduced model as

$$SS_R(\beta_2) = 9663.694$$

and this sum of squares has 2 degrees of freedom (the model has two predictors).

Therefore the extra sum of squares for testing the significance of the quadratic terms is just the difference between the regression sums of squares for the full and reduced models, or

$$SS_R(\beta_1|\beta_2) = SS_R(\beta) - SS_R(\beta_2)$$
$$= 9708.738 - 9663.694$$
$$= 45.044$$

with $5 - 2 = 3$ degrees of freedom. These three degrees of freedom correspond to the three additional terms in the second-order model. The test statistic from Eq. (3.38) is

$$F_0 = \frac{SS_R(\beta_1|\beta_2)/r}{\hat{\sigma}^2}$$
$$= \frac{45.044/3}{56.29}$$
$$= 0.267.$$

This F-statistic is very small, so there is no evidence against the null hypothesis.

Furthermore, from Table 3.5, we observe that the individual t-statistics for the second-order terms are very small and have large P-values, so there is no reason to believe that the model would be improved by adding any of the second-order terms.

It is also interesting to compare the R^2 and R^2_{Adj} statistics for the two models. From Table 3.3, we find that $R^2 = 0.897$ and $R^2_{Adj} = 0.887$ for the original two-variable model, and from Table 3.5, we find that $R^2 = 0.901$ and $R^2_{Adj} = 0.875$ for the quadratic model. Adding the quadratic terms caused the ordinary R^2 to increase slightly (it will never decrease when additional predictors are inserted into the model), but the adjusted R^2 statistic decreased. This decrease in the adjusted R^2 is an indication that the additional variables did not contribute to the explanatory power of the model.

3.3.3 Confidence Intervals on Individual Regression Coefficients

It is often necessary to construct confidence interval (CI) estimates for the parameters in a linear regression and for other quantities of interest from the regression model. The procedure for obtaining these confidence intervals requires that we assume that the model errors are normally and independently distributed with mean zero and variance σ^2, the same assumption made in the two previous sections on hypothesis testing.

Because the least squares estimator $\hat{\beta}$ is a linear combination of the observations, it follows that $\hat{\beta}$ is normally distributed with mean vector β and covariance matrix $V(\hat{\beta}) = \sigma^2(\mathbf{X}'\mathbf{X})^{-1}$. Then each of the statistics

$$\frac{\hat{\beta}_j - \beta_j}{\sqrt{\hat{\sigma}^2 C_{jj}}}, \quad j = 0, 1, \dots, k \tag{3.39}$$

is distributed as t with $n - p$ degrees of freedom, where C_{jj} is the (jj)th element of the $(\mathbf{X}'\mathbf{X})^{-1}$ matrix, and $\hat{\sigma}^2$ is the estimate of the error variance, obtained from Eq. (3.34). Therefore a $100(1 - \alpha)$ percent confidence interval for an individual regression coefficient $\beta_j, j = 0, 1, \dots, k$, is

$$\hat{\beta}_j - t_{\alpha/2, n-p}\sqrt{\hat{\sigma}^2 C_{jj}} \leq \beta_j \leq \hat{\beta}_j + t_{\alpha/2, n-p}\sqrt{\hat{\sigma}^2 C_{jj}}. \tag{3.40}$$

This CI could also be written as

$$\hat{\beta}_j - t_{\alpha/2, n-p}se(\hat{\beta}_j) \leq \beta_j \leq \hat{\beta}_j + t_{\alpha/2, n-p}se(\hat{\beta}_j)$$

because $se(\hat{\beta}_j) = \sqrt{\hat{\sigma}^2 C_{jj}}$.

Example 3.4 We will find a 95% CI on the regression for patient age in the patient satisfaction data regression model. From the JMP output in

Table 3.3, we find that $\hat{\beta}_1 = -1.0311$ and $se(\hat{\beta}_1) = 0.1156$. Therefore the 95% CI is

$$\hat{\beta}_j - t_{\alpha/2,n-p}se(\hat{\beta}_j) \le \beta_j \le \hat{\beta}_j + t_{\alpha/2,n-p}se(\hat{\beta}_j)$$

$$-1.0311 - (2.074)(0.1156) \le \beta_1 \le -1.0311 + (2.074)(0.1156)$$

$$-1.2709 \le \beta_1 \le -0.7913.$$

This confidence interval does not include zero; this is equivalent to rejecting (at the 0.05 level of significance) the null hypothesis that the regression coefficient $\beta_1 = 0$.

3.3.4 Confidence Intervals on the Mean Response

We may also obtain a confidence interval on the mean response at a particular combination of the predictor or regressor variables, say, $x_{01}, x_{02}, \dots,$ x_{0k}. We first define a vector that represents this point expanded to model form. Since the standard multiple linear regression model contains the k predictors and an intercept term, this vector is

$$\mathbf{x}_0 = \begin{bmatrix} 1 \\ x_{01} \\ \vdots \\ x_{0k} \end{bmatrix}$$

The mean response at this point is

$$E[y(\mathbf{x}_0)] = \mu_{y|\mathbf{x}_0} = \mathbf{x}'_0\boldsymbol{\beta}.$$

The estimator of the mean response at this point is found by substituting $\hat{\boldsymbol{\beta}}$ for $\boldsymbol{\beta}$

$$\hat{y}(\mathbf{x}_0) = \hat{\mu}_{y|\mathbf{x}_0} = \mathbf{x}'_0\hat{\boldsymbol{\beta}} \tag{3.41}$$

This estimator is normally distributed because $\hat{\boldsymbol{\beta}}$ is normally distributed and it is also unbiased because $\hat{\boldsymbol{\beta}}$ is an unbiased estimator of $\boldsymbol{\beta}$. The variance of $\hat{y}(\mathbf{x}_0)$ is

$$\text{Var}\,[\hat{y}(\mathbf{x}_0)] = \sigma^2\mathbf{x}'_0(\mathbf{X}'\mathbf{X})^{-1}\mathbf{x}_0. \tag{3.42}$$

Therefore, a $100(1 - \alpha)$ percent CI on the mean response at the point x_{01}, $x_{02}, \ldots, x_{0k}$ is

$$\hat{y}(\mathbf{x}_0) - t_{\alpha/2,n-p}\sqrt{\hat{\sigma}^2\mathbf{x}'_0(\mathbf{X}'\mathbf{X})^{-1}\mathbf{x}_0} \leq \mu_{y|\mathbf{x}_0} \leq \hat{y}(\mathbf{x}_0) + t_{\alpha/2,n-p}\sqrt{\hat{\sigma}^2\mathbf{x}'_0(\mathbf{X}'\mathbf{X})^{-1}\mathbf{x}_0},$$

$$(3.43)$$

where $\hat{\sigma}^2$ is the estimate of the error variance, obtained from Eq. (3.34). Note that the length of this confidence interval will depend on the location of the point $\mathbf{x}_0$ through the term $\mathbf{x}'_0(\mathbf{X}'\mathbf{X})^{-1}\mathbf{x}_0$ in the confidence interval formula. Generally, the length of the CI will increase as the point $\mathbf{x}_0$ moves further from the center of the predictor variable data.

The quantity

$$\sqrt{\mathrm{Var}\,[\hat{y}(\mathbf{x}_0)]} = \sqrt{\hat{\sigma}^2\mathbf{x}'_0(\mathbf{X}'\mathbf{X})^{-1}\mathbf{x}_0}$$

used in the confidence interval calculations in Eq. (3.43) is sometimes called the standard error of the fitted response. JMP will calculate and display these standard errors for each individual observation in the sample used to fit the model and for other non-sample points of interest. The next-to-last column of Table 3.6 displays the standard error of the fitted response for the patient satisfaction data. These standard errors can be used to compute the CI in Eq. (3.43).

Example 3.5   Suppose that we want to find a confidence interval on mean patient satisfaction for the point where x_1 = patient age = 55 and x_2 = severity = 50. This is the first observation in the sample, so refer to Table 3.6, the JMP output for the patient satisfaction regression model. For this observation, JMP reports that the "SE Fit" is 1.51 rounded to two decimal places, or in our notation, $\sqrt{\mathrm{Var}\,[\hat{y}(\mathbf{x}_0)]} = 1.51$. Therefore, if we want to find a 95% CI on the mean patient satisfaction for the case where x_1 = patient age = 55 and x_2 = severity = 50, we would proceed as follows:

$$\hat{y}(\mathbf{x}_0) - t_{\alpha/2,n-p}\sqrt{\hat{\sigma}^2\mathbf{x}'_0(\mathbf{X}'\mathbf{X})^{-1}\mathbf{x}_0} \leq \mu_{y|\mathbf{x}_0} \leq \hat{y}(\mathbf{x}_0) + t_{\alpha/2,n-p}\sqrt{\hat{\sigma}^2\mathbf{x}'_0(\mathbf{X}'\mathbf{X})^{-1}\mathbf{x}_0}$$

$$58.96 - 2.074(1.51) \leq \mu_{y|\mathbf{x}_0} \leq 58.96 + 2.074(1.51)$$

$$55.83 \leq \mu_{y|\mathbf{x}_0} \leq 62.09.$$

From inspection of Table 3.6, note that the standard errors for each observation are different. This reflects the fact that the length of the CI on

TABLE 3.6 JMP Calculations of the Standard Errors of the Fitted Values and Predicted Responses for the Patient Satisfaction Data

Observation	Age	Severity	Satisfaction	Predicted	Residual	SE (Fit)	SE (Predicted)
1	55	50	68	58.96218	9.037816	1.507444	7.275546
2	46	24	77	82.69865	−5.69865	2.98856	7.719631
3	30	46	96	86.96267	9.03733	2.803259	7.6498
4	35	48	80	80.69533	−0.69533	2.446294	7.526323
5	59	58	43	50.38967	−7.38967	1.962621	7.383296
6	61	60	44	47.21548	−3.21548	2.128407	7.429084
7	74	65	26	31.0316	−5.0316	2.89468	7.683772
8	38	42	88	80.93839	7.061606	1.919979	7.372076
9	27	42	75	92.27998	−17.28	2.895873	7.684221
10	51	50	57	63.0864	−6.0864	1.517813	7.277701
11	53	38	56	67.69674	−11.6967	1.856609	7.355826
12	41	30	88	84.51769	3.482312	2.275784	7.472641
13	37	31	88	88.08586	−0.08586	2.260863	7.468111
14	24	34	102	99.82144	2.178556	2.994326	7.721863
15	42	30	88	83.48663	4.513366	2.277152	7.473058
16	50	48	70	65.22953	4.770474	1.462421	7.266351
17	58	61	52	49.75261	2.247393	2.214287	7.454143
18	60	71	43	42.13012	0.869878	3.21204	7.808866
19	62	62	46	45.07236	0.927644	2.296	7.478823
20	68	38	56	52.23094	3.769057	3.038105	7.738945
21	70	41	59	48.50072	10.49928	2.97766	7.715416
22	79	66	26	25.3203	0.679703	3.24021	7.820495
23	63	31	52	61.27847	−9.27847	3.28074	7.837374
24	39	42	83	79.90734	3.092659	1.849178	7.353954
25	49	40	75	70.70888	4.291118	1.58171	7.291295
—	75	60	—	32.78074	—	2.78991	7.644918
—	60	60	—	48.24654	—	2.120899	7.426937

133

the mean response depends on the location of the observation. Generally, the standard error increases as the distance of the point from the center of the predictor variable data increases.

In the case where the point of interest $\mathbf{x}_0$ is not one of the observations in the sample, it is necessary to calculate the standard error for that point $\sqrt{\text{Var}\,[\hat{y}(\mathbf{x}_0)]} = \sqrt{\hat{\sigma}^2 \mathbf{x}'_0(\mathbf{X}'\mathbf{X})^{-1}\mathbf{x}_0}$, which involves finding $\mathbf{x}'_0(\mathbf{X}'\mathbf{X})^{-1}\mathbf{x}_0$ for the observation $\mathbf{x}_0$. This is not too difficult (you can do it in Excel), but it is not necessary, because JMP will provide the CI at any point that you specify. For example, if you want to find a 95% CI on the mean patient satisfaction for the point where x_1 = patient age = 60 and x_2 = severity = 60 (this is not a sample observation), then in the last row of Table 3.6 JMP reports that the estimate of the mean patient satisfaction at the point x_1 = patient age = 60 and x_2 = severity = 60 as $\hat{y}(\mathbf{x}_0)$ = 48.25, and the standard error of the fitted response as $\sqrt{\text{Var}\,[\hat{y}(\mathbf{x}_0)]} = \sqrt{\hat{\sigma}^2 \mathbf{x}'_0(\mathbf{X}'\mathbf{X})^{-1}\mathbf{x}_0}$ = 2.12. Consequently, the 95% CI on the mean patient satisfaction at that point is

$$43.85 \le \mu_{y|\mathbf{x}_0} \le 52.65.$$

3.4 PREDICTION OF NEW OBSERVATIONS

A regression model can be used to predict future observations on the response y corresponding to a particular set of values of the predictor or regressor variables, say, $x_{01}, x_{02}, \ldots, x_{0k}$. Let $\mathbf{x}_0$ represent this point, expanded to model form. That is, if the regression model is the standard multiple regression model, then $\mathbf{x}_0$ contains the coordinates of the point of interest and unity to account for the intercept term, so $\mathbf{x}'_0 = [1, x_{01}, x_{02}, \ldots, x_{0k}]$. A point estimate of the future observation $y(\mathbf{x}_0)$ at the point $x_{01}, x_{02}, \ldots, x_{0k}$ is computed from

$$\hat{y}(\mathbf{x}_0) = \mathbf{x}'_0 \hat{\beta} \tag{3.44}$$

The prediction error in using $\hat{y}(\mathbf{x}_0)$ to estimate $y(\mathbf{x}_0)$ is $y(\mathbf{x}_0) - \hat{y}(\mathbf{x}_0)$. Because $\hat{y}(\mathbf{x}_0)$ and $y(\mathbf{x}_0)$ are independent, the variance of this prediction error is

$$\text{Var}\,[y(\mathbf{x}_0) - \hat{y}(\mathbf{x}_0)] = \text{Var}\,[y(\mathbf{x}_0)] + \text{Var}\,[\hat{y}(\mathbf{x}_0)] = \sigma^2 + \sigma^2 \mathbf{x}'_0(\mathbf{X}'\mathbf{X})^{-1}\mathbf{x}_0$$
$$= \sigma^2 \left[1 + \mathbf{x}'_0(\mathbf{X}'\mathbf{X})^{-1}\mathbf{x}_0\right]. \tag{3.45}$$

If we use $\hat{\sigma}^2$ from Eq. (3.34) to estimate the error variance σ^2, then the ratio

$$\frac{y(\mathbf{x}_0) - \hat{y}(\mathbf{x}_0)}{\sqrt{\hat{\sigma}^2 \left[1 + \mathbf{x}'_0(\mathbf{X}'\mathbf{X})^{-1}\mathbf{x}_0\right]}}$$

has a t distribution with $n - p$ degrees of freedom. Consequently, we can write the following probability statement:

$$P\left(-t_{\alpha/2,n-p} \le \frac{y(\mathbf{x}_0) - \hat{y}(\mathbf{x}_0)}{\sqrt{\hat{\sigma}^2 \left[1 + \mathbf{x}'_0(\mathbf{X}'\mathbf{X})^{-1}\mathbf{x}_0\right]}} \le t_{\alpha/2,n-p}\right) = 1 - \alpha$$

This probability statement can be rearranged as follows:

$$P\left(\begin{array}{l} \hat{y}(\mathbf{x}_0) - t_{\alpha/2,n-p}\sqrt{\hat{\sigma}^2 \left[1 + \mathbf{x}'_0(\mathbf{X}'\mathbf{X})^{-1}\mathbf{x}_0\right]} \le y(\mathbf{x}_0) \\ \le \hat{y}(\mathbf{x}_0) + t_{\alpha/2,n-p}\sqrt{\hat{\sigma}^2 \left[1 + \mathbf{x}'_0(\mathbf{X}'\mathbf{X})^{-1}\mathbf{x}_0\right]} \end{array}\right) = 1 - \alpha.$$

Therefore, the probability is $1 - \alpha$ that the future observation falls in the interval

$$\hat{y}(\mathbf{x}_0) - t_{\alpha/2,n-p}\sqrt{\hat{\sigma}^2 \left[1 + \mathbf{x}'_0(\mathbf{X}'\mathbf{X})^{-1}\mathbf{x}_0\right]} \le y(\mathbf{x}_0)$$

$$\le \hat{y}(\mathbf{x}_0) + t_{\alpha/2,n-p}\sqrt{\hat{\sigma}^2 \left[1 + \mathbf{x}'_0(\mathbf{X}'\mathbf{X})^{-1}\mathbf{x}_0\right]} \qquad (3.46)$$

This statement is called a $100(1 - \alpha)$ percent **prediction interval (PI)** for the future observation $y(\mathbf{x}_0)$ at the point $x_{01}, x_{02}, \ldots, x_{0k}$. The expression in the square tool in Eq. (3.46) is often called the standard error of the predicted response.

The PI formula in Eq. (3.46) looks very similar to the formula for the CI on the mean, Eq. (3.43). The difference is the "1" in the variance of the prediction error under the square root. This will make PI longer than the corresponding CI at the same point. It is reasonable that the PI should be longer, as the CI is an interval estimate on the mean of the response distribution at a specific point, while the PI is an interval estimate on a single future observation from the response distribution at that point. There should be more variability associated with an individual observation than with an estimate of the mean, and this is reflected in the additional length of the PI.

Example 3.6   JMP will compute the standard errors of the predicted response so it is easy to construct the prediction interval in Eq. (3.46). To illustrate, suppose that we want a 95% PI on a future observation of patient satisfaction for a patient whose age is 75 and with severity of illness 60. In the next to last row of Table 3.6 JMP predicted value of satisfaction at this new observation as $\hat{y}(\mathbf{x}_0) = 32.78$, and the standard error of the predicted response is 7.65. Then from Eq. (3.46) the prediction interval is

$$16.93 \leq y(\mathbf{x}_0) \leq 48.64.$$

This example provides us with an opportunity to compare prediction and confidence intervals. First, note that from Table 3.6 the standard error of the fit at this point is smaller than the standard error of the prediction. Therefore, the PI is longer than the corresponding CI. Now compare the length of the CI and the PI for this point with the length of the CI and the PI for the point x_1 = patient age = 60 and x_2 = severity = 60 from Example 3.4. The intervals are longer for the point in this example because this point with x_1 = patient age = 75 and x_2 = severity = 60 is further from the center of the predictor variable data than the point in Example 3.4, where x_1 = patient age = 60 and x_2 = severity = 60.

3.5 MODEL ADEQUACY CHECKING

3.5.1 Residual Plots

An important part of any data analysis and model-building procedure is checking the adequacy of the model. We know that all models are wrong, but a model that is a reasonable fit to the data used to build it and that does not seriously ignore or violate any of the underlying model-building assumptions can be quite useful. Model adequacy checking is particularly important in building regression models for purposes of forecasting, because forecasting will almost always involve some extrapolation or projection of the model into the future, and unless the model is reasonable the forecasting process is almost certainly doomed to failure.

Regression model residuals, originally defined in Eq. (2.2), are very useful in model adequacy checking and to get some sense of how well the regression model assumptions of normally and independently distributed model errors with constant variance are satisfied. Recall that if y_i is the observed value of the response variable and if the corresponding fitted value from the model is $\hat{y}_i$, then the residuals are

$$e_i = y_i - \hat{y}_i, \quad i = 1, 2, \ldots, n.$$

Residual plots are the primary approach to model adequacy checking. The simplest way to check the adequacy of the normality assumption on the model errors is to construct a normal probability plot of the residuals. In Section 2.6.1 we introduced and used the normal probability plot of forecast errors to check for the normality of forecast errors. The use of the normal probability plot for regression residuals follows the same approach. To check the assumption of constant variance, plot the residuals versus the fitted values from the model. If the constant variance assumption is satisfied, this plot should exhibit a random scatter of residuals around zero. Problems with the equal variance assumption usually show up as a **pattern** on this plot. The most common pattern is an outward-opening funnel or megaphone pattern, indicating that the variance of the observations is increasing as the mean increases. Data **transformations** (see Section 2.4.1) are useful in stabilizing the variance. The log transformation is frequently useful in forecasting applications. It can also be helpful to plot the residuals against each of the predictor or regressor variables in the model. Any deviation from random scatter on these plots can indicate how well the model fits a particular predictor.

When the data are a time series, it is also important to plot the residuals versus **time order**. As usual, the anticipated pattern on this plot is random scatter. Trends, cycles, or other patterns in the plot of residuals versus time indicate model inadequacies, possibly due to missing terms or some other model specification issue. A funnel-shaped pattern that increases in width with time is an indication that the variance of the time series is increasing with time. This happens frequently in economic time series data, and in data that span a long period of time. Log transformations are often useful in stabilizing the variance of these types of time series.

Example 3.7   Table 3.3 presents the residuals for the regression model for the patient satisfaction data from Example 3.1. Figure 3.1 presents plots of these residuals. The plot in the upper left-hand portion of the display is a normal probability plot of the residuals. The residuals lie generally along a straight line, so there is no obvious reason to be concerned with

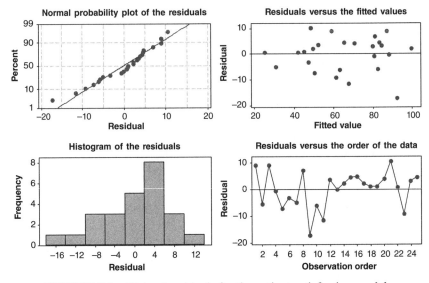

FIGURE 3.1 Plots of residuals for the patient satisfaction model.

the normality assumption. There is a very mild indication that one of the residuals (in the lower tail) may be slightly larger than expected, so this could be an indication of an outlier (a very mild one). The lower left plot is a histogram of the residuals. Histograms are more useful for large samples of data than small ones, so since there are only 25 residuals, this display is probably not as reliable as the normal probability plot. However, the histogram does not give any serious indication of nonnormality. The upper right is a plot of residuals versus the fitted values. This plot indicates essentially random scatter in the residuals, the ideal pattern. If this plot had exhibited a funnel shape, it could indicate problems with the equality of variance assumption. The lower right is a plot of the observations in the order of the data. If this was the order in which the data were collected, or if the data were a time series, this plot could reveal information about how the data may be changing over time. For example, a funnel shape on this plot might indicate that the variability of the observations was changing with time.

In addition to residual plots, other model diagnostics are frequently useful in regression. The following sections introduce and briefly illustrate some of these procedures. For more complete presentations, see Montgomery, Peck, and Vining (2012) and Myers (1990).

3.5.2 Scaled Residuals and PRESS

Standardized Residuals   Many regression model builders prefer to work with **scaled residuals** in contrast to the ordinary least squares (OLS) residuals. These scaled residuals frequently convey more information than do the ordinary residuals. One type of scaled residual is the **standardized residual**,

$$d_i = \frac{e_i}{\hat{\sigma}}, \tag{3.47}$$

where we generally use $\hat{\sigma} = \sqrt{MS_E}$ in the computation. The standardized residuals have mean zero and approximately unit variance; consequently, they are useful in looking for **outliers**. Most of the standardized residuals should lie in the interval $-3 \le d_i \le +3$, and any observation with a standardized residual outside this interval is potentially unusual with respect to its observed response. These outliers should be carefully examined because they may represent something as simple as a data-recording error or something of more serious concern, such as a region of the predictor or regressor variable space where the fitted model is a poor approximation to the true response.

Studentized Residuals   The standardizing process in Eq. (3.47) scales the residuals by dividing them by their approximate average standard deviation. In some data sets, residuals may have standard deviations that differ greatly. We now present a scaling that takes this into account. The vector of fitted values $\hat{y}_i$ that corresponds to the observed values y_i is

$$\hat{\mathbf{y}} = \mathbf{X}\hat{\boldsymbol{\beta}} = \mathbf{X}(\mathbf{X}'\mathbf{X})^{-1}\mathbf{X}'\mathbf{y} = \mathbf{H}\mathbf{y}. \tag{3.48}$$

The $n \times n$ matrix $\mathbf{H} = \mathbf{X}(\mathbf{X}'\mathbf{X})^{-1}\mathbf{X}'$ is usually called the "hat" matrix because it maps the vector of observed values into a vector of fitted values. The hat matrix and its properties play a central role in regression analysis.

The residuals from the fitted model may be conveniently written in matrix notation as

$$\mathbf{e} = \mathbf{y} - \hat{\mathbf{y}} = \mathbf{y} - \mathbf{H}\mathbf{y} = (\mathbf{I} - \mathbf{H})\mathbf{y} \tag{3.49}$$

and the covariance matrix of the residuals is

$$\text{Cov}(\mathbf{e}) = V[(\mathbf{I} - \mathbf{H})\mathbf{y}] = \sigma^2(\mathbf{I} - \mathbf{H}).$$

The matrix $\mathbf{I} - \mathbf{H}$ is in general not diagonal, so the residuals from a linear regression model have different variances and are correlated. The variance of the ith residual is

$$V(e_i) = \sigma^2(1 - h_{ii}), \tag{3.50}$$

where h_{ii} is the ith diagonal element of the hat matrix $\mathbf{H}$. Because $0 \leq h_{ii} \leq 1$ using the mean squared error MS_E to estimate the variance of the residuals actually overestimates the true variance. Furthermore, it turns out that h_{ii} is a measure of the location of the ith point in the predictor variable or x-space; the variance of the residual e_i depends on where the point $\mathbf{x}_i$ lies. As h_{ii} increases, the observation $\mathbf{x}_i$ lies further from the center of the region containing the data. Therefore residuals near the center of the $\mathbf{x}$-space have larger variance than do residuals at more remote locations. Violations of model assumptions are more likely at remote points, so these violations may be hard to detect from inspection of the ordinary residuals e_i (or the standardized residuals d_i) because their residuals will usually be smaller.

We recommend taking this inequality of variance into account when scaling the residuals. We suggest plotting the **studentized residuals as:**

$$r_i = \frac{e_i}{\sqrt{\hat{\sigma}^2(1 - h_{ii})}}, \tag{3.51}$$

with $\hat{\sigma}^2 = MS_E$ instead of the ordinary residuals or the standardized residuals. The studentized residuals have unit variance (i.e., $V(r_i) = 1$) regardless of the location of the observation $\mathbf{x}_i$ when the form of the regression model is correct. In many situations the variance of the residuals stabilizes, particularly for large data sets. In these cases, there may be little difference between the standardized and studentized residuals. Thus standardized and studentized residuals often convey equivalent information. However, because any point with a large residual and a large hat diagonal h_{ii} is potentially highly influential on the least squares fit, examination of the studentized residuals is generally recommended.

Table 3.7 displays the residuals, the studentized residuals, hat diagonals h_{ii}, and several other diagnostics for the regression model for the patient satisfaction data in Example 3.1. These quantities were computer generated using JMP. To illustrate the calculations, consider the first observation.

TABLE 3.7   Residuals and Other Diagnostics for the Regression Model for the Patient Satisfaction Data in Example 3.1

Observation	Residuals	Studentized Residuals	R-Student	h_{ii}	Cook's Distance
1	9.0378	1.29925	1.32107	0.044855	0.026424
2	−5.6986	−0.88216	−0.87754	0.176299	0.055521
3	9.0373	1.38135	1.41222	0.155114	0.116772
4	−0.6953	−0.10403	−0.10166	0.118125	0.000483
5	−7.3897	−1.08009	−1.08440	0.076032	0.031999
6	−3.2155	−0.47342	−0.46491	0.089420	0.007337
7	−5.0316	−0.77380	−0.76651	0.165396	0.039553
8	7.0616	1.03032	1.03183	0.072764	0.027768
9	−17.2800	−2.65767	−3.15124	0.165533	0.467041
10	−6.0864	−0.87524	−0.87041	0.045474	0.012165
11	−11.6967	−1.70227	−1.78483	0.068040	0.070519
12	3.4823	0.51635	0.50757	0.102232	0.010120
13	−0.0859	−0.01272	−0.01243	0.100896	0.000006
14	2.1786	0.33738	0.33048	0.176979	0.008159
15	4.5134	0.66928	0.66066	0.102355	0.017026
16	4.7705	0.68484	0.67634	0.042215	0.006891
17	2.2474	0.33223	0.32541	0.096782	0.003942
18	0.8699	0.13695	0.13386	0.203651	0.001599
19	0.9276	0.13769	0.13458	0.104056	0.000734
20	3.7691	0.58556	0.57661	0.182192	0.025462
21	10.4993	1.62405	1.69133	0.175015	0.186511
22	0.6797	0.10725	0.10481	0.207239	0.001002
23	−9.2785	−1.46893	−1.51118	0.212456	0.194033
24	3.0927	0.44996	0.44165	0.067497	0.004885
25	4.2911	0.61834	0.60945	0.049383	0.006621

The studentized residual is calculated as follows:

$$r_1 = \frac{e_1}{\sqrt{\hat{\sigma}^2(1 - h_{11})}}$$

$$= \frac{e_1}{\hat{\sigma}\sqrt{(1 - h_{11})}}$$

$$= \frac{9.0378}{7.11767\sqrt{1 - 0.044855}}$$

$$= 1.2992$$

which agrees approximately with the value reported by JMP in Table 3.7. Large values of the studentized residuals are usually an indication of potential unusual values or outliers in the data. Absolute values of the studentized residuals that are larger than three or four indicate potentially problematic observations. Note that none of the studentized residuals in Table 3.7 is this large. The largest studentized residual, -2.65767, is associated with observation 9. This observation does show up on the normal probability plot of residuals in Figure 3.1 as a very mild outlier, but there is no indication of a significant problem with this observation.

PRESS   Another very useful residual scaling can be based on the prediction error sum of squares or PRESS. To calculate PRESS, we select an observation—for example, i. We fit the regression model to the remaining $n-1$ observations and use this equation to predict the withheld observation y_i. Denoting this predicted value by $\hat{y}_{(i)}$, we may now find the prediction error for the ith observation as

$$e_{(i)} = y_i - \hat{y}_{(i)} \tag{3.52}$$

The prediction error is often called the ith PRESS residual. Note that the prediction error for the ith observation differs from the ith residual because observation i was not used in calculating the ith prediction value $\hat{y}_{(i)}$. This procedure is repeated for each observation $i = 1, 2, \ldots, n$, producing a set of n PRESS residuals $e_{(1)}, e_{(2)}, \ldots, e_{(n)}$. Then the PRESS statistic is defined as the sum of squares of the n PRESS residuals or

$$\text{PRESS} = \sum_{i=1}^{n} e_{(i)}^2 = \sum_{i=1}^{n} [y_i - \hat{y}_{(i)}]^2 \tag{3.53}$$

Thus PRESS is a form of **data splitting** (discussed in Chapter 2), since it uses each possible subset of $n-1$ observations as an estimation data set, and every observation in turn is used to form a prediction data set. Generally, small values of PRESS imply that the regression model will be useful in predicting new observations. To get an idea about how well the model will predict new data, we can calculate an R^2-like statistic called the R^2 for prediction

$$R^2_{\text{Prediction}} = 1 - \frac{\text{PRESS}}{SS_{\text{T}}} \tag{3.54}$$

Now PRESS will always be larger than the residual sum of squares and, because the ordinary $R^2 = 1 - (SS_E/SS_T)$, if the value of the $R^2_{\text{Prediction}}$ is not much smaller than the ordinary R^2, this is a good indication about potential model predictive performance.

It would initially seem that calculating PRESS requires fitting n different regressions. However, it is possible to calculate PRESS from the results of a single least squares fit to all n observations. It turns out that the ith PRESS residual is

$$e_{(i)} = \frac{e_i}{1 - h_{ii}}, \tag{3.55}$$

where e_i is the OLS residual. The hat matrix diagonals are directly calculated as a routine part of solving the least squares normal equations. Therefore PRESS is easily calculated as

$$\text{PRESS} = \sum_{i=1}^{n} \frac{e_i^2}{1 - h_{ii}} \tag{3.56}$$

JMP will calculate the PRESS statistic for a regression model and the R^2 for prediction based on PRESS from Eq. (3.54). The value of PRESS is PRESS = 1484.93 and the R^2 for prediction is

$$R^2_{\text{Prediction}} = 1 - \frac{\text{PRESS}}{SS_T}$$
$$= 1 - \frac{1484.93}{10778.2}$$
$$= 0.8622.$$

That is, this model would be expected to account for about 86.22% of the variability in new data.

R-Student   The studentized residual r_i discussed earlier is often considered an outlier diagnostic. It is customary to use the mean squared error MS_E as an estimate of σ^2 in computing r_i. This is referred to as **internal scaling** of the residual because MS_E is an internally generated estimate of σ^2 obtained from fitting the model to all n observations. Another approach would be to use an estimate of σ^2 based on a data set with the ith observation removed. We denote the estimate of σ^2 so obtained by $S_{(i)}^2$. We can show that

$$S_{(i)}^2 = \frac{(n-p)MS_E - e_i^2/(1-h_{ii})}{n-p-1} \tag{3.57}$$

The estimate of σ^2 in Eq. (3.57) is used instead of MS_E to produce an **externally studentized residual**, usually called **R-student**, given by

$$t_i = \frac{e_i}{\sqrt{S_{(i)}^2(1-h_{ii})}} \tag{3.58}$$

In many situations, t_i will differ little from the studentized residual r_i. However, if the ith observation is influential, then $S_{(i)}^2$ can differ significantly from MS_E, and consequently the R-student residual will be more sensitive to this observation. Furthermore, under the standard assumptions, the R-student residual t_i has a t-distribution with $n - p - 1$ degrees of freedom. Thus R-student offers a more formal procedure for investigating potential outliers by comparing the absolute magnitude of the residual t_i to an appropriate percentage point of t_{n-p-1}.

JMP will compute the R-student residuals. They are shown in Table 3.7 for the regression model for the patient satisfaction data. The largest value of R-student is for observation 9, $t_9 = -3.15124$. This is another indication that observation 9 is a very mild outlier.

3.5.3 Measures of Leverage and Influence

In building regression models, we occasionally find that a small subset of the data exerts a disproportionate influence on the fitted model. That is, estimates of the model parameters or predictions may depend more on the influential subset than on the majority of the data. We would like to locate these influential points and assess their impact on the model. If these influential points really are "bad" values, they should be eliminated. On the other hand, there may be nothing wrong with these points, but if they control key model properties, we would like to know it because it could affect the use of the model. In this section we describe and illustrate some useful measures of influence.

The disposition of points in the predictor variable space is important in determining many properties of the regression model. In particular, remote observations potentially have disproportionate leverage on the parameter estimates, predicted values, and the usual summary statistics.

The hat matrix $\mathbf{H} = \mathbf{X}(\mathbf{X'X})^{-1}\mathbf{X'}$ is very useful in identifying influential observations. As noted earlier, $\mathbf{H}$ determines the variances and covariances of the predicted response and the residuals because

$$\text{Var }(\hat{\mathbf{y}}) = \sigma^2\mathbf{H} \quad \text{and} \quad \text{Var }(\mathbf{e}) = \sigma^2(\mathbf{I} - \mathbf{H})$$

The elements h_{ij} of the hat matrix $\mathbf{H}$ may be interpreted as the amount of leverage exerted by the observation y_j on the predicted value $\hat{y}_i$. Thus inspection of the elements of $\mathbf{H}$ can reveal points that are potentially influential by virtue of their location in x-space.

Attention is usually focused on the diagonal elements of the hat matrix h_{ii}. It can be shown that $\sum_{i=1}^{n} h_{ii} = \text{rank}(\mathbf{H}) = \text{rank}(\mathbf{X}) = p$, so the average size of the diagonal elements of the $\mathbf{H}$ matrix is p/n. A widely used rough guideline is to compare the diagonal elements h_{ii} to twice their average value $2p/n$, and if any hat diagonal exceeds this value to consider that observation as a high-leverage point.

JMP will calculate and save the values of the hat diagonals. Table 3.7 displays the hat diagonals for the regression model for the patient satisfaction data in Example 3.1. Since there are $p = 3$ parameters in the model and $n = 25$ observations, twice the average size of a hat diagonal for this problem is

$$2p/n = 2(3)/25 = 0.24.$$

The largest hat diagonal, 0.212456, is associated with observation 23. This does not exceed twice the average size of a hat diagonal, so there are no high-leverage observations in these data.

The hat diagonals will identify points that are potentially influential due to their location in x-space. It is desirable to consider both the location of the point and the response variable in measuring influence. Cook (1977, 1979) has suggested using a measure of the squared distance between the least squares estimate based on all n points $\hat{\beta}$ and the estimate obtained by deleting the ith point, say, $\hat{\beta}_{(i)}$. This distance measure can be expressed as

$$D_i = \frac{(\hat{\beta} - \hat{\beta}_{(i)})'\mathbf{X'X}(\hat{\beta} - \hat{\beta}_{(i)})}{pMS_E}, \quad i = 1, 2, \ldots, n \quad (3.59)$$

A reasonable cutoff for D_i is unity. That is, we usually consider observations for which $D_i > 1$ to be influential. Cook's distance statistic D_i is actually calculated from

$$D_i = \frac{r_i^2}{p} \frac{V[\hat{y}(\mathbf{x}_i)]}{V(e_i)} = \frac{r_i^2}{p} \frac{h_{ii}}{1 - h_{ii}} \qquad (3.60)$$

Note that, apart from the constant p, D_i is the product of the square of the ith studentized residual and the ratio $h_{ii}/(1 - h_{ii})$. This ratio can be shown to be the distance from the vector $\mathbf{x}_i$ to the centroid of the remaining data. Thus D_i is made up of a component that reflects how well the regression model fits the ith observation y_i and a component that measures how far that point is from the rest of the data. Either component (or both) may contribute to a large value of D_i.

JMP will calculate and save the values of Cook's distance statistic D_i. Table 3.7 displays the values of Cook's distance statistic for the regression model for the patient satisfaction data in Example 3.1. The largest value, 0.467041, is associated with observation 9. This value was calculated from Eq. (3.60) as follows:

$$
\begin{aligned}
D_i &= \frac{r_i^2}{p} \frac{h_{ii}}{1 - h_{ii}} \\
&= \frac{(-2.65767)^2}{3} \frac{0.165533}{1 - 0.165533} \\
&= 0.467041.
\end{aligned}
$$

This does not exceed twice the cutoff of unity, so there are no influential observations in these data.

3.6 VARIABLE SELECTION METHODS IN REGRESSION

In our treatment of regression we have concentrated on fitting the full regression model. Actually, in most applications of regression the analyst will have a very good idea about the general form of the model he/she wishes to fit, but there may be uncertainty about the exact structure of the model. For example, we may not know if all of the predictor variables are really necessary. These applications of regression frequently involve a moderately large or large set of **candidate predictors**, and the objective of the analyst here is to fit a regression model to the "best subset" of these candidates. This can be a complex problem, as these data sets frequently have outliers, strong correlations between subsets of the variables, and other complicating features.

There are several techniques that have been developed for selecting the best subset regression model. Generally, these methods are either **stepwise-type** variable selection methods or **all possible regressions**. Stepwise-type methods build a regression model by either adding or removing a predictor variable to the basic model at each step. The forward selection version of the procedure begins with a model containing none of the candidate predictor variables and sequentially inserts variables into the model one-at-a-time until a final equation is produced. The criterion for entering a variable into the equation is that the t-statistic for that variable must be significant. The process is continued until there are no remaining candidate predictors that qualify for entry into the equation. In backward elimination, the procedure begins with all of the candidate predictor variables in the equation, and then variables are removed one-at-a-time to produce a final equation. The criterion for removing a variable is usually based on the t-statistic, with the variable having the smallest t-statistic considered for removal first. Variables are removed until all of the predictors remaining in the model have significant t-statistics. Stepwise regression usually consists of a combination of forward and backward stepping. There are many variations of the basic procedures.

In all possible regressions with K candidate predictor variables, the analyst examines all 2^K possible regression equations to identify the ones with potential to be a useful model. Obviously, as K becomes even moderately large, the number of possible regression models quickly becomes formidably large. Efficient algorithms have been developed that implicitly rather than explicitly examine all of these equations. Typically, only the equations that are found to be "best" according to some criterion (such as minimum MS_E or AICc) at each subset size are displayed. For more discussion of variable selection methods, see textbooks on regression such as Montgomery, Peck, and Vining (2012) or Myers (1990).

Example 3.8   Table 3.8 contains an expanded set of data for the hospital patient satisfaction data introduced in Example 3.1. In addition to the patient age and illness severity data, there are two additional regressors, an indicator of whether the patent is a surgical patient (1) or a medical patient (0), and an index indicating the patient's anxiety level. We will use these data to illustrate how variable selection methods in regression can be used to help the analyst build a regression model.

TABLE 3.8   Expanded Patient Satisfaction Data

Observation	Age	Severity	Surgical–Medical	Anxiety	Satisfaction
1	55	50	0	2.1	68
2	46	24	1	2.8	77
3	30	46	1	3.3	96
4	35	48	1	4.5	80
5	59	58	0	2.0	43
6	61	60	0	5.1	44
7	74	65	1	5.5	26
8	38	42	1	3.2	88
9	27	42	0	3.1	75
10	51	50	1	2.4	57
11	53	38	1	2.2	56
12	41	30	0	2.1	88
13	37	31	0	1.9	88
14	24	34	0	3.1	102
15	42	30	0	3.0	88
16	50	48	1	4.2	70
17	58	61	1	4.6	52
18	60	71	1	5.3	43
19	62	62	0	7.2	46
20	68	38	0	7.8	56
21	70	41	1	7.0	59
22	79	66	1	6.2	26
23	63	31	1	4.1	52
24	39	42	0	3.5	83
25	49	40	1	2.1	75

We will illustrate the forward selection procedure first. The JMP output that results from applying forward selection to these data is shown in Table 3.9. We used the AICc criterion for selecting the best model. The forward selection algorithm inserted the predictor patient age first, then severity, then anxiety, and finally surg-med was inserted into the equation. The best model based on the minimum value of AICc contained age and severity.

Table 3.10 presents the results of applying the JMP backward elimination procedure to the patient satisfaction data. Once again the AICc criterion was chosen to select the final model. The procedure begins with all four predictors in the model, then the surgical–medical indicator variable was removed, followed by the anxiety predictor, followed by severity.

TABLE 3.9   JMP Forward Selection for the Patient Satisfaction Data in Table 3.8

Stepwise Fit for Satisfaction

Stepwise Regression Control

Stopping rule: Minimum AICc

Direction: Forward

SSE	DFE	RMSE	RSquare	RSquare Adj	Cp	p	AICc	BIC
1114.5459	22	7.1176667	0.8966	0.8872	2.4553073	3	175.8801	178.7556

Current Estimates

Lock	Entered	Parameter	Estimate	nDF	SS	"F Ratio"	"Prob>F"
☑	☑	Intercept	143.472012	1	0	0.000	1
☐	☑	Age	−1.0310534	1	4029.379	79.536	9.28e-9
☐	☑	Severity	−0.5560378	1	907.0377	17.904	0.00034
☐	☐	Surg-Med	0	1	0.162962	0.003	0.95633
☐	☐	Anxiety	0	1	74.611	1.507	0.23323

Step History

Step	Parameter	Action	"Sig Prob"	Seq SS	RSquare	Cp	p	AICc	BIC	
1	Age	Entered	0.0000	8756.656	0.8124	17.916	2	187.909	190.423	○
2	Severity	Entered	0.0003	907.0377	0.8966	2.4553	3	175.88	178.756	○
3	Anxiety	Entered	0.2332	74.611	0.9035	3.019	4	177.306	180.242	○
4	Surg-Med	Entered	0.8917	0.988332	0.9036	5	5	180.791	183.437	○
5	Best	Specific	.	.	0.8966	2.4553	3	175.88	178.756	◉

However, removing severity causes an increase in AICc so it is added back to the model. The algorithm concluded with both patient age and severity in the model. Note that in this example, the forward selection procedure produced the same model as the backward elimination procedure. This does not always happen, so it is usually a good idea to investigate different model-building techniques for a problem.

Table 3.11 is the JMP stepwise regression algorithm applied to the patient satisfaction data, JMP calls the stepwise option "mixed" variable selection. The default significance levels of 0.25 to enter or remove variables from the model were used. At the first step, patient age is entered in the model. Then severity is entered as the second variable. This is followed by anxiety as the third variable. At that point, none of the remaining predictors met the 0.25 significance level criterion to enter the model, so stepwise regression terminated with age, severity and anxiety as the model predictors. This is not the same model found by backwards elimination and forward selection.

TABLE 3.10 JMP Backward Elimination for the Patient Satisfaction Data in Table 3.8

Stepwise Fit for Satisfaction									

Stepwise Regression Control

Stopping rule: Minimum AICc

Direction: Backward

SSE	DFE	RMSE	RSquare	RSquare Adj	Cp	p	AICc	BIC
1114.5459	22	7.1176667	0.8966	0.8872	2.4553073	3	175.8801	178.7556

Current Estimates

Lock	Entered	Parameter	Estimate	nDF	SS	"F Ratio"	"Prob>F"
☑	☑	Intercept	143.472012	1	0	0.000	1
☐	☑	Age	−1.0310534	1	4029.379	79.536	9.28e-9
☐	☑	Severity	−0.5560378	1	907.0377	17.904	0.00034
☐	☐	Surg-Med	0	1	0.162962	0.003	0.95633
☐	☐	Anxiety	0	1	74.611	1.507	0.23323

Step History

Step	Parameter	Action	"Sig Prob"	Seq SS	RSquare	Cp	p	AICc	BIC	
1	Age	Entered	0.0000	8756.656	0.8124	17.916	2	187.909	190.423	○
2	Severity	Entered	0.0003	907.0377	0.8966	2.4553	3	175.88	178.756	○
3	Surg-Med	Entered	0.9563	0.162962	0.8966	4.4522	4	179.034	181.971	○
4	Anxiety	Entered	0.2422	75.43637	0.9036	5	5	180.791	183.437	○
5	Surg-Med	Removed	0.8917	0.988332	0.9035	3.019	4	177.306	180.242	○
6	Anxiety	Removed	0.2332	74.611	0.8966	2.4553	3	175.88	178.756	○
7	Severity	Removed	0.0003	907.0377	0.8124	17.916	2	187.909	190.423	○
8	Best	Specific	.	.	0.8966	2.4553	3	175.88	178.756	◉

Table 3.12 shows the results of applying the JMP all possible regressions algorithm to the patient satisfaction data. Since there are $k = 4$ predictors, there are 16 possible regression equations. JMP shows the best four of each subset size, along with the full (four-variable) model. For each model, JMP presents the value of R^2, the square root of the mean squared error (RMSE), and the AICc and BIC statistics.

The model with the smallest value of AICc and BIC is the two-variable model with age and severity. The model with the smallest value of the mean squared error (or its square root, RMSE) is the three-variable model with age, severity, and anxiety. Both of these models were found using the stepwise-type algorithms. Either one of these models is likely to be a good regression model describing the effects of the predictor variables on patient satisfaction.

TABLE 3.11 JMP Stepwise (Mixed) Variable Selection for the Patient Satisfaction Data in Table 3.8

Stepwise Fit for Satisfaction
Stepwise Regression Control

Stopping rule: P-Value Threshold
                Prob to Enter 0.25
                Prob to Leave 0.25

Direction: Mixed

SSE	DFE	RMSE	RSquare	RSquare Adj	Cp	p	AICc	BIC
1039.935	21	7.0370954	0.9035	0.8897	3.0190257	4	177.3058	180.2422

Current Estimates

Lock	Entered	Parameter	Estimate	nDF	SS	"F Ratio"	"Prob>F"
[X]		Intercept	143.895206	1	0	0.000	1
[ ]	[X]	Age	-1.1135376	1	3492.683	70.530	3.75e-8
[ ]	[X]	Severity	-0.5849193	1	971.8361	19.625	0.00023
[ ]	[ ]	Surg-Med	0	1	0.988332	0.019	0.89167
[ ]	[X]	Anxiety	1.2961695	1	74.611	1.507	0.23323

Step History

Step	Parameter	Action	"Sig Prob"	Seq SS	RSquare	Cp	p	AICc	BIC	
1	Age	Entered	0.0000	8756.656	0.8124	17.916	2	187.909	190.423	( )
2	Severity	Entered	0.0003	907.0377	0.8966	2.4553	3	175.88	178.756	( )
3	Anxiety	Entered	0.2332	74.611	0.9035	3.019	4	177.306	180.242	(X)

TABLE 3.12 JMP All Possible Models Regression for the Patient Satisfaction Data in Table 3.8

All Possible Models

Ordered up to best 4 models up to 4 terms per model.

Model	Number	RSquare	RMSE	AICc	BIC	
Age	1	0.8124	9.3752	187.909	190.423	O
Severity	1	0.5227	14.9549	211.257	213.771	O
Anxiety	1	0.2876	18.2709	221.271	223.785	O
Surg-Med	1	0.0547	21.0469	228.343	230.857	O
Age, severity	2	0.8966	7.1177	175.880	178.756	O
Age, anxiety	2	0.8133	9.5626	190.644	193.520	O
Age, surg-Med	2	0.8126	9.5817	190.744	193.619	O
Severity, anxiety	2	0.5795	14.3537	210.951	213.827	O
Age, severity, anxiety	3	0.9035	7.0371	177.306	180.242	O
Age, severity, surg-Med	3	0.8966	7.2846	179.034	181.971	O
Age, surg-Med, anxiety	3	0.8135	9.7844	193.785	196.722	O
Severity, surg-Med, anxiety	3	0.5893	14.5186	213.518	216.454	O
Age, severity, surg-Med, anxiety	4	0.9036	7.2074	180.791	183.437	O

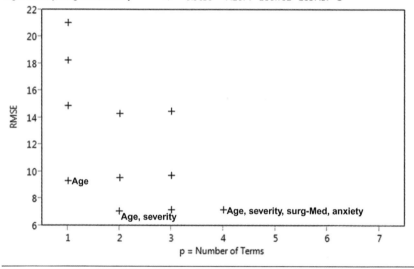

3.7 GENERALIZED AND WEIGHTED LEAST SQUARES

In Section 3.4 we discussed methods for checking the adequacy of a linear regression model. Analysis of the model residuals is the basic methodology. A common defect that shows up in fitting regression models is nonconstant variance. That is, the variance of the observations is not constant but changes in some systematic way with each observation. This problem is often identified from a plot of residuals versus the fitted values. Transformation of the response variable is a widely used method for handling the inequality of variance problem.

Another technique for dealing with nonconstant error variance is to fit the model using the method of **weighted least squares** (WLS). In this method of estimation the deviation between the observed and expected values of y_i is multiplied by a **weight** w_i that is inversely proportional to the variance of y_i. For the case of simple linear regression, the WLS function is

$$L = \sum_{i=1}^{n} w_i (y_i - \beta_0 - \beta_1 x_i)^2, \tag{3.61}$$

where $w_i = 1/\sigma_i^2$ and σ_i^2 is the variance of the ith observation y_i. The resulting least squares normal equations are

$$\hat{\beta}_0 \sum_{i=1}^{n} w_i + \hat{\beta}_1 \sum_{i=1}^{n} w_i x_i = \sum_{i=1}^{n} w_i y_i$$

$$\hat{\beta}_0 \sum_{i=1}^{n} w_i x_i + \hat{\beta}_1 \sum_{i=1}^{n} w_i x_i^2 = \sum_{i=1}^{n} w_i x_i y_i \tag{3.62}$$

Solving Eq. (3.62) will produce WLS estimates of the model parameters β_0 and β_1.

In this section we give a development of WLS for the multiple regression model. We begin by considering a slightly more general situation concerning the structure of the model errors.

3.7.1 Generalized Least Squares

The assumptions that we have made concerning the linear regression model $\mathbf{y} = \mathbf{X}\boldsymbol{\beta} + \boldsymbol{\varepsilon}$ are that $E(\boldsymbol{\varepsilon}) = \mathbf{0}$ and Var $(\boldsymbol{\varepsilon}) = \sigma^2 \mathbf{I}$; that is, the errors have expected value zero and constant variance, and they are uncorrelated. For testing hypotheses and constructing confidence and prediction intervals we also assume that the errors are normally distributed, in which case they are also independent. As we have observed, there are situations where these assumptions are unreasonable. We will now consider the modifications that are necessary to the OLS procedure when $E(\boldsymbol{\varepsilon}) = \mathbf{0}$ and Var $(\boldsymbol{\varepsilon}) = \sigma^2 \mathbf{V}$, where $\mathbf{V}$ is a known $n \times n$ matrix. This situation has a simple interpretation; if $\mathbf{V}$ is diagonal but with unequal diagonal elements, then the observations

y are **uncorrelated** but have **unequal variances**, while if some of the off-diagonal elements of **V** are nonzero, then the observations are **correlated**. When the model is

$$\mathbf{y} = \mathbf{X}\boldsymbol{\beta} + \boldsymbol{\varepsilon}$$
$$E(\boldsymbol{\varepsilon}) = \mathbf{0} \quad \text{and} \quad \text{Var}(\boldsymbol{\varepsilon}) = \sigma^2 \mathbf{I} \tag{3.63}$$

the OLS estimator $\hat{\boldsymbol{\beta}} = (\mathbf{X'X})^{-1}\mathbf{X'y}$ is no longer appropriate. The OLS estimator is unbiased because

$$E(\hat{\boldsymbol{\beta}}) = E[(\mathbf{X'X})^{-1}\mathbf{X'y}] = (\mathbf{X'X})^{-1}\mathbf{X'}E(\mathbf{y}) = (\mathbf{X'X})^{-1}\mathbf{X'}\boldsymbol{\beta} = \boldsymbol{\beta}$$

but the covariance matrix of $\hat{\boldsymbol{\beta}}$ is not $\sigma^2(\mathbf{X'X})^{-1}$. Instead, the covariance matrix is

$$\begin{aligned}\text{Var}\,(\hat{\boldsymbol{\beta}}) &= \text{Var}\,[(\mathbf{X'X})^{-1}\mathbf{X'y}] \\ &= (\mathbf{X'X})^{-1}\mathbf{X'}V(\mathbf{y})\mathbf{X}(\mathbf{X'X})^{-1} \\ &= \sigma^2(\mathbf{X'X})^{-1}\mathbf{X'X}(\mathbf{X'X})^{-1}\end{aligned}$$

Practically, this implies that the variances of the regression coefficients are larger than we expect them to be.

This problem can be avoided if we estimate the model parameters with a technique that takes the correct variance structure in the errors into account. We will develop this technique by transforming the model to a new set of observations that satisfy the standard least squares assumptions. Then we will use OLS on the transformed observations.

Because $\sigma^2 \mathbf{V}$ is the covariance matrix of the errors, **V** must be nonsingular and positive definite, so there exists an $n \times n$ nonsingular symmetric matrix **K** defined such that

$$\mathbf{K'K} = \mathbf{KK} = \mathbf{V}$$

The matrix **K** is often called the **square root** of **V**. Typically, the error variance σ^2 is unknown, in which case **V** represents the known (or assumed) structure of the variances and covariances among the random errors apart from the constant σ^2.

Define the new variables

$$\mathbf{z} = \mathbf{K}^{-1}\mathbf{y}, \quad \mathbf{B} = \mathbf{K}^{-1}\mathbf{X}, \quad \text{and} \quad \boldsymbol{\delta} = \mathbf{K}^{-1}\boldsymbol{\varepsilon} \tag{3.64}$$

so that the regression model $\mathbf{y} = \mathbf{X}\beta + \varepsilon$ becomes, upon multiplication by $\mathbf{K}^{-1}$,

$$\mathbf{K}^{-1}\mathbf{y} = \mathbf{K}^{-1}\mathbf{X}\beta + \mathbf{K}^{-1}\varepsilon$$

or

$$\mathbf{z} = \mathbf{B}\beta + \delta \tag{3.65}$$

The errors in the transformed model Eq. (3.65) have zero expectation because $E(\delta) = E(\mathbf{K}^{-1}\varepsilon) = \mathbf{K}^{-1}E(\varepsilon) = \mathbf{0}$. Furthermore, the covariance matrix of δ is

$$\begin{aligned}
\text{Var}\,(\delta) &= V(\mathbf{K}^{-1}\varepsilon) \\
&= \mathbf{K}^{-1}V(\varepsilon)\mathbf{K}^{-1} \\
&= \sigma^2 \mathbf{K}^{-1}\mathbf{V}\mathbf{K}^{-1} \\
&= \sigma^2 \mathbf{K}^{-1}\mathbf{K}\mathbf{K}\mathbf{K}^{-1} \\
&= \sigma^2 \mathbf{I}
\end{aligned}$$

Thus the elements of the vector of errors δ have mean zero and constant variance and are uncorrelated. Since the errors δ in the model in Eq. (3.65) satisfy the usual assumptions, we may use OLS to estimate the parameters. The least squares function is

$$\begin{aligned}
L &= \delta'\delta \\
&= (\mathbf{K}^{-1}\varepsilon)'\mathbf{K}^{-1}\varepsilon \\
&= \varepsilon'\mathbf{K}^{-1}\mathbf{K}^{-1}\varepsilon \\
&= \varepsilon'\mathbf{V}^{-1}\varepsilon \\
&= (\mathbf{y} - \mathbf{X}\beta)'\mathbf{V}^{-1}(\mathbf{y} - \mathbf{X}\beta)
\end{aligned}$$

The corresponding normal equations are

$$(\mathbf{X}'\mathbf{V}^{-1}\mathbf{X})\hat{\beta}_{\text{GLS}} = \mathbf{X}'\mathbf{V}^{-1}\mathbf{y} \tag{3.66}$$

In Equation (3.66) $\hat{\beta}_{\text{GLS}}$ is the **generalized least squares (GLS) estimator** of the model parameters β. The solution to the GLS normal equations is

$$\hat{\beta}_{\text{GLS}} = (\mathbf{X}'\mathbf{V}^{-1}\mathbf{X})^{-1}\mathbf{X}'\mathbf{V}^{-1}\mathbf{y} \tag{3.67}$$

The GLS estimator is an unbiased estimator for the model parameters β, and the covariance matrix of $\hat{\beta}_{GLS}$ is

$$\text{Var}\,(\hat{\beta}_{GLS}) = \sigma^2(\mathbf{X}'\mathbf{V}^{-1}\mathbf{X})^{-1} \tag{3.68}$$

The GLS estimator is a best linear unbiased estimator of the model parameters β, where "best" means minimum variance.

3.7.2 Weighted Least Squares

Weighted least squares or WLS is a special case of GLS where the n response observations y_i do not have the same variances but are uncorrelated. Therefore the matrix $\mathbf{V}$ is

$$\mathbf{V} = \begin{bmatrix} \sigma_1^2 & 0 & \cdots & 0 \\ 0 & \sigma_2^2 & 0 & 0 \\ \vdots & \vdots & \ddots & \vdots \\ 0 & 0 & \cdots & \sigma_n^2 \end{bmatrix},$$

where σ_i^2 is the variance of the ith observation y_i, $i = 1, 2, \ldots, n$. Because the weight for each observation should be the reciprocal of the variance of that observation, it is convenient to define a diagonal matrix of weights $\mathbf{W} = \mathbf{V}^{-1}$. Clearly, the weights are the main diagonals of the matrix $\mathbf{W}$. Therefore the WLS criterion is

$$L = (\mathbf{y} - \mathbf{X}\beta)'\mathbf{W}(\mathbf{y} - \mathbf{X}\beta) \tag{3.69}$$

and the WLS normal equations are

$$(\mathbf{X}'\mathbf{W}\mathbf{X})\hat{\beta}_{WLS} = \mathbf{X}'\mathbf{W}\mathbf{y}. \tag{3.70}$$

The WLS estimator is

$$\hat{\beta}_{WLS} = (\mathbf{X}'\mathbf{W}\mathbf{X})^{-1}\mathbf{X}'\mathbf{W}\mathbf{y}. \tag{3.71}$$

The WLS estimator is an unbiased estimator for the model parameters β, and the covariance matrix of $\hat{\beta}_{WLS}$ is

$$\text{Var}\,(\hat{\beta}_{WLS}) = (\mathbf{X}'\mathbf{W}\mathbf{X})^{-1}. \tag{3.72}$$

To use WLS, the weights w_i must be known. Sometimes prior knowledge or experience or information from an underlying theoretical model can be used to determine the weights. For example, suppose that a significant source of error is measurement error and different observations are measured by different instruments of unequal but known or well-estimated accuracy. Then the weights could be chosen inversely proportional to the variances of measurement error.

In most practical situations, however, the analyst learns about the inequality of variance problem from the residual analysis for the original model that was fit using OLS. For example, the plot of the OLS residuals e_i versus the fitted values $\hat{y}_i$ may exhibit an outward-opening funnel shape, suggesting that the variance of the observations is increasing with the mean of the response variable y. Plots of the OLS residuals versus the predictor variables may indicate that the variance of the observations is a function of one of the predictors. In these situations we can often use estimates of the weights. There are several approaches that could be used to estimate the weights. We describe two of the most widely used methods.

Estimation of a Variance Equation   In the first method, suppose that analysis of the OLS residuals indicates that the variance of the ith observation is a function of one or more predictors or the mean of y. The squared OLS residual e_i^2 is an estimator of the variance of the ith observation σ_i^2 if the form of the regression model is correct. Furthermore, the absolute value of the residual $|e_i|$ is an estimator of the standard deviation σ_i (because $\sigma_i = |\sqrt{\sigma_i^2}|$). Consequently, we can find a **variance equation** or a regression model relating σ_i^2 to appropriate predictor variables by the following process:

1. Fit the model relating y to the predictor variables using OLS and find the OLS residuals.
2. Use residual analysis to determine potential relationships between σ_i^2 and either the mean of y or some of the predictor variables.
3. Regress the squared OLS residuals on the appropriate predictors to obtain an equation for predicting the variance of each observation, say, $\hat{s}_i^2 = f(x)$ or $\hat{s}_i^2 = f(y)$.
4. Use the fitted values from the estimated variance function to obtain estimates of the weights, $w_i = 1/\hat{s}_i^2$, $i = 1, 2, \ldots, n$.
5. Use the estimated weights as the diagonal elements of the matrix **W** in the WLS procedure.

As an alternative to estimating a variance equation in step 3 above, we could use the absolute value of the OLS residual and fit an equation that relates the standard deviation of each observation to the appropriate regressors. This is the preferred approach if there are potential outliers in the data, because the absolute value of the residuals is less affected by outliers than the squared residuals.

When using the five-step procedure outlined above, it is a good idea to compare the estimates of the model parameters obtained from the WLS fit to those obtained from the original OLS fit. Because both methods produce unbiased estimators, we would expect to find that the point estimates of the parameters from both analyses are very similar. If the WLS estimates differ significantly from their OLS counterparts, it is usually a good idea to use the new WLS residuals and reestimate the variance equation to produce a new set of weights and a revised set of WLS estimates using these new weights. This procedure is called **iteratively reweighted least squares** (IRLS). Usually one or two iterations are all that is required to produce stable estimates of the model parameters.

Using Replicates or Nearest Neighbors   The second approach to estimating the weights makes use of **replicate observations** or **nearest neighbors**. Exact replicates are sample observations that have exactly the same values of the predictor variables. Suppose that there are replicate observations at each of the combination of levels of the predictor variables. The weights w_i can be estimated directly as the reciprocal of the sample variances at each combination of these levels. Each observation in a replicate group would receive the same weight. This method works best when there are a moderately large number of observations in each replicate group, because small samples do not produce reliable estimates of the variance.

Unfortunately, it is fairly unusual to find groups of replicate observations in most regression-modeling situations. It is especially unusual to find them in time series data. An alternative is to look for observations with **similar** x-levels, which can be thought of as a nearest-neighbor group of observations. The observations in a nearest-neighbor group can be considered as pseudoreplicates and the sample variance for all of the observations in each nearest-neighbor group can be computed. The reciprocal of a sample variance would be used as the weight for all observations in the nearest-neighbor group.

Sometimes these nearest-neighbor groups can be identified visually by inspecting the scatter plots of y versus the predictor variables or from plots of the predictor variables versus each other. Analytical methods can also be used to find these nearest-neighbor groups. One nearest-neighbor algorithm is described in Montgomery, Peck, and Vining (2012). These authors also present a complete example showing how the nearest-neighbor approach can be used to estimate the weights for a WLS analysis.

Statistical Inference in WLS In WLS the variances σ_i^2 are almost always unknown and must be estimated. Since statistical inference on the model parameters as well as confidence intervals and prediction intervals on the response are usually necessary, we should consider the effect of using estimated weights on these procedures. Recall that the covariance matrix of the model parameters in WLS was given in Eq. (3.72). This covariance matrix plays a central role in statistical inference. Obviously, when estimates of the weights are substituted into Eq. (3.72) an estimated covariance matrix is obtained. Generally, the impact of using estimated weights is modest, provided that the sample size is not very small. In these situations, statistical tests, confidence intervals, and prediction intervals should be considered as approximate rather than exact.

Example 3.9 Table 3.13 contains 28 observations on the strength of a connector and the age in weeks of the glue used to bond the components of the connector together. A scatter plot of the strength versus age, shown in Figure 3.2, suggests that there may be a linear relationship between strength and age, but there may also be a problem with nonconstant variance in the data. The regression model that was fit to these data is

$$\hat{y} = 25.936 + 0.3759x,$$

where x = weeks.

The residuals from this model are shown in Table 3.13. Figure 3.3 is a plot of the residuals versus weeks. The pronounced outward-opening funnel shape on this plot confirms the inequality of variance problem. Figure 3.4 is a plot of the absolute value of the residuals from this model versus week. There is an indication that a linear relationship may exist between the absolute value of the residuals and weeks, although there is evidence of one outlier in the data. Therefore it seems reasonable to fit a model relating the absolute value of the residuals to weeks. Since the absolute value of a residual is the residual standard deviation, the predicted values from this equation could be used to determine weights for the regression model relating strength to weeks. This regression model is

$$\hat{s}_i = -5.854 + 0.29852x.$$

TABLE 3.13   Connector Strength Data

Observation	Weeks	Strength	Residual	Absolute Residual	Weights
1	20	34	0.5454	0.5454	73.9274
2	21	35	1.1695	1.1695	5.8114
3	23	33	−1.5824	1.5824	0.9767
4	24	36	1.0417	1.0417	0.5824
5	25	35	−0.3342	0.3342	0.3863
6	28	34	−2.4620	2.4620	0.1594
7	29	37	0.1621	0.1621	0.1273
8	30	34	−3.2139	3.2139	0.1040
9	32	42	4.0343	4.0343	0.0731
10	33	35	−3.3416	3.3416	0.0626
11	35	33	−6.0935	6.0935	0.0474
12	37	46	6.1546	6.1546	0.0371
13	38	43	2.7787	2.7787	0.0332
14	40	32	−8.9731	8.9731	0.0270
15	41	37	−4.3491	4.3491	0.0245
16	43	50	7.8991	7.8991	0.0205
17	44	34	−8.4769	8.4769	0.0189
18	45	54	11.1472	11.1472	0.0174
19	46	49	5.7713	5.7713	0.0161
20	48	55	11.0194	11.0194	0.0139
21	50	40	−4.7324	4.7324	0.0122
22	51	33	−12.1084	12.1084	0.0114
23	52	56	10.5157	10.5157	0.0107
24	55	58	11.3879	11.3879	0.0090
25	56	45	−1.9880	1.9880	0.0085
26	57	33	−14.3639	14.3639	0.0080
27	59	60	11.8842	11.8842	0.0072
28	60	35	−13.4917	13.4917	0.0069

The weights would be equal to the inverse of the square of the fitted value for each s_i. These weights are shown in Table 3.13. Using these weights to fit a new regression model to strength using WLS results in

$$\hat{y} = 27.545 + 0.32383x$$

Note that the weighted least squares model does not differ very much from the OLS model. Because the parameter estimates did not change very much, this is an indication that it is not necessary to iteratively reestimate the standard deviation model and obtain new weights.

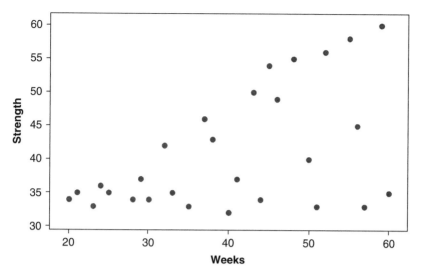

FIGURE 3.2   Scatter diagram of connector strength versus age from Table 3.12.

3.7.3 Discounted Least Squares

Weighted least squares is typically used in situations where the variance of the observations is not constant. We now consider a different situation where a WLS-type procedure is also appropriate. Suppose that the predictor variables in the regression model are only functions of time. As

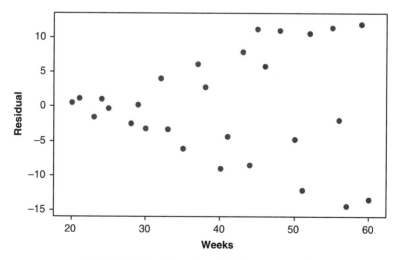

FIGURE 3.3   Plot of residuals versus weeks.

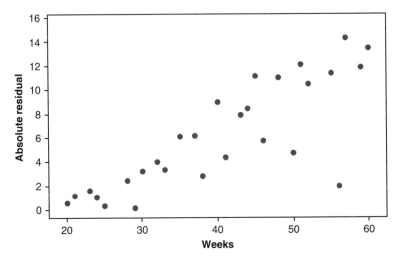

FIGURE 3.4   Scatter plot of absolute residuals versus weeks.

an illustration, consider the linear regression model with a linear trend in time:

$$y_t = \beta_0 + \beta_1 t + \varepsilon, \quad t = 1, 2, \dots, T \tag{3.73}$$

This model was introduced to illustrate trend adjustment in a time series in Section 2.4.2 and Example 3.2. As another example, the regression model

$$y_t = \beta_0 + \beta_1 \sin \frac{2\pi}{d} t + \beta_2 \cos \frac{2\pi}{d} t + \varepsilon \tag{3.74}$$

describes the relationship between a response variable y that varies cyclically or periodically with time where the cyclic variation is modeled as a simple sine wave. A very general model for these types of situations could be written as

$$y_t = \beta_0 + \beta_1 x_1(t) + \cdots + \beta_k x_k(t) + \varepsilon_t, \quad t = 1, 2, \dots, T, \tag{3.75}$$

where the predictors $x_1(t), x_2(t), \dots, x_k(t)$ are mathematical functions of time, t. In these types of models it is often logical to believe that older observations are of less value in predicting the future observations at periods $T + 1, T + 2, \dots$, than are the observations that are close to the current time period, T. In other words, if you want to predict the value of y at time

$T + 1$ given that you are at the end of time period T (or $\hat{y}_{T+1}(T)$), it is logical to assume that the more recent observations such as y_T, y_{T-1}, and y_{T-2} carry much more useful information than do older observations such as y_{T-20}. Therefore it seems reasonable to weight the observations in the regression model so that recent observations are weighted more heavily than older observations. A very useful variation of WLS, called **discounted least squares**, can be used to do this. Discounted least squares also lead to a relatively simple way to update the estimates of the model parameters after each new observation in the time series.

Suppose that the model for observation y_t is given by Eq. (3.75):

$$y_t = \beta_1 x_1(t) + \cdots + \beta_p x_p(t) + \varepsilon_t$$
$$= \mathbf{x}(t)'\boldsymbol{\beta}, \quad t = 1, 2, \ldots, T,$$

where $\mathbf{x}(t)' = [x_1(t), x_2(t), \ldots, x_p(t)]$ and $\boldsymbol{\beta}' = [\beta_1, \beta_2, \ldots, \beta_p]$. This model could have an intercept term, in which case $x_1(t) = 1$ and the final model term could be written as $\beta_k x_k(t)$ as in Eq. (3.75). In matrix form, Eq. (3.75) is

$$\mathbf{y} = \mathbf{X}(T)\boldsymbol{\beta} + \boldsymbol{\varepsilon}, \tag{3.76}$$

where $\mathbf{y}$ is a $T \times 1$ vector of the observations, $\boldsymbol{\beta}$ is a $p \times 1$ vector of the model parameters, $\boldsymbol{\varepsilon}$ is a $T \times 1$ vector of the errors, and $\mathbf{X}(T)$ is the $T \times p$ matrix

$$\mathbf{X}(T) = \begin{bmatrix} x_1(1) & x_2(1) & \cdots & x_p(1) \\ x_1(2) & x_2(2) & \cdots & x_p(2) \\ \vdots & \vdots & \vdots & \vdots \\ x_1(T) & x_2(T) & \cdots & x_p(T) \end{bmatrix}$$

Note that the tth row of $\mathbf{X}(T)$ contains the values of the predictor variables that correspond to the tth observation of the response, y_t.

We will estimate the parameters in Eq. (3.76) using WLS. However, we are going to choose the weights so that they decrease in magnitude with time. Specifically, let the weight for observation y_{T-j} be θ^j, where $0 < \theta < 1$. We are also going to shift the origin of time with each new observation so that T is the current time period. Therefore the WLS criterion is

$$
\begin{aligned}
L &= \sum_{j=0}^{T-1} w_j \left[y_{T-j} - (\beta_1(T)x_1(-j) + \cdots + \beta_p(T)x_k(-j)) \right]^2 \\
&= \sum_{j=0}^{T-1} w_j \left[y_{T-j} - \mathbf{x}(-j)\boldsymbol{\beta}(T) \right]^2,
\end{aligned}
\tag{3.77}
$$

where $\boldsymbol{\beta}(T)$ indicates that the vector of regression coefficients is estimated at the end of time period T, and $\mathbf{x}(-j)$ indicates that the predictor variables, which are just mathematical functions of time, are evaluated at $-j$. This is just WLS with a $T \times T$ diagonal weight matrix

$$
\mathbf{W} = \begin{bmatrix}
\theta^{T-1} & 0 & 0 & \cdots & 0 \\
0 & \theta^{T-2} & 0 & \cdots & 0 \\
\vdots & & \ddots & \vdots & \vdots \\
0 & \cdots & & \theta & 0 \\
0 & 0 & \cdots & 0 & 1
\end{bmatrix}
$$

By analogy with Eq. (3.70), the WLS normal equations are

$$
\mathbf{X}(T)'\mathbf{W}\mathbf{X}(T)\hat{\boldsymbol{\beta}}(T) = \mathbf{X}(T)'\mathbf{W}\mathbf{y}
$$

or

$$
\mathbf{G}(T)\hat{\boldsymbol{\beta}}(T) = \mathbf{g}(T),
\tag{3.78}
$$

where

$$
\begin{aligned}
\mathbf{G}(T) &= \mathbf{X}(T)'\mathbf{W}\mathbf{X}(T) \\
\mathbf{g}(T) &= \mathbf{X}(T)'\mathbf{W}\mathbf{y}
\end{aligned}
\tag{3.79}
$$

The solution to the WLS normal equations is

$$
\hat{\boldsymbol{\beta}}(T) = \mathbf{G}(T)^{-1}\mathbf{g}(T),
\tag{3.80}
$$

$\hat{\boldsymbol{\beta}}(T)$ is called the **discounted least squares estimator** of β.

In many important applications, the discounted least squares estimator can be simplified considerably. Assume that the predictor variables $x_i(t)$ in the model are functions of time that have been chosen so that their values at time period $t + 1$ are linear combinations of their values at the previous time period. That is,

$$x_i(t + 1) = L_{i1}x_1(t) + L_{i2}x_2(t) + \cdots + L_{ip}x_p(t), \quad i = 1, 2, \ldots, p \quad (3.81)$$

In matrix form,

$$\mathbf{x}(t + 1) = \mathbf{L}\mathbf{x}(t), \quad (3.82)$$

where $\mathbf{L}$ is the $p \times p$ matrix of the constants L_{ij} in Eq. (3.81). The transition property in Eq. (3.81) holds for polynomial, trigonometric, and certain exponential functions of time. This transition relationship implies that

$$\mathbf{x}(t) = \mathbf{L}^t\mathbf{x}(0) \quad (3.83)$$

Consider the matrix $\mathbf{G}(T)$ in the normal equations (3.78). We can write

$$\mathbf{G(T)} = \sum_{j=0}^{T-1} \theta^j \mathbf{x}(-j)\mathbf{x}(-j)'$$

$$= \mathbf{G}(T - 1) + \theta^{T-1}\mathbf{x}(-(T - 1))\mathbf{x}(-(T - 1))'$$

If the predictor variables $x_i(t)$ in the model are polynomial, trigonometric, or certain exponential functions of time, the matrix $\mathbf{G}(T)$ approaches a steady-state limiting value $\mathbf{G}$, where

$$\mathbf{G} = \sum_{j=0}^{\infty} \theta^j \mathbf{x}(-j)\mathbf{x}(-j)' \quad (3.84)$$

Consequently, the inverse of $\mathbf{G}$ would only need to be computed once. The right-hand side of the normal equations can also be simplified. We can write

$$\mathbf{g(T)} = \sum_{j=0}^{T-1} \theta^j y_{T-j}\mathbf{x}(-j)$$

$$= y_T\mathbf{x}(0) + \sum_{j=1}^{T-1} \theta^j y_{T-j}\mathbf{x}(-j)$$

$$= y_T\mathbf{x}(0) + \theta \sum_{j=1}^{T-1} \theta^{j-1} y_{T-j}\mathbf{L}^{-1}\mathbf{x}(-j + 1)$$

$$= y_T\mathbf{x}(0) + \theta\mathbf{L}^{-1} \sum_{k=0}^{T-2} \theta^k y_{T-1-k}\mathbf{x}(-k)$$

$$= y_T\mathbf{x}(0) + \theta\mathbf{L}^{-1}\mathbf{g}(T - 1)$$

So the discounted least squares estimator can be written as

$$\hat{\beta}(T) = \mathbf{G}^{-1}\mathbf{g}(T)$$

This can also be simplified. Note that

$$
\begin{aligned}
\hat{\beta}(T) &= \mathbf{G}^{-1}\mathbf{g}(T) \\
&= \mathbf{G}^{-1}[y_T\mathbf{x}(0) + \theta\mathbf{L}^{-1}\mathbf{g}(T-1)] \\
&= \mathbf{G}^{-1}[y_T\mathbf{x}(0) + \theta\mathbf{L}^{-1}\mathbf{G}\hat{\beta}(T-1)] \\
&= y_T\mathbf{G}^{-1}\mathbf{x}(0) + \theta\mathbf{G}^{-1}\mathbf{L}^{-1}\mathbf{G}\hat{\beta}(T-1)
\end{aligned}
$$

or

$$\hat{\beta}(T) = \mathbf{h}y_T + \mathbf{Z}\hat{\beta}(T-1) \tag{3.85}$$

where

$$\mathbf{h} = \mathbf{G}^{-1}\mathbf{x}(0) \tag{3.86}$$

and

$$\mathbf{Z} = \theta\mathbf{G}^{-1}\mathbf{L}^{-1}\mathbf{G} \tag{3.87}$$

The right-hand side of Eq. (3.85) can still be simplified because

$$
\begin{aligned}
\mathbf{L}^{-1}\mathbf{G} &= \mathbf{L}^{-1}\mathbf{G}(\mathbf{L}')^{-1}\mathbf{L}' \\
&= \sum_{j=0}^{\infty} \theta^j \mathbf{L}^{-1}\mathbf{x}(-j)\mathbf{x}(-j)'(\mathbf{L}')^{-1}\mathbf{L}' \\
&= \sum_{j=0}^{\infty} \theta^j [\mathbf{L}^{-1}\mathbf{x}(-j)][\mathbf{L}^{-1}\mathbf{x}(-j)]'\mathbf{L}' \\
&= \sum_{j=0}^{\infty} \theta^j \mathbf{x}(-j-1)\mathbf{x}(-j-1)'\mathbf{L}'
\end{aligned}
$$

and letting $k = j + 1$,

$$
\begin{aligned}
\mathbf{L}^{-1}\mathbf{G} &= \theta^{-1}\sum_{k=1}^{\infty} \theta^k \mathbf{x}(-k)\mathbf{x}(-k)'\mathbf{L}' \\
&= \theta^{-1}[\mathbf{G} - \mathbf{x}(0)\mathbf{x}(0)']\mathbf{L}'
\end{aligned}
$$

Substituting for $\mathbf{L}^{-1}\mathbf{G}$ on the right-hand side of Eq. (3.87) results in

$$
\begin{aligned}
\mathbf{Z} &= \theta\mathbf{G}^{-1}\theta^{-1}[\mathbf{G} - \mathbf{x}(0)\mathbf{x}(0)']\mathbf{L}' \\
&= [\mathbf{I} - \mathbf{G}^{-1}\mathbf{x}(0)\mathbf{x}(0)']\mathbf{L}' \\
&= \mathbf{L}' - \mathbf{h}\mathbf{x}(0)\mathbf{L}' \\
&= \mathbf{L}' - \mathbf{h}[\mathbf{L}\mathbf{x}(0)]' \\
&= \mathbf{L}' - \mathbf{h}\mathbf{x}(1)'
\end{aligned}
$$

Now the vector of discounted least squares parameter estimates at the end of time period T in Eq. (3.85) is

$$
\begin{aligned}
\hat{\beta}(T) &= \mathbf{h}y_T + \mathbf{Z}\hat{\beta}(T-1) \\
&= \mathbf{h}y_T + [\mathbf{L}' - \mathbf{h}\mathbf{x}(1)']\hat{\beta}(T-1) \\
&= \mathbf{L}'\hat{\beta}(T-1) + \mathbf{h}[y_T - \mathbf{x}(1)'\hat{\beta}(T-1)].
\end{aligned}
$$

But $\mathbf{x}(1)'\beta(T-1) = \hat{y}_T(T-1)$ is the forecast of y_T computed at the end of the previous time period, $T-1$, so the discounted least squares vector of parameter estimates computed at the end of time period t is

$$
\begin{aligned}
\hat{\beta}(\text{T}) &= \mathbf{L}'\hat{\beta}(T-1) + \mathbf{h}[y_T - \hat{y}_T(T-1)] \\
&= \mathbf{L}'\hat{\beta}(T-1) + \mathbf{h}e_t(1).
\end{aligned}
\tag{3.88}
$$

The last line in Eq. (3.88) is an extremely important result; it states that in discounted least squares the vector of parameter estimates computed at the end of time period T can be computed as a simple linear combination of the estimates made at the end of the previous time period $T-1$ and the one-step-ahead forecast error for the observation in period T. Note that there are really two things going on in estimating β by discounted least squares: the origin of time is being shifted to the end of the current period, and the estimates of the model parameters are being modified to reflect the forecast error in the current time period. The first and second terms on the right-hand side of Eq. (3.88) accomplish these objectives, respectively.

When discounted least squares estimation is started up, an initial estimate of the parameters is required at time period zero, say, $\hat{\beta}(0)$. This could be found by a standard least squares (or WLS) analysis of historical data.

Because the origin of time is shifted to the end of the current time period, forecasting is easy with discounted least squares. The forecast of the observation at a future time period $T + \tau$, made at the end of time period T, is

$$\hat{y}_{T+\tau}(T) = \hat{\boldsymbol{\beta}}(T)'\mathbf{x}(\tau)$$
$$= \sum_{j=1}^{p} \hat{\beta}_j(T)x_j(\tau). \tag{3.89}$$

Example 3.10 Discounted Least Squares and the Linear Trend Model
To illustrate the discounted least squares procedure, let us consider the linear trend model:

$$y_t = \beta_0 + \beta_1 t + \varepsilon_t, \quad t = 1, 2, \dots, T$$

To write the parameter estimation equations in Eq. (3.88), we need the transition matrix $\mathbf{L}$. For the linear trend model, this matrix is

$$\mathbf{L} = \begin{bmatrix} 1 & 0 \\ 1 & 1 \end{bmatrix}$$

Therefore the parameter estimation equations are

$$\hat{\boldsymbol{\beta}}(T) = \mathbf{L}'\hat{\boldsymbol{\beta}}(T-1) + \mathbf{h}e_T(1)$$

$$\begin{bmatrix} \hat{\beta}_0(T) \\ \hat{\beta}_1(T) \end{bmatrix} = \begin{bmatrix} 1 & 1 \\ 0 & 1 \end{bmatrix} \begin{bmatrix} \hat{\beta}_0(T-1) \\ \hat{\beta}_1(T-1) \end{bmatrix} + \begin{bmatrix} h_1 \\ h_2 \end{bmatrix} e_T(1)$$

or

$$\hat{\beta}_0(T) = \hat{\beta}_0(T-1) + \hat{\beta}_1(T-1) + h_1 e_1(T)$$
$$\hat{\beta}_1(T) = \hat{\beta}_1(T-1) + h_2 e_T(1) \tag{3.90}$$

The elements of the vector $\mathbf{h}$ are found from Eq. (3.86):

$$\mathbf{h} = \mathbf{G}^{-1}\mathbf{x}(0)$$
$$= \mathbf{G}^{-1} \begin{bmatrix} 1 \\ 0 \end{bmatrix}$$

The steady-state matrix $\mathbf{G}$ is found as follows:

$$\mathbf{G}(T) = \sum_{j=0}^{T-1} \theta^j \mathbf{x}(-j)\mathbf{x}(-j)'$$

$$= \sum_{j=0}^{T-1} \theta^j \begin{bmatrix} 1 \\ -j \end{bmatrix} \begin{bmatrix} 1 & -j \end{bmatrix}$$

$$= \sum_{j=0}^{T-1} \theta^j \begin{bmatrix} 1 & -j \\ -j & +j^2 \end{bmatrix}$$

$$= \begin{bmatrix} \sum_{j=0}^{T-1} \theta^j & -\sum_{j=0}^{T-1} j\theta^j \\ -\sum_{j=0}^{T-1} j\theta^j & \sum_{j=0}^{T-1} j^2\theta^j \end{bmatrix}$$

$$= \begin{bmatrix} \dfrac{1 - \theta^T}{1 - \theta} & -\dfrac{\theta(1 - \theta^T)}{1 - \theta} \\ -\dfrac{\theta(1 - \theta^T)}{1 - \theta} & \dfrac{\theta(1 + \theta)(1 - \theta^T)}{(1 - \theta)^3} \end{bmatrix}$$

The steady-state value of $\mathbf{G}(T)$ is found by taking the limit as $T \to \infty$, which results in

$$\mathbf{G} = \lim_{T \to \infty} \mathbf{G}(T)$$

$$= \begin{bmatrix} \dfrac{1}{1 - \theta} & -\dfrac{\theta}{1 - \theta} \\ -\dfrac{\theta}{1 - \theta} & \dfrac{\theta(1 + \theta)}{(1 - \theta)^3} \end{bmatrix}$$

The inverse of $\mathbf{G}$ is

$$\mathbf{G}^{-1} = \begin{bmatrix} 1 - \theta^2 & (1 - \theta)^2 \\ (1 - \theta)^2 & \dfrac{(1 - \theta)^2}{\theta} \end{bmatrix}.$$

Therefore, the vector **h** is

$$\mathbf{h} = \mathbf{G}^{-1}\mathbf{x}(0)$$

$$= \mathbf{G}^{-1} \begin{bmatrix} 1 \\ 0 \end{bmatrix}$$

$$= \begin{bmatrix} 1 - \theta^2 & (1-\theta)^2 \\ (1-\theta)^2 & \dfrac{(1-\theta)^2}{\theta} \end{bmatrix} \begin{bmatrix} 1 \\ 0 \end{bmatrix}$$

$$= \begin{bmatrix} 1 - \theta^2 \\ (1-\theta)^2 \end{bmatrix}.$$

Substituting the elements of the vector **h** into Eq. (3.90) we obtain the parameter estimating equations for the linear trend model as

$$\hat{\beta}_0(T) = \hat{\beta}_0(T-1) + \hat{\beta}_1(T-1) + (1 - \theta^2)e_T(1)$$
$$\hat{\beta}_1(T) = \hat{\beta}_1(T-1) + (1 - \theta)^2 e_T(1)$$

Inspection of these equations illustrates the twin aspects of discounted least squares; shifting the origin of time, and updating the parameter estimates. In the first equation, the updated intercept at time T consists of the old intercept plus the old slope (this shifts the origin of time to the end of the current period T), plus a fraction of the current forecast error (this revises or updates the estimate of the intercept). The second equation revises the slope estimate by adding a fraction of the current period forecast error to the previous estimate of the slope.

To illustrate the computations, suppose that we are forecasting a time series with a linear trend and we have initial estimates of the slope and intercept at time $t = 0$ as

$$\hat{\beta}_0(0) = 50 \quad \text{and} \quad \hat{\beta}_1(0) = 1.5$$

These estimates could have been obtained by regression analysis of historical data.

Assume that $\theta = 0.9$, so that $1 - \theta^2 = 1 - (0.9)^2 = 0.19$ and $(1 - \theta)^2 = (1 - 0.9)^2 = 0.01$. The forecast for time period $t = 1$, made at the end of time period $t = 0$, is computed from Eq. (3.89):

$$\hat{y}_1(0) = \hat{\beta}(0)'\mathbf{x}(1)$$
$$= \hat{\beta}_0(0) + \hat{\beta}_1(0)$$
$$= 50 + 1.5$$
$$= 51.5$$

Suppose that the actual observation in time period 1 is $y_1 = 52$. The forecast error in time period 1 is

$$e_1(1) = y_1 - \hat{y}_1(0)$$
$$= 52 - 51.5$$
$$= 0.5.$$

The updated estimates of the model parameter computed at the end of time period 1 are now

$$\hat{\beta}_0(1) = \hat{\beta}_0(0) + \hat{\beta}_1(0) + 0.19e_1(0)$$
$$= 50 + 1.5 + 0.19(0.5)$$
$$= 51.60$$

and

$$\hat{\beta}_1(1) = \hat{\beta}_1(0) + 0.01e_1(0)$$
$$= 1.5 + 0.01(0.5)$$
$$= 1.55$$

The origin of time is now $T = 1$. Therefore the forecast for time period 2 made at the end of period 1 is

$$\hat{y}_2(1) = \hat{\beta}_0(1) + \hat{\beta}_1(1)$$
$$= 51.6 + 1.55$$
$$= 53.15.$$

If the observation in period 2 is $y_2 = 55$, we would update the parameter estimates exactly as we did at the end of time period 1. First, calculate the forecast error:

$$e_2(1) = y_2 - \hat{y}_2(1)$$
$$= 55 - 53.15$$
$$= 1.85$$

Second, revise the estimates of the model parameters:

$$\hat{\beta}_0(2) = \hat{\beta}_0(1) + \hat{\beta}_1(1) + 0.19e_2(1)$$
$$= 51.6 + 1.55 + 0.19(1.85)$$
$$= 53.50$$

and

$$\hat{\beta}_1(2) = \hat{\beta}_1(1) + 0.01e_2(1)$$
$$= 1.55 + 0.01(1.85)$$
$$= 1.57$$

The forecast for period 3, made at the end of period 2, is

$$\hat{y}_3(2) = \hat{\beta}_0(2) + \hat{\beta}_1(2)$$
$$= 53.50 + 1.57$$
$$= 55.07.$$

Suppose that a forecast at a longer lead time than one period is required. If a forecast for time period 5 is required at the end of time period 2, then because the forecast lead time is $\tau = 5 - 2 = 3$, the desired forecast is

$$\hat{y}_5(2) = \hat{\beta}_0(2) + \hat{\beta}_1(2)3$$
$$= 53.50 + 1.57(3)$$
$$= 58.21.$$

In general, the forecast for any lead time τ, computed at the current origin of time (the end of time period 2), is

$$\hat{y}_5(2) = \hat{\beta}_0(2) + \hat{\beta}_1(2)\tau$$
$$= 53.50 + 1.57\tau.$$

When the discounted least squares procedure is applied to a linear trend model as in Example 3.9, the resulting forecasts are equivalent to the forecasts produced by a method called **double exponential smoothing**. Exponential smoothing is a popular and very useful forecasting technique and will be discussed in detail in Chapter 4.

Discounted least squares can be applied to more complex models. For example, suppose that the model is a polynomial of degree k. The transition matrix for this model is a square $(k + 1) \times (k + 1)$ matrix in which the diagonal elements are unity, the elements immediately to the left of the diagonal are also unity, and all other elements are zero. In this polynomial, the term of degree r is written as

$$\beta_r \binom{t}{r} = \beta_r \frac{t!}{(t - r)!r!}$$

In the next example we illustrate discounted least squares for a simple seasonal model.

Example 3.11 A Simple Seasonal Model   Suppose that a time series can be modeled as a linear trend with a superimposed sine wave to represent a seasonal pattern that is observed monthly. The model is a variation of the one shown in Eq. (3.3):

$$y_t = \beta_0 + \beta_1 t + \beta_2 \sin \frac{2\pi}{d} t + \beta_3 \cos \frac{2\pi}{d} t + \varepsilon \qquad (3.91)$$

Since this model represents monthly data, $d = 12$, Eq. (3.91) becomes

$$y_t = \beta_0 + \beta_1 t + \beta_2 \sin \frac{2\pi}{12} t + \beta_3 \cos \frac{2\pi}{12} t + \varepsilon \qquad (3.92)$$

The transition matrix **L** for this model, which contains a mixture of polynomial and trigonometric terms, is

$$\mathbf{L} = \begin{bmatrix} 1 & 0 & 0 & 0 \\ 1 & 1 & 0 & 0 \\ 0 & 0 & \cos \dfrac{2\pi}{12} & \sin \dfrac{2\pi}{12} \\ 0 & 0 & -\sin \dfrac{2\pi}{12} & \cos \dfrac{2\pi}{12} \end{bmatrix}.$$

Note that **L** has a **block diagonal structure**, with the first block containing the elements for the polynomial portion of the model and the second block containing the elements for the trigonometric terms, and the remaining elements of the matrix are zero. The parameter estimation equations for this model are

$$\hat{\beta}(T) = \mathbf{L}'\hat{\beta}(T-1) + \mathbf{h}e_T(1)$$

$$
\begin{bmatrix} \hat{\beta}_0(T) \\ \hat{\beta}_1(T) \\ \hat{\beta}_2(T) \\ \hat{\beta}_3(T) \end{bmatrix} =
\begin{bmatrix}
1 & 0 & 0 & 0 \\
1 & 1 & 0 & 0 \\
0 & 0 & \cos\dfrac{2\pi}{12} & \sin\dfrac{2\pi}{12} \\
0 & 0 & -\sin\dfrac{2\pi}{12} & \cos\dfrac{2\pi}{12}
\end{bmatrix}
\begin{bmatrix} \hat{\beta}_0(T-1) \\ \hat{\beta}_1(T-1) \\ \hat{\beta}_2(T-1) \\ \hat{\beta}_3(T-1) \end{bmatrix} +
\begin{bmatrix} h_1 \\ h_2 \\ h_3 \\ h_4 \end{bmatrix} e_T(1)
$$

or

$$\hat{\beta}_0(T) = \hat{\beta}_0(T-1) + \hat{\beta}_1(T-1) + h_1 e_T(1)$$
$$\hat{\beta}_1(T) = \hat{\beta}_1(T-1) + h_2 e_T(1)$$
$$\hat{\beta}_2(T) = \cos\frac{2\pi}{12}\hat{\beta}_2(T-1) - \sin\frac{2\pi}{12}\hat{\beta}_3(T-1) + h_3 e_T(1)$$
$$\hat{\beta}_3(T) = \sin\frac{2\pi}{12}\hat{\beta}_2(T-1) + \cos\frac{2\pi}{12}\hat{\beta}_3(T-1) + h_4 e_T(1)$$

and since $2\pi/12 = 30°$, these equations become

$$\hat{\beta}_0(T) = \hat{\beta}_0(T-1) + \hat{\beta}_1(T-1) + h_1 e_T(1)$$
$$\hat{\beta}_1(T) = \hat{\beta}_1(T-1) + h_2 e_T(1)$$
$$\hat{\beta}_2(T) = 0.866\hat{\beta}_2(T-1) - 0.5\hat{\beta}_3(T-1) + h_3 e_T(1)$$
$$\hat{\beta}_3(T) = 0.5\hat{\beta}_2(T-1) + 0.866\hat{\beta}_3(T-1) + h_4 e_T(1)$$

The steady-state **G** matrix for this model is

$$
\mathbf{G} =
\begin{bmatrix}
\displaystyle\sum_{k=0}^{\infty} \theta^k & -\displaystyle\sum_{k=0}^{\infty} k\theta^k & -\displaystyle\sum_{k=0}^{\infty} \theta^k \sin\omega k & \displaystyle\sum_{k=0}^{\infty} \theta^k \cos\omega k \\[2em]
& \displaystyle\sum_{k=0}^{\infty} k^2\theta^k & \displaystyle\sum_{k=0}^{\infty} k\theta^k \sin\omega k & -\displaystyle\sum_{k=0}^{\infty} k\theta^k \cos\omega k \\[2em]
& & \displaystyle\sum_{k=0}^{\infty} \theta^k \sin\omega k \sin\omega k & -\displaystyle\sum_{k=0}^{\infty} \theta^k \sin\omega k \cos\omega k \\[2em]
& & & \displaystyle\sum_{k=0}^{\infty} \theta^k \cos\omega k \cos\omega k
\end{bmatrix}
$$

where we have let $\omega = 2\pi/12$. Because **G** is symmetric, we only need to show the upper half of the matrix. It turns out that there are closed-form expressions for all of the entries in **G**. We will evaluate these expressions for $\theta = 0.9$. This gives the following:

$$\sum_{k=0}^{\infty} \theta^k = \frac{1}{1-\theta} = \frac{1}{1-0.9} = 10$$

$$\sum_{k=0}^{\infty} k\theta^k = \frac{\theta}{(1-\theta)^2} = \frac{0.9}{(1-0.9)^2} = 90$$

$$\sum_{k=0}^{\infty} k^2\theta^k = \frac{\theta(1+\theta)}{(1-\theta)^3} = \frac{0.9(1+0.9)}{(1-0.9)^3} = 1710$$

for the polynomial terms and

$$\sum_{k=0}^{\infty} \theta^k \sin \omega k = \frac{\theta \sin \omega}{1 - 2\theta \cos \omega + \theta^2} = \frac{(0.9)0.5}{1 - 2(0.9)0.866 + (0.9)^2} = 1.79$$

$$\sum_{k=0}^{\infty} \theta^k \cos \omega k = \frac{1 - \theta \cos \omega}{1 - 2\theta \cos \omega + \theta^2} = \frac{1 - (0.9)0.866}{1 - 2(0.9)0.866 + (0.9)^2} = 0.8824$$

$$\sum_{k=0}^{\infty} k\theta^k \sin \omega k = \frac{\theta(1 - \theta^2) \sin \omega}{(1 - 2\theta \cos \omega + \theta^2)^2} = \frac{0.9[1 - (0.9)^2]0.5}{[1 - 2(0.9)0.866 + (0.9)^2]^2}$$
$$= 1.368$$

$$\sum_{k=0}^{\infty} k\theta^k \cos \omega k = \frac{2\theta^2 - \theta(1 + \theta^2) \cos \omega}{(1 - 2\theta \cos \omega + \theta^2)^2} = \frac{2(0.9)^2 - 0.9[1 + (0.9)^2]0.866}{[1 - 2(0.9)0.866 + (0.9)^2]^2}$$
$$= 3.3486$$

$$\sum_{k=0}^{\infty} \theta^k \sin \omega k \sin \omega k = -\frac{1}{2}\left[\frac{1 - \theta \cos(2\omega)}{1 - 2\theta \cos(2\omega) + \theta^2} - \frac{1 - \theta \cos(0)}{1 - 2\theta \cos(0) + \theta^2}\right]$$
$$-\frac{1}{2}\left[\frac{1 - 0.9(0.5)}{1 - 2(0.9)0.5 + (0.9)^2} - \frac{1 - 0.9(1)}{1 - 2(0.9)(1) + (0.9)^2}\right]$$
$$= 4.7528$$

$$\sum_{k=0}^{\infty} \theta^k \sin \omega k \cos \omega k = \frac{1}{2} \left[\frac{\theta \sin(2\omega)}{1 - 2\theta \cos(2\omega) + \theta^2} - \frac{\theta \sin(0)}{1 - 2\theta \cos(0) + \theta^2} \right]$$

$$= \frac{1}{2} \left[\frac{0.9(0.866)}{1 - 2(0.9)0.5 + (0.9)^2} + \frac{0.9(0)}{1 - 2(0.9)1 + (0.9)^2} \right]$$

$$= 0.4284$$

$$\sum_{k=0}^{\infty} \theta^k \cos \omega k \cos \omega k = \frac{1}{2} \left[\frac{1 - \theta \cos(2\omega)}{1 - 2\theta \cos(2\omega) + \theta^2} + \frac{1 - \theta \cos(0)}{1 - 2\theta \cos(0) + \theta^2} \right]$$

$$= \frac{1}{2} \left[\frac{1 - 0.9(0.5)}{1 - 2(0.9)0.5 + (0.9)^2} + \frac{1 - 0.9(1)}{1 - 2(0.9)(1) + (0.9)^2} \right]$$

$$= 5.3022$$

for the trignometric terms. Therefore the **G** matrix is

$$\mathbf{G} = \begin{bmatrix} 10 & -90 & -1.79 & 0.8824 \\ & 1740 & 1.368 & -3.3486 \\ & & 4.7528 & -0.4284 \\ & & & 5.3022 \end{bmatrix}$$

and $\mathbf{G}^{-1}$ is

$$\mathbf{G}^{-1} = \begin{bmatrix} 0.214401 & 0.01987 & 0.075545 & -0.02264 \\ 0.01987 & 0.001138 & 0.003737 & -0.00081 \\ 0.075545 & 0.003737 & 0.238595 & 0.009066 \\ -0.02264 & -0.00081 & 0.009066 & 0.192591 \end{bmatrix}$$

where we have shown the entire matrix. The **h** vector is

$$\mathbf{h} = \mathbf{G}^{-1}\mathbf{x}(0)$$

$$= \begin{bmatrix} 0.214401 & 0.01987 & 0.075545 & -0.02264 \\ 0.01987 & 0.001138 & 0.003737 & -0.00081 \\ 0.075545 & 0.003737 & 0.238595 & 0.009066 \\ -0.02264 & -0.00081 & 0.009066 & 0.192591 \end{bmatrix} \begin{bmatrix} 1 \\ 0 \\ 0 \\ 1 \end{bmatrix}$$

$$= \begin{bmatrix} 0.191762 \\ 0.010179 \\ 0.084611 \\ 0.169953 \end{bmatrix}$$

Therefore the discounted least squares parameter estimation equations are

$$\hat{\beta}_0(T) = \hat{\beta}_0(T-1) + \hat{\beta}_1(T-1) + 0.191762e_T(1)$$

$$\hat{\beta}_1(T) = \hat{\beta}_1(T-1) + 0.010179e_T(1)$$

$$\hat{\beta}_2(T) = \cos\frac{2\pi}{12}\hat{\beta}_2(T-1) - \sin\frac{2\pi}{12}\hat{\beta}_3(T-1) + 0.084611e_T(1)$$

$$\hat{\beta}_3(T) = \sin\frac{2\pi}{12}\hat{\beta}_2(T-1) + \cos\frac{2\pi}{12}\hat{\beta}_3(T-1) + 0.169953e_T(1)$$

3.8 REGRESSION MODELS FOR GENERAL TIME SERIES DATA

Many applications of regression in forecasting involve both predictor and response variables that are time series. Regression models using time series data occur relatively often in economics, business, and many fields of engineering. The assumption of uncorrelated or independent errors that is typically made for cross-section regression data is often not appropriate for time series data. Usually the errors in time series data exhibit some type of autocorrelated structure. You might find it useful at this point to review the discussion of autocorrelation in time series data from Chapter 2.

There are several **sources** of autocorrelation in time series regression data. In many cases, the cause of autocorrelation is the failure of the analyst to include one or more important predictor variables in the model. For example, suppose that we wish to regress the annual sales of a product in a particular region of the country against the annual advertising expenditures for that product. Now the growth in the population in that region over the period of time used in the study will also influence the product sales. If population size is not included in the model, this may cause the errors in the model to be positively autocorrelated, because if the per capita demand for the product is either constant or increasing with time, population size is positively correlated with product sales.

The presence of autocorrelation in the errors has several effects on the OLS regression procedure. These are summarized as follows:

1. The OLS regression coefficients are still unbiased, but they are no longer minimum-variance estimates. We know this from our study of GLS in Section 3.7.

2. When the errors are positively autocorrelated, the residual mean square may seriously underestimate the error variance σ^2.

Consequently, the standard errors of the regression coefficients may be too small. As a result, confidence and prediction intervals are shorter than they really should be, and tests of hypotheses on individual regression coefficients may be misleading, in that they may indicate that one or more predictor variables contribute significantly to the model when they really do not. Generally, underestimating the error variance σ^2 gives the analyst a false impression of precision of estimation and potential forecast accuracy.

3. The confidence intervals, prediction intervals, and tests of hypotheses based on the t and F distributions are, strictly speaking, no longer exact procedures.

There are three approaches to dealing with the problem of autocorrelation. If autocorrelation is present because of one or more omitted predictors and if those predictor variable(s) can be identified and included in the model, the observed autocorrelation should disappear. Alternatively, the WLS or GLS methods discussed in Section 3.7 could be used if there were sufficient knowledge of the autocorrelation structure. Finally, if these approaches cannot be used, the analyst must turn to a model that specifically incorporates the autocorrelation structure. These models usually require special parameter estimation techniques. We will provide an introduction to these procedures in Section 3.8.2.

3.8.1 Detecting Autocorrelation: The Durbin–Watson Test

Residual plots can be useful for the detection of autocorrelation. The most useful display is the plot of residuals versus time. If there is positive autocorrelation, residuals of identical sign occur in clusters: that is, there are not enough changes of sign in the pattern of residuals. On the other hand, if there is negative autocorrelation, the residuals will alternate signs too rapidly.

Various **statistical tests** can be used to detect the presence of autocorrelation. The test developed by Durbin and Watson (1950, 1951, 1971) is a very widely used procedure. This test is based on the assumption that the errors in the regression model are generated by a **first-order autoregressive process** observed at equally spaced time periods; that is,

$$\varepsilon_t = \phi\varepsilon_{t-1} + a_t, \tag{3.93}$$

where ε_t is the error term in the model at time period t, a_t is an NID$(0, \sigma_a^2)$ random variable, and ϕ is a parameter that defines the relationship between successive values of the model errors ε_t and ε_{t-1}. We will require that $|\phi| < 1$, so that the model error term in time period t is equal to a fraction of the error experienced in the immediately preceding period plus a normally and independently distributed random shock or disturbance that is unique to the current period. In time series regression models ϕ is sometimes called the **autocorrelation parameter**. Thus a simple linear regression model with **first-order autoregressive errors** would be

$$y_t = \beta_0 + \beta_1 x_t + \varepsilon_t, \quad \varepsilon_t = \phi \varepsilon_{t-1} + a_t, \tag{3.94}$$

where y_t and x_t are the observations on the response and predictor variables at time period t.

When the regression model errors are generated by the first-order autoregressive process in Eq. (3.93), there are several interesting properties of these errors. By successively substituting for $\varepsilon_t, \varepsilon_{t-1}, \ldots$ on the right-hand side of Eq. (3.93) we obtain

$$\varepsilon_t = \sum_{j=0}^{\infty} \phi^j a_{t-j}$$

In other words, the error term in the regression model for period t is just a linear combination of all of the current and previous realizations of the NID$(0, \sigma^2)$ random variables a_t. Furthermore, we can show that

$$E(\varepsilon_t) = 0$$

$$\text{Var}(\varepsilon_t) = \sigma^2 = \sigma_a^2 \left(\frac{1}{1 - \phi^2} \right) \tag{3.95}$$

$$\text{Cov}(\varepsilon_t, \varepsilon_{t \pm j}) = \phi^j \sigma_a^2 \left(\frac{1}{1 - \phi^2} \right)$$

That is, the errors have zero mean and constant variance but have a nonzero covariance structure unless $\phi = 0$.

The **autocorrelation** between two errors that are one period apart, or the **lag one autocorrelation**, is

$$\rho_1 = \frac{\text{Cov}(\varepsilon_t, \varepsilon_{t+1})}{\sqrt{\text{Var}(\varepsilon_t)} \sqrt{\text{Var}(\varepsilon_t)}}$$

$$= \frac{\phi \sigma_a^2 \left(\frac{1}{1-\phi^2}\right)}{\sqrt{\sigma_a^2 \left(\frac{1}{1-\phi^2}\right)} \sqrt{\sigma_a^2 \left(\frac{1}{1-\phi^2}\right)}}$$

$$= \phi$$

The autocorrelation between two errors that are k periods apart is

$$\rho_k = \phi^k, \quad i = 1, 2, \dots$$

This is called the **autocorrelation function** (refer to Section 2.3.2). Recall that we have required that $|\phi| < 1$. When ϕ is positive, all error terms are positively correlated, but the magnitude of the correlation decreases as the errors grow further apart. Only if $\phi = 0$ are the model errors uncorrelated.

Most time series regression problems involve data with positive auto-correlation. The Durbin–Watson test is a statistical test for the presence of positive autocorrelation in regression model errors. Specifically, the hypotheses considered in the Durbin–Watson test are

$$\begin{aligned} H_0 &: \phi = 0 \\ H_1 &: \phi > 0 \end{aligned} \tag{3.96}$$

The Durbin–Watson **test statistic** is

$$d = \frac{\sum\limits_{t=2}^{T} (e_t - e_{t-1})^2}{\sum\limits_{t=1}^{T} e_t^2} = \frac{\sum\limits_{t=2}^{T} e_t^2 + \sum\limits_{t=2}^{T} e_{t-1}^2 - 2 \sum\limits_{t=2}^{T} e_t e_{t-1}}{\sum\limits_{t=1}^{T} e_t^2} \approx 2(1 - r_1), \tag{3.97}$$

where the e_t, $t = 1, 2, \dots, T$ are the residuals from an OLS regression of y_t on x_t. In Eq. (3.97) r_1 is the lag one autocorrelation between the residuals, so for uncorrelated errors the value of the Durbin–Watson statistic should be approximately 2. Statistical testing is necessary to determine just how far away from 2 the statistic must fall in order for us to conclude that the assumption of uncorrelated errors is violated. Unfortunately, the distribution of the Durbin–Watson test statistic d depends on the $\mathbf{X}$ matrix, and this makes critical values for a statistical test difficult to obtain. However, Durbin and Watson (1951) show that d lies between lower and upper bounds, say, d_L and d_U, such that if d is outside these limits, a conclusion regarding the hypotheses in Eq. (3.96) can be reached. The decision procedure is as follows:

$$\text{If } d < d_L \text{ reject } H_0 : \phi = 0$$

$$\text{If } d > d_U \text{ do not reject } H_0 : \phi = 0$$

$$\text{If } d_L \leq d \leq d_U \text{ the test is inconclusive}$$

Table A.5 in Appendix A gives the bounds d_L and d_U for a range of sample sizes, various numbers of predictors, and three type I error rates ($\alpha = 0.05$, $\alpha = 0.025$, and $\alpha = 0.01$). It is clear that small values of the test statistic d imply that $H_0 : \phi = 0$ should be rejected because positive autocorrelation indicates that successive error terms are of similar magnitude, and the differences in the residuals $e_t - e_{t-1}$ will be small. Durbin and Watson suggest several procedures for resolving inconclusive results. A reasonable approach in many of these inconclusive situations is to analyze the data as if there were positive autocorrelation present to see if any major changes in the results occur.

Situations where negative autocorrelation occurs are not often encountered. However, if a test for negative autocorrelation is desired, one can use the statistic $4 - d$, where d is defined in Eq. (3.97). Then the decision rules for testing the hypotheses $H_0 : \phi = 0$ versus $H_1 : \phi < 0$ are the same as those used in testing for positive autocorrelation. It is also possible to test a two-sided alternative hypothesis ($H_0 : \phi = 0$ versus $H_1 : \phi \neq 0$) by using both of the one-sided tests simultaneously. If this is done, the two-sided procedure has type I error 2α, where α is the type I error used for each individual one-sided test.

Example 3.12   Montgomery, Peck, and Vining (2012) present an example of a regression model used to relate annual regional advertising expenses to annual regional concentrate sales for a soft drink company. Table 3.14 presents the 20 years of these data used by Montgomery, Peck, and Vining (2012). The authors assumed that a straight-line relationship was appropriate and fit a simple linear regression model by OLS. The Minitab output for this model is shown in Table 3.15 and the residuals are shown in the last column of Table 3.14. Because these are time series data, there is a possibility that autocorrelation may be present. The plot of residuals versus time, shown in Figure 3.5, has a pattern indicative of potential autocorrelation; there is a definite upward trend in the plot, followed by a downward trend.

TABLE 3.14   Soft Drink Concentrate Sales Data

Year	Sales (Units)	Expenditures (10^3 Dollars)	Residuals
1	3083	75	−32.3298
2	3149	78	−26.6027
3	3218	80	2.2154
4	3239	82	−16.9665
5	3295	84	−1.1484
6	3374	88	−2.5123
7	3475	93	−1.9671
8	3569	97	11.6691
9	3597	99	−0.5128
10	3725	104	27.0324
11	3794	109	−4.4224
12	3959	115	40.0318
13	4043	120	23.5770
14	4194	127	33.9403
15	4318	135	−2.7874
16	4493	144	−8.6060
17	4683	153	0.5753
18	4850	161	6.8476
19	5005	170	−18.9710
20	5236	182	−29.0625

We will use the Durbin–Watson test for

$$H_0 : \phi = 0$$
$$H_1 : \phi > 0$$

The test statistic is calculated as follows:

$$d = \frac{\sum_{t=2}^{20}(e_t - e_{t-1})^2}{\sum_{t=1}^{20} e_t^2}$$

$$= \frac{[-26.6027 - (-32.3298)]^2 + [2.2154 - (-26.6027)]^2 + \cdots + [-29.0625 - (-18.9710)]^2}{(-32.3298)^2 + (-26.6027)^2 + \cdots + (-29.0625)^2}$$

$$= 1.08$$

Minitab will also calculate and display the Durbin–Watson statistic. Refer to the Minitab output in Table 3.15. If we use a significance level

TABLE 3.15   Minitab Output for the Soft Drink Concentrate Sales Data

Regression Analysis: Sales Versus Expenditures

```
The regression equation is
Sales = 1609 + 20.1 Expenditures

Predictor        Coef  SE Coef        T       P
Constant      1608.51    17.02    94.49   0.000
Expenditures  20.0910   0.1428   140.71   0.000

S = 20.5316   R-Sq = 99.9%   R-Sq(adj) = 99.9%

Analysis of Variance

Source            DF        SS       MS         F       P
Regression         1   8346283  8346283  19799.11   0.000
Residual Error    18      7588      422
Total             19   8353871

Unusual Observations

Obs  Expenditures    Sales      Fit  SE Fit  Residual  St Resid
 12           115  3959.00  3918.97    4.59     40.03     2.00R

R denotes an observation with a large standardized residual.

Durbin-Watson statistic = 1.08005
```

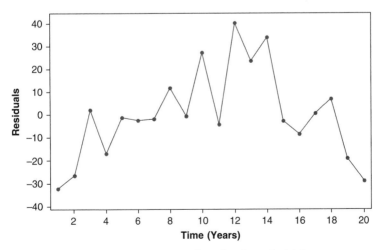

FIGURE 3.5   Plot of residuals versus time for the soft drink concentrate sales model.

of 0.05, Table A.5 in Appendix A gives the critical values corresponding to one predictor variable and 20 observations as $d_L = 1.20$ and $d_U = 1.41$. Since the calculated value of the Durbin–Watson statistic $d = 1.08$ is less than $d_L = 1.20$, we reject the null hypothesis and conclude that the errors in the regression model are positively autocorrelated.

3.8.2 Estimating the Parameters in Time Series Regression Models

A significant value of the Durbin–Watson statistic or a suspicious residual plot indicates a potential problem with auto correlated model errors. This could be the result of an actual time dependence in the errors or an "artificial" time dependence caused by the omission of one or more important predictor variables. If the apparent autocorrelation results from missing predictors and if these missing predictors can be identified and incorporated into the model, the apparent autocorrelation problem may be eliminated. This is illustrated in the following example.

Example 3.13 Table 3.16 presents an expanded set of data for the soft drink concentrate sales problem introduced in Example 3.12. Because it is reasonably likely that regional population affects soft drink sales, Montgomery, Peck, and Vining (2012) provided data on regional population for each of the study years. Table 3.17 is the Minitab output for a regression model that includes as the predictor variables advertising expenditures and population. Both of these predictor variables are highly significant. The last column of Table 3.16 shows the residuals from this model. Minitab calculates the Durbin–Watson statistic for this model as $d = 3.05932$, and the 5% critical values are $d_L = 1.10$ and $d_U = 1.54$, and since d is greater than d_U, we conclude that there is no evidence to reject the null hypothesis. That is, there is no indication of autocorrelation in the errors.

Figure 3.6 is a plot of the residuals from this regression model in time order. This plot shows considerable improvement when compared to the plot of residuals from the model using only advertising expenditures as the predictor. Therefore, we conclude that adding the new predictor population size to the original model has eliminated an apparent problem with autocorrelation in the errors.

The Cochrane–Orcutt Method When the observed autocorrelation in the model errors cannot be removed by adding one or more new predictor variables to the model, it is necessary to take explicit account of the autocorrelative structure in the model and use an appropriate parameter

TABLE 3.16   Expanded Soft Drink Concentrate Sales Data for Example 3.13

Year	Sales (Units)	Expenditures (10^3 Dollars)	Population	Residuals
1	3083	75	825,000	−4.8290
2	3149	78	830,445	−3.2721
3	3218	80	838,750	14.9179
4	3239	82	842,940	−7.9842
5	3295	84	846,315	5.4817
6	3374	88	852,240	0.7986
7	3475	93	860,760	−4.6749
8	3569	97	865,925	6.9178
9	3597	99	871,640	−11.5443
10	3725	104	877,745	14.0362
11	3794	109	886,520	−23.8654
12	3959	115	894,500	17.1334
13	4043	120	900,400	−0.9420
14	4194	127	904,005	14.9669
15	4318	135	908,525	−16.0945
16	4493	144	912,160	−13.1044
17	4683	153	917,630	1.8053
18	4850	161	922,220	13.6264
19	5005	170	925,910	−3.4759
20	5236	182	929,610	0.1025

estimation method. A very good and widely used approach is the procedure devised by Cochrane and Orcutt (1949).

We will describe the Cochrane–Orcutt method for the simple linear regression model with first-order autocorrelated errors given in Eq. (3.94). The procedure is based on transforming the response variable so that $y_t' = y_t - \phi y_{t-1}$. Substituting for y_t and y_{t-1}, the model becomes

$$
\begin{aligned}
y_t' &= y_t - \phi y_{t-1} \\
&= \beta_0 + \beta_1 x_t + \varepsilon_t - \phi(\beta_0 + \beta_1 x_{t-1} + \varepsilon_{t-1}) \\
&= \beta_0(1 - \phi) + \beta_1(x_t - \phi x_{t-1}) + \varepsilon_t - \phi \varepsilon_{t-1} \\
&= \beta_0' + \beta_1 x_t' + a_t,
\end{aligned}
\tag{3.98}
$$

where $\beta_0' = \beta_0(1 - \phi)$ and $x_t' = x_t - \phi x_{t-1}$. Note that the error terms a_t in the transformed or reparameterized model are independent random variables. Unfortunately, this new reparameterized model contains an unknown

TABLE 3.17 Minitab Output for the Soft Drink Concentrate Data in Example 3.13

Regression Analysis: Sales Versus Expenditures, Population

```
The regression equation is
Sales = 320 + 18.4 Expenditures + 0.00168 Population


Predictor          Coef     SE Coef      T      P
Constant          320.3       217.3   1.47  0.159
Expenditures    18.4342      0.2915  63.23  0.000
Population     0.0016787   0.0002829   5.93  0.000


S = 12.0557    R-Sq = 100.0%    R-Sq(adj) = 100.0%


Analysis of Variance

Source           DF        SS       MS         F      P
Regression        2   8351400  4175700  28730.40  0.000
Residual Error   17      2471      145
Total            19   8353871


Source          DF    Seq SS
Expenditures     1   8346283
Population       1      5117


Unusual Observations

Obs  Expenditures   Sales      Fit  SE Fit  Residual  St Resid
 11           109  3794.00  3817.87    4.27    -23.87     -2.12R

R denotes an observation with a large standardized residual.


Durbin-Watson statistic = 3.05932
```

parameter ϕ and it is also no longer linear in the unknown parameters because it involves products of ϕ, β_0, and β_1. However, the first-order autoregressive process $\varepsilon_t = \phi\varepsilon_{t-1} + a_t$ can be viewed as a simple linear regression through the origin and the parameter ϕ can be estimated by obtaining the residuals of an OLS regression of y_t on x_t and then regressing e_t on e_{t-1}. The OLS regression of e_t on e_{t-1} results in

$$\hat{\phi} = \frac{\sum_{t=2}^{T} e_t e_{t-1}}{\sum_{y=1}^{T} e_t^2} \tag{3.99}$$

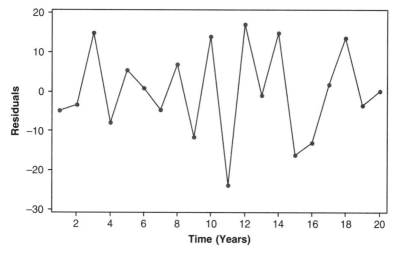

FIGURE 3.6 Plot of residuals versus time for the soft drink concentrate sales model in Example 3.13.

Using $\hat{\phi}$ as an estimate of ϕ, we can calculate the transformed response and predictor variables as

$$y'_t = y_t - \hat{\phi}y_{t-1}$$
$$x'_t = x_t - \hat{\phi}x_{t-1}$$

Now apply OLS to the transformed data. This will result in estimates of the transformed slope $\hat{\beta}'_0$, the intercept $\hat{\beta}_1$, and a new set of residuals. The Durbin–Watson test can be applied to these new residuals from the reparameterized model. If this test indicates that the new residuals are uncorrelated, then no additional analysis is required. However, if positive autocorrelation is still indicated, then another iteration is necessary. In the second iteration ϕ is estimated with new residuals that are obtained by using the regression coefficients from the reparameterized model with the original regressor and response variables. This iterative procedure may be continued as necessary until the residuals indicate that the error terms in the reparameterized model are uncorrelated. Usually only one or two iterations are sufficient to produce uncorrelated errors.

Example 3.14 Montgomery, Peck, and Vining (2012) give data on the market share of a particular brand of toothpaste for 30 time periods and the corresponding selling price per pound. These data are shown in

TABLE 3.18   Toothpaste Market Share Data

Time	Market Share	Price	Residuals	y'_t	x'_t	Residuals
1	3.63	0.97	0.281193			
2	4.20	0.95	0.365398	2.715	0.553	−0.189435
3	3.33	0.99	0.466989	1.612	0.601	0.392201
4	4.54	0.91	−0.266193	3.178	0.505	−0.420108
5	2.89	0.98	−0.215909	1.033	0.608	−0.013381
6	4.87	0.90	−0.179091	3.688	0.499	−0.058753
7	4.90	0.89	−0.391989	2.908	0.522	−0.268949
8	5.29	0.86	−0.730682	3.286	0.496	−0.535075
9	6.18	0.85	−0.083580	4.016	0.498	0.244473
10	7.20	0.82	0.207727	4.672	0.472	0.256348
11	7.25	0.79	−0.470966	4.305	0.455	−0.531811
12	6.09	0.83	−0.659375	3.125	0.507	−0.423560
13	6.80	0.81	−0.435170	4.309	0.471	−0.131426
14	8.65	0.77	0.443239	5.869	0.439	0.635804
15	8.43	0.76	−0.019659	4.892	0.445	−0.192552
16	8.29	0.80	0.811932	4.842	0.489	0.847507
17	7.18	0.83	0.430625	3.789	0.503	0.141344
18	7.90	0.79	0.179034	4.963	0.451	0.027093
19	8.45	0.76	0.000341	5.219	0.437	−0.063744
20	8.23	0.78	0.266136	4.774	0.469	0.284026

Table 3.18. A simple linear regression model is fit to these data, and the resulting Minitab output is in Table 3.19. The residuals are shown in Table 3.18. The Durbin–Watson statistic for the residuals from this model is $d = 1.13582$ (see the Minitab output), and the 5% critical values are $d_{\text{L}} = 1.20$ and $d_{\text{U}} = 1.41$, so there is evidence to support the conclusion that the residuals are positively autocorrelated.

We will use the Cochrane–Orcutt method to estimate the model parameters. The autocorrelation coefficient can be estimated using the residuals in Table 3.18 and Eq. (3.99) as follows:

$$\hat{\phi} = \frac{\sum_{t=2}^{T} e_t e_{t-1}}{\sum_{y=1}^{T} e_t^2}$$

$$= \frac{1.3547}{3.3083}$$

$$= 0.409$$

TABLE 3.19 Minitab Regression Results for the Toothpaste Market Share Data

Regression Analysis: Market Share Versus Price

```
The regression equation is
Market Share = 26.9 - 24.3 Price


Predictor      Coef  SE Coef        T       P
Constant     26.910    1.110    24.25   0.000
Price       -24.290    1.298   -18.72   0.000


S = 0.428710    R-Sq = 95.1%    R-Sq(adj)  = 94.8%


Analysis of Variance

Source           DF      SS       MS        F       P
Regression        1  64.380   64.380   350.29   0.000
Residual Error   18   3.308    0.184
Total            19  67.688


Durbin-Watson statistic = 1.13582
```

The transformed variables are computed according to

$$y'_t = y_t - 0.409y_{t-1}$$
$$x'_t = x_t - 0.409x_{t-1}$$

for $t = 2, 3, \ldots, 20$. These transformed variables are also shown in Table 3.18. The Minitab results for fitting a regression model to the transformed data are summarized in Table 3.20. The residuals from the transformed model are shown in the last column of Table 3.18. The Durbin–Watson statistic for the transformed model is $d = 2.15671$, and the 5% critical values from Table A.5 in Appendix A are $d_L = 1.18$ and $d_U = 1.40$, so we conclude that there is no problem with autocorrelated errors in the transformed model. The Cochrane–Orcutt method has been effective in removing the autocorrelation.

The slope in the transformed model β'_1 is equal to the slope in the original model, β_1. A comparison of the slopes in the two models in Tables 3.19 and 3.20 shows that the two estimates are very similar. However, if the standard errors are compared, the Cochrane–Orcutt method produces an estimate of the slope that has a larger standard error than the standard error of the OLS estimate. This reflects the fact that if the errors are autocorrelated and

TABLE 3.20   Minitab Regression Results for Fitting the Transformed Model to the Toothpaste Sales Data

Regression Analysis: y' Versus x'

```
The regression equation is
y-prime = 16.1 - 24.8 x-prime


Predictor      Coef    SE Coef        T       P
Constant    16.1090     0.9610    16.76   0.000
x-prime     -24.774      1.934   -12.81   0.000


S = 0.390963    R-Sq = 90.6%    R-Sq(adj) = 90.1%


Analysis of Variance

Source            DF      SS       MS        F      P
Regression         1  25.080   25.080   164.08  0.000
Residual Error    17   2.598    0.153
Total             18  27.679


Unusual Observations

Obs  x-prime  y-prime     Fit   SE Fit   Residual   St Resid
  2    0.601   1.6120  1.2198   0.2242     0.3922      1.22 X
  4    0.608   1.0330  1.0464   0.2367    -0.0134     -0.04 X
 15    0.489   4.8420  3.9945   0.0904     0.8475      2.23R


R denotes an observation with a large standardized residual.
X denotes an observation whose X value gives it large influence.


Durbin-Watson statistic = 2.15671
```

OLS is used, the standard errors of the model coefficients are likely to be underestimated.

The Maximum Likelihood Approach

There are other alternatives to the Cochrane–Orcutt method. A popular approach is to use the method of **maximum likelihood** to estimate the parameters in a time series regression model. We will concentrate on the simple linear regression model with first-order autoregressive errors

$$y_t = \beta_0 + \beta_1 x_t + \varepsilon_t, \quad \varepsilon_t = \phi \varepsilon_{t-1} + a_t \qquad (3.100)$$

One reason that the method of maximum likelihood is so attractive is that, unlike the Cochrane–Orcutt method, it can be used in situations where the autocorrelative structure of the errors is more complicated than first-order autoregressive.

For readers unfamiliar with maximum likelihood estimation, we will present a simple example. Consider the time series model

$$y_t = \mu + a_t, \tag{3.101}$$

where a_t is $N(0, \sigma^2)$ and μ is unknown. This is a time series model for a process that varies randomly around a fixed level (μ) and for which there is no autocorrelation. We will estimate the unknown parameter μ using the method of maximum likelihood.

Suppose that there are T observations available, $y_1, y_2, \ldots, y_T$. The probability distribution of any observation is normal, that is,

$$f(y_t) = \frac{1}{\sigma\sqrt{2\pi}} e^{-[(y_t - \mu)/\sigma]^2/2}$$

$$= \frac{1}{\sigma\sqrt{2\pi}} e^{-(a_t/\sigma)^2/2}$$

The **likelihood function** is just the joint probability density function of the sample. Because the observations $y_1, y_2, \ldots, y_T$ are independent, the likelihood function is just the product of the individual density functions, or

$$l(y_t, \mu) = \prod_{t=1}^{T} f(y_t)$$

$$= \prod_{t=1}^{T} \frac{1}{\sigma\sqrt{2\pi}} e^{-(a_t/\sigma)^2/2} \tag{3.102}$$

$$= \left(\frac{1}{\sigma\sqrt{2\pi}}\right)^T \exp\left(-\frac{1}{2\sigma^2} \sum_{t=1}^{T} a_t^2\right)$$

The **maximum likelihood estimator** of μ is the value of the parameter that maximizes the likelihood function. It is often easier to work with the log-likelihood, and this causes no problems because the value of μ that maximizes the likelihood function also maximizes the log-likelihood.

The log-likelihood is

$$\ln l(y_t, \mu) = -\frac{T}{2} \ln(2\pi) - T \ln \sigma - \frac{1}{2\sigma^2} \sum_{t=1}^{T} a_t^2$$

Suppose that σ^2 is known. Then to maximize the log-likelihood we would choose the estimate of μ that minimizes

$$\sum_{t=1}^{T} a_t^2 = \sum_{t=1}^{T} (y_t - \mu)^2$$

Note that this is just the error sum of squares from the model in Eq. (3.101). So, in the case of normally distributed errors, the maximum likelihood estimator of μ is identical to the least squares estimator of μ. It is easy to show that this estimator is just the sample average; that is,

$$\hat{\mu} = \bar{y}$$

Suppose that the mean of the model in Eq. (3.101) is a linear regression function of time, say,

$$\mu = \beta_0 + \beta_1 t$$

so that the model is

$$y_t = \mu + a_t = \beta_0 + \beta_1 t + a_t$$

with independent and normally distributed errors. The likelihood function for this model is identical to Eq. (3.102), and, once again, the maximum likelihood estimators of the model parameters β_0 and β_1 are found by minimizing the error sum of squares from the model. Thus when the errors are normally and independently distributed, the maximum likelihood estimators of the model parameters β_0 and β_1 in the linear regression model are identical to the least squares estimators.

Now let us consider the simple linear regression model with first-order autoregressive errors, first introduced in Eq. (3.94), and repeated for convenience below:

$$y_t = \beta_0 + \beta_1 x_t + \varepsilon_t, \quad \varepsilon_t = \phi \varepsilon_{t-1} + a_t$$

Recall that the a's are normally and independently distributed with mean zero and variance σ_a^2 and ϕ is the autocorrelation parameter. Write this equation for y_{t-1} and subtract ϕy_{t-1} from y_t. This results in

$$y_t - \phi y_{t-1} = (1 - \phi)\beta_0 + \beta_1(x_t - \phi x_{t-1}) + a_t$$

or

$$
\begin{aligned}
y_t &= \phi y_{t-1} + (1 - \phi)\beta_0 + \beta_1(x_t - \phi x_{t-1}) + a_t \\
&= \mu(\mathbf{z}_t, \boldsymbol{\theta}) + a_t,
\end{aligned}
\tag{3.103}
$$

where $\mathbf{z}'_t = [y_{t-1}, x_t]$ and $\boldsymbol{\theta}' = [\phi, \beta_0, \beta_1]$. We can think of $\mathbf{z}_t$ as a vector of predictor variables and $\boldsymbol{\theta}$ as the vector of regression model parameters. Since y_{t-1} appears on the right-hand side of the model in Eq. (3.103), the index of time must run from 2, 3, ..., T. At time period $t = 2$, we treat y_1 as an observed predictor.

Because the a's are normally and independently distributed, the joint probability density of the a's is

$$
\begin{aligned}
f(a_2, a_3, \ldots, a_T) &= \prod_{t=2}^{T} \frac{1}{\sigma_a \sqrt{2\pi}} e^{-(a_t/\sigma_a)^2/2} \\
&= \left(\frac{1}{\sigma_a \sqrt{2\pi}} \right)^{T-1} \exp\left(-\frac{1}{2\sigma_a^2} \sum_{t=1}^{T} a_t^2 \right)
\end{aligned}
$$

and the likelihood function is obtained from this joint distribution by substituting for the a's:

$$
\begin{aligned}
l(y_t, \phi, \beta_0, \beta_1) &= \left(\frac{1}{\sigma_a \sqrt{2\pi}} \right)^{T-1} \\
&\quad \exp\left(-\frac{1}{2\sigma_a^2} \sum_{t=2}^{T} \{y_t - [\phi y_{t-1} + (1 - \phi)\beta_0 + \beta_1(x_t - \phi x_{t-1})]\}^2 \right)
\end{aligned}
$$

The log-likelihood is

$$
\begin{aligned}
\ln l(y_t, \phi, \beta_0, \beta_1) &= -\frac{T-1}{2} \ln(2\pi) - (T-1)\ln\sigma_a \\
&\quad -\frac{1}{2\sigma_a^2} \sum_{t=2}^{T} \{y_t - [\phi y_{t-1} + (1 - \phi)\beta_0 + \beta_1(x_t - \phi x_{t-1})]\}^2
\end{aligned}
$$

This log-likelihood is maximized with respect to the parameters ϕ, β_0, and β_1 by minimizing the quantity

$$SS_E = \sum_{t=2}^{T} \{y_t - [\phi y_{t-1} + (1 - \phi)\beta_0 + \beta_1(x_t - \phi x_{t-1})]\}^2, \quad (3.104)$$

which is the error sum of squares for the model. Therefore the maximum likelihood estimators of ϕ, β_0, and β_1 are also least squares estimators.

There are two important points about the maximum likelihood (or least squares) estimators. First, the sum of squares in Eq. (3.104) is conditional on the initial value of the time series, y_1. Therefore the maximum likelihood (or least squares) estimators found by minimizing this conditional sum of squares are conditional maximum likelihood (or conditional least squares) estimators. Second, because the model involves products of the parameters ϕ and β_0, the model is no longer linear in the unknown parameters. That is, it is not a linear regression model and consequently we cannot give an explicit closed-form solution for the parameter estimators. Iterative methods for fitting nonlinear regression models must be used. These procedures work by linearizing the model about a set of initial guesses for the parameters, solving the linearized model to obtain improved parameter estimates, then using the improved estimates to define a new linearized model, which leads to new parameter estimates and so on. The details of fitting nonlinear models by least squares are discussed in Montgomery, Peck, and Vining (2012).

Suppose that we have obtained a set of parameter estimates, say, $\hat{\theta}' = [\hat{\phi}, \hat{\beta}_0, \hat{\beta}_1]$. The maximum likelihood estimate of σ_a^2 is computed as

$$\hat{\sigma}_a^2 = \frac{SS_E(\hat{\theta})}{n - 1}, \quad (3.105)$$

where $SS_E(\hat{\theta})$ is the error sum of squares in Eq. (3.104) evaluated at the conditional maximum likelihood (or conditional least squares) parameter estimates $\hat{\theta}' = [\hat{\phi}, \hat{\beta}_0, \hat{\beta}_1]$. Some authors (and computer programs) use an adjusted number of degrees of freedom in the denominator to account for the number of parameters that have been estimated. If there are k predictors, then the number of estimated parameters will be $p = k + 3$, and the formula for estimating σ_a^2 is

$$\hat{\sigma}_a^2 = \frac{SS_E(\hat{\theta})}{n - p - 1} = \frac{SS_E(\hat{\theta})}{n - k - 4} \quad (3.106)$$

In order to test hypotheses about the model parameters and to find confidence intervals, standard errors of the model parameters are needed. The standard errors are usually found by expanding the nonlinear model in a first-order Taylor series around the final estimates of the parameters $\hat{\theta}' = [\hat{\phi}, \hat{\beta}_0, \hat{\beta}_1]$. This results in

$$y_t \approx \mu(\mathbf{z}_t, \hat{\theta}) + (\theta - \hat{\theta})' \frac{\partial \mu(\mathbf{z}_t, \theta)}{\partial \theta}\bigg|_{\theta=\hat{\theta}} + a_t$$

The column vector of derivatives, $\partial \mu(\mathbf{z}_t, \theta)/\partial \theta$, is found by differentiating the model with respect to each parameter in the vector $\theta' = [\phi, \beta_0, \beta_1]$. This vector of derivatives is

$$\frac{\partial \mu(\mathbf{z}_t, \theta)}{\partial \theta} = \begin{bmatrix} 1 - \phi \\ x_t - \phi x_{t-1} \\ y_{t-1} - \beta_0 - \beta_1 x_{t-1} \end{bmatrix}$$

This vector is evaluated for each observation at the set of conditional maximum likelihood parameter estimates $\hat{\theta}' = [\hat{\phi}, \hat{\beta}_0, \hat{\beta}_1]$ and assembled into an $\mathbf{X}$ matrix. Then the covariance matrix of the parameter estimates is found from

$$\text{Cov}(\hat{\theta}) = \sigma_a^2 (\mathbf{X}'\mathbf{X})^{-1}$$

When σ_a^2 is replaced by the estimate $\hat{\sigma}_a^2$ from Eq. (3.106) an estimate of the covariance matrix results, and the standard errors of the model parameters are the main diagonals of the covariance matrix.

Example 3.15   We will fit the regression model with time series errors in Eq. (3.104) to the toothpaste market share data originally analyzed in Example 3.14. We will use a widely available software package, SAS (the Statistical Analysis System). The SAS procedure for fitting regression models with time series errors is SAS PROC AUTOREG. Table 3.21 contains the output from this software program for the toothpaste market share data. Note that the autocorrelation parameter (or the lag one autocorrelation) is estimated to be 0.4094, which is very similar to the value obtained by the Cochrane–Orcutt method. The overall R^2 for this model is 0.9601, and we can show that the residuals exhibit no autocorrelative structure, so this is likely a reasonable model for the data.

There is, of course, some possibility that a more complex autocorrelation structure than first-order may exist. SAS PROC AUTOREG can fit

TABLE 3.21 SAS PROC AUTOREG Output for the Toothpaste Market Share Data, Assuming First-Order Autoregressive Errors

```
                          The SAS System

                        The AUTOREG Procedure

                     Dependent Variable     y


                 Ordinary Least Squares Estimates

SSE                3.30825739    DFE                       18
MSE                   0.18379    Root MSE             0.42871
SBC               26.762792      AIC               24.7713275
Regress R-Square       0.9511    Total R-Square        0.9511
Durbin-Watson          1.1358    Pr < DW               0.0098
Pr > DW                0.9902
NOTE: Pr<DW is the p-value for testing positive autocorrelation, and Pr>DW is
the p-value for testing negative autocorrelation.
```

Standard Variable	DF	Estimate	Error	t Value	Approx Pr > |t|	Variable Label
Intercept	1	26.9099	1.1099	24.25	<.0001	
x	1	-24.2898	1.2978	-18.72	<.0001	x

```
                    Estimates of Autocorrelations


Lag    Covariance    Correlation    -1 9 8 7 6 5 4 3 2 1 0 1 2 3 4 5 6 7 8 9 1
0        0.1654       1.000000    |                    |********************|
1        0.0677       0.409437    |                    |********            |


                    Preliminary MSE       0.1377

                Estimates of Autoregressive Parameters

Standard
Lag       Coefficient          Error     t Value

1         -0.409437         0.221275      -1.85

Algorithm converged.
```

more complex patterns. Since there is obviously first-order autocorrelation present, an obvious possibility is that the autocorrelation might be second-order autoregressive, as in

$$\varepsilon_t = \phi_1 \varepsilon_{t-1} + \phi_2 \varepsilon_{t-2} + a_t,$$

where the parameters ϕ_1 and ϕ_2 are autocorrelations at lags one and two, respectively. The output from SAS PROC AUTOREG for this model is in Table 3.22. The t-statistic for the lag two autocorrelation is not significant so there is no reason to believe that this more complex autocorrelative

TABLE 3.21   (*Continued*)

```
                          The SAS System

                       The AUTOREG Procedure

                    Maximum Likelihood Estimates
```

SSE	2.69864377	DFE		17		
MSE	0.15874	Root MSE		0.39843		
SBC	25.8919447	AIC		22.9047479		
Regress R-Square	0.9170	Total R-Square		0.9601		
Durbin-Watson	1.8924	Pr < DW		0.3472		
Pr > DW	0.6528					

NOTE: Pr<DW is the p-value for testing positive autocorrelation, and Pr>DW is
the p-value for testing negative autocorrelation.

Standard Variable	DF	Estimate	Error	t Value	Approx Pr > |t|	Variable Label
Intercept	1	26.3322	1.4777	17.82	<.0001	
x	1	-23.5903	1.7222	-13.70	<.0001	x
AR1	1	-0.4323	0.2203	-1.96	0.0663	

```
                 Autoregressive parameters assumed given.
```

Standard Variable	DF	Approx Estimate	Variable Error	t Value	Pr > |t|	Label
Intercept	1	26.3322	1.4776	17.82	<.0001	
x	1	-23.5903	1.7218	-13.70	<.0001	x

structure is necessary to adequately model the data. The model with first-order autoregressive errors is satisfactory.

Forecasting and Prediction Intervals   We now consider how to obtain forecasts at any lead time using a time series model. It is very tempting to ignore the autocorrelation in the data when forecasting, and simply substitute the conditional maximum likelihood estimates into the regression equation:

$$\hat{y}_t = \hat{\beta}_0 + \hat{\beta}_1 x_t$$

Now suppose that we are at the end of the current time period, T, and we wish to obtain a forecast for period $T + 1$. Using the above equation, this results in

$$\hat{y}_{T+1}(T) = \hat{\beta}_0 + \hat{\beta}_1 x_{T+1},$$

assuming that the value of the predictor variable in the next time period x_{T+1} is known.

TABLE 3.22 SAS PROC AUTOREG Output for the Toothpaste Market Share Data, Assuming Second-Order Autoregressive Errors

```
                          The SAS System

                       The AUTOREG Procedure

                    Dependent Variable     y
                                           y


                 Ordinary Least Squares Estimates
```

SSE	3.30825739	DFE		18
MSE	0.18379	Root MSE		0.42871
SBC	26.762792	AIC		24.7713275
Regress R-Square	0.9511	Total R-Square		0.9511
Durbin-Watson	1.1358	Pr < DW		0.0098
Pr > DW	0.9902			

NOTE: Pr<DW is the p-value for testing positive autocorrelation, and Pr>DW is the p-value for testing negative autocorrelation.

Variable	DF	Approx Estimate	Variable Error	t Value	Pr > |t|	Label
Intercept	1	26.9099	1.1099	24.25	<.0001	
x	1	-24.2898	1.2978	-18.72	<.0001	x

```
                   Estimates of Autocorrelations
```

Lag	Covariance	Correlation	-1 9 8 7 6 5 4 3 2 1 0 1 2 3 4 5 6 7 8 9 1
0	0.1654	1.000000	| |********************|
1	0.0677	0.409437	| |******** |
2	0.0223	0.134686	| |*** |

```
                 Preliminary MSE       0.1375


              Estimates of Autoregressive Parameters
```

Lag	Coefficient	Standard Error	t Value
1	-0.425646	0.249804	-1.70
2	0.039590	0.249804	0.16

Algorithm converged.

TABLE 3.22 *(Continued)*

```
                          The SAS System

                       The AUTOREG Procedure

                    Maximum Likelihood Estimates


SSE                    2.69583958    DFE                        16
MSE                       0.16849    Root MSE              0.41048
SBC                    28.8691217    AIC               24.8861926
Regress R-Square          0.9191    Total R-Square        0.9602
Durbin-Watson             1.9168    Pr < DW               0.3732
Pr > DW                   0.6268
NOTE: Pr<DW is the p-value for testing positive autocorrelation, and Pr>DW is
the p-value for testing negative autocorrelation.


Standard                     Approx    Variable
Variable        DF        Estimate       Error    t Value    Pr > |t|    Label

Intercept        1         26.3406      1.5493      17.00      <.0001
x                1        -23.6025      1.8047     -13.08      <.0001     x
AR1              1         -0.4456      0.2562      -1.74      0.1012
AR2              1          0.0297      0.2617       0.11      0.9110


                 Autoregressive parameters assumed given.

Standard                     Approx    Variable
Variable        DF        Estimate       Error    t Value    Pr > |t|    Label

Intercept        1         26.3406      1.5016      17.54      <.0001
x                1        -23.6025      1.7502     -13.49      <.0001     x
```

Unfortunately, this naive approach is not correct. From Eq. (3.103), we know that the observation at time period t is

$$y_t = \phi y_{t-1} + (1-\phi)\beta_0 + \beta_1(x_t - \phi x_{t-1}) + a_t \qquad (3.107)$$

So at the end of the current time period T the next observation is

$$y_{T+1} = \phi y_T + (1-\phi)\beta_0 + \beta_1(x_{T+1} - \phi x_T) + a_{T+1}$$

Assume that the future value of the regressor variable x_{T+1} is known. Obviously, at the end of the current time period, both y_T and x_T are known. The random error at time $T+1$, a_{T+1}, has not been observed yet, and because we have assumed that the expected value of the errors is zero, the best estimate we can make of a_{T+1} is $a_{T+1} = 0$. This suggests that a reasonable forecast of the observation in time period $T+1$ that we can make at the end of the current time period T is

$$\hat{y}_{T+1}(T) = \hat{\phi}y_T + (1 - \hat{\phi})\hat{\beta}_0 + \hat{\beta}_1(x_{T+1} - \hat{\phi}x_T) \qquad (3.108)$$

Note that this forecast is likely to be very different from the naive forecast obtained by ignoring the autocorrelation.

To find a **prediction interval** on the forecast, we need to find the variance of the prediction error. The one-step-ahead forecast error is

$$y_{T+1} - \hat{y}_{T+1}(T) = a_{T+1},$$

assuming that all of the parameters in the forecasting model are known. The variance of the one-step-ahead forecast error is

$$\text{Var}\,(a_{T+1}) = \sigma_a^2$$

Using the variance of the one-step-ahead forecast error, we can construct a $100(1 - \alpha)$ percent prediction interval for the lead-one forecast from Eq. (3.107). The PI is

$$\hat{y}_{T+1}(T) \pm z_{\alpha/2}\sigma_a,$$

where $z_{\alpha/2}$ is the upper $\alpha/2$ percentage point of the standard normal distribution. To actually compute an interval, we must replace σ_a by an estimate, resulting in

$$\hat{y}_{T+1}(T) \pm z_{\alpha/2}\hat{\sigma}_a \qquad (3.109)$$

as the PI. Because σ_a and the model parameters in the forecasting equation have been replaced by estimates, the probability level on the PI in Eq. (3.109) is only approximate.

Now suppose that we want to forecast two periods ahead assuming that we are at the end of the current time period, T. Using Eq. (3.107), we can write the observation at time period $T + 2$ as

$$y_{T+2} = \phi y_{T+1} + (1 - \phi)\beta_0 + \beta_1(x_{T+2} - \phi x_{T+1}) + a_{T+2}$$
$$= \phi[\phi y_T + (1 - \phi)\beta_0 + \beta_1(x_{T+1} - \phi x_T) + a_{T+1}] + (1 - \phi)\beta_0$$
$$+ \beta_1(x_{T+2} - \phi x_{T+1}) + a_{T+2}$$

Assume that the future value of the regressor variables x_{T+1} and x_{T+2} are known. At the end of the current time period, both y_T and x_T are known. The random errors at time $T+1$ and $T+2$ have not been observed yet, and because we have assumed that the expected value of the errors is zero, the best estimate we can make of both a_{T+1} and a_{T+2} is zero. This suggests that the forecast of the observation in time period $T+2$ made at the end of the current time period T is

$$\hat{y}_{T+2}(T) = \hat{\phi}\hat{y}_{T+1}(T) + (1 - \hat{\phi})\hat{\beta}_0 + \hat{\beta}_1(x_{T+2} - \hat{\phi}x_{T+1})$$
$$= \hat{\phi}[\hat{\phi}y_T + (1 - \hat{\phi})\hat{\beta}_0 + \hat{\beta}_1(x_{T+1} - \hat{\phi}x_T)] \qquad (3.110)$$
$$+ (1 - \hat{\phi})\hat{\beta}_0 + \hat{\beta}_1(x_{T+2} - \hat{\phi}x_{T+1})$$

The two-step-ahead forecast error is

$$y_{T+2} - \hat{y}_{T+2}(T) = a_{T+2} + \phi a_{T+1},$$

assuming that all estimated parameters are actually known. The variance of the two-step-ahead forecast error is

$$\text{Var}(a_{T+2} + \phi a_{T+1}) = \sigma_a^2 + \phi^2\sigma_a^2$$
$$= (1 + \phi^2)\sigma_a^2$$

Using the variance of the two-step-ahead forecast error, we can construct a $100(1 - \alpha)$ percent PI for the lead-one forecast from Eq. (3.107):

$$\hat{y}_{T+2}(T) \pm z_{\alpha/2}(1 + \phi^2)^{1/2}\sigma_a$$

To actually compute the PI, both σ_a and ϕ must be replaced by estimates, resulting in

$$\hat{y}_{T+2}(T) \pm z_{\alpha/2}(1 + \hat{\phi}^2)^{1/2}\hat{\sigma}_a \qquad (3.111)$$

as the PI. Because σ_a and ϕ have been replaced by estimates, the probability level on the PI in Eq. (3.111) is only approximate.

In general, if we want to forecast τ periods ahead, the forecasting equation is

$$\hat{y}_{T+\tau}(T) = \hat{\phi}\hat{y}_{T+\tau-1}(T) + (1 - \hat{\phi})\hat{\beta}_0 + \hat{\beta}_1(x_{T+\tau} - \hat{\phi}x_{T+\tau-1}) \qquad (3.112)$$

The τ-step-ahead forecast error is (assuming that the estimated model parameters are known)

$$y_{T+\tau} - \hat{y}_{T+\tau}(T) = a_{T+\tau} + \phi a_{T+\tau-1} + \cdots + \phi^{\tau-1} a_{T+1}$$

and the variance of the τ-step-ahead forecast error is

$$V(a_{T+\tau} + \phi a_{T+\tau-1} + \cdots + \phi^{\tau-1} a_{T+1}) = (1 + \phi^2 + \cdots + \phi^{2(\tau-1)})\sigma_a^2$$
$$= \frac{1 - \phi^{2\tau}}{1 + \phi^2}\sigma_a^2$$

A $100(1 - \alpha)$ percent PI for the lead-τ forecast from Eq. (3.112) is

$$\hat{y}_{T+\tau}(T) \pm z_{\alpha/2} \left(\frac{1 - \phi^{2\tau}}{1 + \phi^2} \right)^{1/2} \sigma_a$$

Replacing σ_a and ϕ by estimates, the approximate $100(1 - \alpha)$ percent PI is actually computed from

$$\hat{y}_{T+\tau}(T) \pm z_{\alpha/2} \left(\frac{1 - \hat{\phi}^{2\tau}}{1 + \hat{\phi}^2} \right)^{1/2} \hat{\sigma}_a \qquad (3.113)$$

The Case Where the Predictor Variable Must Also Be Forecast

In the preceding discussion, we assumed that in order to make forecasts, any necessary values of the predictor variable in future time periods $T + \tau$ are known. This is often (probably usually) an unrealistic assumption. For example, if you are trying to forecast how many new vehicles will be registered in the state of Arizona in some future year $T + \tau$ as a function of the state population in year $T + \tau$, it is pretty unlikely that you will know the state population in that future year.

A straightforward solution to this problem is to replace the required future values of the predictor variable in future time periods $T + \tau$ by forecasts of these values. For example, suppose that we are forecasting one period ahead. From Eq. (3.108) we know that the forecast for y_{T+1} is

$$\hat{y}_{T+1}(T) = \hat{\phi} y_T + (1 - \hat{\phi})\hat{\beta}_0 + \hat{\beta}_1(x_{T+1} - \hat{\phi} x_T)$$

But the future value of x_{T+1} is not known. Let $\hat{x}_{T+1}(T)$ be an unbiased forecast of x_{T+1}, made at the end of the current time period T. Now the forecast for y_{T+1} is

$$\hat{y}_{T+1}(T) = \hat{\phi} y_T + (1 - \hat{\phi})\hat{\beta}_0 + \hat{\beta}_1[\hat{x}_{T+1}(T) - \hat{\phi} x_T] \qquad (3.114)$$

If we assume that the model parameters are known, the one-step-ahead forecast error is

$$y_{T+1} - \hat{y}_{T+1}(T) = a_{T+1} + \beta_1[x_{T+1} - \hat{x}_{T+1}(T)]$$

and the variance of this forecast error is

$$\text{Var}\,(a_{T+1}) = \sigma_a^2 + \beta_1^2 \sigma_x^2(1), \qquad (3.115)$$

where $\sigma_x^2(1)$ is the variance of the one-step-ahead forecast error for the predictor variable x and we have assumed that the random error a_{T+1} in period $T+1$ is independent of the error in forecasting the predictor variable. Using the variance of the one-step-ahead forecast error from Eq. (3.115), we can construct a $100(1 - \alpha)$ percent prediction interval for the lead-one forecast from Eq. (3.114). The PI is

$$\hat{y}_{T+1}(T) \pm z_{\alpha/2}\left[\sigma_a^2 + \beta_1^2 \sigma_x^2(1)\right]^{1/2},$$

where $z_{\alpha/2}$ is the upper $\alpha/2$ percentage point of the standard normal distribution. To actually compute an interval, we must replace the parameters β_1, σ_a^2, and $\sigma_x^2(1)$ by estimates, resulting in

$$\hat{y}_{T+1}(T) \pm z_{\alpha/2}\left[\hat{\sigma}_a^2 + \hat{\beta}_1^2 \hat{\sigma}_x^2(1)\right]^{1/2} \qquad (3.116)$$

as the PI. Because the parameters have been replaced by estimates, the probability level on the PI in Eq. (3.116) is only approximate.

In general, if we want to forecast τ periods ahead, the forecasting equation is

$$\hat{y}_{T+\tau}(T) = \hat{\phi}\hat{y}_{T+\tau-1}(T) + (1 - \hat{\phi})\hat{\beta}_0 + \hat{\beta}_1[\hat{x}_{T+\tau}(T) - \hat{\phi}\hat{x}_{T+\tau-1}(T)] \quad (3.117)$$

The τ-step-ahead forecast error is, assuming that the model parameters are known,

$$y_{T+\tau} - \hat{y}_{T+\tau}(T) = a_{T+\tau} + \phi a_{T+\tau-1} + \cdots + \phi^{\tau-1} a_{T+1} + \beta_1[x_{T+\tau} - \hat{x}_{T+\tau}(T)]$$

and the variance of the τ-step-ahead forecast error is

$$
\begin{aligned}
\text{Var } (a_{T+\tau} &+ \phi a_{T+\tau-1} + \cdots + \phi^{\tau-1} a_{T+1} + \beta_1 [x_{T+\tau} - \hat{x}_{T+\tau}(t)]) \\
&= (1 + \phi^2 + \cdots + \phi^{2(\tau-1)}) \sigma_a^2 + \beta_1^2 \sigma_x^2(\tau) \\
&= \frac{1 - \phi^{2\tau}}{1 + \phi^2} \sigma_a^2 + \beta_1^2 \sigma_x^2(\tau),
\end{aligned}
$$

where $\sigma_x^2(\tau)$ is the variance of the τ-step-ahead forecast error for the predictor variable x. A $100(1 - \alpha)$ percent PI for the lead-τ forecast from Eq. (3.117) is

$$
\hat{y}_{T+\tau}(T) \pm z_{\alpha/2} \left(\frac{1 - \phi^{2\tau}}{1 + \phi^2} \sigma_a^2 + \beta_1^2 \sigma_x^2(\tau) \right)^{1/2}
$$

Replacing all of the unknown parameters by estimates, the approximate $100(1 - \alpha)$ percent PI is actually computed from

$$
\hat{y}_{T+\tau}(T) \pm z_{\alpha/2} \left(\frac{1 - \hat{\phi}^{2\tau}}{1 + \hat{\phi}^2} \hat{\sigma}_a^2 + \hat{\beta}_1^2 \hat{\sigma}_x^2(\tau) \right)^{1/2} \tag{3.118}
$$

Alternate Forms of the Model   The regression model with autocorrelated errors

$$
y_t = \phi y_{t-1} + (1 - \phi)\beta_0 + \beta_1(x_t - \phi x_{t-1}) + a_t
$$

is a very useful model for forecasting time series regression data. However, when using this model there are two alternatives that should be considered. The first of these is

$$
y_t = \phi y_{t-1} + \beta_0 + \beta_1 x_t + \beta_2 x_{t-1} + a_t \tag{3.119}
$$

This model removes the requirement that the regression coefficient for the lagged predictor variable x_{t-1} be equal to $-\beta_1\phi$. An advantage of this model is that it can be fit by OLS. Another alternative model to consider is to simply drop the lagged value of the predictor variable from Eq. (3.119), resulting in

$$
y_t = \phi y_{t-1} + \beta_0 + \beta_1 x_t + a_t \tag{3.120}
$$

Often just including the lagged value of the response variable is sufficient and Eq. (3.120) will be satisfactory.

The choice between models should always be a data-driven decision. The different models can be fit to the available data, and model selection can be based on the criteria that we have discussed previously, such as model adequacy checking and residual analysis, and (if enough data are available to do some data splitting) forecasting performance over a test or trial period of data.

Example 3.16   Reconsider the toothpaste market share data originally presented in Example 3.14 and modeled with a time series regression model with first-order autoregressive errors in Example 3.15. First we will try fitting the model in Eq. (3.119). This model simply relaxes the restriction that the regression coefficient for the lagged predictor variable x_{t-1} (price in this example) be equal to $-\beta_1 \phi$. Since this is just a linear regression model, we can fit it using Minitab. Table 3.23 contains the Minitab results.

This model is a good fit to the data. The Durbin–Watson statistic is $d = 2.04203$, which indicates no problems with autocorrelation in the residuals. However, note that the t-statistic for the lagged predictor variable (price) is not significant ($P = 0.217$), indicating that this variable could be removed from the model. If x_{t-1} is removed, the model becomes the one in Eq. (3.120). The Minitab output for this model is in Table 3.24.

This model is also a good fit to the data. Both predictors, the lagged variable y_{t-1} and x_t, are significant. The Durbin–Watson statistic does not indicate any significant problems with autocorrelation. It seems that either of these models would be reasonable for the toothpaste market share data. The advantage of these models relative to the time series regression model with autocorrelated errors is that they can be fit by OLS. In this example, including a lagged response variable and a lagged predictor variable has essentially eliminated any problems with autocorrelated errors.

3.9 ECONOMETRIC MODELS

The field of **econometrics** involves the unified study of economics, economic data, mathematics, and statistical models. The term econometrics is generally credited to the Norwegian economist Ragnar Frisch (1895–1973) who was one of the founders of the Econometric Society and the founding editor of the important journal *Econometrica* in 1933. Frisch was a co-winner of the first Nobel Prize in Economic Sciences in 1969. For

TABLE 3.23   Minitab Results for Fitting Model (3.119) to the Toothpaste Market Share Data

---

Regression Analysis: y Versus y_{t-1}, x, x_{t-1}

```
The regression equation is
y = 16.1 + 0.425 y(t-1) - 22.2 x + 7.56 x(t-1)


Predictor       Coef  SE Coef       T      P
Constant      16.100    6.095    2.64  0.019
y(t-1)        0.4253   0.2239    1.90  0.077
x            -22.250    2.488   -8.94  0.000
x(t-1)         7.562    5.872    1.29  0.217


S = 0.402205   R-Sq = 96.0%   R-Sq(adj) = 95.2%


Analysis of Variance

Source           DF      SS      MS       F      P
Regression        3  58.225  19.408  119.97  0.000
Residual Error   15   2.427   0.162
Total            18  60.651

Source   DF  Seq SS
y(t-1)    1  44.768
x         1  13.188
x(t-1)    1   0.268


Durbin-Watson statistic = 2.04203
```

---

introductory books on econometrics, see Greene (2011) and Woodridge (2011).

Econometric models assume that the quantities being studied are random variables and regression modeling techniques are widely used in the field to describe the relationships between these quantities. Typically, an analyst may want to quantify the impact of one set of variables on another variable. For example, one may want to investigate the effect of education on income; that is, what is the change in earnings that result from increasing a worker's education, while holding other variables such as age and gender constant. Large-scale, comprehensive econometric models of macroeconomic relationships are used by government agencies and central banks to evaluate economic activity and to provide guidance on economic

TABLE 3.24 Minitab Results for Fitting Model (3.120) to the Toothpaste Market Share Data

Regression Analysis: y Versus y_{t-1}, x

```
The regression equation is
y = 23.3 + 0.162 y(t-1) - 21.2 x

Predictor      Coef   SE Coef      T       P
Constant     23.279     2.515   9.26   0.000
y(t-1)       0.16172   0.09238   1.75   0.099
x           -21.181     2.394  -8.85   0.000

S = 0.410394   R-Sq = 95.6%   R-Sq(adj) = 95.0%

Analysis of Variance

Source           DF      SS      MS       F       P
Regression        2  57.956  28.978  172.06   0.000
Residual Error   16   2.695   0.168
Total            18  60.651

Source   DF  Seq SS
y(t-1)    1  44.768
x         1  13.188

Durbin-Watson statistic = 1.61416
```

policies. For example, the United States Federal Reserve Bank has maintained macroeconometric models for forecasting and quantitative policy and macroeconomic analysis for over 40 years. The Fed focuses on both the US economy and the global economy.

There are several types of data used in econometric modeling. **Time-series data** are used in many applications. Typical examples include aggregates of economic quantities, such as GDP, asset or commodity prices, and interest rates. As we have discussed earlier in this chapter, time series such as these are characterized by serial correlation. A lot of aggregate economic data are only available at a relatively low sampling frequency, such as monthly, quarterly, or in some cases annually. One exception is financial data, which may be available at very high frequency, such as hourly, daily, or even by individual transaction. **Cross-sectional data** consist of observations taken at the same point in time. In econometric work, surveys are a typical source of cross-sectional data. In typical applications, the surveys are conducted on individuals, households, business organizations, or other economic entities. The early part of this chapter described regression modeling of cross-section data. **Panel data** typically contain both cross-section and time-series data. These data sets consist of a collection of individuals, households, or corporations that are surveyed repeatedly over time. As an example of a simple econometric model involving time series data, suppose that we wish to develop a model for forecasting monthly consumer spending. A plausible model might be

$$y_t = \beta_0 + \beta_1 x_{t-1} + \beta_2 y_{t-1} + \varepsilon_t,$$

where y_t is consumer spending in month t, x_{t-1} is income in month $t-1$, y_{t-1} is consumer spending in month $t-1$, and ε_t is the random error term. This is a lagged-variable regression model of the type discussed earlier in this chapter.

The consumer spending example above is an example of a very simple single-equation econometric model. Many econometric models involve several equations and the predictor variables in these models can be involved in complex interrelationships with each other and with the dependent variables. For example, consider the following econometric model:

Sales $= f_1$(GNP, price, number of competitors, advertising expenditures)

However, price is likely to be a function of other variables, say

Price $= f_2$(production costs, distribution costs, overhead costs, material cost, packaging costs)

and

Production costs $= f_3$(production volume, labor costs, material costs, inventory costs)

Advertising expenditures $= f_4$(sales, number of competitors)

Notice the interrelationships between these variables. Advertising expenditures certainly influence sales, but the level of sales and the number of competitors will influence the money spent on advertising. Furthermore, different levels of sales will have an impact on production costs.

Constructing and maintaining these models is a complex task. One could (theoretically at least) write a very large number of interrelated equations, but data availability and model estimation issues are practical restrictions. The SAS© software package has good capability for this type of simultaneous equation modeling and is widely used in econometrics. However, forecast accuracy does not necessarily increase with the complexity of the models. Often, a relatively simple time series model will outperform a complex econometric model from a pure forecast accuracy point of view. Econometric models are most useful for providing understanding about the way an economic system works, and for evaluating in a broad sense how different economic policies will perform, and the effect that will have on the economy. This is why their use is largely confined to government entities and some large corporations. There are commercial services that offer econometric models that could be useful to smaller organizations, and free alternatives are available from central banks and other government organizations.

3.10 R COMMANDS FOR CHAPTER 3

Example 3.1   The patient satisfaction data are in the sixth column of the array called patsat.data in which the second and third columns are the age and the severity. Note that we can use the "lm" function to fit the linear model. But as in the example, we will show how to obtain the regression coefficients using the matrix notation.

```
nrow<-dim(patsat.data)[1]
X<-cbind(matrix(1,nrow,1),patsat.data[,2:3])
y<-patsat.data[,6]
beta<-solve(t(X)%*%X)%*%t(X)%*%y
beta
                    [,1]
          143.4720118
  Age        -1.0310534
  Severity   -0.5560378
```

Example 3.3   For this example we will use the "lm" function.

```
satisfaction2.fit<-lm(Satisfaction~Age+Severity+Age:Severity+I(Age^2)+
I(Severity^2), data=patsat)
summary(satisfaction2.fit)
 Call:
 lm(formula = Satisfaction ~ Age + Severity + Age:Severity + I(Age^2) +
     I(Severity^2), data = patsat)

 Residuals:
     Min     1Q  Median     3Q     Max
 -16.915  -3.642   2.015  4.000   9.677
```

```
Coefficients:
                Estimate Std. Error t value Pr(> |t|)
(Intercept)    127.527542  27.912923   4.569  0.00021 ***
Age             -0.995233   0.702072  -1.418  0.17251
Severity         0.144126   0.922666   0.156  0.87752
I(Age^2)        -0.002830   0.008588  -0.330  0.74534
I(Severity^2)   -0.011368   0.013533  -0.840  0.41134
Age:Severity     0.006457   0.016546   0.390  0.70071
—
Signif. codes:  0 '***' 0.001 '**' 0.01 '*' 0.05 '.' 0.1 ' ' 1

Residual standard error: 7.503 on 19 degrees of freedom
Multiple R-squared:  0.9008,     Adjusted R-squared:  0.8747
F-statistic:  34.5 on 5 and 19 DF,  p-value: 6.76e-09
```

```
anova(satisfaction2.fit)
```

```
Analysis of Variance Table

Response: Satisfaction
               Df Sum Sq Mean Sq  F value     Pr(> F)
Age             1 8756.7  8756.7 155.5644 1.346e-10 ***
Severity        1  907.0   907.0  16.1138 0.0007417 ***
I(Age^2)        1    1.4     1.4   0.0252 0.8756052
I(Severity^2)   1   35.1    35.1   0.6228 0.4397609
Age:Severity    1    8.6     8.6   0.1523 0.7007070
Residuals      19 1069.5    56.3
—
Signif. codes:  0 '***' 0.001 '**' 0.01 '*' 0.05 '.' 0.1 ' ' 1
```

Example 3.4   We use the "lm" function again and obtain the linear model. Then we use "confint" function to obtain the confidence intervals of the model parameters. Note that the default confidence level is 95%.

```
satisfaction1.fit<-lm(Satisfaction Age+Severity, data=patsat)
confint(satisfaction1.fit,level=.95)
                  2.5 %      97.5 %
(Intercept) 131.122434  155.8215898
Age          -1.270816   -0.7912905
Severity     -0.828566   -0.2835096
```

Example 3.5   This example refers to the linear model (Satisfaction1.fit). We obtain the confidence and prediction intervals for a new data point for which age = 60 and severity = 60.

```
new <- data.frame(Age = 60, Severity=60)
pred.sat1.clim<-predict(Satisfaction1.fit, newdata=new, se.fit =
TRUE, interval = "confidence")
pred.sat1.plim<-predict(Satisfaction1.fit, newdata=new, se.fit =
TRUE, interval = "prediction")
pred.sat1.clim$fit
        fit       lwr       upr
 1 48.24654 43.84806 52.64501
pred.sat1.plim$fit
        fit       lwr       upr
 1 48.24654 32.84401 63.64906
```

Example 3.6   We simply repeat Example 3.5 for age = 75 and severity = 60.

```
new <- data.frame(Age = 75, Severity=60)
pred.sat1.clim<-predict(Satisfaction1.fit, newdata=new, se.fit =
TRUE, interval = "confidence")
pred.sat1.plim<-predict(Satisfaction1.fit, newdata=new, se.fit =
TRUE, interval = "prediction")
pred.sat1.clim$fit
        fit       lwr       upr
 1 32.78074 26.99482 38.56666
pred.sat1.plim$fit
        fit       lwr       upr
 1 32.78074 16.92615 48.63533
```

Example 3.7   The residual plots for Satisfaction1.fit can be obtained using the following commands:

```
par(mfrow=c(2,2),oma=c(0,0,0,0))
qqnorm(Satisfaction1.fit$res,datax=TRUE,pch=16,xlab=
'Residual',main=")
qqline(Satisfaction1.fit$res,datax=TRUE)
plot(Satisfaction1.fit$fit, Satisfaction1.fit$res,pch=16,
 xlab='Fitted Value',ylab='Residual')
abline(h=0)
hist(Satisfaction1.fit$res,col="gray",xlab='Residual',main='')
plot(Satisfaction1.fit$res,type="l",xlab='Observation
Order',ylab='Residual')
points(Satisfaction1.fit$res,pch=16,cex=.5)
abline(h=0)
```

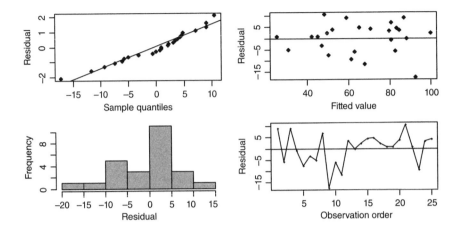

Example 3.8 In R, one can do stepwise regression using step function which allows for stepwise selection of variables in forward, backward, or both directions. Note that step needs to be applied to a model with 4 input variables as indicated in the example. Therefore we first fit that model and apply the step function. Also note that the variable selection is done based on AIC and therefore we get in the forward selection slightly different results than the one provided in the textbook.

```
satisfaction3.fit<-lm(Satisfaction~Age+Severity+Surg.Med+Anxiety,
data=patsat)

step.for<-step(satisfaction3.fit,direction='forward')

 Start:  AIC=103.18
 Satisfaction~Age + Severity + Surg.Med + Anxiety

step.back<-step(satisfaction3.fit,direction='backward')

 Start:  AIC=103.18
 Satisfaction~Age + Severity + Surg.Med + Anxiety

             Df Sum of Sq    RSS    AIC
 - Surg.Med  1       1.0 1039.9 101.20
 - Anxiety   1      75.4 1114.4 102.93
 <none>                  1038.9 103.18
 - Severity  1     971.5 2010.4 117.68
 - Age       1    3387.7 4426.6 137.41

 Step:  AIC=101.2
 Satisfaction~Age + Severity + Anxiety
```

```
            Df Sum of Sq    RSS     AIC
- Anxiety    1      74.6 1114.5 100.93
<none>                    1039.9 101.20
- Severity   1     971.8 2011.8 115.70
- Age        1    3492.7 4532.6 136.00

Step:  AIC=100.93
Satisfaction~Age + Severity

            Df Sum of Sq    RSS     AIC
<none>                    1114.5 100.93
- Severity   1     907.0 2021.6 113.82
- Age        1    4029.4 5143.9 137.17
step.both<-step(satisfaction3.fit,direction='both')

Start:  AIC=103.18
Satisfaction~Age + Severity + Surg.Med + Anxiety

            Df Sum of Sq    RSS     AIC
- Surg.Med   1       1.0 1039.9 101.20
- Anxiety    1      75.4 1114.4 102.93
<none>                    1038.9 103.18
- Severity   1     971.5 2010.4 117.68
- Age        1    3387.7 4426.6 137.41

Step:  AIC=101.2
Satisfaction~Age + Severity + Anxiety

            Df Sum of Sq    RSS     AIC
- Anxiety    1      74.6 1114.5 100.93
<none>                    1039.9 101.20
+ Surg.Med   1       1.0 1038.9 103.18
- Severity   1     971.8 2011.8 115.70
- Age        1    3492.7 4532.6 136.00

Step:  AIC=100.93
Satisfaction~Age + Severity

            Df Sum of Sq    RSS     AIC
<none>                    1114.5 100.93
+ Anxiety    1      74.6 1039.9 101.20
+ Surg.Med   1       0.2 1114.4 102.93
- Severity   1     907.0 2021.6 113.82
- Age        1    4029.4 5143.9 137.17
```

In R, one can do best subset regression using leaps function from leaps package. We first upload the leaps package.

```
library(leaps)
```

```
step.best<-regsubsets(Satisfaction~Age+Severity+Surg.Med+Anxiety,
data=patsat)
```

```
summary(step.best)
```

```
 Subset selection object
 Call: regsubsets.formula(Satisfaction~Age+Severity+Surg.Med+
     Anxiety, data = patsat)
 4 Variables  (and intercept)
           Forced in Forced out
 Age           FALSE       FALSE
 Severity      FALSE       FALSE
 Surg.Med      FALSE       FALSE
 Anxiety       FALSE       FALSE
 1 subsets of each size up to 4
 Selection Algorithm: exhaustive
          Age Severity Surg.Med Anxiety
 1  ( 1 )  "*"  " "      " "      " "
 2  ( 1 )  "*"  "*"      " "      " "
 3  ( 1 )  "*"  "*"      " "      "*"
 4  ( 1 )  "*"  "*"      "*"      "*"
```

Example 3.9   The connector strength data are in the second column of the array called strength.data in which the first column is the Weeks. We start with fitting the linear model and plot the residuals vs. Weeks.

```
strength1.fit<-lm(Strength~Weeks, data=strength.data)
summary(strength1.fit)
 Call:
 lm(formula = Strength~Weeks, data = strength.data)
 Residuals:
      Min      1Q    Median      3Q      Max
 -14.3639  -4.4449  -0.0861   5.8671  11.8842

 Coefficients:
             Estimate Std. Error t value Pr(> |t|)
 (Intercept)  25.9360     5.1116   5.074 2.76e-05 ***
 Weeks         0.3759     0.1221   3.078  0.00486 **
 —
 Signif. codes:  0 '***' 0.001 '**' 0.01 '*' 0.05 '.' 0.1 ' ' 1

 Residual standard error: 7.814 on 26 degrees of freedom
 Multiple R-squared:  0.2671,    Adjusted R-squared:  0.2389
 F-statistic: 9.476 on 1 and 26 DF,  p-value: 0.004863

plot(strength.data[,1],strength1.fit$res, pch=16,cex=.5,
xlab='Weeks',ylab='Residual')
abline(h=0)
```

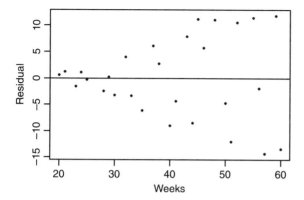

We then fit a linear model to the absolute value of the residuals and obtain the weights as the inverse of the square of the fitted values.

```
res.fit<-lm(abs(strength1.fit$res)~Weeks, data=strength.data)
weights.strength<-1/(res.fit$fitted^2)
```

We then fit a linear model to the absolute value of the residuals and obtain the weights as the inverse of the square of the fitted values.

```
strength2.fit<-lm(Strength~Weeks, data=strength.data,
weights=weights.strength)
summary(strength2.fit)
```

```
 Call:
 lm(formula = Strength~Weeks, data = strength.data, weights =
 weights.strength)

 Weighted Residuals:
     Min      1Q  Median      3Q     Max
 -1.9695 -1.0450 -0.1231  1.1507  1.5785

 Coefficients:
              Estimate Std. Error t value Pr(> |t|)
 (Intercept) 27.54467    1.38051  19.953 <   2e-16 ***
 Weeks        0.32383    0.06769   4.784 5.94e-05 ***
 —

 Signif. codes:  0 '***' 0.001 '**' 0.01 '*' 0.05 '.' 0.1 ' ' 1

 Residual standard error: 1.119 on 26 degrees of freedom
 Multiple R-squared:  0.4682,    Adjusted R-squared:  0.4477
 F-statistic: 22.89 on 1 and 26 DF,  p-value: 5.942e-05
```

Example 3.12   Different packages such as car, lmtest, and bstats offer functions for Durbin–Watson test. We will use function "dwt" in package car. Note that dwt function allows for two-sided or one-sided tests. As in the example we will test for positive autocorrelation. The data are given in softsales.data where the columns are Year, Sales, Expenditures and Population (to be used in the next example).

```
library(car)
soft1.fit<-lm(Sales~Expenditures, data=softsales.data)
dwt(soft1.fit, alternative="positive")
    lag Autocorrelation D-W Statistic p-value
     1        0.3354445       1.08005   0.007
    Alternative hypothesis: rho >  0
```

Since the *p*-value is too small the null hypothesis is rejected concluding that the errors are positively correlated.

Example 3.13   We repeat Example 3.12 with the model expanded to include Population as well.

```
soft2.fit<-lm(Sales~Expenditures+Population, data=softsales.data)
dwt(soft2.fit, alternative="positive")
    lag Autocorrelation D-W Statistic p-value
     1       -0.534382       3.059322  0.974
    Alternative hypothesis: rho >  0
```

As concluded in the example, adding the input variable Population seems to resolve the autocorrelation issue resulting in large *p*-value for the test for autocorrelation.

Example 3.14   The Cochrane–Orcutt method can be found in package orcutt. The function to be used is "cochran.orcutt". The data are given in toothsales.data where the columns are Share and Price.

```
library(orcutt)
tooth1.fit<-lm(Share~Price, data=toothsales.data)
dwt(tooth1.fit, alternative="positive")
        lag Autocorrelation D-W Statistic p-value
     1        0.4094368       1.135816  0.005
    Alternative hypothesis: rho >  0
tooth2.fit<-cochrane.orcutt(tooth1.fit)
```

```
tooth2.fit
 $Cochrane.Orcutt

 Call:
 lm(formula = YB ~ XB - 1)

 Residuals:
      Min        1Q   Median        3Q      Max
 -0.55508 -0.25069 -0.05506  0.25007  0.83017

 Coefficients:
                Estimate Std. Error t value Pr(> |t|)
 XB(Intercept)    26.722      1.633   16.36 7.71e-12 ***
 XBPrice         -24.084      1.938  -12.42 5.90e-10 ***
 —
 Signif. codes:  0 '***' 0.001 '**' 0.01 '*' 0.05 '.' 0.1 ' ' 1

 Residual standard error: 0.3955 on 17 degrees of freedom
 Multiple R-squared:  0.991,     Adjusted R-squared:  0.9899
 F-statistic: 932.7 on 2 and 17 DF,  p-value: <   2.2e-16

 $rho
 [1] 0.4252321
 $number.interaction
 [1] 8
```

The results are not exactly the same as the ones given in the example. This should be due to the difference in the approaches. Where the book uses a two-step procedure, the function Cochrane.Orcutt "estimates both autocorrelation and beta coefficients recursively until we reach the convergence (8th decimal)".

Example 3.15   For this example we will use "gls" function in package nlme. From Example 3.14 we know that there is autocorrelation in the residuals of the linear model. We first assume that the first-order model will be sufficient to model the autocorrelation as shown in the book.

```
tooth3.fit <- gls(Share ~ Price, data = toothsales.data,
correlation=corARMA(p=1), method="ML")
summary(tooth3.fit)
 Generalized least squares fit by maximum likelihood
   Model: Share ~ Price
   Data: toothsales.data
        AIC       BIC     logLik
   24.90475 28.88767 -8.452373

 Correlation Structure: AR(1)
  Formula:  ~ 1
```

```
Parameter estimate(s):
     Phi
0.4325871

Coefficients:
              Value Std.Error   t-value p-value
(Intercept) 26.33218  1.436305  18.33328       0
Price       -23.59030  1.673740 -14.09436       0
 Correlation:
      (Intr)
Price -0.995

Standardized residuals:
       Min          Q1         Med         Q3         Max
-1.85194806 -0.85848738  0.08945623  0.69587678  2.03734437

Residual standard error: 0.4074217
Degrees of freedom: 20 total; 18 residual

intervals(tooth3.fit)

Approximate 95% confidence intervals

 Coefficients:
              lower       est.       upper
(Intercept)  23.31462  26.33218  29.34974
Price       -27.10670 -23.59030 -20.07390
attr(,"label")

[1] "Coefficients:"

 Correlation structure:
        lower       est.      upper
Phi -0.04226294 0.4325871 0.7480172
attr(,"label")

[1] "Correlation structure:"

 Residual standard error:
   lower       est.      upper
0.2805616 0.4074217 0.5916436

predict(tooth3.fit)
```

The second-order autoregressive model for the errors can be fitted using,

```
tooth4.fit <- gls(Share ~ Price, data = toothsales.data,
correlation=corARMA(p=2), method="ML")
```

Example 3.16   To create the lagged version of the variables and also adjust for the number of observations, we use the following commands:

```
T<-length(toothsales.data$Share)
yt<-toothsales.data$Share[2:T]
yt.lag1<- toothsales.data$Share[1:(T-1)]
xt<-toothsales.data$Price[2:T]
xt.lag1<- toothsales.data$Price[1:(T-1)]
tooth5.fit<-lm(yt ~ yt.lag1+xt+xt.lag1)
summary(tooth5.fit)
 Call:
 lm(formula = yt ~ yt.lag1 + xt + xt.lag1)
 Residuals:
     Min       1Q    Median       3Q      Max
 -0.59958 -0.23973 -0.02918  0.26351  0.66532

 Coefficients:
             Estimate Std. Error t value Pr(> |t|)
 (Intercept)  16.0675     6.0904   2.638    0.0186 *
 yt.lag1       0.4266     0.2237   1.907    0.0759 .
 xt          -22.2532     2.4870  -8.948 2.11e-07 ***
 xt.lag1       7.5941     5.8697   1.294    0.2153
 —
 Signif. codes:  0 '***' 0.001 '**' 0.01 '*' 0.05 '.' 0.1 ' ' 1

 Residual standard error: 0.402 on 15 degrees of freedom
 Multiple R-squared:   0.96,    Adjusted R-squared:  0.952
 F-statistic: 120.1 on 3 and 15 DF,  p-value: 1.037e-10
```

EXERCISES

3.1 An article in the journal *Air and Waste* (Update on Ozone Trends in California's South Coast Air Basin, Vol. 43, 1993) investigated the ozone levels in the South Coast Air Basin of California for the years 1976–1991. The author believes that the number of days the ozone levels exceeded 0.20 ppm (the response) depends on the seasonal meteorological index, which is the seasonal average 850-millibar Temperature (the predictor). Table E3.1 gives the data.

a. Construct a scatter diagram of the data.

b. Estimate the prediction equation.

c. Test for significance of regression.

d. Calculate the 95% CI and PI on for a seasonal meteorological index value of 17. Interpret these quantities.

e. Analyze the residuals. Is there evidence of model inadequacy?

f. Is there any evidence of autocorrelation in the residuals?

TABLE E3.1   Days that Ozone Levels Exceed 20 ppm and Seasonal Meteorological Index

Year	Days	Index
1976	91	16.7
1977	105	17.1
1978	106	18.2
1979	108	18.1
1980	88	17.2
1981	91	18.2
1982	58	16.0
1983	82	17.2
1984	81	18.0
1985	65	17.2
1986	61	16.9
1987	48	17.1
1988	61	18.2
1989	43	17.3
1990	33	17.5
1991	36	16.6

3.2 Montgomery, Peck, and Vining (2012) present data on the number of pounds of steam used per month at a plant. Steam usage is thought to be related to the average monthly ambient temperature. The past year's usages and temperatures are shown in Table E3.2.

TABLE E3.2   Monthly Steam Usage and Average Ambient Temperature

Month	Temperature (°F)	Usage/1000	Month	Temperature (°F)	Usage/1000
January	21	185.79	July	68	621.55
February	24	214.47	August	74	675.06
March	32	288.03	September	62	562.03
April	47	424.84	October	50	452.93
May	50	454.68	November	41	369.95
June	59	539.03	December	30	273.98

a. Fit a simple linear regression model to the data.

b. Test for significance of regression.

c. Analyze the residuals from this model.

d. Plant management believes that an increase in average ambient temperature of one degree will increase average monthly steam consumption by 10,000 lb. Do the data support this statement?

e. Construct a 99% prediction interval on steam usage in a month with average ambient temperature of 58°F.

3.3 On March 1, 1984, the *Wall Street Journal* published a survey of television advertisements conducted by Video Board Tests, Inc., a New York ad-testing company that interviewed 4000 adults. These people were regular product users who were asked to cite a commercial they had seen for that product category in the past week. In this case, the response is the number of millions of retained impressions per week. The predictor variable is the amount of money spent by the firm on advertising. The data are in Table E3.3.

TABLE E3.3   Number of Retained Impressions and Advertising Expenditures

Firm	Amount Spent (Millions)	Retained Impressions per Week (Millions)
Miller Lite	50.1	32.1
Pepsi	74.1	99.6
Stroh's	19.3	11.7
Federal Express	22.9	21.9
Burger King	82.4	60.8
Coca-Cola	40.1	78.6
McDonald's	185.9	92.4
MCI	26.9	50.7
Diet Cola	20.4	21.4
Ford	166.2	40.1
Levi's	27	40.8
Bud Lite	45.6	10.4
ATT Bell	154.9	88.9
Calvin Klein	5	12
Wendy's	49.7	29.2
Polaroid	26.9	38
Shasta	5.7	10
Meow Mix	7.6	12.3
Oscar Meyer	9.2	23.4
Crest	32.4	71.1
Kibbles N Bits	6.1	4.4

a. Fit the simple linear regression model to these data.

b. Is there a significant relationship between the amount that a company spends on advertising and retained impressions? Justify your answer statistically.

c. Analyze the residuals from this model.

d. Construct the 95% confidence intervals on the regression coefficients.

e. Give the 95% confidence and prediction intervals for the number of retained impressions for MCI.

3.4 Suppose that we have fit the straight-line regression model $\hat{y} = \hat{\beta}_0 + \hat{\beta}_1 x_1$, but the response is affected by a second variable x_2 such that the true regression function is

$$E(y) = \beta_0 + \beta_1 x_1 + \beta_2 x_2$$

a. Is the least squares estimator of the slope in the original simple linear regression model unbiased?

b. Show the bias in $\hat{\beta}_1$.

3.5 Suppose that we are fitting a straight line and wish to make the standard error of the slope as small as possible. Suppose that the "region of interest" for x is $-1 \leq x \leq 1$. Where should the observations $x_1, x_2, \ldots, x_n$ be taken? Discuss the practical aspects of this data collection plan.

3.6 Consider the simple linear regression model

$$y = \beta_0 + \beta_1 x + \varepsilon,$$

where the intercept β_0 is known.

a. Find the least squares estimator of β_1 for this model. Does this answer seem reasonable?

b. What is the variance of the slope $(\hat{\beta}_1)$ for the least squares estimator found in part a?

c. Find a $100(1 - \alpha)$ percent CI for β_1. Is this interval narrower than the estimator for the case where both slope and intercept are unknown?

3.7 The quality of Pinot Noir wine is thought to be related to the properties of clarity, aroma, body, flavor, and oakiness. Data for 38 wines are given in Table E3.4.

TABLE E3.4 Wine Quality Data[a] (Found in Minitab)

Clarity, x_1	Aroma, x_2	Body, x_3	Flavor, x_4	Oakiness, x_5	Quality, y	Region
1	3.3	2.8	3.1	4.1	9.8	1
1	4.4	4.9	3.5	3.9	12.6	1
1	3.9	5.3	4.8	4.7	11.9	1
1	3.9	2.6	3.1	3.6	11.1	1
1	5.6	5.1	5.5	5.1	13.3	1
1	4.6	4.7	5	4.1	12.8	1
1	4.8	4.8	4.8	3.3	12.8	1
1	5.3	4.5	4.3	5.2	12	1
1	4.3	4.3	3.9	2.9	13.6	3
1	4.3	3.9	4.7	3.9	13.9	1
1	5.1	4.3	4.5	3.6	14.4	3
0.5	3.3	5.4	4.3	3.6	12.3	2
0.8	5.9	5.7	7	4.1	16.1	3
0.7	7.7	6.6	6.7	3.7	16.1	3
1	7.1	4.4	5.8	4.1	15.5	3
0.9	5.5	5.6	5.6	4.4	15.5	3
1	6.3	5.4	4.8	4.6	13.8	3
1	5	5.5	5.5	4.1	13.8	3
1	4.6	4.1	4.3	3.1	11.3	1
0.9	3.4	5	3.4	3.4	7.9	2
0.9	6.4	5.4	6.6	4.8	15.1	3
1	5.5	5.3	5.3	3.8	13.5	3
0.7	4.7	4.1	5	3.7	10.8	2
0.7	4.1	4	4.1	4	9.5	2
1	6	5.4	5.7	4.7	12.7	3
1	4.3	4.6	4.7	4.9	11.6	2
1	3.9	4	5.1	5.1	11.7	1
1	5.1	4.9	5	5.1	11.9	2
1	3.9	4.4	5	4.4	10.8	2
1	4.5	3.7	2.9	3.9	8.5	2
1	5.2	4.3	5	6	10.7	2
0.8	4.2	3.8	3	4.7	9.1	1
1	3.3	3.5	4.3	4.5	12.1	1
1	6.8	5	6	5.2	14.9	3
0.8	5	5.7	5.5	4.8	13.5	1
0.8	3.5	4.7	4.2	3.3	12.2	1
0.8	4.3	5.5	3.5	5.8	10.3	1
0.8	5.2	4.8	5.7	3.5	13.2	1

[a] The wine here is Pinot Noir. Region refers to distinct geographic regions.

a. Fit a multiple linear regression model relating wine quality to these predictors. Do not include the "Region" variable in the model.

b. Test for significance of regression. What conclusions can you draw?

c. Use t-tests to assess the contribution of each predictor to the model. Discuss your findings.

d. Analyze the residuals from this model. Is the model adequate?

e. Calculate R^2 and the adjusted R^2 for this model. Compare these values to the R^2 and adjusted R^2 for the linear regression model relating wine quality to only the predictors "Aroma" and "Flavor." Discuss your results.

f. Find a 95% CI for the regression coefficient for "Flavor" for both models in part e. Discuss any differences.

3.8 Reconsider the wine quality data in Table E3.4. The "Region" predictor refers to three distinct geographical regions where the wine was produced. Note that this is a categorical variable.

a. Fit the model using the "Region" variable as it is given in Table E3.4. What potential difficulties could be introduced by including this variable in the regression model using the three levels shown in Table E3.4?

b. An alternative way to include the categorical variable "Region" would be to introduce two indicator variables x_1 and x_2 as follows:

Region	x_1	x_2
1	0	0
2	1	0
3	0	1

Why is this approach better than just using the codes 1, 2, and 3?

c. Rework Exercise 3.7 using the indicator variables defined in part b for "Region."

3.9 Table B.6 in Appendix B contains data on the global mean surface air temperature anomaly and the global CO_2 concentration. Fit a regression model to these data, using the global CO_2 concentration as the predictor. Analyze the residuals from this model. Is there evidence of autocorrelation in these data? If so, use one iteration of the Cochrane–Orcutt method to estimate the parameters.

3.10 Table B.13 in Appendix B contains hourly yield measurements from a chemical process and the process operating temperature. Fit a regression model to these data, using the temperature as the predictor. Analyze the residuals from this model. Is there evidence of autocorrelation in these data?

3.11 The data in Table E3.5 give the percentage share of market of a particular brand of canned peaches (y_t) for the past 15 months and the relative selling price (x_t).

TABLE E3.5 Market Share and Price of Canned Peaches

t	x_t	y_t	t	x_t	y_t
1	100	15.93	9	85	16.60
2	98	16.26	10	83	17.16
3	100	15.94	11	81	17.77
4	89	16.81	12	79	18.05
5	95	15.67	13	90	16.78
6	87	16.47	14	77	18.17
7	93	15.66	15	78	17.25
8	82	16.94			

a. Fit a simple linear regression model to these data. Plot the residuals versus time. Is there any indication of autocorrelation?

b. Use the Durbin–Watson test to determine if there is positive autocorrelation in the errors. What are your conclusions?

c. Use one iteration of the Cochrane–Orcutt procedure to estimate the regression coefficients. Find the standard errors of these regression coefficients.

d. Is there positive autocorrelation remaining after the first iteration? Would you conclude that the iterative parameter estimation technique has been successful?

3.12 The data in Table E3.6 give the monthly sales for a cosmetics manufacturer (y_t) and the corresponding monthly sales for the entire industry (x_t). The units of both variables are millions of dollars.

a. Build a simple linear regression model relating company sales to industry sales. Plot the residuals against time. Is there any indication of autocorrelation?

TABLE E3.6   Cosmetic Sales Data for Exercise 3.12

t	x_t	y_t	t	x_t	y_t
1	5.00	0.318	10	6.16	0.650
2	5.06	0.330	11	6.22	0.655
3	5.12	0.356	12	6.31	0.713
4	5.10	0.334	13	6.38	0.724
5	5.35	0.386	14	6.54	0.775
6	5.57	0.455	15	6.68	0.78
7	5.61	0.460	16	6.73	0.796
8	5.80	0.527	17	6.89	0.859
9	6.04	0.598	18	6.97	0.88

    b. Use the Durbin–Watson test to determine if there is positive autocorrelation in the errors. What are your conclusions?

    c. Use one iteration of the Cochrane–Orcutt procedure to estimate the model parameters. Compare the standard error of these regression coefficients with the standard error of the least squares estimates.

    d. Test for positive autocorrelation following the first iteration. Has the procedure been successful?

3.13 Reconsider the data in Exercise 3.12. Define a new set of transformed variables as the first difference of the original variables, $y'_t = y_t - y_{t-1}$ and $x'_t = x_t - x_{t-1}$. Regress y'_t on x'_t through the origin. Compare the estimate of the slope from this first-difference approach with the estimate obtained from the iterative method in Exercise 3.12.

3.14 Show that an equivalent way to perform the test for significance of regression in multiple linear regression is to base the test on R^2 as follows. To test $H_0 : \beta_1 = \beta_2 = \cdots = \beta_k$ versus H_1: at least one $\beta_j \neq 0$, calculate

$$F_0 = \frac{R^2(n-p)}{k(1-R^2)}$$

and reject H_0 if the computed value of F_0 exceeds $F_{a,k,n-p}$, where $p = k + 1$.

3.15 Suppose that a linear regression model with $k = 2$ regressors has been fit to $n = 25$ observations and $R^2 = 0.90$.

a. Test for significance of regression at $\alpha = 0.05$. Use the results of the Exercise 3.14.

b. What is the smallest value of R^2 that would lead to the conclusion of a significant regression if $\alpha = 0.05$? Are you surprised at how small this value of R^2 is?

3.16 Consider the simple linear regression model $y_t = \beta_0 + \beta_1 x + \varepsilon_t$, where the errors are generated by the second-order autoregressive process

$$\varepsilon_t = \rho_1 \varepsilon_{t-1} + \rho_2 \varepsilon_{t-2} + a_t$$

Discuss how the Cochrane–Orcutt iterative procedure could be used in this situation. What transformations would be used on the variables y_t and x_t? How would you estimate the parameters ρ_1 and ρ_2?

3.17 Show that an alternate computing formula for the regression sum of squares in a linear regression model is

$$SS_R = \sum_{i=1}^{n} \hat{y}_i^2 - n\bar{y}^2$$

3.18 An article in *Quality Engineering* (The Catapult Problem: Enhanced Engineering Modeling Using Experimental Design, Vol. 4, 1992) conducted an experiment with a catapult to determine the effects of hook (x_1), arm length (x_2), start angle (x_3), and stop angle (x_4) on the distance that the catapult throws a ball. They threw the ball three times for each setting of the factors. Table E3.7 summarizes the experimental results.

TABLE E3.7   Catapult Experiment Data for Exercise 3.18

x_1	x_2	x_3	x_4	y		
−1	−1	−1	−1	28.0	27.1	26.2
−1	−1	1	1	46.5	43.5	46.5
−1	1	−1	1	21.9	21.0	20.1
−1	1	1	−1	52.9	53.7	52.0
1	−1	−1	1	75.0	73.1	74.3
1	−1	1	−1	127.7	126.9	128.7
1	1	−1	−1	86.2	86.5	87.0
1	1	1	1	195.0	195.9	195.7

    a. Fit a regression model to the data and perform a residual analysis for the model.

    b. Use the sample variances as the basis for WLS estimation of the original data (not the sample means).

    c. Fit an appropriate model to the sample variances. Use this model to develop the appropriate weights and repeat part b.

3.19 Consider the simple linear regression model $y_i = \beta_0 + \beta_1 x_i + \varepsilon_i$, where the variance of ε_i is proportional to x_i^2; that is, $\text{Var}(\varepsilon_i) = \sigma^2 x_i^2$.

    a. Suppose that we use the transformations $y' = y/x$ and $x' = 1/x$. Is this a variance-stabilizing transformation?

    b. What are the relationships between the parameters in the original and transformed models?

    c. Suppose we use the method of WLS with $w_i = 1/x_i^2$. Is this equivalent to the transformation introduced in part a?

3.20 Consider the WLS normal equations for the case of simple linear regression where time is the predictor variable, Eq. (3.62). Suppose that the variances of the errors are proportional to the index of time such that $w_t = 1/t$. Simplify the normal equations for this situation. Solve for the estimates of the model parameters.

3.21 Consider the simple linear regression model where time is the predictor variable. Assume that the errors are uncorrelated and have constant variance σ^2. Show that the variances of the model parameter estimates are

$$V(\hat{\beta}_0) = \sigma^2 \frac{2(2T+1)}{T(T-1)}$$

and

$$V(\hat{\beta}_1) = \sigma^2 \frac{12}{T(T^2-1)}$$

3.22 Analyze the regression model in Exercise 3.1 for leverage and influence. Discuss your results.

3.23 Analyze the regression model in Exercise 3.2 for leverage and influence. Discuss your results.

3.24 Analyze the regression model in Exercise 3.3 for leverage and influence. Discuss your results.

3.25 Analyze the regression model for the wine quality data in Exercise 3.7 for leverage and influence. Discuss your results.

3.26 Consider the wine quality data in Exercise 3.7. Use variable selection techniques to determine an appropriate regression model for these data.

3.27 Consider the catapult data in Exercise 3.18. Use variable selection techniques to determine an appropriate regression model for these data. In determining the candidate variables, consider all of the two-factor cross-products of the original four variables.

3.28 Table B.10 in Appendix B presents monthly data on airline miles flown in the United Kingdom. Fit an appropriate regression model to these data. Analyze the residuals and comment on model adequacy.

3.29 Table B.11 in Appendix B presents data on monthly champagne sales. Fit an appropriate regression model to these data. Analyze the residuals and comment on model adequacy.

3.30 Consider the data in Table E3.5. Fit a time series regression model with autocorrected errors to these data. Compare this model with the results you obtained in Exercise 3.11 using the Cochrane–Orcutt procedure.

3.31 Consider the data in Table E3.5. Fit the lagged variables regression models shown in Eqs. (3.119) and (3.120) to these data. Compare these models with the results you obtained in Exercise 3.11 using the Cochrane–Orcutt procedure, and with the time series regression model from Exercise 3.30.

3.32 Consider the data in Table E3.5. Fit a time series regression model with autocorrected errors to these data. Compare this model with the results you obtained in Exercise 3.13 using the Cochrane–Orcutt procedure.

3.33 Consider the data in Table E3.6. Fit the lagged variables regression models shown in Eqs. (3.119) and (3.120) to these data. Compare these models with the results you obtained in Exercise 3.13 using the Cochrane–Orcutt procedure, and with the time series regression model from Exercise 3.32.

3.34 Consider the global surface air temperature anomaly data and the CO_2 concentration data in Table B.6 in Appendix B. Fit a time series regression model to these data, using global surface air temperature anomaly as the response variable. Is there any indication of auto-correlation in the residuals? What corrective action and modeling strategies would you recommend?

3.35   Table B.20 in Appendix B contains data on tax refund amounts and population. Fit an OLS regression model to these data.

   a. Analyze the residuals and comment on model adequacy.

   b. Fit the lagged variables regression models shown in Eqs. (3.119) and (3.120) to these data. How do these models compare with the OLS model in part a?

3.36   Table B.25 contains data from the National Highway Traffic Safety Administration on motor vehicle fatalities from 1966 to 2012, along with several other variables. These data are used by a variety of governmental and industry groups, as well as research organizations.

   a. Plot the fatalities data. Comment on the graph.

   b. Construct a scatter plot of fatalities versus number of licensed drivers. Comment on the apparent relationship between these two factors.

   c. Fit a simple linear regression model to the fatalities data, using the number of licensed drivers as the predictor variable. Discuss the summary statistics from this model.

   d. Analyze the residuals from the model in part c. Discuss the adequacy of the fitted model.

   e. Calculate the Durbin–Watson test statistic for the model in part c. Is there evidence of autocorrelation in the residuals? Is a time series regression model more appropriate than an OLS model for these data?

3.37   Consider the motor vehicle fatalities data in Appendix Table B.25 and the simple linear regression model from Exercise 3.36. There are several candidate predictors that could be added to the model. Add the number of registered motor vehicles to the model that you fit in Exercise 3.36. Has the addition of another predictor improved the model?

3.38   Consider the motor vehicle fatalities data in Appendix Table B.25. There are several candidate predictors that could be added to the model. Use stepwise regression to find an appropriate subset of predictors for the fatalities data. Analyze the residuals from the model, including the Durbin–Watson test, and comment on model adequacy.

3.39 Consider the motor vehicle fatalities data in Appendix Table B.25. There are several candidate predictors that could be added to the model. Use an all-possible-models approach to find an appropriate subset of predictors for the fatalities data. Analyze the residuals from the model, including the Durbin–Watson test, and comment on model adequacy. Compare this model to the one you obtained through stepwise model fitting in Exercise 3.38.

3.40 Appendix Table B.26 contains data on monthly single-family residential new home sales from 1963 through 2014. The number of building permits issued is also given in the table.

    a. Plot the home sales data. Comment on the graph.

    b. Construct a scatter plot of home sales versus number of building permits. Comment on the apparent relationship between these two factors.

    c. Fit a simple linear regression model to the home sales data, using the number of building permits as the predictor variable. Discuss the summary statistics from this model.

    d. Analyze the residuals from the model in part c. Discuss the adequacy of the fitted model.

    e. Calculate the Durbin–Watson test statistic for the model in part c. Is there evidence of autocorrelation in the residuals? Is a time series regression model more appropriate than an OLS model for these data?

CHAPTER 4

EXPONENTIAL SMOOTHING METHODS

If you have to forecast, forecast often.

EDGAR R. FIEDLER, *American economist*

4.1 INTRODUCTION

We can often think of a data set as consisting of two distinct components: **signal** and **noise**. Signal represents any pattern caused by the intrinsic dynamics of the process from which the data are collected. These patterns can take various forms from a simple constant process to a more complicated structure that cannot be extracted visually or with any basic statistical tools. The constant process, for example, is represented as

$$y_t = \mu + \varepsilon_t, \tag{4.1}$$

where μ represents the underlying constant level of system response and ε_t is the noise at time t. The ε_t is often assumed to be uncorrelated with mean 0 and constant variance σ_ε^2.

We have already discussed some basic data smoothers in Section 2.2.2. **Smoothing** can be seen as a technique to separate the signal and the noise

Introduction to Time Series Analysis and Forecasting, Second Edition.
Douglas C. Montgomery, Cheryl L. Jennings and Murat Kulahci.
© 2015 John Wiley & Sons, Inc. Published 2015 by John Wiley & Sons, Inc.

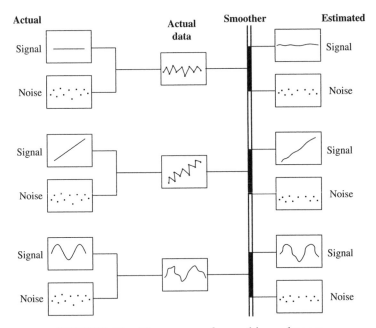

FIGURE 4.1   The process of smoothing a data set.

as much as possible and in that a smoother acts as a filter to obtain an "estimate" for the signal. In Figure 4.1, we give various types of signals that with the help of a smoother can be "reconstructed" and the underlying pattern of the signal is to some extent recovered. The smoothers that we will discuss in this chapter achieve this by simply relating the current observation to the previous ones. For a given data set, one can devise forward and/or backward looking smoothers but in this chapter we will only consider backward looking smoothers. That is, at any given T, the observation y_T will be replaced by a combination of observations at and before T. It does then intuitively make sense to use some sort of an "average" of the current and the previous observations to smooth the data. An obvious choice is to replace the current observation with the average of the observations at $T, T-1, \ldots, 1$. In fact this is the "best" choice in the least squares sense for a constant process given in Eq. (4.1).

A constant process can be smoothed by replacing the current observation with the best estimate for μ. Using the least squares criterion, we define the error sum of squares, SS_E, for the constant process as

$$SS_E = \sum_{t=1}^{T} (y_t - \mu)^2.$$

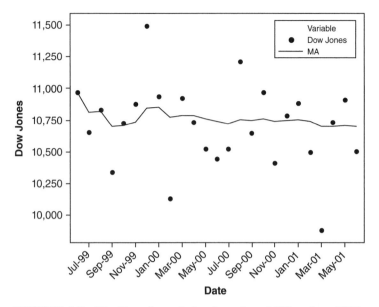

FIGURE 4.2   The Dow Jones Index from June 1999 to June 2001.

The least squares estimate of μ can be found by setting the derivative of SS with respect to μ to 0. This gives

$$\hat{\mu} = \frac{1}{T} \sum_{t=1}^{T} y_t, \qquad (4.2)$$

where $\hat{\mu}$ is the least squares estimate of μ. Equation (4.2) shows that the least squares estimate of μ is indeed the average of observations up to time T.

Figure 4.2 shows the monthly data for the Dow Jones Index from June 1999 to June 2001. Visual inspection suggests that a constant model can be used to describe the general pattern of the data.[1] To further confirm this claim, we use the smoother described in Eq. (4.2) for each data point by taking the average of the available data up to that point in time. The smoothed observations are shown by the line segments in Figure 4.2. It can be seen that the smoother in Eq. (4.2) indeed extracts the main pattern

---

[1] Please note that for this data the independent errors assumption in the constant process in Eq. (4.1) may have been violated. Remedies to check and handle such violations will be provided in the following chapters.

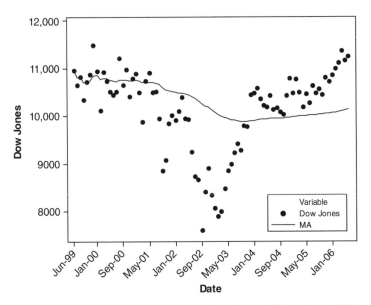

FIGURE 4.3   The Dow Jones Index from June 1999 to June 2006.

in the data and leads to the conclusion that during the 2-year period from June 1999 to June 2001, the Dow Jones Index was quite stable.

As we can see, for the constant process the smoother in Eq. (4.2) is quite effective in providing a clear picture of the underlying pattern. What happens if the process is not constant but exhibits a more complicated pattern? Consider again, for example, the Dow Jones Index from June 1999 to June 2006 given in Figure 4.3 (the complete data set is in Table 4.1). It is clear that the data do not follow the behavior typical of a constant behavior during this period. In Figure 4.3, we can also see the pattern that the smoother in Eq. (4.2) extracts for the same period. As the process changes, this smoother is having trouble keeping up with the process. What could be the reason for the poor performance after June 2001? The answer is quite simple: the constant process assumption is no longer valid. However, as time goes on, the smoother in Eq. (4.2) accumulates more and more data points and gains some sort of "inertia". So when there is a change in the process, it becomes increasingly more difficult for this smoother to react to it.

How often is the constant process assumption violated? The answer to this question is provided by the Second Law of Thermodynamics, which in the most simplistic way states that if left on its own (free of external influences) any system will deteriorate. Thus the constant process is not

TABLE 4.1 Dow Jones Index at the End of the Month from June 1999 to June 2006

Date	Dow Jones	Date	Dow Jones	Date	Dow Jones	Date	Dow Jones
Jun-99	10,970.8	Apr-01	10,735	Feb-03	7891.08	Dec-04	10,783
Jul-99	10,655.2	May-01	10,911.9	Mar-03	7992.13	Jan-05	10,489.9
Aug-99	10,829.3	Jun-01	10,502.4	Apr-03	8480.09	Feb-05	10,766.2
Sep-99	10,337	Jul-01	10,522.8	May-03	8850.26	Mar-05	10,503.8
Oct-99	10,729.9	Aug-01	9949.75	Jun-03	8985.44	Apr-05	10,192.5
Nov-99	10,877.8	Sep-01	8847.56	Jul-03	9233.8	May-05	10,467.5
Dec-99	11,497.1	Oct-01	9075.14	Aug-03	9415.82	Jun-05	10,275
Jan-00	10,940.5	Nov-01	9851.56	Sep-03	9275.06	Jul-05	10,640.9
Feb-00	10,128.3	Dec-01	10,021.6	Oct-03	9801.12	Aug-05	10,481.6
Mar-00	10,921.9	Jan-02	9920	Nov-03	9782.46	Sep-05	10,568.7
Apr-00	10,733.9	Feb-02	10,106.1	Dec-03	10,453.9	Oct-05	10,440.1
May-00	10,522.3	Mar-02	10,403.9	Jan-04	10488.1	Nov-05	10,805.9
Jun-00	10,447.9	Apr-02	9946.22	Feb-04	10,583.9	Dec-05	10,717.5
Jul-00	10,522	May-02	9925.25	Mar-04	10,357.7	Jan-06	10,864.9
Aug-00	11,215.1	Jun-02	9243.26	Apr-04	10,225.6	Feb-06	10,993.4
Sep-00	10,650.9	Jul-02	8736.59	May-04	10,188.5	Mar-06	11,109.3
Oct-00	10,971.1	Aug-02	8663.5	Jun-04	10,435.5	Apr-06	11,367.1
Nov-00	10,414.5	Sep-02	7591.93	Jul-04	10,139.7	May-06	11,168.3
Dec-00	10,788	Oct-02	8397.03	Aug-04	10,173.9	Jun-06	11,247.9
Jan-01	10,887.4	Nov-02	8896.09	Sep-04	10,080.3		
Feb-01	10,495.3	Dec-02	8341.63	Oct-04	10,027.5		
Mar-01	9878.78	Jan-03	8053.81	Nov-04	10,428		

the norm but at best an exception. So what can we do to deal with this issue? Recall that the problem with the smoother in Eq. (4.2) was that it reacted too slowly to process changes because of its inertia. In fact, when there is a change in the process, earlier data no longer carry the information about the change in the process, yet they contribute to this inertia at an equal proportion compared to the more recent (and probably more useful) data. The most obvious choice is to somehow discount the older data. Also recall that in a simple average, as in Eq. (4.2), all the observations are weighted equally and hence have the same amount of influence on the average. Thus, if the weights of each observation are changed so that earlier observations are weighted less, a faster reacting smoother should be obtained. As mentioned in Section 2.2.2, a common solution is to use the **simple moving average** given in Eq. (2.3):

$$M_T = \frac{y_T + y_{T-1} + \cdots + y_{T-N+1}}{N} = \frac{1}{N} \sum_{t=T-N+1}^{N} y_t.$$

The most crucial issue in simple moving averages is the choice of the **span**, N. A simple moving average will react faster to the changes if N is small. However, we know from Section 2.2.2 that the variance of the simple moving average with uncorrelated observations with variance σ^2 is given as

$$\text{Var}(M_T) = \frac{\sigma^2}{N}.$$

This means that as N gets small, the variance of the moving average gets bigger. This creates a dilemma in the choice of N. If the process is expected to be constant, a large N can be used whereas a small N is preferred if the process is changing. In Figure 4.4, we show the effect of going from a span of 10 observations to 5 observations. While the latter exhibits a more jittery behavior, it nevertheless follows the actual data more closely. A more thorough analysis on the choice of N can be performed based on the prediction error. We will explore this for exponential smoothers in Section 4.6.1, where we will discuss forecasting using exponential smoothing.

A final note on the moving average is that even if the individual observations are independent, the moving averages will be autocorrelated as two successive moving averages contain the same $N-1$ observations. In fact,

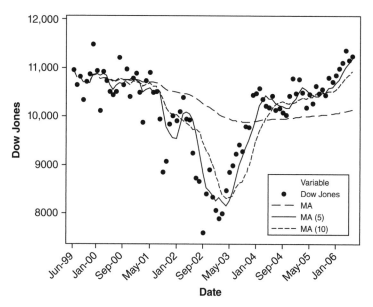

FIGURE 4.4 The Dow Jones Index from June 1999 to June 2006 with moving averages of span 5 and 10.

the autocorrelation function (ACF) of the moving averages that are k-lags apart is given as

$$\rho k = \begin{cases} 1 - \dfrac{|k|}{N}, & k < N \\ 0, & k \geq N \end{cases}.$$

4.2 FIRST-ORDER EXPONENTIAL SMOOTHING

Another approach to obtain a smoother that will react to process changes faster is to give geometrically decreasing weights to the past observations. Hence an exponentially weighted smoother is obtained by introducing a discount factor θ as

$$\sum_{t=0}^{T-1} \theta^t y_{T-t} = y_T + \theta y_{T-1} + \theta^2 y_{T-2} + \cdots + \theta^{T-1} y_1. \qquad (4.3)$$

Please note that if the past observations are to be discounted in a geometrically decreasing manner, then we should have $|\theta| < 1$. However, the smoother in Eq. (4.3) is not an *average* as the sum of the weights is

$$\sum_{t=0}^{T-1} \theta^t = \frac{1 - \theta^T}{1 - \theta} \tag{4.4}$$

and hence does not necessarily add up to 1. For that we can adjust the smoother in Eq. (4.3) by multiplying it by $(1-\theta)/(1-\theta^T)$. However, for large T values, θ^T goes to zero and so the exponentially weighted average will have the following form:

$$\tilde{y}_T = (1 - \theta) \sum_{t=0}^{T-1} \theta^t y_{T-t} \tag{4.5}$$

$$= (1 - \theta)(y_T + \theta y_{T-1} + \theta^2 y_{T-2} + \cdots + \theta^{T-1} y_1)$$

This is called a **simple** or **first-order exponential smoother**. There is an extensive literature on exponential smoothing. For example, see the books by Brown (1963), Abraham and Ledolter (1983), and Montgomery et al. (1990), and the papers by Brown and Meyer (1961), Chatfield and Yar (1988), Cox (1961), Gardner (1985), Gardner and Dannenbring (1980), and Ledolter and Abraham (1984).

An alternate expression in a recursive form for simple exponential smoothing is given by

$$\tilde{y}_T = (1 - \theta)y_T + (1 - \theta)(\theta y_{T-1} + \theta^2 y_{T-2} + \cdots + \theta^{T-1} y_1)$$

$$= (1 - \theta)y_T + \theta \underbrace{(1 - \theta)(y_{T-1} + \theta^1 y_{T-2} + \cdots + \theta^{T-2} y_1)}_{\tilde{y}_{T-1}} \tag{4.6}$$

$$= (1 - \theta)y_T + \theta \tilde{y}_{T-1}.$$

The recursive form in Eq. (4.6) shows that first-order exponential smoothing can also be seen as the linear combination of the current observation and the smoothed observation at the previous time unit. As the latter contains the data from all previous observations, the smoothed observation at time T is in fact the linear combination of the current observation and the discounted sum of all previous observations. The simple exponential smoother is often represented in a different form by setting $\lambda = 1-\theta$,

$$\tilde{y}_T = \lambda y_T + (1 - \lambda)\tilde{y}_{T-1} \tag{4.7}$$

In this representation the **discount factor**, λ, represents the weight put on the last observation and $(1-\lambda)$ represents the weight put on the smoothed value of the previous observations.

Analogous to the size of the span in moving average smoothers, an important issue for the exponential smoothers is the choice of the discount factor, λ. Moreover, from Eq. (4.7), we can see that the calculation of $\tilde{y}_1$ would require us to know $\tilde{y}_0$. We will discuss these issues in the next two sections.

4.2.1 The Initial Value, $\tilde{y}_0$

Since $\tilde{y}_0$ is needed in the recursive calculations that start with $\tilde{y}_1 = \lambda y_1 + (1 - \lambda)\tilde{y}_0$, its value needs to be estimated. But from Eq. (4.7) we have

$$\tilde{y}_1 = \lambda y_1 + (1 - \lambda)\tilde{y}_0$$
$$\tilde{y}_2 = \lambda y_2 + (1 - \lambda)\tilde{y}_1 = \lambda y_2 + (1 - \lambda)(\lambda y_1 + (1 - \lambda)\tilde{y}_0)$$
$$= \lambda(y_2 + (1 - \lambda)y_1) + (1 - \lambda)^2\tilde{y}_0$$
$$\tilde{y}_3 = \lambda(y_3 + (1 - \lambda)y_2 + (1 - \lambda)^2 y_1) + (1 - \lambda)^3\tilde{y}_0$$
$$\vdots$$
$$\tilde{y}_T = \lambda(y_T + (1 - \lambda)y_{T-1} + \cdots + (1 - \lambda)^{T-1}y_1) + (1 - \lambda)^T\tilde{y}_0,$$

which means that as T gets large and hence $(1 - \lambda)^T$ gets small, the contribution of $\tilde{y}_0$ to $\tilde{y}_T$ becomes negligible. Thus for large data sets, the estimation of $\tilde{y}_0$ has little relevance. Nevertheless, two commonly used estimates for $\tilde{y}_0$ are the following.

1. Set $\tilde{y}_0 = y_1$. If the changes in the process are expected to occur early and fast, this choice for the starting value for $\tilde{y}_T$ is reasonable.
2. Take the average of the available data or a subset of the available data, $\bar{y}$, and set $\tilde{y}_0 = \bar{y}$. If the process is at least at the beginning locally constant, this starting value may be preferred.

4.2.2 The Value of λ

In Figures 4.5 and 4.6, respectively, we have two simple exponential smoothers for the Dow Jones Index data with $\lambda = 0.2$ and $\lambda = 0.4$. It can be seen that in the latter the smoothed values follow the original observations more closely. In general, as λ gets closer to 1, and more emphasis is put on the last observation, the smoothed values will approach the original observations. Two extreme cases will be when $\lambda = 0$ and $\lambda = 1$. In the former, the smoothed values will all be equal to a constant, namely, y_0.

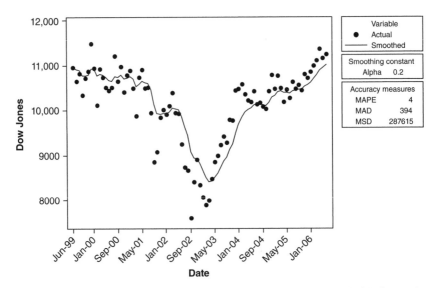

FIGURE 4.5   The Dow Jones Index from June 1999 to June 2006 with first-order exponential smoothing with $\lambda = 0.2$.

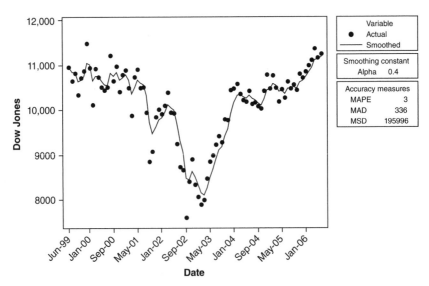

FIGURE 4.6   The Dow Jones Index from June 1999 to June 2006 with first-order exponential smoothing with $\lambda = 0.4$.

We can think of the constant line as the "smoothest" version of whatever pattern the actual time series follows. For $\lambda = 1$, we have $\tilde{y}_T = y_T$ and this will represent the "least" smoothed (or unsmoothed) version of the original time series. We can accordingly expect the variance of the simple exponential smoother to vary between 0 and the variance of the original time series based on the choice of λ. Note that under the independence and constant variance assumptions we have

$$
\begin{aligned}
\text{Var}(\tilde{y}_T) &= \text{Var}\left(\lambda \sum_{t=0}^{\infty} (1 - \lambda)^t y_{T-t}\right) \\
&= \lambda^2 \sum_{t=0}^{\infty} (1 - \lambda)^{2t} \text{Var}(y_{T-t}) \\
&= \lambda^2 \sum_{t=0}^{\infty} (1 - \lambda)^{2t} \text{Var}(y_T) \\
&= \text{Var}(y_T) \lambda^2 \sum_{t=0}^{\infty} (1 - \lambda)^{2t} \\
&= \frac{\lambda}{(2 - \lambda)} \text{Var}(y_T).
\end{aligned}
\tag{4.8}
$$

Thus the question will be how much smoothing is needed. In the literature, λ values between 0.1 and 0.4 are often recommended and do indeed perform well in practice. A more rigorous method of finding the right λ value will be discussed in Section 4.6.1.

Example 4.1 Consider the Dow Jones Index from June 1999 to June 2006 given in Figure 4.3. For first-order exponential smoothing we would need to address two issues as stated in the previous sections: how to pick the initial value y_0 and the smoothing constant λ. Following the recommendation in Section 4.2.2, we will consider the smoothing constants 0.2 and 0.4. As for the initial value, we will consider the first recommendation in Section 4.2.1 and set $\tilde{y}_0 = y_1$. Figures 4.5 and 4.6 show the smoothed and actual data obtained from Minitab with smoothing constants 0.2 and 0.4, respectively.

Note that Minitab reports several measures of accuracy; MAPE, MAD, and MSD. Mean absolute percentage error (MAPE) is the average absolute percentage change between the predicted value that is $\tilde{y}_{t-1}$ for a one-step-ahead forecast and the true value, given as

$$
\text{MAPE} = \frac{\sum_{t=1}^{T} |(y_t - \tilde{y}_{t-1})/y_t|}{T} \times 100 \quad (y_t \neq 0).
$$

Mean absolute deviation (MAD) is the average absolute difference between the predicted and the true values, given as

$$\text{MAD} = \frac{\sum\limits_{t=1}^{T} |(y_t - \tilde{y}_{t-1})|}{T}.$$

Mean squared deviation (MSD) is the average squared difference between the predicted and the true values, given as

$$\text{MSD} = \frac{\sum\limits_{t=1}^{T} (y_t - \tilde{y}_{t-1})^2}{T}.$$

It should also be noted that the smoothed data with $\lambda = 0.4$ follows the actual data closer. However, in both cases, when there is an apparent linear trend in the data (e.g., from February 2003 to February 2004) the smoothed values consistently underestimate the actual data. We will discuss this issue in greater detail in Section 4.3.

As an alternative estimate for the initial value, we can also use the average of the data between June 1999 and June 2001, since during this period the time series data appear to be stable. Figures 4.7 and 4.8 show

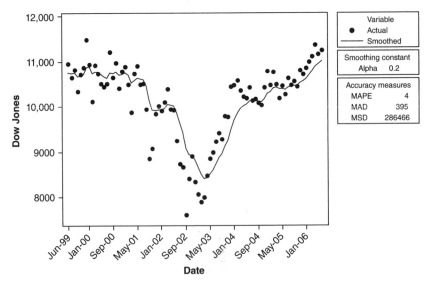

FIGURE 4.7   The Dow Jones Index from June 1999 to June 2006 with first-order exponential smoothing with $\lambda = 0.2$ and $\tilde{y}_0 = (\sum_{t=1}^{25} y_t / 25)$ (i.e., initial value equal to the average of the first 25 observations).

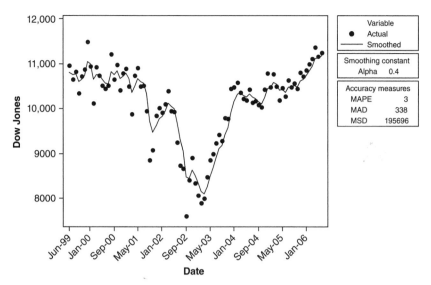

FIGURE 4.8   The Dow Jones Index from June 1999 to June 2006 with first-order exponential smoothing with $\lambda = 0.4$ and $\tilde{y}_0 = (\sum_{t=1}^{25} y_t/25)$ (i.e., initial value equal to the average of the first 25 observations).

the single exponential smoothing with the initial value equal to the average of the first 25 observations corresponding to the period between June 1999 and June 2001. Note that the choice of the initial value has very little effect on the smoothed values as time goes on.

4.3 MODELING TIME SERIES DATA

In Section 4.1, we considered the constant process where the time series data are expected to vary around a constant level with random fluctuations, which are usually characterized by uncorrelated errors with mean 0 and constant variance σ_ε^2. In fact the constant process represents a very special case in a more general set of models often used in modeling time series data as a function of time. The general class of models can be represented as

$$y_t = f(t; \beta) + \varepsilon_t, \tag{4.9}$$

where β is the vector of unknown parameters and ε_t represents the uncorrelated errors. Thus as a member of this general class of models, the constant process can be represented as

$$y_t = \beta_0 + \varepsilon_t, \tag{4.10}$$

where β_0 is equal to μ in Eq. (4.1). We have seen in Chapter 3 how to estimate and make inferences about the regression coefficients. The same principles apply to the class of models in Eq. (4.9). However, we have seen in Section 4.1 that the least squares estimates for β_0 at any given time T will be very slow to react to changes in the level of the process. For that, we suggested to use either the moving average or simple exponential smoothing.

As mentioned earlier, smoothing techniques are effective in illustrating the underlying pattern in the time series data. We have so far focused particularly on exponential smoothing techniques. For the class of models given in Eq. (4.9), we can find another use for the exponential smoothers: model estimation. Indeed for the constant process, we can see the simple exponential smoother as the estimate of the process level, or in regards to Eq. (4.10) an estimate of β_0. To show this in greater detail we need to introduce the sum of weighted squared errors for the constant process. Remember that the sum of squared errors for the constant process is given by

$$SS_E = \sum_{t=1}^{T} (y_t - \mu)^2.$$

If we argue that not all observations should have equal influence on the sum and decide to introduce a string of weights that are geometrically decreasing in time, the sum of squared errors becomes

$$SS_E^* = \sum_{t=0}^{T-1} \theta^t (y_{T-1} - \beta_0)^2, \tag{4.11}$$

where $|\theta|\ 1 < 1$. To find the least squares estimate for β_0, we take the derivative of Eq. (4.11) with respect to β_0 and set it to zero:

$$\frac{dSS_E^*}{d\beta_0}\bigg|_{\beta_0} = -2\sum_{t=0}^{T-1} \theta^t (y_{T-t} - \hat{\beta}_0) = 0. \tag{4.12}$$

The solution to Eq. (4.12), $\hat{\beta}_0$, which is the least squares estimate of β_0, is

$$\hat{\beta}_0 \sum_{t=0}^{T-1} \theta^t = \sum_{t=0}^{T-1} \theta^t y_{T-t}. \tag{4.13}$$

From Eq. (4.4), we have

$$\hat{\beta}_0 = \frac{1 - \theta}{1 - \theta^T} \sum_{t=0}^{T-1} \theta^t y_{T-t}. \tag{4.14}$$

Once again for large T, θ^T goes to zero. We then have

$$\hat{\beta}_0 = (1 - \theta) \sum_{t=0}^{T-1} \theta^t y_{T-t}. \tag{4.15}$$

We can see from Eqs. (4.5) and (4.15) that $\hat{\beta}_0 = \tilde{y}_T$. Thus the simple exponential smoothing procedure does in fact provide a weighted least squares estimate of $\hat{\beta}_0$ in the constant process with weights that are exponentially decreasing in time.

Now we return to our general class of models given in Eq. (4.9) and note that $f(t; \beta)$ can in fact be any function of t. For practical purposes it is usually more convenient to consider the polynomial family for nonseasonal time series. For seasonal time series, we will consider other forms of $f(t; \beta)$ that fit the data and exhibit a certain periodicity better. In the polynomial family, the constant process is indeed the simplest model we can consider. We will now consider the next obvious choice: the linear trend model.

4.4 SECOND-ORDER EXPONENTIAL SMOOTHING

We will now return to our Dow Jones Index data but consider only the subset of the data from February 2003 to February 2004 as given in Figure 4.9. Evidently for that particular time period it was a bullish market and correspondingly the Dow Jones Index exhibits an upward linear trend as indicated with the dashed line.

For this time period, an appropriate model in time from the polynomial family should be the linear trend model given as

$$y_t = \beta_0 + \beta_1 t + \varepsilon_t, \tag{4.16}$$

where the ε_t is once again assumed to be uncorrelated with mean 0 and constant variance σ^2_ε. Based on what we have learned so far, we may attempt to smooth/model this linear trend using the simple exponential smoothing procedure. The actual and fitted values for the simple exponential smoothing procedure are given in Figure 4.10. For the exponential

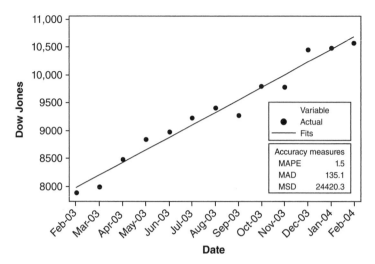

FIGURE 4.9   The Dow Jones Index from February 2003 to February 2004.

smoother, without any loss of generality, we used $\tilde{y}_0 = y_1$ and $\lambda = 0.3$. From Figure 4.10, we can see that while the simple exponential smoother was to some extent able to capture the slope of the linear trend, it also exhibits some bias. That is, the fitted values based on the exponential smoother are consistently underestimating the actual data. More interestingly, the amount of underestimation is more or less constant for all observations.

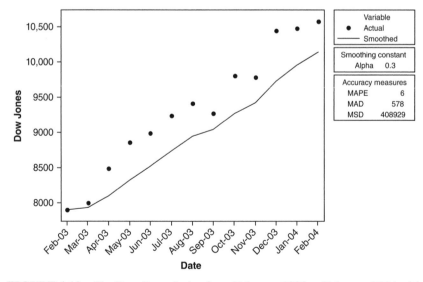

FIGURE 4.10   The Dow Jones Index from February 2003 to February 2004 with simple exponential smoothing with $\lambda = 0.3$.

In fact similar behavior for the simple exponential smoother can be observed in Figure 4.5 for the entire data from June 1999 to June 2006. Whenever the data exhibit a linear trend, the simple exponential smoother seems to over- or underestimates the actual data consistently. To further explore this, we will consider the expected value of $\tilde{y}_T$,

$$E(\tilde{y}_T) = E\left(\lambda \sum_{t=0}^{\infty} (1 - \lambda)^t y_{T-t}\right)$$

$$= \lambda \sum_{t=0}^{\infty} (1 - \lambda)^t E(y_{T-t}).$$

For the linear trend model in Eq. (4.16), $E(y_t) = \beta_0 + \beta_1 t$. So we have

$$E(\tilde{y}_T) = \lambda \sum_{t=0}^{\infty} (1 - \lambda)^t(\beta_0 + \beta_1(T - t))$$

$$= \lambda \sum_{t=0}^{\infty} (1 - \lambda)^t(\beta_0 + \beta_1 T) - \lambda \sum_{t=0}^{\infty} (1 - \lambda)^t(\beta_1 t)$$

$$= (\beta_0 + \beta_1 T)\lambda \sum_{t=0}^{\infty} (1 - \lambda)^t - \lambda\beta_1 \sum_{t=0}^{\infty} (1 - \lambda)^t t.$$

But for the infinite sums we have

$$\sum_{t=0}^{\infty} (1 - \lambda)^t = \frac{1}{1 - (1 - \lambda)} = \frac{1}{\lambda} \text{ and } \sum_{t=0}^{\infty} (1 - \lambda)^t t = \frac{1 - \lambda}{\lambda^2}.$$

Hence the expected value of the simple exponential smoother for the linear trend model is

$$E(\tilde{y}_T) = (\beta_0 + \beta_1 T) - \frac{1 - \lambda}{\lambda}\beta_1$$

$$= E(y_T) - \frac{1 - \lambda}{\lambda}\beta_1.$$

(4.17)

This means that the simple exponential smoother is a biased estimator for the linear trend model and the amount of bias is $-[(1 - \lambda)/\lambda]\beta_1$. This indeed explains the underestimation in Figure 4.10. One solution will be to use a large λ value since $(1 - \lambda)/\lambda \to 0$ as $\lambda \to 1$. In Figure 4.11, we show two simple exponential smoothers with $\lambda = 0.3$ and $\lambda = 0.99$. It can be

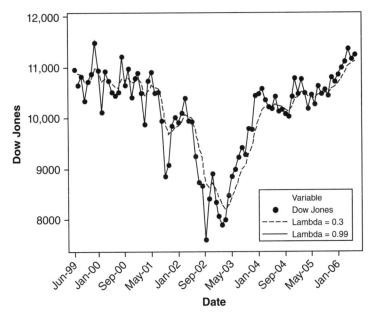

FIGURE 4.11 The Dow Jones Index from June 1999 to June 2006 using exponential smoothing with $\lambda = 0.3$ and 0.99.

seen that the latter does a better job in capturing the linear trend. However, it should also be noted that as the smoother with $\lambda = 0.99$ follows the actual observations very closely, it fails to smooth out the constant pattern during the first 2 years of the data. A method based on adaptive updating of the discount factor, λ, following the changes in the process is given in Section 4.6.4. In this section to model a linear trend model we will instead introduce the second-order exponential smoothing by applying simple exponential smoothing on $\tilde{y}_T$ as

$$\tilde{y}_T^{(2)} = \lambda \tilde{y}_T^{(1)} + (1 - \lambda)\tilde{y}_{T-1}^{(2)}, \qquad (4.18)$$

where $\tilde{y}_T^{(1)}$ and $\tilde{y}_T^{(2)}$ denote the first- and second-order smoothed exponentials, respectively. Of course, in Eq. (4.18) we can use a different λ than in Eq. (4.7). However, for the derivations that follow, we will assume that the same λ is used in the calculations of both $\tilde{y}_T^{(1)}$ and $\tilde{y}_T^{(2)}$.

From Eq. (4.17), we can see that the first-order exponential smoother introduces bias in estimating a linear trend. It can also be seen in Figure 4.7 that the first-order exponential smoother for the linear trend model exhibits a linear trend as well. Hence the second-order smoother—that is, a first-order exponential smoother of the original first-order exponential smoother—should also have a bias. We can represent this as

$$E\left(\tilde{y}_T^{(2)}\right) = E\left(\tilde{y}_T^{(1)}\right) - \frac{1-\lambda}{\lambda}\beta_1. \tag{4.19}$$

From Eq. (4.19), an estimate for β_1 at time T is

$$\hat{\beta}_{1,T} = \frac{\lambda}{1-\lambda}\left(\tilde{y}_T^1 - \tilde{y}_T^2\right) \tag{4.20}$$

and for an estimate of β_0 at time T, we have from Eq. (4.17)

$$\tilde{y}_T^{(1)} = \left(\hat{\beta}_{0,T} + \hat{\beta}_{1,T}T\right) - \frac{1-\lambda}{\lambda}\hat{\beta}_{1,T}$$

$$\Rightarrow \hat{\beta}_{0,T} = \tilde{y}_T^{(1)} - T\hat{\beta}_{1,T} + \frac{1-\lambda}{\lambda}\hat{\beta}_{1,T}. \tag{4.21}$$

In terms of the first- and second-order exponential smoothers, we have

$$\hat{\beta}_{0,T} = \tilde{y}_T^{(1)} - T\frac{\lambda}{1-\lambda}\left(\tilde{y}_T^{(1)} - \tilde{y}_T^{(2)}\right) + \frac{1-\lambda}{\lambda}\left(\frac{\lambda}{1-\lambda}\left(\tilde{y}_T^{(1)} - \tilde{y}_T^{(2)}\right)\right)$$

$$= \tilde{y}_T^{(1)} - T\frac{\lambda}{1-\lambda}\left(\tilde{y}_T^{(1)} - \tilde{y}_T^{(2)}\right) + \left(\tilde{y}_T^{(1)} - \tilde{y}_T^{(2)}\right) \tag{4.22}$$

$$= \left(2 - T\frac{\lambda}{1-\lambda}\right)\tilde{y}_T^{(1)} - \left(1 - T\frac{\lambda}{1-\lambda}\right)\tilde{y}_T^{(2)}.$$

Finally, combining Eq. (4.20) and (4.22), we have a predictor for y_T as

$$\tilde{y}_T = \hat{\beta}_{0,T} + \hat{\beta}_{1,T}T$$

$$= 2\tilde{y}_T^{(1)} - \tilde{y}_T^{(2)}. \tag{4.23}$$

It can easily be shown that $\hat{y}_T$ is an unbiased predictor of y_T. In Figure 4.12, we use Eq. (4.23) to estimate the Dow Jones Index from February 2003 to February 2004. From Figures 4.10 and 4.12, we can clearly see that the second-order exponential smoother is doing a much better job in modeling the linear trend compared to the simple exponential smoother.

As in the simple exponential smoothing, we have the same two issues to deal with: initial values for the smoothers and the discount factors. The

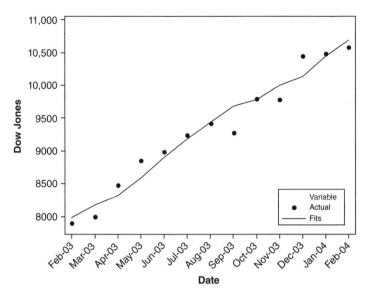

FIGURE 4.12 The Dow Jones Index from February 2003 to February 2004 with second-order exponential smoother with discount factor of 0.3.

latter will be discussed in Section 4.6.1. For the former we will combine Eqs. (4.17) and (4.19) as the following:

$$\tilde{y}_0^{(1)} = \hat{\beta}_{0,0} - \frac{1-\lambda}{\lambda}\hat{\beta}_{1,0}$$

$$\tilde{y}_0^{(2)} = \hat{\beta}_{0,0} - 2\left(\frac{1-\lambda}{\lambda}\right)\hat{\beta}_{1,0}. \tag{4.24}$$

The initial estimates of the model parameters are usually obtained by fitting the linear trend model to the entire or a subset of the available data. The least squares estimates of the parameter estimates are then used for $\hat{\beta}_{0,0}$ and $\hat{\beta}_{1,0}$.

Example 4.2   Consider the US Consumer Price Index (CPI) from January 1995 to December 2004 in Table 4.2. Figure 4.13 clearly shows that the data exhibits a linear trend. To smooth the data, following the recommendation in Section 4.2, we can use single exponential smoothing with $\lambda = 0.3$ as given in Figure 4.14.

As we expected, the exponential smoother does a very good job in capturing the general trend in the data and provides a less jittery (smooth) version of it. However, we also notice that the smoothed values are

TABLE 4.2 Consumer Price Index from January 1995 to December 2004

Month-Year	CPI	Month-Year	CPI	Month-Year	CPI	Month-Year	CPI	Month-Year	CPI
Jan-1995	150.3	Jan-1997	159.1	Jan-1999	164.3	Jan-2001	175.1	Jan-2003	181.7
Feb-1995	150.9	Feb-1997	159.6	Feb-1999	164.5	Feb-2001	175.8	Feb-2003	183.1
Mar-1995	151.4	Mar-1997	160	Mar-1999	165	Mar-2001	176.2	Mar-2003	184.2
Apr-1995	151.9	Apr-1997	160.2	Apr-1999	166.2	Apr-2001	176.9	Apr-2003	183.8
May-1995	152.2	May-1997	160.1	May-1999	166.2	May-2001	177.7	May-2003	183.5
Jun-1995	152.5	Jun-1997	160.3	Jun-1999	166.2	Jun-2001	178	Jun-2003	183.7
Jul-1995	152.5	Jul-1997	160.5	Jul-1999	166.7	Jul-2001	177.5	Jul-2003	183.9
Aug-1995	152.9	Aug-1997	160.8	Aug-1999	167.1	Aug-2001	177.5	Aug-2003	184.6
Sep-1995	153.2	Sep-1997	161.2	Sep-1999	167.9	Sep-2001	178.3	Sep-2003	185.2
Oct-1995	153.7	Oct-1997	161.6	Oct-1999	168.2	Oct-2001	177.7	Oct-2003	185
Nov-1995	153.6	Nov-1997	161.5	Nov-1999	168.3	Nov-2001	177.4	Nov-2003	184.5
Dec-1995	153.5	Dec-1997	161.3	Dec-1999	168.3	Dec-2001	176.7	Dec-2003	184.3
Jan-1996	154.4	Jan-1998	161.6	Jan-2000	168.8	Jan-2002	177.1	Jan-2004	185.2
Feb-1996	154.9	Feb-1998	161.9	Feb-2000	169.8	Feb-2002	177.8	Feb-2004	186.2
Mar-1996	155.7	Mar-1998	162.2	Mar-2000	171.2	Mar-2002	178.8	Mar-2004	187.4
Apr-1996	156.3	Apr-1998	162.5	Apr-2000	171.3	Apr-2002	179.8	Apr-2004	188
May-1996	156.6	May-1998	162.8	May-2000	171.5	May-2002	179.8	May-2004	189.1
Jun-1996	156.7	Jun-1998	163	Jun-2000	172.4	Jun-2002	179.9	Jun-2004	189.7
Jul-1996	157	Jul-1998	163.2	Jul-2000	172.8	Jul-2002	180.1	Jul-2004	189.4
Aug-1996	157.3	Aug-1998	163.4	Aug-2000	172.8	Aug-2002	180.7	Aug-2004	189.5
Sep-1996	157.8	Sep-1998	163.6	Sep-2000	173.7	Sep-2002	181	Sep-2004	189.9
Oct-1996	158.3	Oct-1998	164	Oct-2000	174	Oct-2002	181.3	Oct-2004	190.9
Nov-1996	158.6	Nov-1998	164	Nov-2000	174.1	Nov-2002	181.3	Nov-2004	191
Dec-1996	158.6	Dec-1998	163.9	Dec-2000	174	Dec-2002	180.9	Dec-2004	190.3

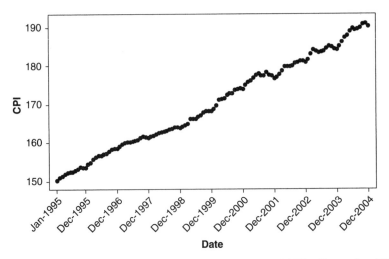

FIGURE 4.13 US Consumer Price Index from January 1995 to December 2004.

consistently below the actual values. Hence there is an apparent bias in our smoothing. To fix this problem we have two choices: use a bigger λ or **second-order** exponential smoothing. The former will lead to less smooth estimates and hence defeat the purpose. For the latter, however, we can use $\lambda = 0.3$ to calculate and $\tilde{y}_T^{(1)}$ and $\tilde{y}_T^{(2)}$ as given in Table 4.3.

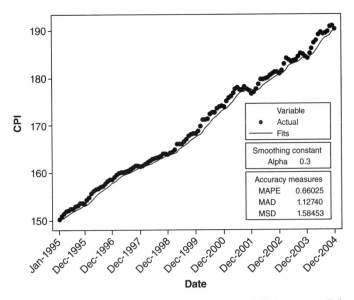

FIGURE 4.14 Single exponential smoothing of the US Consumer Price Index (with $\tilde{y}_0 = y_1$).

TABLE 4.3   Second-Order Exponential Smoothing of the US Consumer Price Index (with $\lambda = 0.3, \tilde{y}_0^{(1)} = y_1$, and $\tilde{y}_0^{(2)} = \tilde{y}_0^{(1)}$)

Date	y_t	$\tilde{y}_T^{(1)}$	$\tilde{y}_T^{(2)}$	$\hat{y}_T = 2\tilde{y}_T^{(1)} - \tilde{y}_T^{(2)}$
Jan-1995	150.3	150.300	150.300	150.300
Feb-1995	150.9	150.480	150.354	150.606
Mar-1995	151.4	150.756	150.475	151.037
Apr-1995	151.9	151.099	150.662	151.536
May-1995	152.2	151.429	150.892	151.967
Nov-2004	191.0	190.041	188.976	191.106
Dec-2004	190.3	190.119	189.319	190.919

Note that we used $\tilde{y}_0^{(1)} = y_1$, and $\tilde{y}_0^{(2)} = \tilde{y}_0^{(1)}$ as the initial values of $\tilde{y}_T^{(1)}$ and $\tilde{y}_T^{(2)}$. A more rigorous approach would involve fitting a linear regression model in time to the available data that give

$$\hat{y}_t = \hat{\beta}_{0,T} + \hat{\beta}_{1,T}t$$
$$= 149.89 + 0.33t,$$

where t goes from 1 to 120. Then from Eq. (4.24) we have

$$\tilde{y}_0^{(1)} = \hat{\beta}_{0,0} - \frac{1-\lambda}{\lambda}\hat{\beta}_{1,0}$$

$$= 149.89 - \frac{1-0.3}{0.3}0.33 = 146.22$$

$$\tilde{y}_0^{(2)} = \hat{\beta}_{0,0} - 2\left(\frac{1-\lambda}{\lambda}\right)\hat{\beta}_{1,0}$$

$$= 149.89 - 2\left(\frac{1-0.3}{0.3}\right)0.33 = 142.56.$$

Figure 4.15 shows the second-order exponential smoothing of the CPI. As we can see, the second-order exponential smoothing not only captures the trend in the data but also does not exhibit any bias.

The calculations for the second-order smoothing for the CPI data are performed using Minitab. We first obtained the first-order exponential smoother for the CPI, $\tilde{y}_T^{(1)}$, using $\lambda = 0.3$ and $\tilde{y}_0^{(1)} = y_1$. Then we obtained $\tilde{y}_T^{(2)}$ by taking the first-order exponential smoother $\tilde{y}_T^{(1)}$ using $\lambda = 0.3$ and $\tilde{y}_0^{(2)} = \tilde{y}_1^{(1)}$. Then using Eq. (4.23) we have $\hat{y}_T = 2\tilde{y}_T^{(1)} - \tilde{y}_T^{(2)}$.

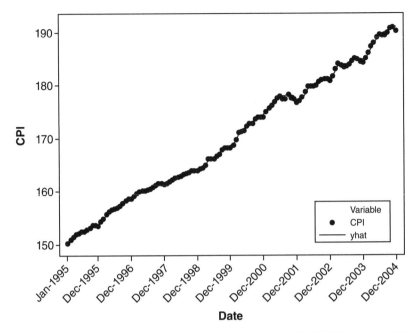

FIGURE 4.15 Second-order exponential smoothing of the US Consumer Price Index (with $\lambda = 0.3$, $\tilde{y}_0^{(1)} = y_1$, and $\tilde{y}_0^{(2)} = \tilde{y}_1^{(1)}$).

The "Double Exponential Smoothing" option available in Minitab is a slightly different approach based on Holt's method (Holt, 1957). This method divides the time series data into two components: the level, L_t, and the trend, T_t. These two components can be calculated from

$$L_t = \alpha y_t + (1 - \alpha)(L_{t-1} + T_{t-1})$$
$$T_t = \gamma(L_t - L_{t-1}) + (1 - \gamma)T_{t-1}$$

Hence for a given set of α and γ, these two components are calculated and L_t is used to obtain the double exponential smoothing of the data at time t. Furthermore, the sum of the level and trend components at time t can be used as the one-step-ahead $(t + 1)$ forecast. Figure 4.16 shows the actual and smoothed data using the double exponential smoothing option in Minitab with $\alpha = 0.3$ and $\gamma = 0.3$.

In general, the initial values for the level and the trend terms can be obtained by fitting a linear regression model to the CPI data with time as

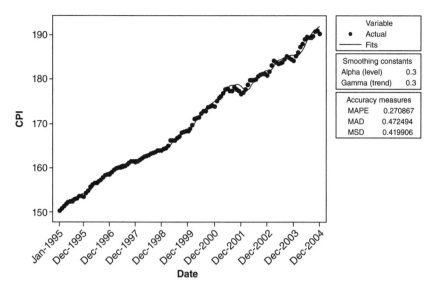

FIGURE 4.16 The double exponential smoothing of the US Consumer Price Index (with $\alpha = 0.3$ and $\gamma = 0.3$).

the regressor. Then the intercept and the slope can be used as the initial values of L_t and T_t respectively.

Example 4.3 For the Dow Jones Index data, we observed that first-order exponential smoothing with low values of λ showed some bias when there were linear trends in the data. We may therefore decide to use the second-order exponential smoothing approach for this data as shown in Figure 4.17. Note that the bias present with first-order exponential smoothing has been eliminated. The calculations for second-order exponential smoothing for the Dow Jones Index are given in Table 4.4.

4.5 HIGHER-ORDER EXPONENTIAL SMOOTHING

So far we have discussed the use of exponential smoothers in estimating the constant and linear trend models. For the former we employed the **simple** or **first-order** exponential smoother and for the latter the **second-order** exponential smoother. It can further be shown that for the general nth-degree polynomial model of the form

$$y_t = \beta_0 + \beta_1 t + \frac{\beta_2}{2!}t^2 + \cdots + \frac{\beta_n}{n!}t^n + \varepsilon_t, \tag{4.25}$$

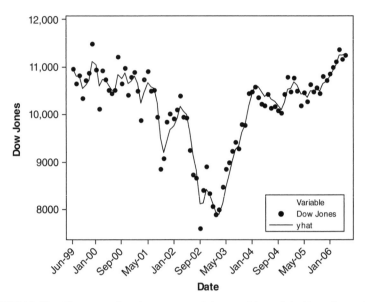

FIGURE 4.17 The second-order exponential smoothing of the Dow Jones Index (with $\lambda = 0.3$, $\tilde{y}_0^{(1)} = y_1$, and $\tilde{y}_0^{(2)} = \tilde{y}_1^{(1)}$).

where the ε_t is assumed to be independent with mean 0 and constant variance σ_ε^2, we employ $(n + 1)$-order exponential smoothers

$$\tilde{y}_T^{(2)} = \lambda y_T + (1 - \lambda)\tilde{y}_{T-1}^{(1)}$$
$$\tilde{y}_T^{(2)} = \lambda\tilde{y}_T^{(1)} + (1 - \lambda)\tilde{y}_{T-1}^{(2)}$$
$$\vdots$$
$$\tilde{y}_T^{(n)} = \lambda\tilde{y}_T^{(n-1)} + (1 - \lambda)\tilde{y}_{T-1}^{(n)}$$

TABLE 4.4 Second-Order Exponential Smoothing of the Dow Jones Index (with $\lambda = 0.3$, $\tilde{y}_0^{(1)} = y_1$, and $\tilde{y}_0^{(2)} = \tilde{y}_1^{(1)}$)

Date	$\tilde{y}_t$	$\tilde{y}_T^1$	$\tilde{y}_T^2$	$\hat{y}_T = 2\tilde{y}_T^{(1)} - \tilde{y}_T^{(2)}$
Jun-1999	10,970.8	10,970.8	10,970.8	10,970.8
Jul-1999	10,655.2	10,876.1	10,942.4	10,809.8
Aug-1999	10,829.3	10,862.1	10,918.3	10,805.8
Sep-1999	10,337.0	10,704.6	10,854.2	10,554.9
Oct-1999	10,729.9	10,712.2	10,811.6	10,612.7
May-2006	11,168.3	11,069.4	10,886.5	11,252.3
Jun-2006	11,247.9	11,123.0	10,957.4	11,288.5

to estimate the model parameters. For even the quadratic model (second-degree polynomial), the calculations get quite complicated. Refer to Montgomery et al. (1990), Brown (1963), and Abraham and Ledolter (1983) for the solutions to higher-order exponential smoothing problems. If a high-order polynomial does seem to be required for the time series, the autoregressive integrated moving average (ARIMA) models and techniques discussed in Chapter 5 can instead be considered.

4.6 FORECASTING

We have so far considered exponential smoothing techniques as either visual aids to point out the underlying patterns in the time series data or to estimate the model parameters for the class of models given in Eq. (4.9). The latter brings up yet another use of exponential smoothing—forecasting future observations. At time T, we may wish to forecast the observation in the next time unit, $T + 1$, or further into the future. For that, we will denote the τ-step-ahead forecast made at time T as $\hat{y}_{T+\tau}(T)$. In the next two sections and without any loss of generality, we will once again consider first- and second-order exponential smoothers as examples for forecasting time series data from the constant and linear trend processes.

4.6.1 Constant Process

In Section 4.2 we discussed first-order exponential smoothing for the constant process in Eq. (4.1) as

$$\tilde{y}_T = \lambda y_T + (1 - \lambda)\tilde{y}_{T-1}.$$

In Section 4.3 we further showed that the constant level in Eq. (4.1), β_0, can be estimated by $\tilde{y}_T$. Since the constant model consists of two parts—β_0 that can be estimated by the first-order exponential smoother and the random error that cannot be predicted—our forecast for the future observation is simply equal to the current value of the exponential smoother

$$\hat{y}_{T+\tau}(T) = \tilde{y}_T = \tilde{y}_T. \tag{4.26}$$

Please note that, for the constant process, the forecast in Eq. (4.26) is the same for all future values. Since there may be changes in the level of the constant process, forecasting all future observations with the same value will most likely be misleading. However, as we start accumulating more observations, we can update our forecast. For example, if the data at $T + 1$ become available, our forecast for the future observations becomes

$$\tilde{y}_{T+1} = \lambda y_{T+1} + (1 - \lambda)\tilde{y}_T$$

or

$$\hat{y}_{T+1+\tau}(T + 1) = \lambda y_{T+1} + (1 - \lambda)\hat{y}_{T+\tau}(T) \qquad (4.27)$$

We can rewrite Eq. (4.27) for $\tau = 1$ as

$$\begin{aligned} \hat{y}_{T+2}(T + 1) &= \hat{y}_{T+1}(T) + \lambda(y_{T+1} - \hat{y}_{T+1}(T)) \\ &= \hat{y}_{T+1}(T) + \lambda e_{T+1}(1), \end{aligned} \qquad (4.28)$$

where $e_{T+1}(1) = y_{T+1} - \hat{y}_{T+1}(T)$ is called the one-step-ahead forecast or prediction error. The interpretation of Eq. (4.28) makes it easier to understand the forecasting process using exponential smoothing: our forecast for the next observation is simply our previous forecast for the current observation plus a fraction of the forecast error we made in forecasting the current observation. The fraction in this summation is determined by λ. Hence how fast our forecast will react to the forecast error depends on the discount factor. A large discount factor will lead to fast reaction to the forecast error but it may also make our forecast react fast to random fluctuations. This once again brings up the issue of the choice of the discount factor.

Choice of λ   We will define the sum of the squared one-step-ahead forecast errors as

$$SS_E(\lambda) = \sum_{t=1}^{T} e_t^2(1). \qquad (4.29)$$

For a given historic data, we can in general calculate SS_E values for various values of λ and pick the value of λ that gives the smallest sum of the squared forecast errors.

Prediction Intervals   Another issue in forecasting is the uncertainty associated with it. That is, we may be interested not only in the "point estimates" but also in the quantification of the prediction uncertainty. This is usually achieved by providing the prediction intervals that are expected at a specific confidence level to contain the future observations. Calculations of the prediction intervals will require the estimation of the variance of the forecast errors. We will discuss two different techniques in estimating prediction error variance in Section 4.6.3. For the constant process, the 100 $(1 - \alpha/2)$ percent prediction intervals for any lead time τ are given as

$$\tilde{y}_T \pm Z_{\alpha/2}\hat{\sigma}_e,$$

where $\tilde{y}_T$ is the first-order exponential smoother, $Z_{\alpha/2}$ is the $100(1 - \alpha/2)$ percentile of the standard normal distribution, and $\hat{\sigma}_e$ is the estimate of the standard deviation of the forecast errors.

It should be noted that the prediction interval is constant for all lead times. This of course can be (and probably is in most cases) quite unrealistic. As it will be more likely that the process goes through some changes as time goes on, we would correspondingly expect to be less and less "sure" about our predictions for large lead times (or large τ values). Hence we would anticipate prediction intervals that are getting wider and wider for increasing lead times. We propose a remedy for this in Section 4.6.3. We will discuss this issue further in Chapter 6.

Example 4.4   We are interested in the average speed on a specific stretch of a highway during nonrush hours. For the past year and a half (78 weeks), we have available weekly averages of the average speed in miles/hour between 10 AM and 3 PM. The data are given in Table 4.5. Figure 4.18 shows that the time series data follow a constant process. To smooth out the excessive variation, however, first-order exponential smoothing can be used. The "best" smoothing constant can be determined by finding the smoothing constant value that minimizes the sum of the squared one-step-ahead prediction errors.

The sum of the squared one-step-ahead prediction errors for various λ values is given in Table 4.6. Furthermore, Figure 4.19 shows that the minimum SS_E is obtained for $\lambda = 0.4$.

Let us assume that we are also asked to make forecasts for the next 12 weeks at week 78. Figure 4.20 shows the smoothed values for the first 78 weeks together with the forecasts for weeks 79–90 with prediction intervals. It also shows the actual weekly speed during that period. Note that since the constant process is assumed, the forecasts for the next 12 weeks are the same. Similarly, the prediction intervals are constant for that period.

TABLE 4.5   The Weekly Average Speed During Nonrush Hours

Week	Speed	Week	Speed	Week	Speed	Week	Speed
1	47.12	26	46.74	51	45.71	76	45.69
2	45.01	27	46.62	52	43.84	77	44.59
3	44.69	28	45.31	53	45.09	78	43.45
4	45.41	29	44.69	54	44.16	79	44.75
5	45.45	30	46.39	55	46.21	80	45.46
6	44.77	31	43.79	56	45.11	81	43.73
7	45.24	32	44.28	57	46.16	82	44.15
8	45.27	33	46.04	58	46.50	83	44.05
9	46.93	34	46.45	59	44.88	84	44.83
10	47.97	35	46.31	60	45.68	85	43.93
11	45.27	36	45.65	61	44.40	86	44.40
12	45.10	37	46.28	62	44.17	87	45.25
13	43.31	38	44.11	63	45.18	88	44.80
14	44.97	39	46.00	64	43.73	89	44.75
15	45.31	40	46.70	65	45.14	90	44.50
16	45.23	41	47.84	66	47.98	91	45.12
17	42.92	42	48.24	67	46.52	92	45.28
18	44.99	43	45.59	68	46.89	93	45.15
19	45.12	44	46.56	69	46.01	94	46.24
20	46.67	45	45.02	70	44.98	95	46.15
21	44.62	46	43.67	71	45.76	96	46.57
22	45.11	47	44.53	72	45.38	97	45.51
23	45.18	48	44.37	73	45.33	98	46.98
24	45.91	49	44.62	74	44.07	99	46.64
25	48.39	50	46.71	75	44.02	100	44.31

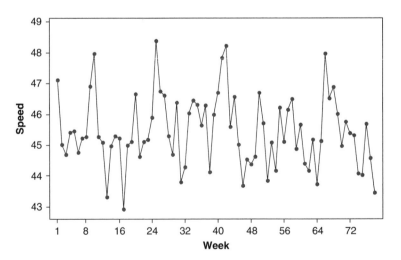

FIGURE 4.18   The weekly average speed during nonrush hours.

TABLE 4.6 SS_E for Different λ Values for the Average Speed Data

| | | λ | | | | | | | | | | | |
| | | 0.1 | | 0.2 | | 0.3 | | 0.4 | | 0.5 | | 0.9 | |
Week	Speed	Forecast	$e(t)$	Forecast	$e(t)$	Forecast	$e(t)$	Forecast	$e(t)$	Forecast	$e(t)$	Forecast	$e(t)$
1	47.12	47.12	0.00	47.12	0.00	47.12	0.00	47.12	0.00	47.12	0.00	47.12	0.00
2	45.01	47.12	−2.11	47.12	−2.11	47.12	−2.11	47.12	−2.11	47.12	−2.U	47.12	−2.11
3	44.69	46.91	−2.22	46.70	−2.01	46.49	−1.80	46.28	−1.59	46.07	−1.38	45.22	−0.53
4	45.41	46.69	−1.28	46.30	−0.89	45.95	−0.54	45.64	−0.23	45.38	0.03	44.74	0.67
5	45.45	46.56	−1.11	46.12	−0.67	45.79	−0.34	45.55	−0.10	45.39	0.06	45.34	0.11
6	44.77	46.45	−1.68	45.99	−1.22	45.69	−0.92	45.51	−0.74	45.42	−0.65	45.44	−0.67
7	45.24	46.28	−1.04	45.74	−0.50	45.41	−0.17	45.21	0.03	45.10	0.14	44.84	0.40
8	45.27	46.18	−0.91	45.64	−0.37	45.36	−0.09	45.22	0.05	45.17	0.10	45.20	0.07
9	46.93	46.09	0.84	45.57	1.36	45.33	1.60	45.24	1.69	45.22	1.71	45.26	1.67
10	47.97	46.17	1.80	45.84	2.13	45.81	2.16	45.92	2.05	46.07	1.90	46.76	1.21
⋮	⋮	⋮		⋮		⋮		⋮		⋮		⋮	
75	44.02	45.42	−1.40	45.30	−1.28	45.12	−1.10	44.93	−0.91	44.75	−0.73	44.20	−0.18
76	45.69	45.28	0.41	45.05	0.64	44.79	0.90	44.56	1.13	44.39	1.30	44.04	1.65
77	44.59	45.32	−0.73	45.18	−0.59	45.06	−0.47	45.01	−0.42	45.04	−0.45	45.52	−0.93
78	43.45	45.25	−1.80	45.06	−1.61	44.92	−1.47	44.84	−1.39	44.81	−1.36	44.68	−1.23
SS_E			124.14		118.88		117.27		116.69		116.95		128.98

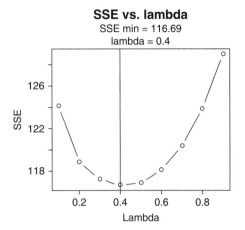

FIGURE 4.19 Plot of SS_E for various λ values for average speed data.

4.6.2 Linear Trend Process

The t-step-ahead forecast for the linear trend model is given by

$$
\begin{aligned}
\hat{y}_{T+\tau}(T) &= \hat{\beta}_{0,T} + \hat{\beta}_{1,T}(T + \tau) \\
&= \hat{\beta}_{0,T} + \hat{\beta}_{1,T}T + \hat{\beta}_{1,T}\tau \\
&= \hat{y}_T + \hat{\beta}_{1,T}\tau.
\end{aligned}
\tag{4.30}
$$

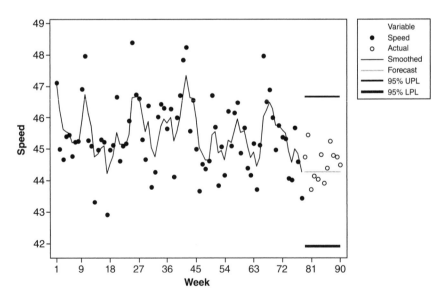

FIGURE 4.20 Forecasts for the weekly average speed data for weeks 79–90.

In terms of the exponential smoothers, we can rewrite Eq. (4.30) as

$$
\begin{aligned}
\hat{y}_{T+\tau}(\tau) &= \left(2\tilde{y}_T^{(1)} - \tilde{y}_T^{(2)}\right) + \tau \frac{\lambda}{1-\lambda}\left(\tilde{y}_T^{(1)} - \tilde{y}_T^{(2)}\right) \\
&= \left(2 + \frac{\lambda}{1-\lambda}\tau\right)\tilde{y}_T^{(1)} - \left(1 + \frac{\lambda}{1-\lambda}\tau\right)\tilde{y}_T^{(2)}.
\end{aligned}
\tag{4.31}
$$

It should be noted that the predictions for the trend model depend on the lead time and, as opposed to the constant model, will be different for different lead times. As we collect more data, we can improve our forecasts by updating our parameter estimates using

$$
\begin{aligned}
\hat{\beta}_{0,T+1} &= \lambda(1+\lambda)y_{T+1} + (1-\lambda)^2(\hat{\beta}_{0,T} + \hat{\beta}_{1,T}) \\
\hat{\beta}_{1,T+1} &= \frac{\lambda}{(2-\lambda)}\left(\hat{\beta}_{0,T+1} - \hat{\beta}_{0,T}\right) + \frac{2(1-\lambda)}{(2-\lambda)}\hat{\beta}_{1,T}
\end{aligned}
\tag{4.32}
$$

Subsequently, we can update our τ-step-ahead forecasts based on Eq. (4.32). As in the constant process, the discount factor, λ, can be estimated by minimizing the sum of the squared one-step-ahead forecast errors given in Eq. (4.29).

In this case, the $100(1 - \alpha/2)$ percent prediction interval for any lead time τ is

$$
\left(2 + \frac{\lambda}{1-\lambda}\tau\right)\hat{y}_T^{(1)} - \left(1 + \frac{\lambda}{1-\lambda}\tau\right)\hat{y}_T^{(2)} \pm Z_{\alpha/2}\frac{c_\tau}{c_1}\hat{\sigma}_e,
$$

where

$$
c_i^2 = 1 + \frac{\lambda}{(2-\lambda)^3}[(10 - 14\lambda + 5\lambda^2) + 2i\lambda(4 - 3\lambda) + 2i^2\lambda^2].
$$

Example 4.5   Consider the CPI data in Example 4.2. Assume that we are currently in December 2003 and would like to make predictions of the CPI for the following year. Although the data from January 1995 to December 2003 clearly exhibit a linear trend, we may still like to consider first-order exponential smoothing first. We will then calculate the "best" λ value that minimizes the sum of the squared one-step-ahead prediction errors. The predictions and prediction errors for various λ values are given in Table 4.7.

Figure 4.21 shows the sum of the squared one-step-ahead prediction errors (SS_E) for various values of λ.

TABLE 4.7 The Predictions and Prediction Errors for Various λ Values for CPI Data

Month-Year	CPI	$\lambda = 0.1$ Prediction	Error	$\lambda = 0.2$ Prediction	Error	$\lambda = 0.3$ Prediction	Error	$\lambda = 0.9$ Prediction	Error	$\lambda = 0.99$ Prediction	Error
Jan-1995	150.3	150.30	0.00	150.30	0.00	150.30	0.00	150.30	0.00	150.30	0.00
Feb-1995	150.9	150.30	0.60	150.30	0.60	150.30	0.60	150.30	0.60	150.30	0.60
Mar-1995	151.4	150.36	1.04	150.42	0.98	150.48	0.92	150.84	0.56	150.89	0.51
Apr-1995	151.9	150.46	1.44	150.62	1.28	150.76	1.14	151.34	0.56	151.39	0.51
⋮	⋮	⋮		⋮		⋮		⋮		⋮	
Nov-2003	184.5	182.29	2.21	183.92	0.58	184.45	0.05	185.01	-0.51	185.00	-0.50
Dec-2003	184.3	182.51	1.79	184.03	0.27	184.46	-0.16	184.55	-0.25	184.51	-0.21
SS_E		1061.50		309.14		153.71		31.90		28.62	

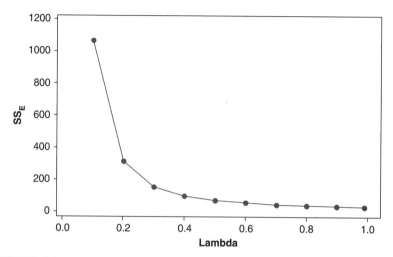

FIGURE 4.21   Scatter plot of the sum of the squared one-step-ahead prediction errors versus λ.

We notice that the SS_E keeps on getting smaller as λ gets bigger. This suggests that the data are highly autocorrelated. This can be clearly seen in the ACF plot in Figure 4.22. In fact if the "best" λ value (i.e., λ value that minimizes SS_E) turns out to be high, it may indeed be better to switch to a higher-order smoothing or use an ARIMA model as discussed in Chapter 5.

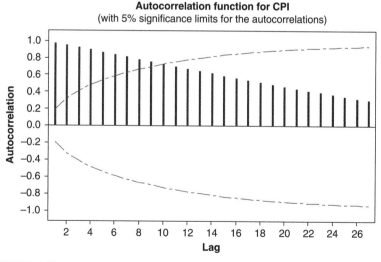

FIGURE 4.22   ACF plot for the CPI data (with 5% significance limits for the autocorrelations).

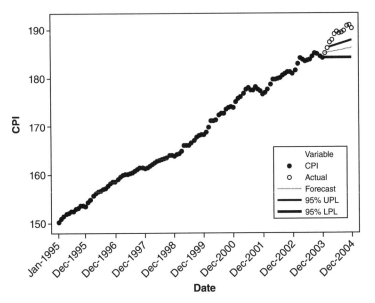

FIGURE 4.23 The 1- to 12-step-ahead forecasts of the CPI data for 2004.

Since the first-order exponential smoothing is deemed inadequate, we will now try the second-order exponential smoothing to forecast next year's monthly CPI values. Usually we have two options:

1. On December 2003, make forecasts for the entire 2004 year; that is, 1-step-ahead, 2-step-ahead, ... , 12-step-ahead forecasts. For that we can use Eq. (4.30) or equivalently Eq. (4.31). Using the double exponential smoothing option in Minitab with $\lambda = 0.3$, we obtain the forecasts given in Figure 4.23.

Note that the forecasts further in the future (for the later part of 2004) are quite a bit off. To remedy this we may instead use the following strategy.

2. In December 2003, make the one-step-ahead forecast for January 2004. When the data for January 2004 becomes available, then make the one-step-ahead forecast for February 2004, and so on. We can see from Figure 4.24 that forecasts when only one-step-ahead forecasts are used and adjusted as actual data becomes available perform better than in the previous case where, for December 2003, forecasts are made for the entire following year.

The JMP software package also has an excellent forecasting capability. Table 4.8 shows output from JMP for the CPI data for double

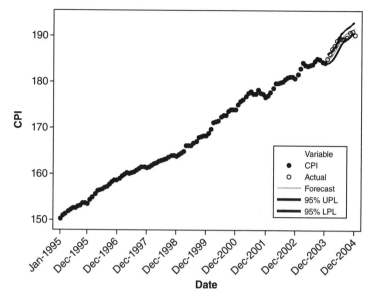

FIGURE 4.24   The one-step-ahead forecasts of the CPI data for 2004.

exponential smoothing. JMP uses the double smoothing procedure that employs a single smoothing constant. The JMP output shows the time series plot and summary statistics including the sample ACF. It also provides a sample partial ACF, which we will discuss in Chapter 5. Then an optimal smoothing constant is chosen by finding the value of λ that

TABLE 4.8   JMP Output for the CPI Data

Time series CPI

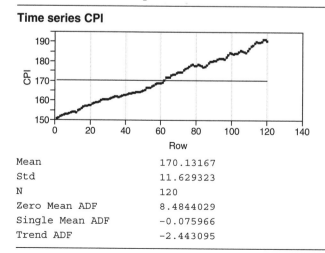

Mean	170.13167
Std	11.629323
N	120
Zero Mean ADF	8.4844029
Single Mean ADF	−0.075966
Trend ADF	−2.443095

(continued)

TABLE 4.8 *(Continued)*

Time series basic diagnostics

Lag	AutoCorr	Plot autocorr	Ljung-box Q	p-Value
0	1.0000			
1	0.9743		116.774	<.0001
2	0.9472		228.081	<.0001
3	0.9203		334.053	<.0001
4	0.8947		435.091	<.0001
5	0.8694		531.310	<.0001
6	0.8436		622.708	<.0001
7	0.8166		709.101	<.0001
8	0.7899		790.659	<.0001
9	0.7644		867.721	<.0001
10	0.7399		940.580	<.0001
11	0.7161		1009.46	<.0001

Lag	AutoCorr	Plot autocorr	Ljung-box Q	p-Value
12	0.6924		1074.46	<.0001
13	0.6699		1135.85	<.0001
14	0.6469		1193.64	<.0001
15	0.6235		1247.84	<.0001
16	0.6001		1298.54	<.0001
17	0.5774		1345.93	<.0001
18	0.5550		1390.14	<.0001
19	0.5324		1431.24	<.0001
20	0.5098		1469.29	<.0001
21	0.4870		1504.36	<.0001
22	0.4637		1536.48	<.0001
23	0.4416		1565.91	<.0001
24	0.4205		1592.87	<.0001
25	0.4000		1617.54	0.0000

Lag	Partial	plot partial
0	1.0000	
1	0.9743	
2	−0.0396	
3	−0.0095	
4	0.0128	
5	−0.0117	
6	−0.0212	
7	−0.0379	
8	−0.0070	
9	0.0074	
10	0.0033	
11	−0.0001	
12	−0.0116	
13	0.0090	
14	−0.0224	
15	−0.0220	
16	−0.0139	
17	−0.0022	
18	−0.0089	
19	−0.0174	
20	−0.0137	
21	−0.0186	
22	−0.0234	
23	0.0074	
24	0.0030	
25	−0.0036	

TABLE 4.8 *(Continued)*

```
Model Comparison
Model                                       DF  Variance        AIC
Double (Brown) Exponential Smoothing  117  0.247119   171.05558
      SBC   RSquare   -2LogLH  AIC  Rank  SBC Rank   MAPE MAE
173.82626    0.998  169.05558    0     0  0.216853  0.376884
```

```
Model: Double (Brown) Exponential Smoothing
Model Summary
```

DF	117
Sum of Squared Errors	28.9129264
Variance Estimate	0.24711903
Standard Deviation	0.49711068
Akaike's 'A' Information Criterion	171.055579
Schwarz's Bayesian Criterion	173.826263
RSquare	0.99812888
RSquare Adj	0.99812888
MAPE	0.21685285
MAE	0.37688362
-2LogLikelihood	169.055579

```
Stable    Yes
Invertible Yes
```

```
Parameter Estimates
Term                    Estimate    Std Error  t Ratio  Prob>|t|
Level Smoothing Weight  0.81402446  0.0919040     8.86    <.0001
```

Forecast

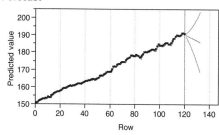

Residuals

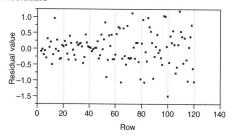

(continued)

TABLE 4.8 (*Continued*)

Lag	AutoCorr plot autocorr	Ljung-box Q	p-Value
0	1.0000	.	.
1	0.0791	0.7574	0.3841
2	−0.3880	19.1302	<.0001
3	−0.2913	29.5770	<.0001
4	−0.0338	29.7189	<.0001
5	0.1064	31.1383	<.0001
6	0.1125	32.7373	<.0001
7	0.1867	37.1819	<.0001
8	−0.1157	38.9063	<.0001
9	−0.3263	52.7344	<.0001
10	−0.1033	54.1324	<.0001
11	0.2149	60.2441	<.0001
12	0.2647	69.6022	<.0001
13	−0.0773	70.4086	<.0001
14	0.0345	70.5705	<.0001
15	−0.1243	72.6937	<.0001
16	−0.1429	75.5304	<.0001
17	0.0602	76.0384	<.0001
18	0.1068	77.6533	<.0001
19	0.0370	77.8497	<.0001
20	−0.0917	79.0656	<.0001
21	−0.0363	79.2579	<.0001
22	−0.0995	80.7177	<.0001
23	−0.0306	80.8570	<.0001
24	0.2602	91.0544	<.0001
25	0.1728	95.6007	<.0001

Lag	Partial plot partial
0	1.0000
1	0.0791
2	−0.3967
3	−0.2592
4	−0.1970
5	−0.1435
6	−0.0775
7	0.1575
8	−0.1144
9	−0.2228
10	−0.1482

Lag	AutoCorr plot autocorr	Ljung-box Q	p-Value
11	−0.0459		
12	0.0368		
13	−0.1335		
14	0.2308		
15	−0.0786		
16	0.0050		
17	0.0390		
18	−0.0903		
19	−0.0918		
20	0.0012		
21	−0.0077		
22	−0.1935		
23	−0.0665		
24	0.1783		
25	0.0785		

minimizes the error sum of squares. The value selected is $\lambda = 0.814$. This relatively large value is not unexpected, because there is a very strong linear trend in the data and considerable autocorrelation. Values of the forecast for the next 12 periods at origin December 2004 and the associated prediction interval are also shown. Finally, the residuals from the model fit are shown along with the sample ACF and sample partial ACF plots of the residuals. The sample ACF indicates that there may be a small amount of structure in the residuals, but it is not enough to cause concern.

4.6.3 Estimation of σ_e^2

In the estimation of the variance of the forecast errors, σ_e^2, it is often assumed that the model (e.g., constant, linear trend) is correct and constant in time. With these assumptions, we have two different ways of estimating σ_e^2:

1. We already defined the one-step-ahead forecast error as $e_T(1) = y_T - \hat{y}_T(T-1)$. The idea is to apply the model to the historic data and obtain the forecast errors to calculate:

$$
\begin{aligned}
\hat{\sigma}_e^2 &= \frac{1}{T}\sum_{t=1}^{T} e_t^2(1) \\
&= \frac{1}{T}\sum_{t=1}^{T} (y_t - \hat{y}_t(t-1))^2
\end{aligned}
\tag{4.33}
$$

It should be noted that in the variance calculations the mean adjustment was not needed, since for the correct model the forecasts are unbiased; that is, the expected value of the forecast errors is 0.

As more data are collected, the variance of the forecast errors can be updated as

$$
\hat{\sigma}_{eT+1}^2 = \frac{1}{T+1}\left(T\hat{\sigma}_{e,T}^2 + e_{T+1}^2(1)\right).
\tag{4.34}
$$

As discussed in Section 4.6.1, it may be counterintuitive to have a constant forecast error variance for all lead times. We can instead define $\sigma_e^2(\tau)$ as the τ-step-ahead forecast error variance and estimate it by

$$
\hat{\sigma}_e^2(\tau) = \frac{1}{T-\tau+1}\sum_{t=\tau}^{T} e_1^2(\tau).
\tag{4.35}
$$

Hence the estimate in Eq. (4.35) can instead be used in the calculations of the prediction interval for the τ-step-ahead forecast.

2. For the second method of estimating σ_e^2 we will first define the *mean absolute deviation* Δ as

$$\Delta = E(|e - E(e)|) \qquad (4.36)$$

and, assuming that the model is correct, calculate its estimate by

$$\hat{\Delta}_T = \delta|e_T(1)| + (1 - \delta)\hat{\Delta}_{T-1}. \qquad (4.37)$$

Then the estimate of the σ_e^2 is given by

$$\hat{\sigma}_{e,T} = 1.25\hat{\Delta}_T. \qquad (4.38)$$

For further details, see Montgomery et al. (1990).

4.6.4 Adaptive Updating of the Discount Factor

In the previous sections we discussed estimation of the "best" discount factor, $\hat{\lambda}$, by minimizing the sum of the squared one-step-ahead forecasts errors. However, as we have seen with the Dow Jones Index data, changes in the underlying time series model will make it difficult for the exponential smoother with fixed discount factor to follow these changes. Hence a need for monitoring and, if necessary, modifying the discount factor arises. By doing so, the discount factor will adapt to the changes in the time series model. For that we will employ the procedure originally described by Trigg and Leach (1967) for single discount factor. As an example we will consider the first-order exponential smoother and modify it as

$$\hat{y}_T = \lambda_T y_T + (1 - \lambda_T)\tilde{y}_{T-1}. \qquad (4.39)$$

Please note that in Eq. (4.39), the discount factor λ_T is given as a function of time and hence it is allowed to adapt to changes in the time series model. We also define the *smoothed error* as

$$Q_T = \delta e_T(1) + (1 - \delta)Q_{T-1}, \qquad (4.40)$$

where δ is a smoothing parameter.

Finally, we define the tracking signal as

$$\frac{Q_T}{\hat{\Delta}_T},\qquad(4.41)$$

where $\hat{\Delta}_T$ is given in Eq. (4.37). This ratio is expected to be close to 0 when the forecasting system performs well and to approach ±1 as it starts to fail. In fact, Trigg and Leach (1967) suggest setting the discount factor to

$$\lambda_T = \left|\frac{Q_T}{\hat{\Delta}_T}\right|\qquad(4.42)$$

Equation (4.42) will allow for automatic updating of the discount factor.

Example 4.6   Consider the Dow Jones Index from June 1999 to June 2006 given in Table 4.1. Figure 4.2 shows that the data do not exhibit a single regime of constant or linear trend behavior. Hence a single exponential smoother with adaptive discount factor as given in Eq. (4.42) can be used. Figure 4.25 shows two simple exponential smoothers for the Dow Jones Index: one with fixed $\lambda = 0.3$ and another one with adaptive updating based on the Trigg–Leach method given in Eq. (4.42).

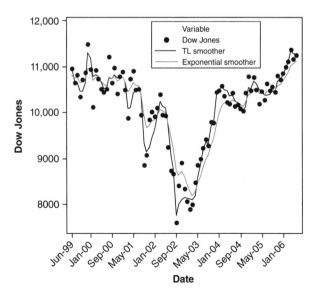

FIGURE 4.25   Time series plot of the Dow Jones Index from June 1999 to June 2006, the simple exponential smoother with $\lambda = 0.3$, and the Trigg–Leach (TL) smoother with $\delta = 0.3$.

TABLE 4.9   The Trigg–Leach Smoother for the Dow Jones Index

Date	Dow Jones	Smoothed	λ	Error	Q_t	D_t
Jun-99	10,970.8	10,970.8	1		0	0
Jul-99	10,655.2	10,655.2	1	−315.6	−94.68	94.68
Aug-99	10,829.3	10,675.835	0.11853	174.1	−14.046	118.506
Sep-99	10,337	10,471.213	0.6039	−338.835	−111.483	184.605
Oct-99	10,729.9	10,471.753	0.00209	258.687	−0.43178	206.83
$\vdots$	$\vdots$	$\vdots$	$\vdots$	$\vdots$	$\vdots$	$\vdots$
May-06	11,168.3	11,283.962	0.36695	−182.705	68.0123	185.346
Jun-06	11,247.9	11,274.523	0.26174	−36.0619	36.79	140.561

This plot shows that a better smoother can be obtained by making automatic updates to the discount factor. The calculations for the Trigg–Leach smoother are given in Table 4.9.

The adaptive smoothing procedure suggested by Trigg and Leach is a useful technique. For other approaches to adaptive adjustment of exponential smoothing parameters, see Chow (1965), Roberts and Reed (1969), and Montgomery (1970).

4.6.5 Model Assessment

If the forecast model performs as expected, the forecast errors should not exhibit any pattern or structure; that is, they should be uncorrelated. Therefore it is always a good idea to verify this. As noted in Chapter 2, we can do so by calculating the sample ACF of the forecast errors from

$$rk = \frac{\sum_{t=k}^{T-1} [e_t(1) - \bar{e}]\left[e_{t-k}(1) - \bar{e}\right]}{\sum_{T=0}^{T-1} [e_t(1) - \bar{e}]^2}, \tag{4.43}$$

where

$$\bar{e} = \frac{1}{n} \sum_{t=1}^{T} e_t(1).$$

If the one-step-ahead forecast errors are indeed uncorrelated, the sample autocorrelations for any lag k should be around 0 with a standard error $1/\sqrt{T}$. Hence a sample autocorrelation for any lag k that lies outside the $\pm 2/\sqrt{T}$ limits will require further investigation of the model.

4.7 EXPONENTIAL SMOOTHING FOR SEASONAL DATA

Some time series data exhibit cyclical or seasonal patterns that cannot be effectively modeled using the polynomial model in Eq. (4.25). Several approaches are available for the analysis of such data. In this chapter we will discuss exponential smoothing techniques that can be used in modeling seasonal time series. The methodology we will focus on was originally introduced by Holt (1957) and Winters (1960) and is generally known as Winters' method, where a seasonal adjustment is made to the linear trend model. Two types of adjustments are suggested—additive and multiplicative.

4.7.1 Additive Seasonal Model

Consider the US clothing sales data given in Figure 4.26. Clearly, for certain months of every year we have high (or low) sales. Hence we can conclude that the data exhibit seasonality. The data also exhibit a linear trend as the sales tend to get higher for the same month as time goes on. As the final observation, we note that the amplitude of the seasonal pattern, that is, the range of the periodic behavior within a year, remains more or

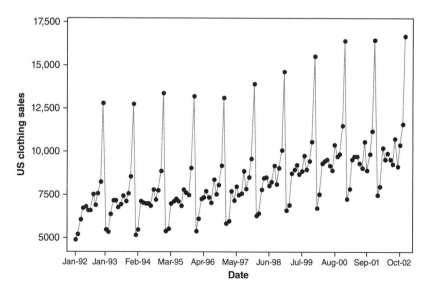

FIGURE 4.26 Time series plot of US clothing sales from January 1992 to December 2003.

less constant in time and remains independent of the average level within a year.

We will for this case assume that the seasonal time series can be represented by the following model:

$$y_t = L_t + S_t + \varepsilon_t, \qquad (4.44)$$

where L_t represents the level or linear trend component and can in turn be represented by $\beta_0 + \beta_1 t$; S_t represents the seasonal adjustment with $S_t = S_{t+s} = S_{t+2s} = \dots$ for $t = 1, \dots, s - 1$, where s is the length of the season (period) of the cycles; and the ε_t are assumed to be uncorrelated with mean 0 and constant variance σ_ε^2. Sometimes the level is called the *permanent component*. One usual restriction on this model is that the seasonal adjustments add to zero during one season,

$$\sum_{t=1}^{s} S_t = 0. \qquad (4.45)$$

In the model given in Eq. (4.44), for forecasting the future observations, we will employ first-order exponential smoothers with different discount factors. The procedure for updating the parameter estimates once the current observation y_T is obtained is as follows.

Step 1. Update the estimate of L_T using

$$\hat{L}_T = \lambda_1(y_T - \hat{S}_{T-s}) + (1 - \lambda_1)\left(\hat{L}_{T-1} + \hat{\beta}_{1,T-1}\right), \qquad (4.46)$$

where $0 < \lambda_1 < 1$. It should be noted that in Eq. (4.46), the first part can be seen as the "current" value for L_T and the second part as the forecast of L_T based on the estimates at $T - 1$.

Step 2. Update the estimate of β_1 using

$$\hat{\beta}_{1,T} = \lambda_2(\hat{L}_T - \hat{L}_{T-1}) + (1 - \lambda_2)\hat{\beta}_{1,T-1}, \qquad (4.47)$$

where $0 < \lambda_2 < 1$. As in Step 1, the estimate of β_1 in Eq. (4.47) can be seen as the linear combination of the "current" value of β_1 and its "forecast" at $T - 1$.

Step 3. Update the estimate of S_t using

$$\hat{S}_T = \lambda_3(y_T - \hat{L}_T) + (1 - \lambda_3)\hat{S}_{T-s}, \qquad (4.48)$$

where $0 < \lambda_3 < 1$.

Step 4. Finally, the τ-step-ahead forecast, $\hat{y}_{T+\tau}(T)$, is

$$\hat{y}_{T+\tau}(T) = \hat{L}_T + \hat{\beta}_{1,T}\tau + \hat{S}_T(\tau - s). \qquad (4.49)$$

As before, estimating the initial values of the exponential smoothers is important. For a given set of historic data with n seasons (hence ns observations), we can use the least squares estimates of the following model:

$$y_t = \beta_0 + \beta_1 t + \sum_{i=1}^{s-1} \gamma_i (I_{t,i} - I_{t,s}) + \varepsilon_t, \qquad (4.50)$$

where

$$I_{t,i} = \begin{cases} 1, & t = i, i+s, i+2s, \dots \\ 0, & \text{otherwise} \end{cases} . \qquad (4.51)$$

The least squares estimates of the parameters in Eq. (4.50) are used to obtain the initial values as

$$\hat{\beta}_{0,0} = \hat{L}_0 = \hat{\beta}_0$$

$$\hat{\beta}_{1,0} = \hat{\beta}_1$$

$$\hat{S}_{j-s} = \hat{Y}_j \quad \text{for } 1 \leq j \leq s - 1$$

$$\hat{S}_0 = -\sum_{j=1}^{s-1} \hat{y}_j$$

These are initial values of the model parameters at the *original* origin of time, $t = 0$. To make forecasts from the correct origin of time the permanent component must be shifted to time T by computing $\hat{L}_T = \hat{L}_0 + ns\hat{\beta}_1$. Alternatively, one could smooth the parameters using equations (4.46)–(4.48) for time periods $t = 1, 2, \dots, T$.

Prediction Intervals As in the nonseasonal smoothing case, the calculations of the prediction intervals would require an estimate for the prediction error variance. The most common approach is to use the relationship between the exponential smoothing techniques and the ARIMA models of Chapter 5 as discussed in Section 4.8, and estimate the prediction error variance accordingly. It can be shown that the seasonal exponential smoothing using the three parameter Holt–Winters method is optimal for an ARIMA $(0, 1, s + 1) \times (0, 1, 0)_s$, process, where s represents the length of the period of the seasonal cycles. For further details, see Yar and Chatfield (1990) and McKenzie (1986).

An alternate approach is to recognize that the additive seasonal model is just a linear regression model and to use the ordinary least squares (OLS) regression procedure for constructing prediction intervals as discussed in Chapter 3. If the errors are correlated, the regression methods for autocorrelated errors could be used instead of OLS.

Example 4.7   Consider the clothing sales data given in Table 4.10. To obtain the smoothed version of this data, we can use the Winters' method option in Minitab. Since the amplitude of the seasonal pattern is constant over time, we decide to use the additive model. Two issues we have encountered in previous exponential smoothers have to be addressed in this case as well—initial values and the choice of smoothing constants. Similar recommendations as in the previous exponential smoothing options can also be made in this case. Of course, the choice of the smoothing constant, in particular, is a bit more concerning since it involves the estimation of three smoothing constants. In this example, we follow our usual recommendation and choose smoothing constants that are all equal to 0.2. For more complicated cases, we recommend seasonal ARIMA models, which we will discuss in Chapter 5.

Figure 4.27 shows the smoothed version of the seasonal clothing sales data. To use this model for forecasting, let us assume that we are currently in December 2002 and we are asked to make forecasts for the following year. Figure 4.28 shows the forecasted sales for 2003 together with the actual data and the 95% prediction limits. Note that the forecast for December 2003 is the 12-step-ahead forecast made in December 2002. Even though the forecast is made further in the future, it still performs well since in the "seasonal" sense it is in fact a one-step-ahead forecast.

4.7.2 Multiplicative Seasonal Model

If the amplitude of the seasonal pattern is proportional to the average level of the seasonal time series, as in the liquor store sales data given in Figure 4.29, the following multiplicative seasonal model will be more appropriate:

$$y_t = L_t S_t + \varepsilon_t, \tag{4.52}$$

where L_t once again represents the permanent component (i.e., $\beta_0 + \beta_1 t$); S_t represents the seasonal adjustment with $S_t = S_{t+s} = S_{t+2s} = \cdots$ for

TABLE 4.10 US Clothing Sales from January 1992 to December 2003

Date	Sales	Date	Sales	Date	Sales	Date	Sales	Date	Sales
Jan-92	4889	Aug-94	7824	Mar-97	7695	Oct-99	9481	May-02	9906
Feb-92	5197	Sep-94	7229	Apr-97	7161	Nov-99	10577	Jun-02	9530
Mar-92	6061	Oct-94	7772	May-97	7978	Dec-99	15552	Jul-02	9298
Apr-92	6720	Nov-94	8873	Jun-97	7506	Jan-00	6726	Aug-02	10,755
May-92	6811	Dec-94	13397	Jul-97	7602	Feb-00	7514	Sep-02	9128
Jun-92	6579	Jan-95	5377	Aug-97	8877	Mar-00	9330	Oct-02	10,408
Jul-92	6598	Feb-95	5516	Sep-97	7859	Apr-00	9472	Nov-02	11,618
Aug-92	7536	Mar-95	6995	Oct-97	8500	May-00	9551	Dec-02	16,721
Sep-92	6923	Apr-95	7131	Nov-97	9594	Jun-00	9203	Jan-03	7891
Oct-92	7566	May-95	7246	Dec-97	13952	Jul-00	8910	Feb-03	7892
Nov-92	8257	Jun-95	7140	Jan-98	6282	Aug-00	10378	Mar-03	9874
Dec-92	12,804	Jul-95	6863	Feb-98	6419	Sep-00	9731	Apr-03	9920
Jan-93	5480	Aug-95	7790	Mar-98	7795	Oct-00	9868	May-03	10,431
Feb-93	5322	Sep-95	7618	Apr-98	8478	Nov-00	11512	Jun-03	9758
Mar-93	6390	Oct-95	7484	May-98	8501	Dec-00	16422	Jul-03	10,003
Apr-93	7155	Nov-95	9055	Jun-98	8044	Jan-01	7263	Aug-03	11,055

TABLE 4.10 *(Continued)*

Date	Sales	Date	Sales	Date	Sales	Date	Sales	Date	Sales
May-93	7175	Dec-95	13,201	Jul-98	8272	Feb-01	7866	Sep-03	9941
Jun-93	6770	Jan-96	5375	Aug-98	9189	Mar-01	9535	Oct-03	10,763
Jul-93	6954	Feb-96	6105	Sep-98	8099	Apr-01	9710	Nov-03	12058
Aug-93	7438	Mar-96	7246	Oct-98	9054	May-01	9711	Dec-03	17535
Sep-93	7144	Apr-96	7335	Nov-98	10,093	Jun-01	9324		
Oct-93	7585	May-96	7712	Dec-98	14668	Jul-01	9063		
Nov-93	8558	Jun-96	7337	Jan-99	6617	Aug-01	10,584		
Dec-93	12,753	Jul-96	7059	Feb-99	6928	Sep-01	8928		
Jan-94	5166	Aug-96	8374	Mar-99	8734	Oct-01	9843		
Feb-94	5464	Sep-96	7554	Apr-99	8973	Nov-01	11,211		
Mar-94	7145	Oct-96	8087	May-99	9237	Dec-01	16,470		
Apr-94	7062	Nov-96	9180	Jun-99	8689	Jan-02	7508		
May-94	6993	Dec-96	13109	Jul-99	8869	Feb-02	8002		
Jun-94	6995	Jan-97	5833	Aug-99	9764	Mar-02	10,203		
Jul-94	6886	Feb-97	5949	Sep-99	8970	Apr-02	9548		

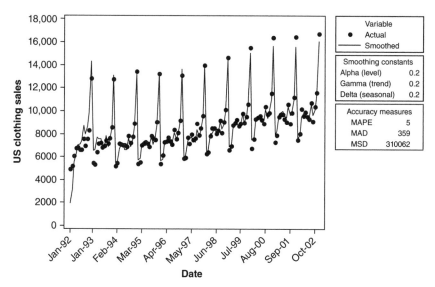

FIGURE 4.27 Smoothed data for the US clothing sales from January 1992 to December 2003 using the additive model.

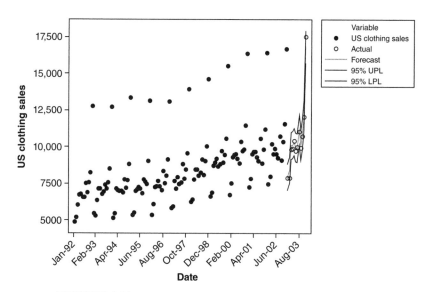

FIGURE 4.28 Forecasts for 2003 for the US clothing sales.

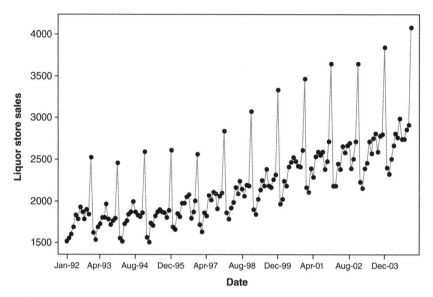

FIGURE 4.29   Time series plot of liquor store sales data from January 1992 to December 2004.

$t = i,\ldots, s - 1$, where s is the length of the period of the cycles; and the ε_t are assumed to be uncorrelated with mean 0 and constant variance σ_ε^2. The restriction for the seasonal adjustments in this case becomes

$$\sum_t^s S_t = s. \tag{4.53}$$

As in the additive model, we will employ three exponential smoothers to estimate the parameters in Eq. (4.52).

Step 1. Update the estimate of L_T using

$$\hat{L}_T = \lambda_1 \frac{y_T}{\hat{S}_{T-s}} + (1 - \lambda_1)(\hat{L}_{T-1} + \hat{\beta}_{1,T-1}), \tag{4.54}$$

where $0 < \lambda_1 < 1$. Similar interpretation as in the additive model can be made for the exponential smoother in Eq. (4.54).

Step 2. Update the estimate of β_1 using

$$\hat{\beta}_{1,T} = \lambda_2(\hat{L}_T - \hat{L}_{T-1}) + (1 - \lambda_2)\hat{\beta}_{1,T-1}, \tag{4.55}$$

where $0 < \lambda_2 < 1$.

Step 3. Update the estimate of S_t using

$$\hat{S}_T = \lambda_3 \frac{y_T}{\hat{L}_T} + (1 - \lambda_3)\hat{S}_{T-s}, \qquad (4.56)$$

where $0 < \lambda_3 < 1$.

Step 4. The τ-step-ahead forecast, $\hat{y}_{T+\tau}(T)$, is

$$\hat{y}_{T+\tau}(T) = (\hat{L}_T + \hat{\beta}_{1,T}\tau)\hat{S}_T(\tau - s). \qquad (4.57)$$

It will almost be necessary to obtain starting values of the model parameters. Suppose that a record consisting of n seasons of data is available. From this set of historical data, the initial values, $\hat{\beta}_{0,0}$, $\hat{\beta}_{1,0}$, and $\hat{S}_0$, can be calculated as

$$\hat{\beta}_{0,0} = \hat{L}_0 = \frac{\bar{y}_n - \bar{y}_1}{(n-1)s},$$

where

$$\bar{y}_i = \frac{1}{s} \sum_{t=(i-1)s+1}^{is} y_t$$

and

$$\hat{\beta}_{1,0} = \bar{y}_1 - \frac{s}{2}\hat{\beta}_{0,0}$$

$$\hat{S}_{j-s} = s\frac{\hat{S}_j^*}{\sum\limits_{i=1}^{s} \hat{S}_i^*} \text{ for } 1 \leq j \leq s,$$

where

$$\hat{S}_j^* = \frac{1}{n}\sum_{t=1}^{n} \frac{y_{(t-1)s+j}}{\bar{y}_t - ((s+1)/2 - j)\hat{\beta}_0}.$$

For further details, please see Montgomery et al. (1990) and Abraham and Ledolter (1983).

Prediction Intervals   Constructing prediction intervals for the multiplicative model is much harder than the additive model as the former is nonlinear. Several authors have considered this problem, including Chatfield and Yar (1991), Sweet (1985), and Gardner (1988). Chatfield and Yar (1991) propose an empirical method in which the length of the prediction interval depends on the point of origin of the forecast and may decrease in length near the low points of the seasonal cycle. They also discuss the case where the error is assumed to be proportional to the seasonal effect rather than constant, which is the standard assumption in Winters' method. Another approach would be to obtain a "linearized" version of Winters' model by expanding it in a first-order Taylor series and use this to find an approximate variance of the predicted value (statisticians call this the delta method). Then this prediction variance could be used to construct prediction intervals much as is done in the linear regression model case.

Example 4.8   Consider the liquor store data given in Table 4.11. In Figure 4.29, we can see that the amplitude of the periodic behavior gets larger as the average level of the seasonal data gets larger due to a linear trend. Hence the multiplicative model will be more appropriate. Figures 4.30 and 4.31 show the smoothed data with additive and multiplicative models, respectively. Based on the performance of the smoothers, it should therefore be clear that the multiplicative model should indeed be preferred.

As for forecasting using the multiplicative model, we can assume as usual that we are currently in December 2003 and are asked to forecast the sales in 2004. Figure 4.32 shows the forecasts together with the actual values and the prediction intervals.

4.8 EXPONENTIAL SMOOTHING OF BIOSURVEILLANCE DATA

Bioterrorism is the use of biological agents in a campaign of aggression. The use of biological agents in warfare is not new; many centuries ago plague and other contagious diseases were employed as weapons. Their use today is potentially catastrophic, so medical and public health officials are designing and implementing biosurveillance systems to monitor populations for potential disease outbreaks. For example, public health officials collect syndrome data from sources such as hospital emergency rooms, outpatient clinics, and over-the-counter medication sales to detect disease outbreaks, such as the onset of the flu season. For an excellent and highly readable introduction to statistical techniques for biosurveillance and syndromic surveillance, see Fricker (2013). Monitoring of syndromic data is also a type of epidemiologic surveillance in a biosurveillance process,

TABLE 4.11 Liquor Store Sales from January 1992 to December 2004

Date	Sales	Date	Sales	Date	Sales	Date	Sales	Date	Sales
Jan-92	1519	Aug-94	1870	Mar-97	1862	Oct-99	2264	May-02	2661
Feb-92	1551	Sep-94	1834	Apr-97	1826	Nov-99	2321	Jun-02	2579
Mar-92	1606	Oct-94	1817	May-97	2071	Dec-99	3336	Jul-02	2667
Apr-92	1686	Nov-94	1857	Jun-97	2012	Jan-00	1963	Aug-02	2698
May-92	1834	Dec-94	2593	Jul-97	2109	Feb-00	2022	Sep-02	2392
Jun-92	1786	Jan-95	1565	Aug-97	2092	Mar-00	2242	Oct-02	2504
Jul-92	1924	Feb-95	1510	Sep-97	1904	Apr-00	2184	Nov-02	2719
Aug-92	1874	Mar-95	1736	Oct-97	2063	May-00	2415	Dec-02	3647
Sep-92	1781	Apr-95	1709	Nov-97	2096	Jun-00	2473	Jan-03	2228
Oct-92	1894	May-95	1818	Dec-97	2842	Jul-00	2524	Feb-03	2153
Nov-92	1843	Jun-95	1873	Jan-98	1863	Aug-00	2483	Mar-03	2395
Dec-92	2527	Jul-95	1898	Feb-98	1786	Sep-00	2419	Apr-03	2460
Jan-93	1623	Aug-95	1872	Mar-98	1913	Oct-00	2413	May-03	2718
Feb-93	1539	Sep-95	1856	Apr-98	1985	Nov-00	2615	Jun-03	2570
Mar-93	1688	Oct-95	1800	May-98	2164	Dec-00	3464	Jul-03	2758
Apr-93	1725	Nov-95	1892	Jun-98	2084	Jan-01	2165	Aug-03	2809

(*continued*)

TABLE 4.11 (Continued)

Date	Sales	Date	Sales	Date	Sales	Date	Sales	Date	Sales
May-93	1807	Dec-95	2616	Jul-98	2237	Feb-01	2107	Sep-03	2597
Jun-93	1804	Jan-96	1690	Aug-98	2146	Mar-01	2390	Oct-03	2785
Jul-93	1962	Feb-96	1662	Sep-98	2058	Apr-01	2292	Nov-03	2803
Aug-93	1788	Mar-96	1849	Oct-98	2193	May-01	2538	Dec-03	3849
Sep-93	1717	Apr-96	1810	Nov-98	2186	Jun-01	2596	Jan-04	2406
Oct-93	1769	May-96	1970	Dec-98	3082	Jul-01	2553	Feb-04	2324
Nov-93	1794	Jun-96	1971	Jan-99	1897	Aug-01	2590	Mar-04	2509
Dec-93	2459	Jul-96	2047	Feb-99	1838	Sep-01	2384	Apr-04	2670
Jan-94	1557	Aug-96	2075	Mar-99	2021	Oct-01	2481	May-04	2809
Feb-94	1514	Sep-96	1791	Apr-99	2136	Nov-01	2717	Jun-04	2764
Mar-94	1724	Oct-96	1870	May-99	2250	Dec-01	3648	Jul-04	2995
Apr-94	1769	Nov-96	2003	Jun-99	2186	Jan-02	2182	Aug-04	2745
May-94	1842	Dec-96	2562	Jul-99	2383	Feb-02	2180	Sep-04	2742
Jun-94	1869	Jan-97	1716	Aug-99	2182	Mar-02	2447	Oct-04	2863
Jul-94	1994	Feb-97	1629	Sep-99	2169	Apr-02	2380	Nov-04	2912
								Dec-04	4085

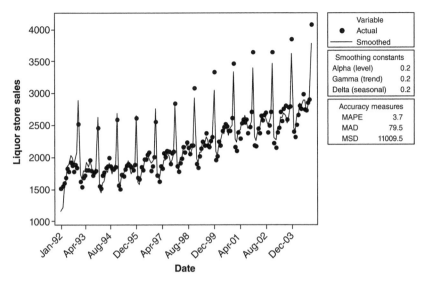

FIGURE 4.30 Smoothed data for the liquor store sales from January 1992 to December 2004 using the additive model.

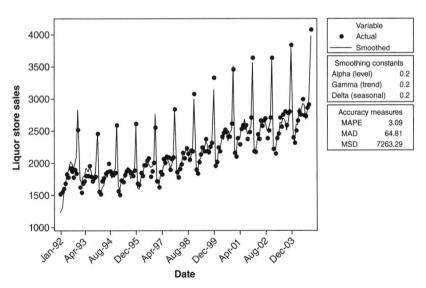

FIGURE 4.31 Smoothed data for the liquor store sales from January 1992 to December 2004 using the multiplicative model.

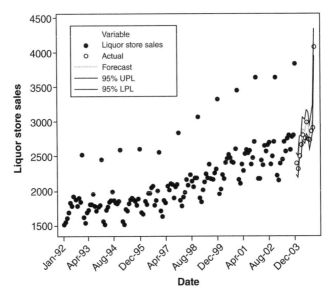

FIGURE 4.32 Forecasts for the liquor store sales for 2004 using the multiplicative model.

where significantly higher than anticipated counts of influenza-like illness might signal a potential bioterrorism attack.

As an example of such syndromic data, Fricker (2013) describes daily counts of respiratory and gastrointestinal complaints for more than $2\frac{1}{2}$ years at several hospitals in a large metropolitan area. Table 4.12 presents the respiratory count data from one of these hospitals. There are 980 observations. Fifty observations were missing from the original data set. The missing values were replaced with the last value that was observed on the same day of the week. This type of data imputation is a variation of "Hot Deck Imputation" discussed in Section 1.4.3 and in Fricker (2013). It is also sometimes called last observation (or Value) carried forward (LOCF). For additional discussion see the web site: http://missingdata.lshtm.ac.uk/.

Figure 4.33 is a time series plot of the respiratory syndrome count data in Table 4.12. This plot was constructed using the Graph Builder feature in JMP. This software package overlays a smoothed curve on the data. The curve is fitted using **locally weighted regression**, often called **loess**. This is a variation of kernel regression that uses a weighted average of the data in a local neighborhood around a specific location to determine the value to plot at that location. Loess usually uses either first-order linear regression or a quadratic regression model for the weighted least squares fit. For more information on kernel regression and loess see Montgomery, et al. (2012).

TABLE 4.12 Counts of Respiratory Complaints at a Metropolitan Hospital

Day	Count	Day	Count	Day	Count	Day	Count	Day	Count	Day	Count	Day	Count	Day	Count	Day	Count	Day	Count
1	17	101	30	201	31	301	12	401	28	501	35	601	26	701	19	801	41	901	29
2	29	102	21	202	23	302	16	402	26	502	27	602	31	702	12	802	50	902	26
3	31	103	32	203	13	303	24	403	28	503	33	603	23	703	17	803	42	903	36
4	34	104	32	204	18	304	21	404	29	504	30	604	24	704	22	804	56	904	31
5	18	105	43	205	36	305	14	405	33	505	30	605	27	705	20	805	36	905	25
6	43	106	25	206	23	306	15	406	36	506	29	606	24	706	22	306	51	906	31
7	34	107	32	207	22	307	23	407	62	507	30	607	31	707	21	807	40	907	32
8	23	108	31	208	23	308	10	403	31	508	22	608	29	708	24	808	29	908	30
9	23	109	33	209	26	309	16	409	30	509	40	609	36	709	16	809	61	909	31
10	39	110	40	210	22	310	11	410	31	510	40	610	31	710	14	810	42	910	29
11	25	111	37	211	21	311	16	411	27	511	41	611	30	711	14	811	56	911	30
12	15	112	34	212	25	312	16	412	35	512	34	612	27	712	30	812	60	912	35
13	29	113	29	213	20	313	12	413	45	513	30	613	27	713	24	813	38	913	24
14	20	114	50	214	18	314	23	414	37	514	33	614	25	714	25	814	52	914	27
15	21	115	27	215	26	315	10	415	23	515	17	615	34	715	17	815	32	915	22
16	22	116	28	216	32	315	15	416	31	516	32	616	33	716	27	816	43	916	33
17	24	117	23	217	41	317	11	417	33	517	40	617	36	717	25	817	54	917	29
18	19	118	27	218	30	318	17	418	27	518	30	618	26	718	14	818	36	913	37
19	28	119	27	219	34	319	13	419	28	519	27	619	20	719	25	819	51	919	29
20	29	120	41	220	38	320	14	420	46	520	30	620	27	720	25	820	57	920	32
21	26	121	29	221	22	321	20	421	39	521	38	621	25	721	26	821	48	921	27
22	22	122	26	222	35	322	10	422	53	522	22	622	36	722	20	822	70	922	22
23	21	123	28	223	36	323	15	423	33	523	27	623	30	723	21	823	48	923	33

(continued)

TABLE 4.12 (Continued)

Day	Count	Day	Count	Day	Count	Day	Count	Day	Count	Day	Count	Day	Count	Day	Count	Day	Count	Day	Count
24	29	124	30	224	37	324	14	424	32	524	19	624	39	724	29	824	54	924	29
25	25	125	49	225	27	325	6	425	45	525	19	625	26	725	16	825	36	925	37
26	20	126	43	226	23	326	17	426	21	526	33	626	20	726	24	826	51	926	29
27	20	127	27	227	31	327	17	427	47	527	45	627	27	727	42	827	52	927	20
28	29	128	32	228	39	328	17	423	23	528	34	628	36	728	44	828	48	928	13
29	29	129	13	229	39	329	23	429	39	529	27	629	43	729	34	829	70	929	27
30	32	130	26	230	31	330	9	430	32	530	31	630	46	730	33	830	48	930	23
31	16	131	34	231	43	331	21	431	27	531	19	631	33	731	26	831	57	931	17
32	25	132	27	232	35	332	13	432	29	532	22	632	26	732	29	832	38	932	26
33	20	133	33	233	41	333	13	433	37	533	23	633	33	733	33	833	44	933	23
34	22	134	42	234	24	334	14	434	32	534	13	634	24	734	34	834	34	934	27
35	27	135	29	235	39	335	25	435	28	535	29	635	23	735	42	835	50	935	28
36	32	136	29	236	44	336	15	436	42	536	13	636	51	736	43	836	39	936	21
37	23	137	29	237	35	337	18	437	33	537	20	637	35	737	33	837	65	937	20
38	31	138	28	238	30	338	21	438	36	538	20	638	26	738	31	838	55	938	25
39	22	139	35	239	29	339	18	439	25	539	23	639	32	739	30	839	46	939	30
40	21	140	33	240	13	340	12	440	19	540	17	640	29	740	35	840	57	940	13
41	27	141	35	241	23	341	10	441	34	541	31	641	24	741	34	841	43	941	19
42	37	142	23	242	19	342	10	442	34	542	21	642	18	742	43	842	50	942	20
43	28	143	28	243	24	343	17	443	33	543	29	643	36	743	21	843	39	943	27
44	41	144	23	244	19	344	12	444	26	544	20	644	15	744	42	844	55	944	14
45	45	145	31	245	27	345	24	445	43	545	21	645	33	745	30	845	38	945	21
46	40	146	29	246	20	345	22	446	31	546	25	646	21	746	29	846	29	946	32
47	32	147	24	247	19	347	14	447	30	547	35	647	25	747	29	347	32	947	18
48	45	148	22	248	28	348	14	448	41	548	24	648	25	748	41	848	27	948	25

49	48	149	30	249	19	349	9	449	15	549	25	649	19	749	35	849	22	949	13
50	51	150	21	250	29	350	19	450	23	550	23	650	23	750	29	850	23	950	25
51	51	151	24	251	24	351	15	451	25	551	27	651	18	751	37	851	25	951	19
52	21	152	21	252	33	352	9	452	27	552	35	652	26	752	31	852	19	952	27
53	43	153	30	253	20	353	18	453	40	553	36	653	27	753	24	853	29	953	27
54	42	154	25	254	29	354	17	454	40	554	33	654	11	754	47	854	34	954	18
55	56	155	17	255	17	355	15	455	34	555	27	655	20	755	3	855	27	955	25
55	51	155	22	255	19	355	21	455	42	555	33	656	13	755	34	355	30	956	26
57	51	157	18	257	23	357	22	457	12	557	25	657	20	757	35	857	30	957	39
58	60	158	19	258	26	358	17	458	24	558	32	658	23	758	39	858	24	958	59
59	35	159	20	259	25	359	21	459	20	559	23	659	19	759	29	859	33	959	34
60	43	160	22	260	32	360	26	460	26	560	42	660	21	760	41	860	29	960	34
61	42	161	39	261	21	361	23	461	46	561	25	661	21	761	36	861	36	961	24
62	55	162	35	262	15	362	20	462	35	562	33	662	29	762	50	862	29	962	25
63	46	163	29	263	20	363	28	463	46	563	19	663	18	763	33	863	27	963	40
64	49	164	24	264	19	364	34	464	33	564	40	664	25	764	38	864	32	964	19
65	40	165	22	265	13	365	23	465	27	565	35	665	24	765	40	865	30	965	35
66	33	166	26	266	25	366	20	466	35	566	36	666	19	766	41	866	23	966	34
67	45	167	27	267	19	367	37	467	33	567	33	667	15	767	34	867	25	967	33
68	37	168	28	268	9	368	22	468	29	568	25	668	23	768	42	868	23	968	29
69	44	169	36	269	20	369	32	469	45	569	33	669	14	769	40	869	29	969	23
70	50	170	31	270	20	370	41	470	18	570	33	670	16	770	50	870	26	970	29
71	37	171	31	271	21	371	35	471	21	571	27	671	16	771	30	871	29	971	25
72	36	172	34	272	21	372	41	472	35	572	33	672	22	772	34	872	22	972	19
73	43	173	19	273	20	373	43	473	39	573	33	673	13	773	28	873	16	973	34
74	49	174	37	274	13	374	33	474	40	574	39	674	19	774	21	874	25	974	37
75	4C	175	39	275	25	375	32	475	33	575	30	675	23	775	24	875	26	975	34

(continued)

TABLE 4.12 (*Continued*)

Day	Count	Day	Count	Day	Count	Day	Count	Day	Count	Day	Count	Day	Count	Day	Count	Day	Count	Day	Count
76	65	176	32	276	29	376	28	476	35	576	33	676	14	776	37	876	25	976	29
77	49	177	36	277	16	377	42	477	20	577	26	677	16	777	44	877	24	977	27
78	49	178	42	278	18	378	27	478	36	578	26	678	10	778	39	878	29	978	18
79	34	179	31	279	32	379	25	479	34	579	23	679	14	779	37	879	34	979	26
80	33	180	28	280	32	380	32	480	35	580	24	680	13	780	35	880	35	980	28
81	29	181	35	281	19	381	27	481	36	581	32	681	15	781	32	881	29		
82	32	182	36	282	24	382	35	482	29	582	24	682	15	782	41	882	39		
83	57	183	35	283	18	383	26	483	19	583	32	683	11	783	41	883	31		
84	43	184	32	284	20	384	32	484	36	584	41	634	11	784	51	884	26		
85	40	185	26	285	20	385	42	485	35	585	26	685	18	785	43	885	24		
86	46	186	29	286	20	386	38	486	31	586	28	686	16	786	35	886	31		
87	33	187	25	287	24	387	36	487	23	587	25	687	18	787	33	887	24		
88	30	188	23	288	15	388	26	488	31	588	29	688	15	788	33	888	29		
89	41	189	29	289	22	389	26	489	29	589	40	689	16	789	31	889	26		
90	38	190	29	290	16	390	24	490	44	590	34	690	11	790	43	890	45		
91	29	191	26	291	14	391	30	491	42	591	41	691	11	791	45	891	36		
92	41	192	18	292	17	392	32	492	31	592	37	692	23	792	43	892	29		
93	28	193	19	293	15	393	14	493	31	593	36	693	20	793	42	893	22		
94	47	194	17	294	8	394	27	494	24	594	26	694	18	794	36	894	31		
95	42	195	22	295	23	395	26	495	30	595	42	695	24	795	34	895	38		
96	34	196	25	296	17	396	25	496	26	596	40	696	14	796	30	896	36		
97	40	197	33	297	13	397	23	497	26	597	34	697	22	797	46	897	33		
98	35	198	10	298	15	398	27	498	39	598	41	698	16	798	54	898	34		
99	40	199	25	299	15	399	36	499	35	599	37	699	26	799	52	899	34		
100	24	200	25	300	13	400	40	500	34	600	36	700	17	800	39	900	25		

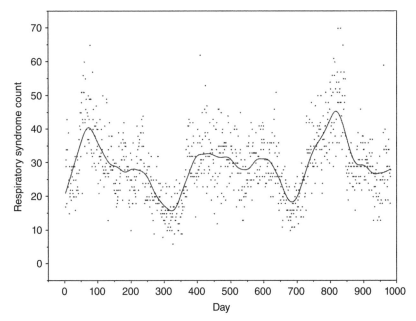

FIGURE 4.33   Time series plot of daily respiratory syndrome count, with kernel-smoothed fitted line. ($\alpha = 0.1$).

Over the 2 $\frac{1}{2}$ year period, the daily counts of the respiratory syndrome appear to follow a weak seasonal pattern, with the highest peak in November–December (late fall), a secondary peak in March–April, and then decreasing to the lowest counts in June–August (summer). The amplitude, or range within a year, seems to vary, but counts do not appear to be increasing or decreasing over time.

Not immediately evident from the time series plots is a potential day effect. The box plots of the residuals from the loess smoothed line in Figure 4.33 are plotted in Figure 4.34 versus day of the week. These plots exhibit variation that indicates slightly higher-than-expected counts on Monday and slightly lower-than-expected counts on Thursday, Friday, and Saturday.

The exponential smoothing procedure in JMP was applied to the respiratory syndrome data. The results of first-order or simple exponential smoothing are summarized in Table 4.13 and Figure 4.35, which plots only the last 100 observations along with the smoothed values. JMP reported the value of the smoothing constant that produced the minimum value of the error sum of squares as $\lambda = 0.21$. This value also minimizes the AIC and BIC criteria, and results in the smallest values of the mean absolute

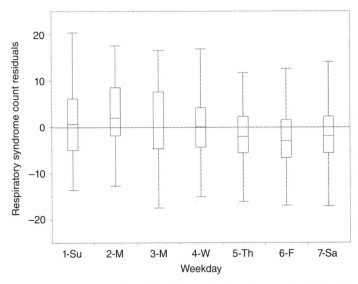

FIGURE 4.34   Box plots of residuals from the kernel-smoothed line fit to daily respiratory syndrome count.

prediction error and the mean absolute, although there is very little difference between the optimal value of $\lambda = 0.21$ and the values $\lambda = 0.1$ and $\lambda = 0.4$.

The results of using second-order exponential smoothing are summarized in Table 4.14 and illustrated graphically for the last 100 observations in Figure 4.36. There is not a lot of difference between the two procedures, although the optimal first-order smoother does perform slightly better and the larger smoothing parameters in the double smoother perform more poorly.

Single and double exponential smoothing do not account for the apparent mild seasonality observed in the original time series plot of the data.

TABLE 4.13   First-Order Simple Exponential Smoothing Applied to the Respiratory Data

Model	Variance	AIC	BIC	MAPE	MAE
First-Order Exponential (min SSE, $\lambda = 0.21$)	52.66	6660.81	6665.70	21.43	5.67
First-Order Exponential ($\lambda = 0.1$)	55.65	6714.67	6714.67	22.23	5.85
First-Order Exponential ($\lambda = 0.4$)	55.21	6705.63	6705.63	21.87	5.82

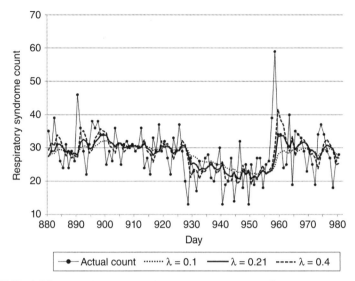

FIGURE 4.35 Respiratory syndrome counts using first-order exponential smoothing with $\lambda = 0.1$, $\lambda = 0.21$ (min SSE), and $\lambda = 0.4$.

We used JMP to fit Winters' additive seasonal model to the respiratory syndrome count data. Because the seasonal patterns are not strong, we investigated seasons of length 3, 7, and 12 periods. The results are summarized in Table 4.15 and illustrated graphically for the last 100 observations in Figure 4.37. The 7-period season works best, probably reflecting the daily seasonal pattern that we observed in Figure 4.34. This is also the best smoother of all the techniques that were investigated. The values of $\lambda = 0$ for the trend and seasonal components in this model are an indication that there is not a significant linear trend in the data and that the seasonal pattern is relatively stable over the period of available data.

TABLE 4.14 **Second-Order Simple Exponential Smoothing Applied to the Respiratory Data**

Model	Variance	AIC	BIC	MAPE	MAE
Second-Order Exponential (min SSE, $\lambda = 0.10$)	54.37	6690.98	6695.86	21.71	5.78
Second-Order Exponential ($\lambda = 0.2$)	58.22	6754.37	6754.37	22.44	5.98
Second-Order Exponential ($\lambda = 0.4$)	74.46	6992.64	6992.64	25.10	6.74

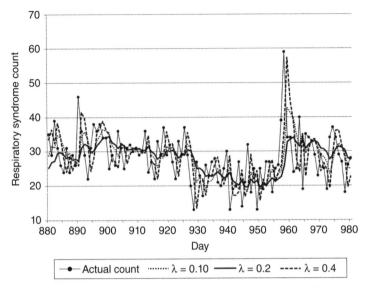

FIGURE 4.36 Respiratory syndrome counts using second-order exponential smoothing with $\lambda = 0.10$ (min SSE), $\lambda = 0.2$, and $\lambda = 0.4$.

TABLE 4.15   Winters' Additive Seasonal Exponential Smoothing Applied to the Respiratory Data

Model	Variance	AIC	BIC	MAPE	MAE
$S = 3$					
Winters Additive (min SSE, $\lambda1 = 0.21, \lambda2 = 0, \lambda3 = 0$)	52.75	6662.75	6677.40	21.70	5.72
Winters Additive ($\lambda1 = 0.2, \lambda2 = 0.1, \lambda3 = 0.1$)	57.56	6731.59	6731.59	22.38	5.94
$S = 7$					
Winters Additive (min SSE, $\lambda1 = 0.22, \lambda2 = 0, \lambda3 = 0$)	49.77	6593.83	6608.47	21.10	5.56
Winters Additive ($\lambda1 = 0.2, \lambda2 = 0.1, \lambda3 = 0.1$)	54.27	6652.57	6652.57	21.47	5.70
$S = 12$					
Winters Additive (min SSE, $\lambda1 = 0.21, \lambda2 = 0, \lambda3 = 0$)	52.74	6635.58	6650.21	22.13	5.84
Winters Additive ($\lambda1 = 0.2, \lambda2 = 0.1, \lambda3 = 0.1$)	58.76	6703.79	6703.79	22.77	6.08

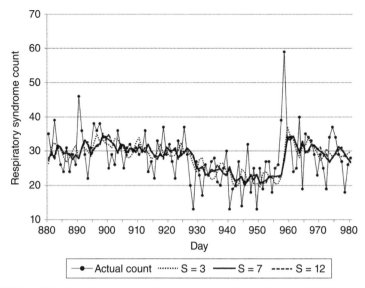

FIGURE 4.37   Respiratory syndrome counts using winters' additive seasonal exponential smoothing with $S = 3$, $S = 7$, and $S = 12$, and smoothing parameters that minimize SSE.

4.9 EXPONENTIAL SMOOTHERS AND ARIMA MODELS

The first-order exponential smoother presented in Section 4.2 is a very effective model in forecasting. The discount factor, λ, makes this smoother fairly flexible in handling time series data with various characteristics. The first-order exponential smoother is particularly good in forecasting time series data with certain specific characteristics.

Recall that the first-order exponential smoother is given as

$$\tilde{y}_T = \lambda y_T + (1 - \lambda)\tilde{y}_{T-1} \qquad (4.58)$$

and the forecast error is defined as

$$e_T = y_T - \hat{y}_{T-1}. \qquad (4.59)$$

Similarly, we have

$$e_{T-1} = y_{T-1} - \hat{y}_{T-2}. \qquad (4.60)$$

By multiplying Eq. (4.60) by $(1 - \lambda)$ and subtracting it from Eq. (4.59), we obtain

$$
\begin{aligned}
e_T - (1 - \lambda)e_{T-1} &= (y_T - \hat{y}_{T-1}) - (1 - \lambda)(y_{T-1} - \hat{y}_{T-2}) \\
&= y_T - y_{T-1} - \hat{y}_{T-1} + \underbrace{\lambda y_{T-1} + (1 - \lambda)\hat{y}_{T-2}}_{=\hat{y}_{T-1}} \\
&= y_T - y_{T-1} - \hat{y}_{T-1} + \hat{y}_{T-1} \\
&= y_T - y_{T-1}.
\end{aligned}
\tag{4.61}
$$

We can rewrite Eq. (4.61) as

$$
y_T - y_{T-1} = e_T - \theta e_{T-1},
\tag{4.62}
$$

where $\theta = 1 - \lambda$. Recall from Chapter 2 the **backshift** *operator,* B, defined as $B(y_t) = y_{t-1}$. Thus Eq. (4.62) becomes

$$
(1 - B)y_T = (1 - \theta B)e_T.
\tag{4.63}
$$

We will see in Chapter 5 that the model in Eq. (4.63) is called the **integrated moving average** model denoted as IMA(1,1), for the backshift operator is used only once on y_T and only once on the error. It can be shown that if the process exhibits the dynamics defined in Eq. (4.63), that is an IMA(1,1) process, the first-order exponential smoother provides minimum mean squared error (MMSE) forecasts (see Muth (1960), Box and Luceno (1997), and Box, Jenkins, and Reinsel (1994)). For more discussion of the equivalence between exponential smoothing techniques and the ARIMA models, see Abraham and Ledolter (1983), Cogger (1974), Goodman (1974), Pandit and Wu (1974), and McKenzie (1984).

4.10   R COMMANDS FOR CHAPTER 4

Example 4.1   The Dow Jones index data are in the second column of the array called dji.data in which the first column is the month of the year. We can use the following simple function to obtain the first-order exponential smoothing

```
firstsmooth<-function(y,lambda,start=y[1]){
      ytilde<-y
      ytilde[1]<-lambda*y[1]+(1-lambda)*start
      for (i in 2:length(y)){
            ytilde[i]<-lambda*y[i]+(1-lambda)*ytilde[i-1]
      }
ytilde
}
```

Note that this function uses the first observation as the starting value by default. One can change this by providing a specific start value when calling the function.

We can then obtain the smoothed version of the data for a specified lambda value and plot the fitted value as the following:

```
dji.smooth1<-firstsmooth(y=dji.data[,2],lambda=0.4)
plot(dji.data[,2],type="p", pch=16,cex=.5,xlab='Date',ylab='Dow
    Jones',xaxt='n')
axis(1, seq(1,85,12), dji.data[seq(1,85,12),1])
lines(dji.smooth1)
```

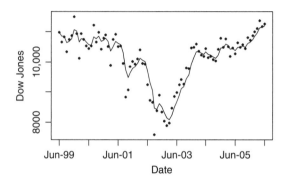

For the first-order exponential smoothing, measures of accuracy such as MAPE, MAD, and MSD can be obtained from the following function:

```
measacc.fs<- function(y,lambda){
            out<- firstsmooth(y,lambda)
            T<-length(y)
            #Smoothed version of the original is the one step
              ahead prediction
            #Hence the predictions (forecasts) are given as
            pred<-c(y[1],out[1:(T-1)])
            prederr<- y-pred
            SSE<-sum(prederr^2)
            MAPE<-100*sum(abs(prederr/y))/T
            MAD<-sum(abs(prederr))/T
            MSD<-sum(prederr^2)/T
            ret1<-c(SSE,MAPE,MAD,MSD)
            names(ret1)<-c("SSE","MAPE","MAD","MSD")
            return(ret1)
}

measacc.fs(dji.data[,2],0.4)
         SSE            MAPE           MAD            MSD
1.665968e+07 3.461342e+00 3.356325e+02 1.959962e+05
```

Note that alternatively we could use the Holt–Winters function from the stats package. The function requires three parameters (alpha, beta, and gamma) to be defined. Providing a specific value for alpha and setting beta and gamma to "FALSE" give the first-order exponential as the following

```
dji1.fit<-HoltWinters(dji.data[,2],alpha=.4, beta=FALSE, gamma=FALSE)
```

Beta corresponds to the second-order smoothing (or the trend term) and gamma is for the seasonal effect.

Example 4.2 The US CPI data are in the second column of the array called cpi.data in which the first column is the month of the year. For this case we use the firstsmooth function twice to obtain the double exponential smoothing as

```
cpi.smooth1<-firstsmooth(y=cpi.data[,2],lambda=0.3)
cpi.smooth2<-firstsmooth(y=cpi.smooth1,lambda=0.3)
cpi.hat<-2*cpi.smooth1-cpi.smooth2 #Equation 4.23
plot(cpi.data[,2],type="p", pch=16,cex=.5,xlab='Date',ylab='CPI',
  xaxt='n')
axis(1, seq(1,120,24), cpi.data[seq(1,120,24),1])
lines(cpi.hat)
```

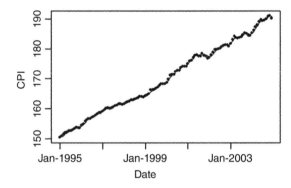

Note that the fitted values are obtained using Eq. (4.23). Also the corresponding command using Holt–Winters function is

```
HoltWinters(cpi.data[,2],alpha=0.3, beta=0.3, gamma=FALSE)
```

Example 4.3   In this example we use the firstsmooth function twice for the Dow Jones Index data to obtain the double exponential smoothing as in the previous example.

```
dji.smooth1<-firstsmooth(y=dji.data[,2],lambda=0.3)
dji.smooth2<-firstsmooth(y=dji.smooth1,lambda=0.3)
dji.hat<-2*dji.smooth1-dji.smooth2 #Equation 4.23
plot(dji.data[,2],type="p", pch=16,cex=.5,xlab='Date',ylab='Dow
  Jones',xaxt='n')
axis(1, seq(1,85,12), cpi.data[seq(1,85,12),1])
lines(dji.hat)
```

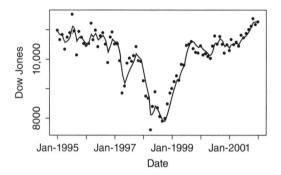

Example 4.4   The average speed data are in the second column of the array called speed.data in which the first column is the index for the week. To find the "best" smoothing constant, we will use the firstsmooth function for various lambda values and obtain the sum of squared one-step-ahead prediction error (SS_E) for each. The lambda value that minimizes the sum of squared prediction errors is deemed the "best" lambda. The obvious option is to apply firstsmooth function in a for loop to obtain SS_E for various lambda values. Even though in this case this may not be an issue, in many cases for loops can slow down the computations in R and are to be avoided if possible. We will do that using sapply function.

```
lambda.vec<-seq(0.1, 0.9, 0.1)
sse.speed<-function(sc){measacc.fs(speed.data[1:78,2],sc)[1]}
sse.vec<-sapply(lambda.vec, sse.speed)
opt.lambda<-lambda.vec[sse.vec == min(sse.vec)]
plot(lambda.vec, sse.vec, type="b", main = "SSE vs. lambda\n",
  xlab='lambda\n',ylab='SSE')
abline(v=opt.lambda, col = 'red')
mtext(text = paste("SSE min = ", round(min(sse.vec),2), "\n lambda
  = ", opt.lambda))
```

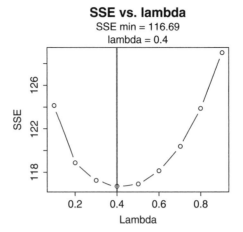

Note that we can also use Holt–Winters function to find the "best" value for the smoothing constant by not specifying the appropriate parameter as the following:

```
HoltWinters(speed.data[,2], beta=FALSE, gamma=FALSE)
```

Example 4.5   We will first try to find the best lambda for the CPI data using first-order exponential smoothing. We will also plot ACF of the data.

Note that we will use the data up to December 2003.

```
lambda.vec<-c(seq(0.1, 0.9, 0.1), .95, .99)
sse.cpi<-function(sc){measacc.fs(cpi.data[1:108,2],sc)[1]}
sse.vec<-sapply(lambda.vec, sse.cpi)
opt.lambda<-lambda.vec[sse.vec == min(sse.vec)]
plot(lambda.vec, sse.vec, type="b", main = "SSE vs. lambda\n",
  xlab='lambda\n',ylab='SSE', pch=16,cex=.5)
acf(cpi.data[1:108,2],lag.max=25)
```

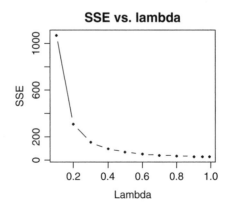

Series cpi.data[1:108, 2]

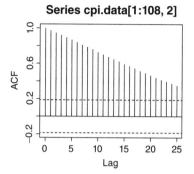

We now use the second-order exponential smoothing with lambda of 0.3. We calculate the forecasts using Eq. (4.31) for the two options suggested in the Example 4.5.

Option 1: On December 2003, make the forecasts for the entire 2004 (1- to 12-step-ahead forecasts).

```
lcpi<-0.3
cpi.smooth1<-firstsmooth(y=cpi.data[1:108,2],lambda=lcpi)
cpi.smooth2<-firstsmooth(y=cpi.smooth1,lambda=lcpi)
cpi.hat<-2*cpi.smooth1-cpi.smooth2
tau<-1:12
T<-length(cpi.smooth1)
cpi.forecast<-(2+tau*(lcpi/(1-lcpi)))*cpi.smooth1[T]-(1+tau*(lcpi/
   (1-lcpi)))*cpi.smooth2[T]
ctau<-sqrt(1+(lcpi/((2-lcpi)^3))*(10-14*lcpi+5*(lcpi^2)+2*tau*lcpi
   *(4-3*lcpi)+2*(tau^2)*(lcpi^2)))
alpha.lev<-.05
sig.est<- sqrt(var(cpi.data[2:108,2]- cpi.hat[1:107]))
cl<-qnorm(1-alpha.lev/2)*(ctau/ctau[1])*sig.est
plot(cpi.data[1:108,2],type="p", pch=16,cex=.5,xlab='Date',
   ylab='CPI',xaxt='n',xlim=c(1,120),ylim=c(150,192))
axis(1, seq(1,120,24), cpi.data[seq(1,120,24),1])
points(109:120,cpi.data[109:120,2])
lines(109:120,cpi.forecast)
lines(109:120,cpi.forecast+cl)
lines(109:120,cpi.forecast-cl)
```

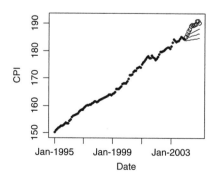

Option 2: On December 2003, make the forecast for January 2004. Then when January 2004 data are available, make the forecast for February 2004 (only one-step-ahead forecasts).

```
lcpi<-0.3
T<-108
tau<-12
alpha.lev<-.05
cpi.forecast<-rep(0,tau)
cl<-rep(0,tau)
cpi.smooth1<-rep(0,T+tau)
cpi.smooth2<-rep(0,T+tau)

for (i in 1:tau) {
    cpi.smooth1[1:(T+i-1)]<-firstsmooth(y=cpi.data[1:(T+i-1),2],
      lambda=lcpi)
    cpi.smooth2[1:(T+i-1)]<-firstsmooth(y=cpi.smooth1[1:(T+i-1)],
      lambda=lcpi)
    cpi.forecast[i]<-(2+(lcpi/(1-lcpi)))*cpi.smooth1[T+i-1]-
      (1+(lcpi/(1-lcpi)))*cpi.smooth2[T+i-1]
    cpi.hat<-2*cpi.smooth1[1:(T+i-1)]-cpi.smooth2[1:(T+i-1)]
    sig.est<- sqrt(var(cpi.data[2:(T+i-1),2]- cpi.hat[1:(T+i-2)]))
    cl[i]<-qnorm(1-alpha.lev/2)*sig.est
    }

plot(cpi.data[1:T,2],type="p", pch=16,cex=.5,xlab='Date',ylab='CPI',
  xaxt='n',xlim=c(1,T+tau),ylim=c(150,192))
axis(1, seq(1,T+tau,24), cpi.data[seq(1,T+tau,24),1])
points((T+1):(T+tau),cpi.data[(T+1):(T+tau),2],cex=.5)
lines((T+1):(T+tau),cpi.forecast)
lines((T+1):(T+tau),cpi.forecast+cl)
lines((T+1):(T+tau),cpi.forecast-cl)
```

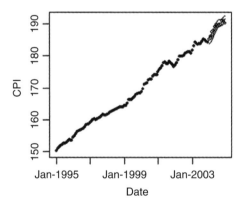

Example 4.6 The function for the Trigg–Leach smoother is given as:

```
tlsmooth<-function(y,gamma,y.tilde.start=y[1],lambda.start=1){
      T<-length(y)

#Initialize the vectors
      Qt<-vector()
      Dt<-vector()
      y.tilde<-vector()
      lambda<-vector()
      err<-vector()
#Set the starting values for the vectors
      lambda[1]=lambda.start
      y.tilde[1]=y.tilde.start
      Qt[1]<-0
      Dt[1]<-0
      err[1]<-0

      for (i in 2:T){
            err[i]<-y[i]-y.tilde[i-1]
            Qt[i]<-gamma*err[i]+(1-gamma)*Qt[i-1]
            Dt[i]<-gamma*abs(err[i])+(1-gamma)*Dt[i-1]
            lambda[i]<-abs(Qt[i]/Dt[i])
            y.tilde[i]=lambda[i]*y[i] + (1-lambda[i])*y.tilde[i-1]
      }
return(cbind(y.tilde,lambda,err,Qt,Dt))
}


#Obtain the TL smoother for Dow Jones Index
out.tl.dji<-tlsmooth(dji.data[,2],0.3)
```

```
#Obtain the exponential smoother for Dow Jones Index
dji.smooth1<-firstsmooth(y=dji.data[,2],lambda=0.4)

#Plot the data together with TL and exponential smoother for
  comparison
plot(dji.data[,2],type="p", pch=16,cex=.5,xlab='Date',ylab='Dow
  Jones',xaxt='n')
axis(1, seq(1,85,12), cpi.data[seq(1,85,12),1])
lines(out.tl.dji[,1])
lines(dji.smooth1,col="grey40")
legend(60,8000,c("Dow Jones","TL Smoother","Exponential Smoother"),
  pch=c(16, NA, NA),lwd=c(NA,.5,.5),cex=.55,col=c("black",
  "black","grey40"))
```

Example 4.7   The clothing sales data are in the second column of the array called closales.data in which the first column is the month of the year. We will use the data up to December 2002 to fit the model and make forecasts for the coming year (2003). We will use Holt–Winters function given in stats package. The model is additive seasonal model with all parameters equal to 0.2.

```
dat.ts = ts(closales.data[,2], start = c(1992,1), freq = 12)
y1<-closales.data[1:132,]
# convert data to ts object
y1.ts<-ts(y1[,2], start = c(1992,1), freq = 12)
clo.hw1<-HoltWinters(y1.ts,alpha=0.2,beta=0.2,gamma=0.2,seasonal
  ="additive")
plot(y1.ts,type="p", pch=16,cex=.5,xlab='Date',ylab='Sales')
lines(clo.hw1$fitted[,1])
```

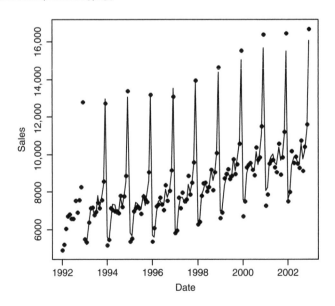

```
#Forecast the the sales for 2003
y2<-closales.data[133:144,]
y2.ts<-ts(y2[,2],start=c(2003,1),freq=12)

y2.forecast<-predict(clo.hw1, n.ahead=12, prediction.interval
  = TRUE)
plot(y1.ts,type="p", pch=16,cex=.5,xlab='Date',ylab='Sales',
  xlim=c(1992,2004))
points(y2.ts)
lines(y2.forecast[,1])
lines(y2.forecast[,2])
lines(y2.forecast[,3])
```

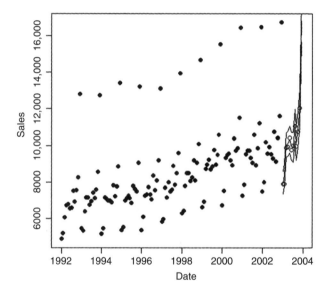

Example 4.8   The liquor store sales data are in the second column of the array called liqsales.data in which the first column is the month of the year. We will first fit additive and multiplicative seasonal models to the entire data to see the difference in the fits. Then we will use the data up to December 2003 to fit the multiplicative model and make forecasts for the coming year (2004). We will once again use Holt–Winters function given in stats package. In all cases we set all parameters to 0.2.

```
y.ts<- ts(liqsales.data[,2], start = c(1992,1), freq = 12)

liq.hw.add<-HoltWinters(y.ts,alpha=0.2,beta=0.2,gamma=0.2,
  seasonal="additive")
plot(y.ts,type="p", pch=16,cex=.5,xlab='Date',ylab='Sales',
  main="Additive Model")
lines(liq.hw.add$fitted[,1])
```

Additive model

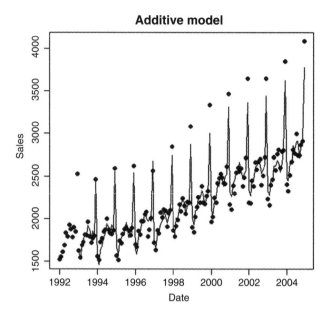

```
liq.hw.mult<-HoltWinters(y.ts,alpha=0.2,beta=0.2,gamma=0.2,
  seasonal="multiplicative")
plot(y.ts,type="p", pch=16,cex=.5,xlab='Date',ylab='Sales',
  main="Multiplicative Model")
lines(liq.hw.mult$fitted[,1])
```

Multiplicative model

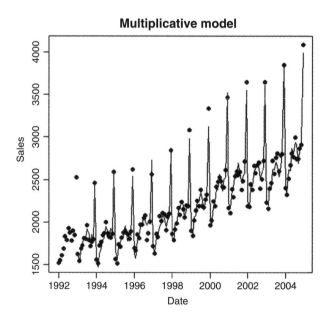

```
y1<-liqsales.data[1:144,]
y1.ts<-ts(y1[,2], start = c(1992,1), freq = 12)
liq.hw1<-HoltWinters(y1.ts,alpha=0.2,beta=0.2,gamma=0.2,
  seasonal="multiplicative")
y2<-liqsales.data[145:156,]
y2.ts<-ts(y2[,2],start=c(2004,1),freq=12)

y2.forecast<-predict(liq.hw1, n.ahead=12, prediction.interval =
  TRUE)
plot(y1.ts,type="p", pch=16,cex=.5,xlab='Date',ylab='Sales',
  xlim=c(1992,2005))
points(y2.ts)
lines(y2.forecast[,1])
lines(y2.forecast[,2])
lines(y2.forecast[,3])
```

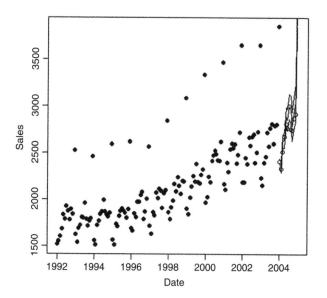

EXERCISES

4.1   Consider the time series data shown in Table E4.1.

   a. Make a time series plot of the data.

   b. Use simple exponential smoothing with $\lambda = 0.2$ to smooth the first 40 time periods of this data. How well does this smoothing procedure work?

   c. Make one-step-ahead forecasts of the last 10 observations. Determine the forecast errors.

TABLE E4.1   Data for Exercise 4.1

Period	y_t	Period	y_t	Period	y_t	Period	y_t	Period	y_t
1	48.7	11	49.1	21	45.3	31	50.8	41	47.9
2	45.8	12	46.7	22	43.3	32	46.4	42	49.5
3	46.4	13	47.8	23	44.6	33	52.3	43	44.0
4	46.2	14	45.8	24	47.1	34	50.5	44	53.8
5	44.0	15	45.5	25	53.4	35	53.4	45	52.5
6	53.8	16	49.2	26	44.9	36	53.9	46	52.0
7	47.6	17	54.8	27	50.5	37	52.3	47	50.6
8	47.0	18	44.7	28	48.1	38	53.0	48	48.7
9	47.6	19	51.1	29	45.4	39	48.6	49	51.4
10	51.1	20	47.3	30	51.6	40	52.4	50	47.7

4.2 Reconsider the time series data shown in Table E4.1.

   a. Use simple exponential smoothing with the optimum value of λ to smooth the first 40 time periods of this data (you can find the optimum value from Minitab). How well does this smoothing procedure work? Compare the results with those obtained in Exercise 4.1.

   b. Make one-step-ahead forecasts of the last 10 observations. Determine the forecast errors. Compare these forecast errors with those from Exercise 4.1. How much has using the optimum value of the smoothing constant improved the forecasts?

4.3 Find the sample ACF for the time series in Table E4.1. Does this give you any insight about the optimum value of the smoothing constant that you found in Exercise 4.2?

4.4 Consider the time series data shown in Table E4.2.

   a. Make a time series plot of the data.

   b. Use simple exponential smoothing with $\lambda = 0.2$ to smooth the first 40 time periods of this data. How well does this smoothing procedure work?

   c. Make one-step-ahead forecasts of the last 10 observations. Determine the forecast errors.

4.5 Reconsider the time series data shown in Table E4.2.

   a. Use simple exponential smoothing with the optimum value of λ to smooth the first 40 time periods of this data (you can find the optimum value from Minitab). How well does this smoothing procedure work? Compare the results with those obtained in Exercise 4.4.

TABLE E4.2   Data for Exercise 4.4

Period	y_t	Period	y_t	Period	y_t	Period	y_t	Period	y_t
1	43.1	11	41.8	21	47.7	31	52.9	41	48.3
2	43.7	12	50.7	22	51.1	32	47.3	42	45.0
3	45.3	13	55.8	23	67.1	33	50.0	43	55.2
4	47.3	14	48.7	24	47.2	34	56.7	44	63.7
5	50.6	15	48.2	25	50.4	35	42.3	45	64.4
6	54.0	16	46.9	26	44.2	36	52.0	46	66.8
7	46.2	17	47.4	27	52.0	37	48.6	47	63.3
8	49.3	18	49.2	28	35.5	38	51.5	48	60.0
9	53.9	19	50.9	29	48.4	39	49.5	49	60.9
10	42.5	20	55.3	30	55.4	40	51.4	50	56.1

    b. Make one-step-ahead forecasts of the last 10 observations. Determine the forecast errors. Compare these forecast errors with those from Exercise 4.4. How much has using the optimum value of the smoothing constant improved the forecasts?

4.6 Find the sample ACF for the time series in Table E4.2. Does this give you any insight about the optimum value of the smoothing constant that you found in Exercise 4.5?

4.7 Consider the time series data shown in Table E4.3.

    a. Make a time series plot of the data.

    b. Use simple exponential smoothing with $\lambda = 0.1$ to smooth the first 30 time periods of this data. How well does this smoothing procedure work?

TABLE E4.3   Data for Exercise 4.7

Period	y_t	Period	y_t	Period	y_t	Period	y_t	Period	y_t
1	275	11	297	21	231	31	255	41	293
2	245	12	235	22	238	32	255	42	284
3	222	13	237	23	251	33	229	43	276
4	169	14	203	24	253	34	286	44	290
5	236	15	238	25	283	35	236	45	250
6	259	16	232	26	283	36	194	46	235
7	268	17	206	27	245	37	228	47	275
8	225	18	295	28	234	38	244	48	350
9	246	19	247	29	273	39	241	49	290
10	263	20	227	30	293	40	284	50	269

    c. Make one-step-ahead forecasts of the last 20 observations. Determine the forecast errors.

    d. Plot the forecast errors on a control chart for individuals. Use a moving range chart to estimate the standard deviation of the forecast errors in constructing this chart. What conclusions can you draw about the forecasting procedure and the time series?

4.8 The data in Table E4.4 exhibit a linear trend.

    a. Verify that there is a trend by plotting the data.

    b. Using the first 12 observations, develop an appropriate procedure for forecasting.

    c. Forecast the last 12 observations and calculate the forecast errors. Does the forecasting procedure seem to be working satisfactorily?

TABLE E4.4   Data for Exercise 4.8

Period	y_t	Period	y_t
1	315	13	460
2	195	14	395
3	310	15	390
4	316	16	450
5	325	17	458
6	335	18	570
7	318	19	520
8	355	20	400
9	420	21	420
10	410	22	580
11	485	23	475
12	420	24	560

4.9 Reconsider the linear trend data in Table E4.4. Take the first difference of this data and plot the time series of first differences. Has differencing removed the trend? Use exponential smoothing on the first 11 differences. Instead of forecasting the original data, forecast the first differences for the remaining data using exponential smoothing and use these forecasts of the first differences to obtain forecasts for the original data.

4.10 Table B.1 in Appendix B contains data on the market yield on US Treasury Securities at 10-year constant maturity.

    a. Make a time series plot of the data.

b. Use simple exponential smoothing with $\lambda = 0.2$ to smooth the data, excluding the last 20 observations. How well does this smoothing procedure work?

c. Make one-step-ahead forecasts of the last 20 observations. Determine the forecast errors.

4.11 Reconsider the US Treasury Securities data shown in Table B.1.

a. Use simple exponential smoothing with the optimum value of λ to smooth the data, excluding the last 20 observations (you can find the optimum value from Minitab). How well does this smoothing procedure work? Compare the results with those obtained in Exercise 4.10.

b. Make one-step-ahead forecasts of the last 10 observations. Determine the forecast errors. Compare these forecast errors with those from Exercise 4.10. How much has using the optimum value of the smoothing constant improved the forecasts?

4.12 Table B.2 contains data on pharmaceutical product sales.

a. Make a time series plot of the data.

b. Use simple exponential smoothing with $\lambda = 0.1$ to smooth this data. How well does this smoothing procedure work?

c. Make one-step-ahead forecasts of the last 10 observations. Determine the forecast errors.

4.13 Reconsider the pharmaceutical sales data shown in Table B.2.

a. Use simple exponential smoothing with the optimum value of λ to smooth the data (you can find the optimum value from either Minitab or JMP). How well does this smoothing procedure work? Compare the results with those obtained in Exercise 4.12.

b. Make one-step-ahead forecasts of the last 10 observations. Determine the forecast errors. Compare these forecast errors with those from Exercise 4.12. How much has using the optimum value of the smoothing constant improved the forecasts?

c. Construct the sample ACF for these data. Does this give you any insight regarding the optimum value of the smoothing constant?

4.14 Table B.3 contains data on chemical process viscosity.

a. Make a time series plot of the data.

b. Use simple exponential smoothing with $\lambda = 0.1$ to smooth this data. How well does this smoothing procedure work?

    c. Make one-step-ahead forecasts of the last 10 observations. Determine the forecast errors.

4.15 Reconsider the chemical process data shown in Table B.3.

    a. Use simple exponential smoothing with the optimum value of λ to smooth the data (you can find the optimum value from either Minitab or JMP). How well does this smoothing procedure work? Compare the results with those obtained in Exercise 4.14.

    b. Make one-step-ahead forecasts of the last 10 observations. Determine the forecast errors. Compare these forecast errors with those from Exercise 4.14. How much has using the optimum value of the smoothing constant improved the forecasts?

    c. Construct the sample ACF for these data. Does this give you any insight regarding the optimum value of the smoothing constant?

4.16 Table B.4 contains data on the annual US production of blue and gorgonzola cheeses. This data have a strong trend.

    a. Verify that there is a trend by plotting the data.

    b. Develop an appropriate exponential smoothing procedure for forecasting.

    c. Forecast the last 10 observations and calculate the forecast errors. Does the forecasting procedure seem to be working satisfactorily?

4.17 Reconsider the blue and gorgonzola cheese data in Table B.4 and Exercise 4.16. Take the first difference of this data and plot the time series of first differences. Has differencing removed the trend? Use exponential smoothing on the first differences. Instead of forecasting the original data, develop a procedure for forecasting the first differences and explain how you would use these forecasts of the first differences to obtain forecasts for the original data.

4.18 Table B.5 shows data for US beverage manufacturer product shipments. Develop an appropriate exponential smoothing procedure for forecasting these data.

4.19 Table B.6 contains data on the global mean surface air temperature anomaly.

    a. Make a time series plot of the data.

    b. Use simple exponential smoothing with $\lambda = 0.2$ to smooth the data. How well does this smoothing procedure work? Do you think this would be a reliable forecasting procedure?

4.20 Reconsider the global mean surface air temperature anomaly data shown in Table B.6 and used in Exercise 4.19.

    a. Use simple exponential smoothing with the optimum value of λ to smooth the data (you can find the optimum value from either Minitab or JMP). How well does this smoothing procedure work? Compare the results with those obtained in Exercise 4.19.

    b. Do you think using the optimum value of the smoothing constant would result in improved forecasts from exponential smoothing?

    c. Take the first difference of this data and plot the time series of first differences. Use exponential smoothing on the first differences. Instead of forecasting the original data, develop a procedure for forecasting the first differences and explain how you would use these forecasts of the first differences to obtain forecasts for the original global mean surface air temperature anomaly.

4.21 Table B.7 contains daily closing stock prices for the Whole Foods Market.

    a. Make a time series plot of the data.

    b. Use simple exponential smoothing with $\lambda = 0.1$ to smooth the data. How well does this smoothing procedure work? Do you think this would be a reliable forecasting procedure?

4.22 Reconsider the Whole Foods Market data shown in Table B.7 and used in Exercise 4.21.

    a. Use simple exponential smoothing with the optimum value of λ to smooth the data (you can find the optimum value from either Minitab or JMP). How well does this smoothing procedure work? Compare the results with those obtained in Exercise 4.21.

    b. Do you think that using the optimum value of the smoothing constant would result in improved forecasts from exponential smoothing?

    c. Use an exponential smoothing procedure for trends on this data. Is this an apparent improvement over the use of simple exponential smoothing with the optimum smoothing constant?

    d. Take the first difference of this data and plot the time series of first differences. Use exponential smoothing on the first differences. Instead of forecasting the original data, develop a procedure for forecasting the first differences and explain how you would use these forecasts of the first differences to obtain forecasts for the stock price.

4.23 Unemployment rate data are given in Table B.8.

   a. Make a time series plot of the data.

   b. Use simple exponential smoothing with $\lambda = 0.2$ to smooth the data. How well does this smoothing procedure work? Do you think that simple exponential smoothing should be used to forecast this data?

4.24 Reconsider the unemployment rate data shown in Table B.8 and used in Exercise 4.23.

   a. Use simple exponential smoothing with the optimum value of λ to smooth the data (you can find the optimum value from either Minitab or JMP). How well does this smoothing procedure work? Compare the results with those obtained in Exercise 4.23.

   b. Do you think that using the optimum value of the smoothing constant would result in improved forecasts from exponential smoothing?

   c. Use an exponential smoothing procedure for trends on this data. Is this an apparent improvement over the use of simple exponential smoothing with the optimum smoothing constant?

   d. Take the first difference of this data and plot the time series of first differences. Use exponential smoothing on the first differences. Is this a reasonable procedure for forecasting the first differences?

4.25 Table B.9 contains yearly data on the international sunspot numbers.

   a. Construct a time series plot of the data.

   b. Use simple exponential smoothing with $\lambda = 0.1$ to smooth the data. How well does this smoothing procedure work? Do you think that simple exponential smoothing should be used to forecast this data?

4.26 Reconsider the sunspot data shown in Table B.9 and used in Exercise 4.25.

   a. Use simple exponential smoothing with the optimum value of λ to smooth the data (you can find the optimum value from either Minitab or JMP). How well does this smoothing procedure work? Compare the results with those obtained in Exercise 4.25.

   b. Do you think that using the optimum value of the smoothing constant would result in improved forecasts from exponential smoothing?

c. Use an exponential smoothing procedure for trends on this data. Is this an apparent improvement over the use of simple exponential smoothing with the optimum smoothing constant?

4.27 Table B.10 contains 7 years of monthly data on the number of airline miles flown in the United Kingdom. This is seasonal data.

a. Make a time series plot of the data and verify that it is seasonal.

b. Use Winters' multiplicative method for the first 6 years to develop a forecasting method for this data. How well does this smoothing procedure work?

c. Make one-step-ahead forecasts of the last 12 months. Determine the forecast errors. How well did your procedure work in forecasting the new data?

4.28 Reconsider the airline mileage data in Table B.10 and used in Exercise 4.27.

a. Use the additive seasonal effects model for the first 6 years to develop a forecasting method for this data. How well does this smoothing procedure work?

b. Make one-step-ahead forecasts of the last 12 months. Determine the forecast errors. How well did your procedure work in forecasting the new data?

c. Compare these forecasts with those found using Winters' multiplicative method in Exercise 4.27.

4.29 Table B.11 contains 8 years of monthly champagne sales data. This is seasonal data.

a. Make a time series plot of the data and verify that it is seasonal. Why do you think seasonality is present in these data?

b. Use Winters' multiplicative method for the first 7 years to develop a forecasting method for this data. How well does this smoothing procedure work?

c. Make one-step-ahead forecasts of the last 12 months. Determine the forecast errors. How well did your procedure work in forecasting the new data?

4.30 Reconsider the monthly champagne sales data in Table B.11 and used in Exercise 4.29.

a. Use the additive seasonal effects model for the first 7 years to develop a forecasting method for this data. How well does this smoothing procedure work?

b. Make one-step-ahead forecasts of the last 12 months. Determine the forecast errors. How well did your procedure work in forecasting the new data?

c. Compare these forecasts with those found using Winters' multiplicative method in Exercise 4.29.

4.31 Montgomery et al. (1990) give 4 years of data on monthly demand for a soft drink. These data are given in Table E4.5.

a. Make a time series plot of the data and verify that it is seasonal. Why do you think seasonality is present in these data?

b. Use Winters' multiplicative method for the first 3 years to develop a forecasting method for this data. How well does this smoothing procedure work?

c. Make one-step-ahead forecasts of the last 12 months. Determine the forecast errors. How well did your procedure work in forecasting the new data?

TABLE E4.5   Soft Drink Demand Data

Period	y_t	Period	y_t	Period	y_t	Period	y_t
1	143	13	189	25	359	37	332
2	191	14	326	26	264	38	244
3	195	15	289	27	315	39	320
4	225	16	293	28	362	40	437
5	175	17	279	29	414	41	544
6	389	18	552	30	647	42	830
7	454	19	674	31	836	43	1011
8	618	20	827	32	901	44	1081
9	770	21	1000	33	1104	45	1400
10	564	22	502	34	874	46	1123
11	327	23	512	35	683	47	713
12	235	24	300	36	352	48	487

4.32 Reconsider the soft drink demand data in Table E4.5 and used in Exercise 4.31.

a. Use the additive seasonal effects model for the first 3 years to develop a forecasting method for this data. How well does this smoothing procedure work?

b. Make one-step-ahead forecasts of the last 12 months. Determine the forecast errors. How well did your procedure work in forecasting the new data?

c. Compare these forecasts with those found using Winters' multiplicative method in Exercise 4.31.

4.33 Table B.12 presents data on the hourly yield from a chemical process and the operating temperature. Consider only the yield data in this exercise.

   a. Construct a time series plot of the data.

   b. Use simple exponential smoothing with $\lambda = 0.2$ to smooth the data. How well does this smoothing procedure work? Do you think that simple exponential smoothing should be used to forecast this data?

4.34 Reconsider the chemical process yield data shown in Table B.12.

   a. Use simple exponential smoothing with the optimum value of λ to smooth the data (you can find the optimum value from either Minitab or JMP). How well does this smoothing procedure work? Compare the results with those obtained in Exercise 4.33.

   b. How much has using the optimum value of the smoothing constant improved the forecasts?

4.35 Find the sample ACF for the chemical process yield data in Table B.12. Does this give you any insight about the optimum value of the smoothing constant that you found in Exercise 4.34?

4.36 Table B.13 presents data on ice cream and frozen yogurt sales. Develop an appropriate exponential smoothing forecasting procedure for this time series.

4.37 Table B.14 presents the CO_2 readings from Mauna Loa.

   a. Use simple exponential smoothing with the optimum value of λ to smooth the data (you can find the optimum value from either Minitab or JMP).

   b. Use simple exponential smoothing with $\lambda = 0.1$ to smooth the data. How well does this smoothing procedure work? Compare the results with those obtained using the optimum smoothing constant. How much has using the optimum value of the smoothing constant improved the exponential smoothing procedure?

4.38 Table B.15 presents data on the occurrence of violent crimes. Develop an appropriate exponential smoothing forecasting procedure for this time series.

4.39 Table B.16 presents data on the US. gross domestic product (GDP). Develop an appropriate exponential smoothing forecasting procedure for the GDP time series.

4.40 Total annual energy consumption is shown in Table B.17. Develop an appropriate exponential smoothing forecasting procedure for the energy consumption time series.

4.41 Table B.18 contains data on coal production. Develop an appropriate exponential smoothing forecasting procedure for the coal production time series.

4.42 Table B.19 contains data on the number of children 0–4 years old who drowned in Arizona.
   a. Plot the data. What type of forecasting model seems appropriate?
   b. Develop a forecasting model for this data?

4.43 Data on tax refunds and population are shown in Table B.20. Develop an appropriate exponential smoothing forecasting procedure for the tax refund time series.

4.44 Table B.21 contains data from the US Energy Information Administration on monthly average price of electricity for the residential sector in Arizona. This data have a strong seasonal component. Use the data from 2001–2010 to develop a multiplicative Winters-type exponential smoothing model for this data. Use this model to simulate one-month-ahead forecasts for the remaining years. Calculate the forecast errors. Discuss the reasonableness of the forecasts.

4.45 Use the electricity price data in Table B.21 from 2010–2010 and an additive Winters-type exponential smoothing procedure to develop a forecasting model.
   a. Use this model to simulate one-month-ahead forecasts for the remaining years. Calculate the forecast errors. Discuss the reasonableness of the forecasts.
   b. Compare the performance of this model with the multiplicative model you developed in Exercise 4.44.

4.46 Table B.22 contains data from the Danish Energy Agency on Danish crude oil production.
   a. Plot the data and comment on any features that you observe from the graph. Calculate and plot the sample ACF and variogram. Interpret these graphs.

b. Use first-order exponential smoothing to develop a forecasting model for crude oil production. Plot the smoothed statistic on the same axes with the original data. How well does first-order exponential smoothing seem to work?

c. Use double exponential smoothing to develop a forecasting model for crude oil production. Plot the smoothed statistic on the same axes with the original data. How well does double exponential smoothing seem to work?

d. Compare the two smoothing models from parts b and c. Which approach seems preferable?

4.47 Apply a first difference to the Danish crude oil production data in Table B.22.

a. Plot the data and comment on any features that you observe from the graph. Calculate and plot the sample ACF and variogram. Interpret these graphs.

b. Use first-order exponential smoothing to develop a forecasting model for crude oil production. Plot the smoothed statistic on the same axes with the original data. How well does first-order exponential smoothing seem to work? How does this compare to the first-order exponential smoothing model you developed in Exercise 4.46 for the original (undifferenced) data?

4.48 Table B.23 shows weekly data on positive laboratory test results for influenza. Notice that these data have a number of missing values. In exercise you were asked to develop and implement a scheme to estimate the missing values. This data have a strong seasonal component. Use the data from 1997–2010 to develop a multiplicative Winters-type exponential smoothing model for this data. Use this model to simulate one-week-ahead forecasts for the remaining years. Calculate the forecast errors. Discuss the reasonableness of the forecasts.

4.49 Repeat Exercise 4.48 using an additive Winters-type model. Compare the performance of the additive and the multiplicative model from Exercise 4.48.

4.50 Data from the Western Regional Climate Center for the monthly mean daily solar radiation (in Langleys) at the Zion Canyon, Utah, station are shown in Table B.24. This data have a strong seasonal component. Use the data from 2003–2012 to develop a multiplicative Winters-type exponential smoothing model for this data. Use this model to simulate one-month-ahead forecasts for the remaining years. Calculate the forecast errors. Discuss the reasonableness of the forecasts.

4.51   Repeat Exercise 4.50 using an additive Winters-type model. Compare the performance of the additive and the multiplicative model from Exercise 4.50.

4.52   Table B.25 contains data from the National Highway Traffic Safety Administration on motor vehicle fatalities from 1966 to 2012. This data are used by a variety of governmental and industry groups, as well as research organizations.

    a. Plot the fatalities data and comment on any features of the data that you see.

    b. Develop a forecasting procedure using first-order exponential smoothing. Use the data from 1966–2006 to develop the model, and then simulate one-year-ahead forecasts for the remaining years. Compute the forecasts errors. How well does this method seem to work?

    c. Develop a forecasting procedure using based on double exponential smoothing. Use the data from 1966–2006 to develop the model, and then simulate one-year-ahead forecasts for the remaining years. Compute the forecasts errors. How well does this method seem to work in comparison to the method based on first-order exponential smoothing?

4.53   Apply a first difference to the motor vehicle fatalities data in Table B.25.

    a. Plot the differenced data and comment on any features of the data that you see.

    b. Develop a forecasting procedure for the first differences based on first-order exponential smoothing. Use the data from 1966–2006 to develop the model, and then simulate one-year-ahead forecasts for the remaining years. Compute the forecasts errors. How well does this method seem to work?

    c. Compare this approach with the two smoothing methods used in Exercise 4.52.

4.54   Appendix Table B.26 contains data on monthly single-family residential new home sales from 1963 through 2014.

    a. Plot the home sales data. Comment on the graph.

b. Develop a forecasting procedure using first-order exponential smoothing. Use the data from 1963–2000 to develop the model, and then simulate one-year-ahead forecasts for the remaining years. Compute the forecasts errors. How well does this method seem to work?

c. Can you explain the unusual changes in sales observed in the data near the end of the graph?

4.55 Appendix Table B.27 contains data on the airline's best on-time arrival and airport performance. The data are given by month from January 1995 through February 2013.

a. Plot the data and comment on any features of the data that you see.

b. Construct the sample ACF and variogram. Comment on these displays.

c. Develop an appropriate exponential smoothing model for these data.

4.56 Data from the US Census Bureau on monthly domestic automobile manufacturing shipments (in millions of dollars) are shown in Table B.28.

a. Plot the data and comment on any features of the data that you see.

b. Construct the sample ACF and variogram. Comment on these displays.

c. Develop an appropriate exponential smoothing model for these data. Note that there is some apparent seasonality in the data. Why does this seasonal behavior occur?

d. Plot the first difference of the data. Now compute the sample ACF and variogram for the differenced data. What impact has differencing had? Is there still some apparent seasonality in the differenced data?

4.57 Suppose that simple exponential smoothing is being used to forecast a process. At the start of period t^*, the mean of the process shifts to a new level $\mu + \delta$. The mean remains at this new level for subsequent time periods. Show that the expected value of the exponentially smoothed statistic is

$$E(\hat{y}_t) = \begin{cases} \mu, & T \le t^* \\ \mu + \delta - \delta(1-\lambda)^{T-t^*+1}, & T \ge t^* \end{cases}$$

4.58 Using the results of Exercise 4.44, determine the number of periods following the step change for the expected value of the exponential smoothing statistic to be within $0.10\,\delta$ of the new time series level $\mu + \delta$. Plot the number of periods as a function of the smoothing constant. What conclusions can you draw?

4.59 Suppose that simple exponential smoothing is being used to forecast the process $y_t = \mu + \varepsilon_t$. At the start of period t^*, the mean of the process experiences a transient; that is, it shifts to a new level $\mu + \delta$, but reverts to its original level y at the start of the next period $t^* + 1$. The mean remains at this level for subsequent time periods. Show that the expected value of the exponentially smoothed statistic is

$$E(\hat{y}_t) = \begin{cases} \mu, & T \leq t^* \\ \mu + \delta\lambda(1 - \lambda)^{T-t^*}, & T \geq t^* \end{cases}.$$

4.60 Using the results of Exercise 4.46, determine the number of periods that it will take following the impulse for the expected value of the exponential smoothing statistic to return to within $0.10\,\delta$ of the original time series level μ. Plot the number of periods as a function of the smoothing constant. What conclusions can you draw?

CHAPTER 5

AUTOREGRESSIVE INTEGRATED MOVING AVERAGE (ARIMA) MODELS

All models are wrong, some are useful.

GEORGE E. P. BOX, *British statistician*

5.1 INTRODUCTION

In the previous chapter, we discussed forecasting techniques that, in general, were based on some variant of exponential smoothing. The general assumption for these models was that any time series data can be represented as the sum of two distinct components: deterministic and stochastic (random). The former is modeled as a function of time whereas for the latter we assumed that some random noise that is added on to the deterministic signal generates the stochastic behavior of the time series. One very important assumption is that the random noise is generated through independent shocks to the process. In practice, however, this assumption is often violated. That is, usually successive observations show serial dependence. Under these circumstances, forecasting methods based on exponential smoothing may be inefficient and sometimes inappropriate

Introduction to Time Series Analysis and Forecasting, Second Edition.
Douglas C. Montgomery, Cheryl L. Jennings and Murat Kulahci.
© 2015 John Wiley & Sons, Inc. Published 2015 by John Wiley & Sons, Inc.

because they do not take advantage of the serial dependence in the observations in the most effective way. To formally incorporate this dependent structure, in this chapter we will explore a general class of models called autoregressive integrated moving average (MA) models or ARIMA models (also known as Box–Jenkins models).

5.2 LINEAR MODELS FOR STATIONARY TIME SERIES

In statistical modeling, we are often engaged in an endless pursuit of finding the ever elusive true relationship between certain inputs and the output. As cleverly put by the quote of this chapter, these efforts usually result in models that are nothing but approximations of the "true" relationship. This is generally due to the choices the analyst makes along the way to ease the modeling efforts. A major assumption that often provides relief in modeling efforts is the linearity assumption. A **linear filter**, for example, is a linear operation from one time series x_t to another time series y_t,

$$y_t = L(x_t) = \sum_{i=-\infty}^{+\infty} \psi_i x_{t-i} \tag{5.1}$$

with $t = \ldots, -1, 0, 1, \ldots$. In that regard the linear filter can be seen as a "process" that converts the input, x_t, into an output, y_t, and that conversion is not instantaneous but involves all (present, past, and future) values of the input in the form of a summation with different "weights", $\{\psi_i\}$, on each x_t. Furthermore, the linear filter in Eq. (5.1) is said to have the following properties:

1. **Time-invariant** as the coefficients $\{\psi_i\}$ do not depend on time.
2. **Physically realizable** if $\psi_i = 0$ for $i < 0$; that is, the output y_t is a linear function of the current and past values of the input: $y_t = \psi_0 x_t + \psi_1 x_{t-1} + \cdots$.
3. **Stable** if $\sum_{i=-\infty}^{+\infty} |\psi_i| < \infty$.

In linear filters, under certain conditions, some properties such as **stationarity** of the input time series are also reflected in the output. We discussed stationarity previously in Chapter 2. We will now give a more formal description of it before proceeding further with linear models for time series.

5.2.1 Stationarity

The **stationarity** of a time series is related to its statistical properties in time. That is, in the more strict sense, a stationary time series exhibits similar "statistical behavior" in time and this is often characterized as a constant probability distribution in time. However, it is usually satisfactory to consider the first two moments of the time series and define stationarity (or **weak stationarity**) as follows: (1) the expected value of the time series does not depend on time and (2) the autocovariance function defined as $\text{Cov}(y_t, y_{t+k})$ for any lag k is only a function of k and not time; that is, $\gamma_y(k) = \text{Cov}(y_t, y_{t+k})$.

In a crude way, the stationarity of a time series can be determined by taking arbitrary "snapshots" of the process at different points in time and observing the general behavior of the time series. If it exhibits "similar" behavior, one can then proceed with the modeling efforts under the assumption of stationarity. Further preliminary tests also involve observing the behavior of the autocorrelation function. A strong and slowly dying ACF will also suggest deviations from stationarity. Better and more methodological tests of stationarity also exist and we will discuss some of them later in this chapter. Figure 5.1 shows examples of stationary and nonstationary time series data.

5.2.2 Stationary Time Series

For a time-invariant and stable linear filter and a stationary input time series x_t with $\mu_x = E(x_t)$ and $\gamma_x(k) = \text{Cov}(x_t, x_{t+k})$, the output time series y_t given in Eq. (5.1) is also a stationary time series with

$$E(y_t) = \mu_y = \sum_{-\infty}^{\infty} \psi_i \mu_x$$

and

$$\text{Cov}(y_t, y_{t+k}) = \gamma_y(k) = \sum_{i=-\infty}^{\infty} \sum_{j=-\infty}^{\infty} \psi_i \psi_j \gamma_x(i - j + k)$$

It is then easy to show that the following stable linear process with white noise time series, ε_t, is also stationary:

$$y_t = \mu + \sum_{i=0}^{\infty} \psi_i \varepsilon_{t-i} \qquad (5.2)$$

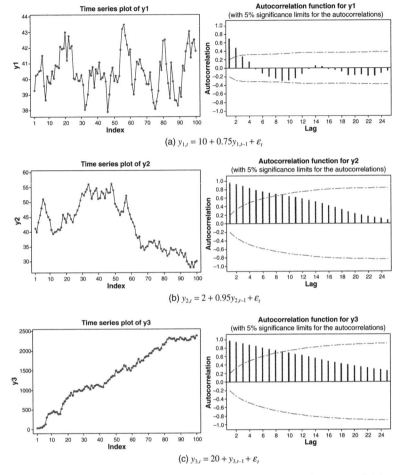

FIGURE 5.1 Realizations of (a) stationary, (b) near nonstationary, and (c) nonstationary processes.

with $E(\varepsilon_t) = 0$, and

$$\gamma_\varepsilon(h) = \begin{cases} \sigma^2 & \text{if } h = 0 \\ 0 & \text{if } h \neq 0 \end{cases}$$

So for the autocovariance function of y_t, we have

$$\gamma_y(k) = \sum_{i=0}^{\infty} \sum_{j=0}^{\infty} \psi_i \psi_j \gamma_\varepsilon(i - j + k)$$

$$= \sigma^2 \sum_{i=0}^{\infty} \psi_i \psi_{i+k} \tag{5.3}$$

We can rewrite the linear process in Eq. (5.2) in terms of the **backshift operator**, B, as

$$
\begin{aligned}
y_t &= \mu + \psi_0 \varepsilon_t + \psi_1 \varepsilon_{t-1} + \psi_2 \varepsilon_{t-2} + \cdots \\
&= \mu + \sum_{i=0}^{\infty} \psi_i B^i \varepsilon_t \\
&= \mu + \underbrace{\left(\sum_{i=0}^{\infty} \psi_i B^i \right)}_{=\Psi(B)} \varepsilon_t \\
&= \mu + \Psi(B) \varepsilon_t
\end{aligned}
\tag{5.4}
$$

This is called the **infinite moving average** and serves as a general class of models for any stationary time series. This is due to a theorem by Wold (1938) and basically states that **any** nondeterministic weakly stationary time series y_t can be represented as in Eq. (5.2), where $\{\psi_i\}$ satisfy $\sum_{i=0}^{\infty} \psi_i^2 < \infty$. A more intuitive interpretation of this theorem is that a stationary time series can be seen as the weighted sum of the present and past random "disturbances." For further explanations see Yule (1927) and Bisgaard and Kulahci (2005, 2011).

The theorem by Wold requires that the random shocks in (5.4) to be white noise which we defined as uncorrelated random shocks with constant variance. Some textbooks discuss independent or strong white noise for random shocks. It should be noted that there is a difference between correlation and independence. Independent random variables are also uncorrelated but the opposite is not always true. Independence between two random variables refers their joint probability distribution function being equal to the product of the marginal distributions. That is, two random variables X and Y are said to be independent if

$$
f(X, Y) = f_X(X) f_Y(Y)
$$

This can also be loosely interpreted as if X and Y are independent, knowing the value of X for example does not provide any information about what the value of Y might be.

For two uncorrelated random variables X and Y, we have their correlation and their covariance equal to zero. That is,

$$
\begin{aligned}
Cov(X, Y) &= E[(X - \mu_X)(Y - \mu_Y)] \\
&= E[XY] - E[X]E[Y] \\
&= 0
\end{aligned}
$$

This implies that if X and Y are uncorrelated, $E[XY] = E[X]E[Y]$.

Clearly if two random variables are independent, they are also uncorrelated since under independence we always have

$$
\begin{aligned}
E[XY] &= \iint xyf(x, y)dxdy \\
&= \iint xyf(x)f(y)dxdy \\
&= \left\{ \int xf(x)dx \right\} \left\{ \int yf(y)dx \right\} \\
&= E[X]E[Y]
\end{aligned}
$$

As we mentioned earlier, the opposite is not always true. To illustrate this with an example, consider X, a random variable with a symmetric probability density function around 0, i.e., $E[X] = 0$. Assume that the second variable Y is equal to $|X|$. Since knowing the value of X also determines the value of Y, these two variables are clearly not independent. However we can show that $E[Y] = 2 \int_0^\infty xf(x)dx$ and $E[XY] = \int_0^\infty x^2f(x)dx - \int_{-\infty}^0 x^2f(x)dx = 0$ and hence $E[XY] = E[X]E[Y]$ This shows that X and Y are uncorrected but not independent.

Wold's decomposition theorem practically forms the foundation of the models we discuss in this chapter. This means that the strong assumption of independence is not necessarily needed except for the discussion on forecasting using ARIMA models in Section 5.8 where we assume the random shocks to be independent.

It can also be seen from Eq. (5.3) that there is a direct relation between the weights $\{\psi_i\}$ and the autocovariance function. In modeling a stationary time series as in Eq. (5.4), it is obviously impractical to attempt to estimate the infinitely many weights given in $\{\psi_i\}$. Although very powerful in providing a general representation of any stationary time series, the infinite moving average model given in Eq. (5.2) is useless in practice except for certain special cases:

1. Finite order moving average (MA) models where, except for a finite number of the weights in $\{\psi_i\}$, they are set to 0.
2. Finite order autoregressive (AR) models, where the weights in $\{\psi_i\}$ are generated using only a finite number of parameters.
3. A mixture of finite order autoregressive and moving average models (ARMA).

We shall now discuss each of these classes of models in great detail.

5.3 FINITE ORDER MOVING AVERAGE PROCESSES

In finite order moving average or MA models, conventionally ψ_0 is set to 1 and the weights that are not set to 0 are represented by the Greek letter θ with a minus sign in front. Hence a moving average process of order q (MA(q)) is given as

$$y_t = \mu + \varepsilon_t - \theta_1 \varepsilon_{t-1} - \cdots - \theta_q \varepsilon_{t-q} \tag{5.5}$$

where $\{\varepsilon_t\}$ is white noise. Since Eq. (5.5) is a special case of Eq. (5.4) with only finite weights, an MA(q) process is **always** stationary regardless of values of the weights. In terms of the backward shift operator, the MA(q) process is

$$
\begin{aligned}
y_t &= \mu + \left(1 - \theta_1 B - \cdots - \theta_q B^q\right)\varepsilon_t \\
&= \mu + \left(1 - \sum_{i=1}^{q} \theta_i B^i\right)\varepsilon_t \\
&= \mu + \Theta(B)\,\varepsilon_t
\end{aligned}
\tag{5.6}
$$

where $\Theta(B) = 1 - \sum_{i=1}^{q} \theta_i B^i$.

Furthermore, since $\{\varepsilon_t\}$ is white noise, the expected value of the MA(q) process is simply

$$
\begin{aligned}
E(y_t) &= E\left(\mu + \varepsilon_t - \theta_1 \varepsilon_{t-1} - \cdots - \theta_q \varepsilon_{t-q}\right) \\
&= \mu
\end{aligned}
\tag{5.7}
$$

and its variance is

$$
\begin{aligned}
\mathrm{Var}(y_t) = \gamma_y(0) &= \mathrm{Var}\left(\mu + \varepsilon_t - \theta_1 \varepsilon_{t-1} - \cdots - \theta_q \varepsilon_{t-q}\right) \\
&= \sigma^2 \left(1 + \theta_1^2 + \cdots + \theta_q^2\right)
\end{aligned}
\tag{5.8}
$$

Similarly, the autocovariance at lag k can be calculated from

$$
\begin{aligned}
\gamma_y(k) &= \mathrm{Cov}(y_t, y_{t+k}) \\
&= E[(\varepsilon_t - \theta_1 \varepsilon_{t-1} - \cdots - \theta_q \varepsilon_{t-q})(\varepsilon_{t+k} - \theta_1 \varepsilon_{t+k-1} - \cdots - \theta_q \varepsilon_{t+k-q})] \\
&= \begin{cases} \sigma^2(-\theta_k + \theta_1 \theta_{k+1} + \cdots + \theta_{q-k}\theta_q), & k = 1, 2, \ldots, q \\ 0, & k > q \end{cases}
\end{aligned}
\tag{5.9}
$$

From Eqs. (5.8) and (5.9), the autocorrelation function of the MA(q) process is

$$
\rho_y(k) = \frac{\gamma_y(k)}{\gamma_y(0)} =
\begin{cases}
\dfrac{-\theta_k + \theta_1\theta_{k+1} + \cdots + \theta_{q-k}\theta_q}{1 + \theta_1^2 + \cdots + \theta_q^2}, & k = 1, 2, \ldots, q \\[2mm]
0, & k > q
\end{cases}
\tag{5.10}
$$

This feature of the ACF is very helpful in identifying the MA model and its appropriate order as it "cuts off" after lag q. In real life applications, however, the sample ACF, $r(k)$, will not necessarily be equal to zero after lag q. It is expected to become very small in absolute value after lag q. For a data set of N observations, this is often tested against $\pm 2/\sqrt{N}$ limits, where $1/\sqrt{N}$ is the approximate value for the standard deviation of the ACF for any lag under the assumption $\rho(k) = 0$ for all k's as discussed in Chapter 2.

Note that a more accurate formula for the standard error of the kth sample autocorrelation coefficient is provided by Bartlett (1946) as

$$
s.e.\,(r(k)) = N^{-1/2}\left(1 + 2\sum_{j=1}^{k-1} r(j)^{*2}\right)^{1/2}
$$

where

$$
r(j)^* =
\begin{cases}
r(j) & \text{for } \rho(j) \neq 0 \\
0 & \text{for } \rho(j) = 0
\end{cases}
$$

A special case would be white noise data for which $\rho(j) = 0$ for all j's. Hence for a white noise process (i.e., no autocorrelation), a reasonable interval for the sample autocorrelation coefficients to fall in would be $\pm 2/\sqrt{N}$ and any indication otherwise may be considered as evidence for serial dependence in the process.

5.3.1 The First-Order Moving Average Process, MA(1)

The simplest finite order MA model is obtained when $q = 1$ in Eq. (5.5):

$$
y_t = \mu + \varepsilon_t - \theta_1\varepsilon_{t-1}
\tag{5.11}
$$

For the first-order moving average or MA(1) model, we have the autocovariance function as

$$\gamma_y(0) = \sigma^2 \left(1 + \theta_1^2\right)$$
$$\gamma_y(1) = -\theta_1 \sigma^2 \qquad\qquad (5.12)$$
$$\gamma_y(k) = 0, \quad k > 1$$

Similarly, we have the autocorrelation function as

$$\rho_y(1) = \frac{-\theta_1}{1 + \theta_1^2}$$
$$\rho_y(k) = 0, \quad k > 1 \qquad\qquad (5.13)$$

From Eq. (5.13), we can see that the first lag autocorrelation in MA(1) is bounded as

$$\left|\rho_y(1)\right| = \frac{|\theta_1|}{1 + \theta_1^2} \leq \frac{1}{2} \qquad\qquad (5.14)$$

and the autocorrelation function cuts off after lag 1.

Consider, for example, the following MA(1) model:

$$y_t = 40 + \varepsilon_t + 0.8\varepsilon_{t-1}$$

A realization of this model with its sample ACF is given in Figure 5.2. A visual inspection reveals that the mean and variance remain stable while there are some short runs where successive observations tend to follow each other for very brief durations, suggesting that there is indeed some positive autocorrelation in the data as revealed in the sample ACF plot.

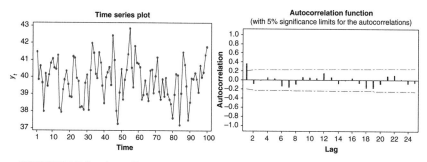

FIGURE 5.2 A realization of the MA(1) process, $y_t = 40 + \varepsilon_t + 0.8\varepsilon_{t-1}$.

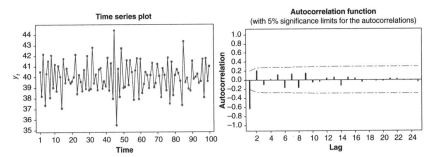

FIGURE 5.3   A realization of the MA(1) process, $y_t = 40 + \varepsilon_t - 0.8\varepsilon_{t-1}$.

We can also consider the following model:

$$y_t = 40 + \varepsilon_t - 0.8\varepsilon_{t-1}$$

A realization of this model is given in Figure 5.3. We can see that observations tend to oscillate successively. This suggests a negative autocorrelation as confirmed by the sample ACF plot.

5.3.2 The Second-Order Moving Average Process, MA(2)

Another useful finite order moving average process is MA(2), given as

$$
\begin{aligned}
y_t &= \mu + \varepsilon_t - \theta_1 \varepsilon_{t-1} - \theta_2 \varepsilon_{t-2} \\
&= \mu + \left(1 - \theta_1 B - \theta_2 B^2\right) \varepsilon_t
\end{aligned}
\tag{5.15}
$$

The autocovariance and autocorrelation functions for the MA(2) model are given as

$$
\begin{aligned}
\gamma_y (0) &= \sigma^2 \left(1 + \theta_1^2 + \theta_2^2\right) \\
\gamma_y (1) &= \sigma^2 (-\theta_1 + \theta_1 \theta_2) \\
\gamma_y (2) &= \sigma^2 (-\theta_2) \\
\gamma_y (k) &= 0, \quad k > 2
\end{aligned}
\tag{5.16}
$$

and

$$
\begin{aligned}
\rho_y (1) &= \frac{-\theta_1 + \theta_1 \theta_2}{1 + \theta_1^2 + \theta_2^2} \\
\rho_y (2) &= \frac{-\theta_2}{1 + \theta_1^2 + \theta_2^2} \\
\rho_y (k) &= 0, \quad k > 2
\end{aligned}
\tag{5.17}
$$

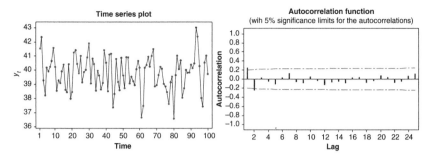

FIGURE 5.4 A realization of the MA(2) process, $y_t = 40 + \varepsilon_t + 0.7\varepsilon_{t-1} - 0.28\varepsilon_{t-2}$.

Figure 5.4 shows the time series plot and the autocorrelation function for a realization of the MA(2) model:

$$y_t = 40 + \varepsilon_t + 0.7\varepsilon_{t-1} - 0.28\varepsilon_{t-2}$$

Note that the sample ACF cuts off after lag 2.

5.4 FINITE ORDER AUTOREGRESSIVE PROCESSES

As mentioned in Section 5.1, while it is quite powerful and important, Wold's decomposition theorem does not help us much in our modeling and forecasting efforts as it implicitly requires the estimation of the infinitely many weights, $\{\psi_i\}$. In Section 5.2 we discussed a special case of this decomposition of the time series by assuming that it can be adequately modeled by only estimating a finite number of weights and setting the rest equal to 0. Another interpretation of the finite order MA processes is that at any given time, of the infinitely many past disturbances, only a finite number of those disturbances "contribute" to the current value of the time series and that the time window of the contributors "moves" in time, making the "oldest" disturbance obsolete for the next observation. It is indeed not too far fetched to think that some processes might have these intrinsic dynamics. However, for some others, we may be required to consider the "lingering" contributions of the disturbances that happened back in the past. This will of course bring us back to square one in terms of our efforts in estimating infinitely many weights. Another solution to this problem is through the autoregressive models in which the infinitely many weights are assumed to follow a distinct pattern and can be successfully represented with only a handful of parameters. We shall now consider some special cases of autoregressive processes.

5.4.1 First-Order Autoregressive Process, AR(1)

Let us first consider again the time series given in Eq. (5.2):

$$y_t = \mu + \sum_{i=0}^{\infty} \psi_i \varepsilon_{t-i}$$

$$= \mu + \sum_{i=0}^{\infty} \psi_i B^i \varepsilon_t$$

$$= \mu + \Psi(B) \varepsilon_t$$

where $\Psi(B) = \sum_{i=0}^{\infty} \psi_i B^i$. As in the finite order MA processes, one approach to modeling this time series is to assume that the contributions of the disturbances that are way in the past should be small compared to the more recent disturbances that the process has experienced. Since the disturbances are independently and identically distributed random variables, we can simply assume a set of infinitely many weights in descending magnitudes reflecting the diminishing magnitudes of contributions of the disturbances in the past. A simple, yet intuitive set of such weights can be created following an exponential decay pattern. For that we will set $\psi_i = \phi^i$, where $|\phi| < 1$ to guarantee the exponential "decay." In this notation, the weights on the disturbances starting from the current disturbance and going back in past will be $1, \phi, \phi^2, \phi^3, \ldots$ Hence Eq. (5.2) can be written as

$$y_t = \mu + \varepsilon_t + \phi \varepsilon_{t-1} + \phi^2 \varepsilon_{t-2} + \cdots$$
$$= \mu + \sum_{i=0}^{\infty} \phi^i \varepsilon_{t-i} \tag{5.18}$$

From Eq. (5.18), we also have

$$y_{t-1} = \mu + \varepsilon_{t-1} + \phi \varepsilon_{t-2} + \phi^2 \varepsilon_{t-3} + \cdots \tag{5.19}$$

We can then combine Eqs. (5.18) and (5.19) as

$$y_t = \mu + \varepsilon_t + \underbrace{\phi \varepsilon_{t-1} + \phi^2 \varepsilon_{t-2} + \cdots}_{=\phi y_{t-1} - \phi \mu}$$
$$= \underbrace{\mu - \phi \mu}_{=\delta} + \phi y_{t-1} + \varepsilon_t \tag{5.20}$$
$$= \delta + \phi y_{t-1} + \varepsilon_t$$

where $\delta = (1 - \phi)\mu$. The process in Eq. (5.20) is called a **first-order autoregressive process**, AR(1), because Eq. (5.20) can be seen as a regression of y_t on y_{t-1} and hence the term **auto**regressive process.

The assumption of $|\phi| < 1$ results in the weights that decay exponentially in time and also guarantees that $\sum_{i=0}^{+\infty} |\psi_i| < \infty$. This means that an AR(1) process is stationary if $|\phi| < 1$. For $|\phi| > 1$, past disturbances will get exponentially increasing weights as time goes on and the resulting time series will be explosive. Box et al. (2008) argue that this type of processes are of little practical interest and therefore only consider cases where $|\phi| = 1$ and $|\phi| < 1$. The solution in (5.18) does indeed not converge for $|\phi| > 1$. We can however rewrite the AR(1) process for y_{t+1}

$$y_{t+1} = \phi y_t + a_{t+1} \tag{5.21}$$

For y_t, we then have

$$
\begin{aligned}
y_t &= -\phi^{-1}\mu + \phi^{-1}y_{t+1} - \phi^{-1}\varepsilon_{t+1} \\
&= -\phi^{-1}\mu + \phi^{-1}\left(-\phi^{-1}\mu + \phi^{-1}y_{t+2} - \phi^{-1}\varepsilon_{t+2}\right) - \phi^{-1}\varepsilon_{t+1} \\
&= -\left(\phi^{-1} + \phi^{-2}\right)\mu + \phi^{-2}y_{t+2} - \phi^{-1}\varepsilon_{t+1} - \phi^{-2}\varepsilon_{t+2} \\
&\vdots \\
&= -\mu \sum_{i=1}^{\infty} \phi^{-1} - \sum_{i=1}^{\infty} \phi^{-1}\varepsilon_{t+1}
\end{aligned}
\tag{5.22}
$$

For $|\phi| > 1$ we have $|\phi^{-1}| < 1$ and therefore the solution for y_t given in (5.22) is stationary. The only problem is that it involves future values of disturbances. This of course is impractical as this type of models requires knowledge about the future to make forecasts about it. These are called non-causal models. Therefore there exists a stationary solution for an AR(1) process when $|\phi| > 1$, however, it results in a non-causal model. Throughout the book when we discuss the stationary autoregressive models, we implicitly refer to the causal autoregressive models. We can in fact show that an AR(I) process is nonstationary if and only if $|\phi| = 1$.

The mean of a stationary AR(1) process is

$$E(y_t) = \mu = \frac{\delta}{1 - \phi} \tag{5.23}$$

The autocovariance function of a stationary AR(1) can be calculated from Eq. (5.18) as

$$\gamma(k) = \sigma^2 \phi^k \frac{1}{1 - \phi^2} \quad \text{for } k = 0, 1, 2, \ldots \tag{5.24}$$

The covariance is then given as

$$\gamma(0) = \sigma^2 \frac{1}{1 - \phi^2} \tag{5.25}$$

Correspondingly, the autocorrelation function for a stationary AR(1) process is given as

$$\rho(k) = \frac{\gamma(k)}{\gamma(0)} = \phi^k \quad \text{for } k = 0, 1, 2, \ldots \tag{5.26}$$

Hence the ACF for an AR(1) process has an exponential decay form.

A realization of the following AR(1) model,

$$y_t = 8 + 0.8y_{t-1} + \varepsilon_t$$

is shown in Figure 5.5. As in the MA(1) model with $\theta = -0.8$, we can observe some short runs during which observations tend to move in the upward or downward direction. As opposed to the MA(1) model, however, the duration of these runs tends to be longer and the trend tends to linger. This can also be observed in the sample ACF plot.

Figure 5.6 shows a realization of the AR(1) model $y_t = 8 - 0.8y_{t-1} + \varepsilon_t$. We observe that instead of lingering runs, the observations exhibit jittery up/down movements because of the negative ϕ value.

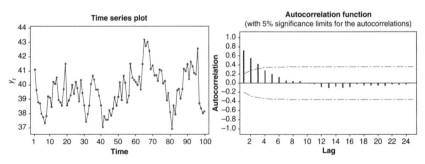

FIGURE 5.5   A realization of the AR(1) process, $y_t = 8 + 0.8y_{t-1} + \varepsilon_t$.

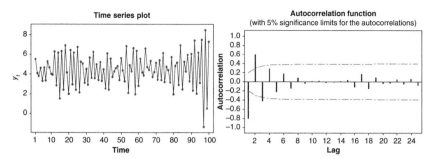

FIGURE 5.6   A realization of the AR(1) process, $y_t = 8 - 0.8y_{t-1} + \varepsilon_t$.

5.4.2 Second-Order Autoregressive Process, AR(2)

In this section, we will first start with the obvious extension of Eq. (5.20) to include the observation y_{t-2} as

$$y_t = \delta + \phi_1 y_{t-1} + \phi_2 y_{t-2} + \varepsilon_t \tag{5.27}$$

We will then show that Eq. (5.27) can be represented in the infinite MA form and provide the conditions of stationarity for y_t in terms of ϕ_1 and ϕ_2. For that we will rewrite Eq. (5.27) as

$$(1 - \phi_1 B - \phi_2 B^2)y_t = \delta + \varepsilon_t \tag{5.28}$$

or

$$\Phi(B)y_t = \delta + \varepsilon_t \tag{5.29}$$

Furthermore, applying $\Phi(B)^{-1}$ to both sides, we obtain

$$
\begin{aligned}
y_t &= \underbrace{\Phi(B)^{-1}\delta}_{=\mu} + \underbrace{\Phi(B)^{-1}}_{=\Psi(B)}\varepsilon_t \\
&= \mu + \Psi(B)\varepsilon_t \\
&= \mu + \sum_{i=0}^{\infty} \psi_i \varepsilon_{t-i} \\
&= \mu + \sum_{i=0}^{\infty} \psi_i B^i \varepsilon_t
\end{aligned}
\tag{5.30}
$$

where

$$\mu = \Phi(B)^{-1} \delta \tag{5.31}$$

and

$$\Phi(B)^{-1} = \sum_{i=0}^{\infty} \psi_i B^i = \Psi(B) \tag{5.32}$$

We can use Eq. (5.32) to obtain the weights in Eq. (5.30) in terms of ϕ_1 and ϕ_2. For that, we will use

$$\Phi(B)\,\Psi(B) = 1 \tag{5.33}$$

That is,

$$(1 - \phi_1 B - \phi_2 B^2)(\psi_0 + \psi_1 B + \psi_2 B^2 + \cdots) = 1$$

or

$$\psi_0 + (\psi_1 - \phi_1 \psi_0)B + (\psi_2 - \phi_1 \psi_1 - \phi_2 \psi_0)B^2$$
$$+ \cdots + (\psi_j - \phi_1 \psi_{j-1} - \phi_2 \psi_{j-2})B^j + \cdots = 1 \tag{5.34}$$

Since on the right-hand side of the Eq. (5.34) there are no backshift operators, for $\Phi(B)\,\Psi(B) = 1$, we need

$$\psi_0 = 1$$
$$(\psi_1 - \phi_1 \psi_0) = 0 \tag{5.35}$$
$$(\psi_j - \phi_1 \psi_{j-1} - \phi_2 \psi_{j-2}) = 0 \quad \text{for all } j = 2, 3, \ldots$$

The equations in (5.35) can indeed be solved for each ψ_j in a futile attempt to estimate infinitely many parameters. However, it should be noted that the ψ_j in Eq. (5.35) satisfy the second-order linear difference equation and that they can be expressed as the solution to this equation in terms of the two roots m_1 and m_2 of the associated polynomial

$$m^2 - \phi_1 m - \phi_2 = 0 \tag{5.36}$$

If the roots obtained by

$$m_1, m_2 = \frac{\phi_1 \pm \sqrt{\phi_1^2 + 4\phi_2}}{2}$$

satisfy $|m_1|, |m_2| < 1$, then we have $\sum_{i=0}^{+\infty} |\psi_i| < \infty$. Hence if the roots m_1 and m_2 are both less than 1 in absolute value, then the AR(2) model is causal and stationary. Note that if the roots of Eq. (5.36) are complex conjugates of the form $a \pm ib$, the condition for stationarity is that $\sqrt{a^2 + b^2} < 1$. Furthermore, under the condition that $|m_1|, |m_2| < 1$, the AR(2) time series, $\{y_t\}$, has an infinite MA representation as in Eq. (5.30).

This implies that for the second-order autoregressive process to be stationary, the parameters ϕ_1 and ϕ_2 must satisfy.

$$\phi_1 + \phi_2 < 1$$
$$\phi_2 - \phi_1 < 1$$
$$|\phi_2| < 1$$

Now that we have established the conditions for the stationarity of an AR(2) time series, let us now consider its mean, autocovariance, and autocorrelation functions. From Eq. (5.27), we have

$$E(y_t) = \delta + \phi_1 E(y_{t-1}) + \phi_2 E(y_{t-2}) + 0$$
$$\mu = \delta + \phi_1 \mu + \phi_2 \mu$$
$$\Rightarrow \mu = \frac{\delta}{1 - \phi_1 - \phi_2} \tag{5.37}$$

Note that for $1 - \phi_1 - \phi_2 = 0$, $m = 1$ is one of the roots for the associated polynomial in Eq. (5.36) and hence the time series is deemed nonstationary. The autocovariance function is

$$\begin{aligned}
\gamma(k) &= \text{Cov}(y_t, y_{t-k}) \\
&= \text{Cov}(\delta + \phi_1 y_{t-1} + \phi_2 y_{t-2} + \varepsilon_t, y_{t-k}) \\
&= \phi_1 \text{Cov}(y_{t-1}, y_{t-k}) + \phi_2 \text{Cov}(y_{t-2}, y_{t-k}) + \text{Cov}(\varepsilon_t, y_{t-k}) \quad (5.38) \\
&= \phi_1 \gamma(k-1) + \phi_2 \gamma(k-2) + \begin{cases} \sigma^2 & \text{if } k = 0 \\ 0 & \text{if } k > 0 \end{cases}
\end{aligned}$$

Thus $\gamma(0) = \phi_1\gamma(1) + \phi_2\gamma(2) + \sigma^2$ and

$$\gamma(k) = \phi_1\gamma(k-1) + \phi_2\gamma(k-2), \quad k = 1, 2, \ldots \tag{5.39}$$

The equations in (5.39) are called the **Yule–Walker** equations for $\gamma(k)$. Similarly, we can obtain the autocorrelation function by dividing Eq. (5.39) by $\gamma(0)$:

$$\rho(k) = \phi_1\rho(k-1) + \phi_2\rho(k-2), \quad k = 1, 2, \ldots \tag{5.40}$$

The Yule–Walker equations for $\rho(k)$ in Eq. (5.40) can be solved recursively as

$$\rho(1) = \phi_1 \underbrace{\rho(0)}_{=1} + \phi_2 \underbrace{\rho(-1)}_{=\rho(1)}$$

$$= \frac{\phi_1}{1 - \phi_2}$$

$$\rho(2) = \phi_1\rho(1) + \phi_2$$

$$\rho(3) = \phi_1\rho(2) + \phi_2\rho(1)$$

$$\vdots$$

A general solution can be obtained through the roots m_1 and m_2 of the associated polynomial $m^2 - \phi_1 m - \phi_2 = 0$. There are three cases.

Case 1. If m_1 and m_2 are distinct, real roots, we then have

$$\rho(k) = c_1 m_1^k + c_2 m_2^k, \quad k = 0, 1, 2, \ldots \tag{5.41}$$

where c_1 and c_2 are particular constants and can, for example, be obtained from $\rho(0)$ and $\rho(1)$. Moreover, since for stationarity we have $|m_1|, |m_2| < 1$, in this case, the autocorrelation function is a **mixture of two exponential decay terms**.

Case 2. If m_1 and m_2 are complex conjugates in the form of $a \pm ib$, we then have

$$\rho(k) = R^k \left[c_1 \cos(\lambda k) + c_2 \sin(\lambda k) \right], \quad k = 0, 1, 2, \ldots \tag{5.42}$$

where $R = |m_i| = \sqrt{a^2 + b^2}$ and λ is determined by $\cos(\lambda) = a/R$, $\sin(\lambda) = b/R$. Hence we have $a \pm ib = R[\cos(\lambda) \pm i \sin(\lambda)]$. Once again c_1 and c_2 are particular constants. The ACF in this case has the form of a **damped sinusoid**, with damping factor R and frequency λ; that is, the period is $2\pi / \lambda$.

Case 3. If there is one real root m_0, $m_1 = m_2 = m_0$, we then have

$$\rho(k) = (c_1 + c_2 k)m_0^k \quad k = 0, 1, 2, \ldots \tag{5.43}$$

In this case, the ACF will exhibit an exponential decay pattern.

In case 1, for example, an AR(2) model can be seen as an "adjusted" AR(1) model for which a single exponential decay expression as in the AR(1) model is not enough to describe the pattern in the ACF, and hence an additional exponential decay expression is "added" by introducing the second lag term, y_{t-2}.

Figure 5.7 shows a realization of the AR(2) process

$$y_t = 4 + 0.4y_{t-1} + 0.5y_{t-2} + \varepsilon_t$$

Note that the roots of the associated polynomial of this model are real. Hence the ACF is a mixture of two exponential decay terms.

Similarly, Figure 5.8 shows a realization of the following AR(2) process

$$y_t = 4 + 0.8y_{t-1} - 0.5y_{t-2} + \varepsilon_t.$$

For this process, the roots of the associated polynomial are complex conjugates. Therefore the ACF plot exhibits a damped sinusoid behavior.

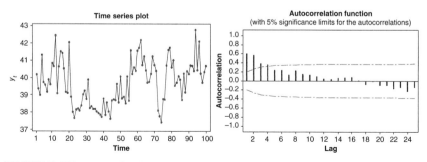

FIGURE 5.7 A realization of the AR(2) process, $y_t = 4 + 0.4y_{t-1} + 0.5y_{t-2} + \varepsilon_t$.

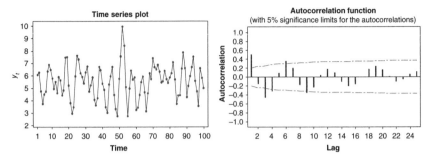

FIGURE 5.8   A realization of the AR(2) process, $y_t = 4 + 0.8y_{t-1} - 0.5y_{t-2} + \varepsilon_t$.

5.4.3 General Autoregressive Process, AR(p)

From the previous two sections, a general, pth-order AR model is given as

$$y_t = \delta + \phi_1 y_{t-1} + \phi_2 y_{t-2} + \cdots + \phi_p y_{t-p} + \varepsilon_t \qquad (5.44)$$

where ε_t is white noise. Another representation of Eq. (5.44) can be given as

$$\Phi(B)y_t = \delta + \varepsilon_t \qquad (5.45)$$

where $\Phi(B) = 1 - \phi_1 B - \phi_2 B^2 - \cdots - \phi_p B^p$.

The AR(p) time series $\{y_t\}$ in Eq. (5.44) is causal and stationary if the roots of the associated polynomial

$$m^p - \phi_1 m^{p-1} - \phi_2 m^{p-2} - \cdots - \phi_p = 0 \qquad (5.46)$$

are less than one in absolute value. Furthermore, under this condition, the AR(p) time series $\{y_t\}$ is also said to have an **absolutely summable** infinite MA representation

$$y_t = \mu + \Psi(B)\,\varepsilon_t = \mu + \sum_{i=0}^{\infty} \psi_i \varepsilon_{t-i} \qquad (5.47)$$

where $\Psi(B) = \Phi(B)^{-1}$ with $\sum_{i=0}^{\infty} |\psi_i| < \infty$.

As in AR(2), the weights of the random shocks in Eq. (5.47) can be obtained from $\Phi(B)\,\Psi(B) = 1$ as

$$\psi_j = 0, \quad j < 0$$
$$\psi_0 = 1 \tag{5.48}$$
$$\psi_j - \phi_1 \psi_{j-1} - \phi_2 \psi_{j-2} - \cdots - \phi_p \psi_{j-p} = 0 \quad \text{for all } j = 1, 2, \ldots$$

We can easily show that, for the stationary AR(p) process

$$E(y_t) = \mu = \frac{\delta}{1 - \phi_1 - \phi_2 - \cdots - \phi_p}$$

and

$$
\begin{aligned}
\gamma(k) &= \mathrm{Cov}(y_t, y_{t-k}) \\
&= \mathrm{Cov}(\delta + \phi_1 y_{t-1} + \phi_2 y_{t-2} + \cdots + \phi_p y_{t-p} + \varepsilon_t, y_{t-k}) \\
&= \sum_{i=1}^{p} \phi_i \mathrm{Cov}(y_{t-i}, y_{t-k}) + \mathrm{Cov}(\varepsilon_t, y_{t-k}) \\
&= \sum_{i=1}^{p} \phi_i \gamma(k-i) + \begin{cases} \sigma^2 & \text{if } k = 0 \\ 0 & \text{if } k > 0 \end{cases}
\end{aligned}
\tag{5.49}
$$

Thus we have

$$\gamma(0) = \sum_{i=1}^{p} \phi_i \gamma(i) + \sigma^2 \tag{5.50}$$

$$\Rightarrow \gamma(0)\left[1 - \sum_{i=1}^{p} \phi_i \rho(i) \right] = \sigma^2 \tag{5.51}$$

By dividing Eq. (5.49) by $\gamma(0)$ for $k > 0$, it can be observed that the ACF of an AR(p) process satisfies the Yule–Walker equations

$$\rho(k) = \sum_{i=1}^{p} \phi_i \rho(k-i), \quad k = 1, 2, \ldots \tag{5.52}$$

The equations in (5.52) are pth-order **linear difference equations**, implying that the ACF for an AR(p) model can be found through the p roots of the associated polynomial in Eq. (5.46). For example, if the roots are all distinct and real, we have

$$\rho(k) = c_1 m_1^k + c_2 m_2^k + \cdots + c_p m_p^k, \quad k = 1, 2, \ldots \tag{5.53}$$

where $c_1, c_2, \ldots, c_p$ are particular constants. However, in general, the roots may not all be distinct or real. Thus the ACF of an AR(p) process can be a **mixture of exponential decay and damped sinusoid** expressions depending on the roots of Eq. (5.46).

5.4.4 Partial Autocorrelation Function, PACF

In Section 5.2, we saw that the ACF is an excellent tool in identifying the order of an MA(q) process, because it is expected to "cut off" after lag q. However, in the previous section, we pointed out that the ACF is not as useful in the identification of the order of an AR(p) process for which it will most likely have a mixture of exponential decay and damped sinusoid expressions. Hence such behavior, while indicating that the process might have an AR structure, fails to provide further information about the order of such structure. For that, we will define and employ the **partial autocorrelation function** (PACF) of the time series. But before that, we discuss the concept of partial correlation to make the interpretation of the PACF easier.

Partial Correlation Consider three random variables X, Y, and Z. Then consider simple linear regression of X on Z and Y on Z as

$$\hat{X} = a_1 + b_1 Z \quad \text{where } b_1 = \frac{\text{Cov}(Z, X)}{\text{Var}(Z)}$$

and

$$\hat{Y} = a_2 + b_2 Z \quad \text{where } b_2 = \frac{\text{Cov}(Z, Y)}{\text{Var}(Z)}$$

Then the errors can be obtained from

$$X^* = X - \hat{X} = X - (a_1 + b_1 Z)$$

and

$$Y^* = Y - \hat{Y} = Y - (a_2 + b_2 Z)$$

Then the **partial correlation** between X and Y after adjusting for Z is defined as the correlation between X^* and Y^*; $\text{corr}(X^*, Y^*) = \text{corr}(X - \hat{X}, Y - \hat{Y})$. That is, partial correlation can be seen as the correlation between two variables after being adjusted for a common factor that may be affecting them. The generalization is of course possible by allowing for adjustment for more than just one factor.

Partial Autocorrelation Function Following the above definition, the **PACF** between y_t and y_{t-k} is the autocorrelation between y_t and y_{t-k} after adjusting for $y_{t-1}, y_{t-2}, \ldots, y_{t-k+1}$. Hence for an AR($p$) model the PACF between y_t and y_{t-k} for $k > p$ should be equal to zero. A more formal definition can be found below.

Consider a stationary time series model $\{y_t\}$ that is not necessarily an AR process. Further consider, for any fixed value of k, the Yule–Walker equations for the ACF of an AR(p) process given in Eq. (5.52) as

$$\rho(j) = \sum_{i=1}^{k} \phi_{ik} \rho(j - i), \quad j = 1, 2, \ldots, k \tag{5.54}$$

or

$$\rho(1) = \phi_{1k} + \phi_{2k}\rho(1) + \cdots + \phi_{kk}\rho(k - 1)$$
$$\rho(2) = \phi_{1k}\rho(1) + \phi_{2k} + \cdots + \phi_{kk}\rho(k - 2)$$
$$\vdots$$
$$\rho(k) = \phi_{1k}\rho(k-) + \phi_{2k}\rho(k - 2) + \cdots + \phi_{kk}$$

Hence we can write the equations in (5.54) in matrix notation as

$$\begin{bmatrix} 1 & \rho(1) & \rho(2) & \cdots & \rho(k-1) \\ \rho(1) & 1 & \rho(3) & \cdots & \rho(k-2) \\ \rho(2) & \rho(1) & 1 & \cdots & \rho(k-3) \\ \vdots & \vdots & \vdots & \ddots & \vdots \\ \rho(k-1) & \rho(k-2) & \rho(k-3) & \cdots & 1 \end{bmatrix} \begin{bmatrix} \phi_{1k} \\ \phi_{2k} \\ \phi_{3k} \\ \vdots \\ \phi_{kk} \end{bmatrix} = \begin{bmatrix} \rho(1) \\ \rho(2) \\ \rho(3) \\ \vdots \\ \rho(k) \end{bmatrix} \tag{5.55}$$

or

$$\mathbf{P}_k \boldsymbol{\phi}_k = \boldsymbol{\rho}_k \tag{5.56}$$

where

$$\mathbf{P}_k = \begin{bmatrix} 1 & \rho(1) & \rho(2) & \cdots & \rho(k-1) \\ \rho(1) & 1 & \rho(3) & \cdots & \rho(k-2) \\ \rho(2) & \rho(1) & 1 & \cdots & \rho(k-3) \\ \vdots & \vdots & \vdots & \ddots & \vdots \\ \rho(k-1) & \rho(k-2) & \rho(k-3) & \cdots & 1 \end{bmatrix},$$

$$\phi_k = \begin{bmatrix} \phi_{1k} \\ \phi_{2k} \\ \phi_{3k} \\ \vdots \\ \phi_{kk} \end{bmatrix}, \quad \text{and} \quad \rho_k = \begin{bmatrix} \rho(1) \\ \rho(2) \\ \rho(3) \\ \vdots \\ \rho(k) \end{bmatrix}.$$

Thus to solve for ϕ_k, we have

$$\phi_k = \mathbf{P}_k^{-1} \rho_k \tag{5.57}$$

For any given k, $k = 1, 2, \ldots$, the last coefficient ϕ_{kk} is called the partial autocorrelation of the process at lag k. Note that for an AR(p) process $\phi_{kk} = 0$ for $k > p$. Hence we say that the PACF cuts off after lag p for an AR(p). This suggests that the PACF can be used in identifying the order of an AR process similar to how the ACF can be used for an MA process.

For sample calculations, $\hat{\phi}_{kk}$, the sample estimate of ϕ_{kk}, is obtained by using the sample ACF, $r(k)$. Furthermore, in a sample of N observations from an AR(p) process, $\hat{\phi}_{kk}$ for $k > p$ is approximately normally distributed with

$$E(\hat{\phi}_{kk}) \approx 0 \quad \text{and} \quad \text{Var}(\hat{\phi}_{kk}) \approx \frac{1}{N} \tag{5.58}$$

Hence the 95% limits to judge whether any $\hat{\phi}_{kk}$ is statistically significantly different from zero are given by $\pm 2/\sqrt{N}$. For further detail see Quenouille (1949), Jenkins (1954, 1956), and Daniels (1956).

Figure 5.9 shows the sample PACFs of the models we have considered so far. In Figure 5.9a we have the sample PACF of the realization of the MA(1) model with $\theta = 0.8$ given in Figure 5.3. It exhibits an exponential decay pattern. Figure 5.9b shows the sample PACF of the realization of the MA(2) model in Figure 5.4 and it also has an exponential decay pattern in absolute value since for this model the roots of the associated polynomial are real. Figures 5.9c and 5.9d show the sample PACFs of the realization of the AR(1) model with $\phi = 0.8$ and $\phi = -0.8$, respectively. In both

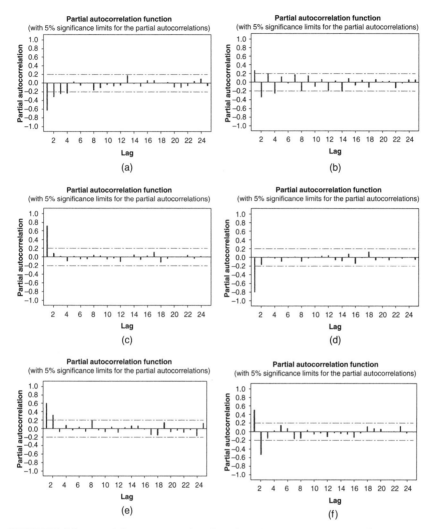

FIGURE 5.9 Partial autocorrelation functions for the realizations of (a) MA(1) process, $y_t = 40 + \varepsilon_t - 0.8\varepsilon_{t-1}$; (b) MA(2) process, $y_t = 40 + \varepsilon_t + 0.7\varepsilon_{t-1} - 0.28\varepsilon_{t-2}$; (c) AR(1) process, $y_t = 8 + 0.8y_{t-1} + \varepsilon_t$; (d) AR(1) process, $y_t = 8 - 0.8y_{t-1} + \varepsilon_t$; (e) AR(2) process, $y_t = 4 + 0.4y_{t-1} + 0.5y_{t-2} + \varepsilon_t$; and (f) AR(2) process, $y_t = 4 + 0.8y_{t-1} - 0.5y_{t-2} + \varepsilon_t$.

cases the PACF "cuts off" after the first lag. That is, the only significant sample PACF value is at lag 1, suggesting that the AR(1) model is indeed appropriate to fit the data. Similarly, in Figures 5.9e and 5.9f, we have the sample PACFs of the realizations of the AR(2) model. Note that the sample PACF cuts off after lag 2.

As we discussed in Section 5.3, finite order MA processes are stationary. On the other hand as in the causality concept we discussed for the autoregressive processes, we will impose some restrictions on the parameters of the MA models as well. Consider for example the MA(1) model in (5.11)

$$y_t = \varepsilon_t - \theta_1 \varepsilon_{t-1} \tag{5.59}$$

Note that for the sake of simplicity, in (5.59) we consider a centered process, i.e. $E(y_t) = 0$.

We can then rewrite (5.59) as

$$
\begin{aligned}
\varepsilon_t &= y_t + \theta_1 \varepsilon_{t-1} \\
&= y_t + \theta_1 \left[y_{t-1} + \theta_1 \varepsilon_{t-2} \right] \\
&= y_t + \theta_1 y_{t-1} + \theta_1^2 \varepsilon_{t-2} \\
&\quad \vdots \\
&= \sum_{i=0}^{\infty} \theta_1^i y_{t-i}
\end{aligned}
\tag{5.60}
$$

It can be seen from (5.60) that for $|\theta_1| < 1$, ε_t is a convergent series of current and past observations and the process is called an **invertible** moving average process. Similar to the causality argument, for $|\theta_1| > 1$, ε_t can be written as a convergent series of future observations and is called noninvertible. When $|\theta_1| = 1$, the MA(1) process is considered noninvertible in a more restricted sense (Brockwell and Davis (1991)).

The direct implication of invertibility becomes apparent in model identification. Consider the MA(1) process as an example. The first lag autocorrelation for that process is given as

$$\rho(1) = \frac{-\theta_1}{1 + \theta_1^2} \tag{5.61}$$

This allows for the calculation of θ_1 for a given $\rho(1)$ by rearranging (5.61) as

$$\theta_1^2 - \frac{\theta_1}{\rho(1)} + 1 = 0 \tag{5.62}$$

and solving for θ_1. Except for the case of a repeated root, this equation has two solutions. Consider for example $\rho(1) = 0.4$ for which both $\theta_1 = 0.5$ and $\theta_1 = 2$ are the solutions for (5.62). Following the above argument, only $\theta_1 = 0.5$ yields the invertible MA(1) process. It can be shown that when there are multiple solutions for possible values of MA parameters, there is only one solution that will satisfy the invertibility condition (Box et al. (2008), Section 6.4.1).

Consider the MA(q) process

$$y_t = \mu + \left(1 - \sum_{i=1}^{q} \theta_i B^i \right) \varepsilon_t$$
$$= \mu + \Theta(B)\, \varepsilon_t$$

After multiplying both sides with $\Theta(B)^{-1}$, we have

$$\Theta(B)^{-1} y_t = \Theta(B)^{-1} \mu + \varepsilon_t$$
$$\Pi(B)\, y_t = \delta + \varepsilon_t \tag{5.63}$$

where $\Pi(B) = 1 - \sum_{i=1}^{\infty} \pi_i B^i = \Theta(B)^{-1}$ and $\Theta(B)^{-1} \mu = \delta$. Hence the infinite AR representation of an MA(q) process is given as

$$y_t - \sum_{i=1}^{\infty} \pi_i y_{t-i} = \delta + \varepsilon_t \tag{5.64}$$

with $\sum_{i=1}^{\infty} |\pi_i| < \infty$. The π_i can be determined from

$$(1 - \theta_1 B - \theta_2 B^2 - \cdots - \theta_q B^q)(1 - \pi_1 B - \pi_2 B^2 + \cdots) = 1 \tag{5.65}$$

which in turn yields

$$\pi_1 + \theta_1 = 0$$
$$\pi_2 - \theta_1 \pi_1 + \theta_2 = 0$$
$$\vdots \tag{5.66}$$
$$\pi_j - \theta_1 \pi_{j-1} - \cdots - \theta_q \pi_{j-q} = 0$$

with $\pi_0 = -1$ and $\pi_j = 0$ for $j < 0$. Hence as in the previous arguments for the stationarity of AR(p) models, the π_i are the solutions to the qth-order linear difference equations and therefore the condition for the invertibility of an MA(q) process turns out to be very similar to the stationarity condition of an AR(p) process: the roots of the associated polynomial given in Eq. (5.66) should be less than 1 in absolute value,

$$m^q - \theta_1 m^{q-1} - \theta_2 m^{q-2} - \cdots - \theta_q = 0 \tag{5.67}$$

An invertible MA(q) process can then be written as an infinite AR process.

Correspondingly, for such a process, adjusting for $y_{t-1}, y_{t-2}, \ldots, y_{t-k+1}$ does not necessarily eliminate the correlation between y_t and y_{t-k} and therefore its PACF will never "cut off." In general, the PACF of an MA(q) process is a **mixture of exponential decay and damped sinusoid** expressions.

The ACF and the PACF do have very distinct and indicative properties for MA and AR models, respectively. Therefore, in model identification, we strongly recommend the use of both the sample ACF and the sample PACF **simultaneously**.

5.5 MIXED AUTOREGRESSIVE–MOVING AVERAGE PROCESSES

In the previous sections we have considered special cases of Wold's decomposition of a stationary time series represented as a weighted sum of infinite random shocks. In an AR(1) process, for example, the weights in the infinite sum are forced to follow an exponential decay form with ϕ as the rate of decay. Since there are no restrictions apart from $\sum_{i=0}^{\infty} \psi_i^2 < \infty$ on the weights (ψ_i), it may not be possible to approximate them by an exponential decay pattern. For that, we will need to increase the order of the AR model to approximate any pattern that these weights may in fact be exhibiting. On some occasions, however, it is possible to make simple adjustments to the exponential decay pattern by adding only a few terms and hence to have a more parsimonious model. Consider, for example, that the weights ψ_i do indeed exhibit an exponential decay pattern with a constant rate except for the fact that ψ_1 is not equal to this rate of decay as it would be in the case of an AR(1) process. Hence instead of increasing the order of the AR model to accommodate for this "anomaly," we can add an MA(1) term that will simply adjust ψ_1 while having no effect on the rate of exponential decay pattern of the rest of the weights. This results in a mixed **autoregressive moving average** or ARMA(1,1) model. In general, an ARMA(p, q) model is given as

$$
\begin{aligned}
y_t &= \delta + \phi_1 y_{t-1} + \phi_2 y_{t-2} + \cdots + \phi_p y_{t-p} + \varepsilon_t - \theta_1 \varepsilon_{t-1} \\
&\quad - \theta_2 \varepsilon_{t-2} - \cdots - \theta_q \varepsilon_{t-q} \\
&= \delta + \sum_{i=1}^{p} \phi_i y_{t-i} + \varepsilon_t - \sum_{i=1}^{q} \theta_i \varepsilon_{t-i}
\end{aligned}
\tag{5.68}
$$

or

$$\Phi(B)\, y_t = \delta + \Theta(B)\, \varepsilon_t \tag{5.69}$$

where ε_t is a white noise process.

5.5.1 Stationarity of ARMA(p, q) Process

The **stationarity** of an ARMA process is related to the AR component in the model and can be checked through the roots of the associated polynomial

$$m^p - \phi_1 m^{p-1} - \phi_2 m^{p-2} - \cdots - \phi_p = 0. \tag{5.70}$$

If all the roots of Eq. (5.70) are less than one in absolute value, then ARMA(p, q) is stationary. This also implies that, under this condition, ARMA(p, q) has an infinite MA representation as

$$y_t = \mu + \sum_{i=0}^{\infty} \psi_i \varepsilon_{t-i} = \mu + \Psi(B)\, \varepsilon_t \tag{5.71}$$

with $\Psi(B) = \Phi(B)^{-1}\,\Theta(B)$. The coefficients in $\Psi(B)$ can be found from

$$\psi_i - \phi_1 \psi_{i-1} - \phi_2 \psi_{i-2} - \cdots - \phi_p \psi_{i-p} = \begin{cases} -\theta_i, & i = 1, \ldots, q \\ 0, & i > q \end{cases} \tag{5.72}$$

and $\psi_0 = 1$.

5.5.2 Invertibility of ARMA(p, q) Process

Similar to the stationarity condition, the **invertibility** of an ARMA process is related to the MA component and can be checked through the roots of the associated polynomial

$$m^q - \theta_1 m^{q-1} - \theta_2 m^{q-2} - \cdots - \theta_q = 0 \tag{5.73}$$

If all the roots of Eq. (5.71) are less than one in absolute value, then ARMA(p, q) is said to be invertible and has an infinite AR representation,

$$\Pi(B)\, y_t = \alpha + \varepsilon_t \tag{5.74}$$

where $\alpha = \Theta(B)^{-1}\,\delta$ and $\Pi(B) = \Theta(B)^{-1}\,\Phi(B)$. The coefficients in $\Pi(B)$ can be found from

$$\pi_i - \theta_1 \pi_{i-1} - \theta_2 \pi_{i-2} - \cdots - \theta_q \pi_{i-q} = \begin{cases} \phi_i, & i = 1, \ldots, p \\ 0, & i > p \end{cases} \tag{5.75}$$

and $\pi_0 = -1$.

In Figure 5.10 we provide realizations of two ARMA(1,1) models:

$$y_t = 16 + 0.6y_{t-1} + \varepsilon_t + 0.8\varepsilon_{t-1} \quad \text{and} \quad y_t = 16 - 0.7y_{t-1} + \varepsilon_t - 0.6\varepsilon_{t-1}.$$

Note that the sample ACFs and PACFs exhibit exponential decay behavior (sometimes in absolute value depending on the signs of the AR and MA coefficients).

5.5.3 ACF and PACF of ARMA(p, q) Process

As in the stationarity and invertibility conditions, the ACF and PACF of an ARMA process are determined by the AR and MA components, respectively. It can therefore be shown that the ACF and PACF of an ARMA(p, q) both exhibit exponential decay and/or damped sinusoid patterns, which makes the identification of the order of the ARMA(p, q) model relatively more difficult. For that, additional sample functions such as the Extended Sample ACF (ESACF), the Generalized Sample PACF (GPACF), the Inverse ACF (IACF), and canonical correlations can be used. For further information see Box, Jenkins, and Reinsel (2008), Wei (2006), Tiao and Box (1981), Tsay and Tiao (1984), and Abraham and Ledolter (1984). However, the availability of sophisticated statistical software packages such as Minitab JMP and SAS makes it possible for the practitioner to consider several different models with various orders and compare them based on the model selection criteria such as AIC, AICC, and BIC as described in Chapter 2 and residual analysis.

The theoretical values of the ACF and PACF for stationary time series are summarized in Table 5.1. The summary of the sample ACFs and PACFs of the realizations of some of the models we have covered in this chapter are given in Table 5.2, Table 5.3, and Table 5.4 for MA, AR, and ARMA models, respectively.

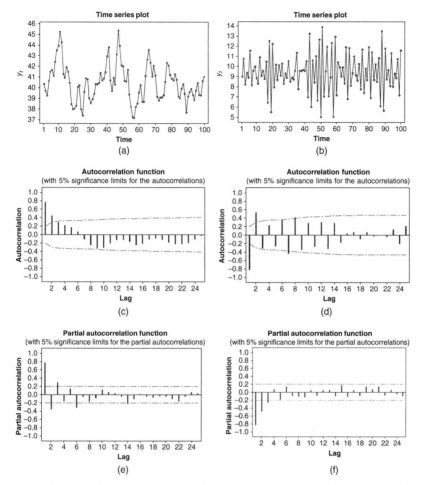

FIGURE 5.10 Two realizations of the ARMA(1,1) model: (a) $y_t = 16 + 0.6y_{t-1} + \varepsilon_t + 0.8\varepsilon_{t-1}$ and (b) $y_t = 16 - 0.7y_{t-1} + \varepsilon_t - 0.6\varepsilon_{t-1}$. (c) The ACF of (a), (d) the ACF of (b), (e) the PACF of (a), and (f) the PACF of (b).

TABLE 5.1   Behavior of Theoretical ACF and PACF for Stationary Processes

Model	ACF	PACF
MA(q)	Cuts off after lag q	Exponential decay and/or damped sinusoid
AR(p)	Exponential decay and/or damped sinusoid	Cuts off after lag p
ARMA(p,q)	Exponential decay and/or damped sinusoid	Exponential decay and/or damped sinusoid

TABLE 5.2   Sample ACFs and PACFs for Some Realizations of MA(1) and MA(2) Models

Model	Sample ACF	Sample PACF

MA(1)

$$y_t = 40 + \varepsilon_t - 0.8\varepsilon_{t-1}$$

$$y_t = 40 + \varepsilon_t + 0.8\varepsilon_{t-1}$$

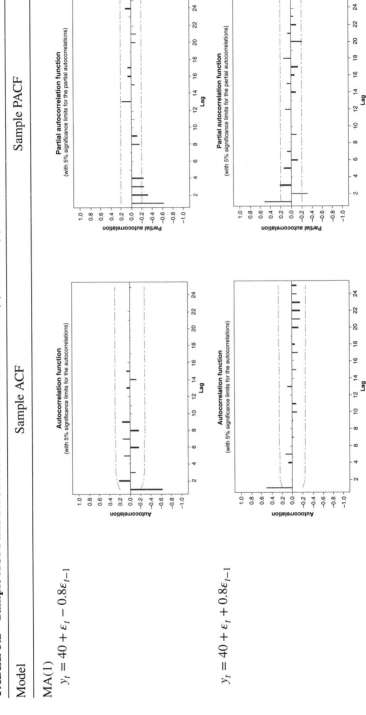

MA(2)

$$y_t = 40 + \varepsilon_t + 0.7\varepsilon_{t-1} - 0.28\varepsilon_{t-2}$$

$$y_t = 40 + \varepsilon_t - 1.1\varepsilon_{t-1} + 0.8\varepsilon_{t-2}$$

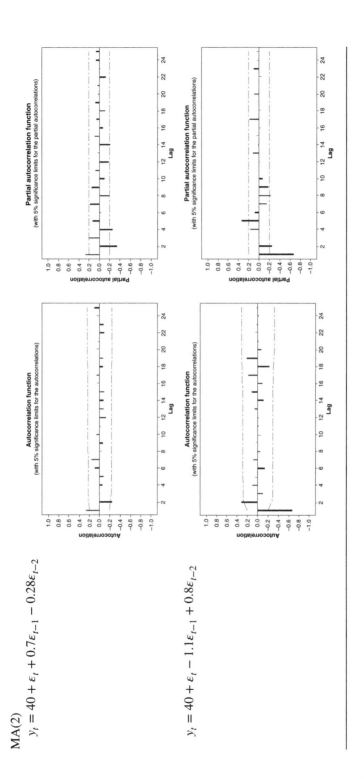

359

TABLE 5.3   Sample ACFs and PACFs for Some Realizations of AR(1) and AR(2) Models

Model	Sample ACF	Sample PACF

AR(1)
$$y_t = 8 + 0.8y_{t-1} + \varepsilon_t$$

$$y_t = 8 - 0.8y_{t-1} + \varepsilon_t$$

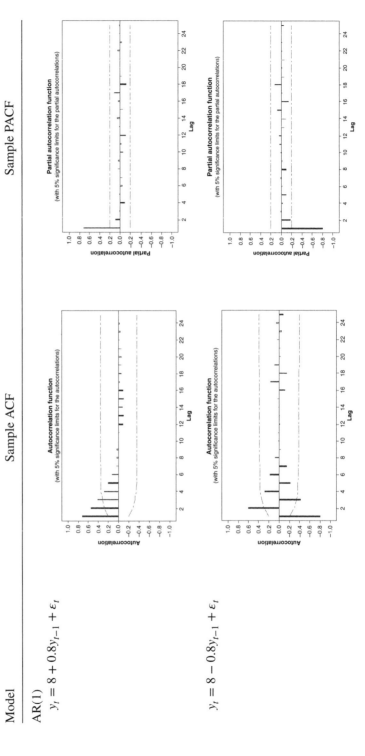

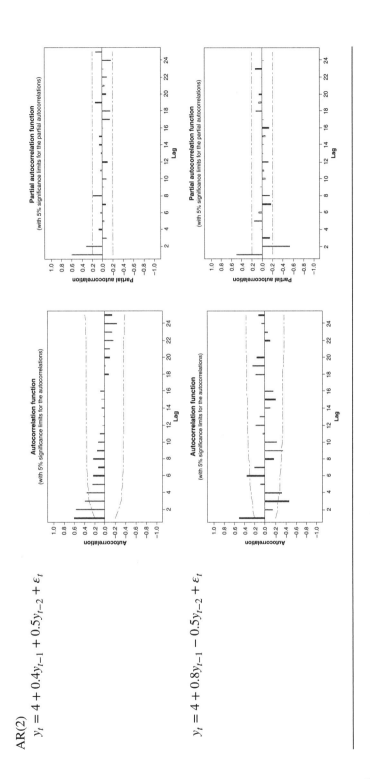

AR(2)

$$y_t = 4 + 0.4y_{t-1} + 0.5y_{t-2} + \varepsilon_t$$

$$y_t = 4 + 0.8y_{t-1} - 0.5y_{t-2} + \varepsilon_t$$

TABLE 5.4 Sample ACFs and PACFs for Some Realizations of ARMA(1,1) Models

Model	Sample ACF	Sample PACF

ARMA(1.1)

$$y_t = 16 + 0.6y_{t-1} + \varepsilon_t + 0.8\varepsilon_{t-1}$$

$$y_t = 16 - 0.7y_{t-1} + \varepsilon_t - 0.6\varepsilon_{t-1}$$

5.6 NONSTATIONARY PROCESSES

It is often the case that while the processes may not have a constant level, they exhibit homogeneous behavior over time. Consider, for example, the linear trend process given in Figure 5.1c. It can be seen that different snapshots taken in time do exhibit similar behavior except for the mean level of the process. Similarly, processes may show nonstationarity in the slope as well. We will call a time series, y_t, homogeneous nonstationary if it is not stationary but its first difference, that is, $w_t = y_t - y_{t-1} = (1 - B) y_t$, or higher-order differences, $w_t = (1 - B)^d y_t$, produce a stationary time series. We will further call y_t an **autoregressive integrated moving average** (ARIMA) process of orders p, d, and q—that is, ARIMA(p, d, q)—if its dth difference, denoted by $w_t = (1 - B)^d y_t$, produces a stationary ARMA(p, q) process. The term integrated is used since, for $d = 1$, for example, we can write y_t as the sum (or "integral") of the w_t process as

$$
\begin{aligned}
y_t &= w_t + y_{t-1} \\
&= w_t + w_{t-1} + y_{t-2} \\
&= w_t + w_{t-1} + \cdots + w_1 + y_0
\end{aligned}
\tag{5.76}
$$

Hence an ARIMA(p, d, q) can be written as

$$
\Phi(B)(1 - B)^d y_t = \delta + \Theta(B) \varepsilon_t
\tag{5.77}
$$

Thus once the differencing is performed and a stationary time series $w_t = (1 - B)^d y_t$ is obtained, the methods provided in the previous sections can be used to obtain the full model. In most applications first differencing $(d = 1)$ and occasionally second differencing $(d = 2)$ would be enough to achieve stationarity. However, sometimes transformations other than differencing are useful in reducing a nonstationary time series to a stationary one. For example, in many economic time series the variability of the observations increases as the average level of the process increases; however, the percentage of change in the observations is relatively independent of level. Therefore taking the logarithm of the original series will be useful in achieving stationarity.

5.6.1 Some Examples of ARIMA(p, d, q) Processes

The **random walk process, ARIMA(0, 1, 0)** is the simplest nonstationary model. It is given by

$$
(1 - B)y_t = \delta + \varepsilon_t
\tag{5.78}
$$

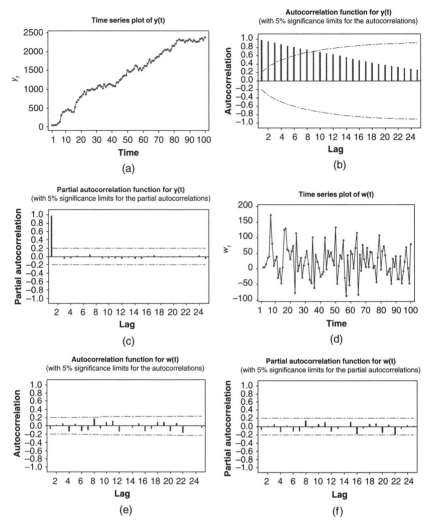

FIGURE 5.11   A realization of the ARIMA(0, 1, 0) model, y_t, its first difference, w_t, and their sample ACFs and PACFs.

suggesting that first differencing eliminates all serial dependence and yields a white noise process.

Consider the process $y_t = 20 + y_{t-1} + \varepsilon_t$. A realization of this process together with its sample ACF and PACF are given in Figure 5.11a–c. We can see that the sample ACF dies out very slowly, while the sample PACF is only significant at the first lag. Also note that the PACF value at the first lag is very close to one. All this evidence suggests that the process is not stationary. The first difference, $w_t = y_t - y_{t-1}$, and its sample ACF and PACF are shown in Figure 5.11d–f. The time series plot of w_t implies that the first difference is stationary. In fact, the sample ACF and PACF do not show any significant values. This further suggests that differencing the original data once "clears out" the autocorrelation. Hence the data can be modeled using the random walk model given in Eq. (5.78).

The **ARIMA(0, 1, 1) process** is given by

$$(1 - B)y_t = \delta + (1 - \theta B)\,\varepsilon_t \tag{5.79}$$

The infinite AR representation of Eq. (5.79) can be obtained from Eq. (5.75)

$$\pi_i - \theta \pi_{i-1} = \begin{cases} 1, & i = 1 \\ 0, & i > 1 \end{cases} \tag{5.80}$$

with $\pi_0 = -1$. Thus we have

$$y_t = \alpha + \sum_{i=1}^{\infty} \pi_i y_{t-i} + \varepsilon_t$$
$$= \alpha + (1 - \theta)(y_{t-1} + \theta y_{t-2} + \cdots) + \varepsilon_t \tag{5.81}$$

This suggests that an ARIMA(0, 1, 1) (a.k.a. IMA(1, 1)) can be written as an exponentially weighted moving average (EWMA) of all past values.

Consider the time series data in Figure 5.12a. It looks like the mean of the process is changing (moving upwards) in time. Yet the change in the mean (i.e., nonstationarity) is not as obvious as in the previous example. The sample ACF plot of the data in Figure 5.12b dies down relatively slowly and the sample PACF of the data in Figure 5.12c shows two significant values at lags 1 and 2. Hence we might be tempted to model this data using an AR(2) model because of the exponentially decaying ACF and significant PACF at the first two lags. Indeed, we might even have a good fit using an AR(2) model. We should nevertheless check the roots of the associated polynomial given in Eq. (5.36) to make sure that its roots are less than 1 in absolute value. Also note that a technically stationary process will behave more and more nonstationary as the roots of the associated polynomial approach unity. For that, observe the realization of the near nonstationary process, $y_t = 2 + 0.95y_{t-1} + \varepsilon_t$, given in Figure 5.1b. Based on the visual inspection, however, we may deem the process nonstationary and proceed with taking the first difference of the data. This is because the ϕ value of the AR(1) model is close to 1. Under these circumstances, where the nonstationarity

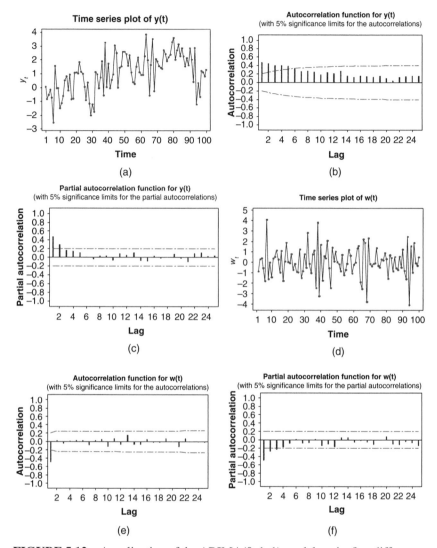

FIGURE 5.12   A realization of the ARIMA(0, 1, 1) model, y_t, its first difference, w_t, and their sample ACFs and PACFs.

of the process is dubious, we strongly recommend that the analyst refer back to basic underlying process knowledge. If, for example, the process mean is expected to wander off as in some financial data, assuming that the process is nonstationary and proceeding with differencing the data would be more appropriate. For the data given in Figure 5.12a, its first difference given in Figure 5.12d looks stationary. Furthermore, its sample ACF and

PACF given in Figures 5.12e and 5.12f, respectively, suggest that an MA(1) model would be appropriate for the first difference since its ACF cuts off after the first lag and the PACF exhibits an exponential decay pattern. Hence the ARIMA(0, 1, 1) model given in Eq. (5.79) can be used for this data.

5.7 TIME SERIES MODEL BUILDING

A three-step iterative procedure is used to build an ARIMA model. First, a tentative model of the ARIMA class is identified through analysis of historical data. Second, the unknown parameters of the model are estimated. Third, through residual analysis, diagnostic checks are performed to determine the adequacy of the model, or to indicate potential improvements. We shall now discuss each of these steps in more detail.

5.7.1 Model Identification

Model identification efforts should start with preliminary efforts in understanding the type of process from which the data is coming and how it is collected. The process' perceived characteristics and sampling frequency often provide valuable information in this preliminary stage of model identification. In today's data rich environments, it is often expected that the practitioners would be presented with "enough" data to be able to generate reliable models. It would nevertheless be recommended that 50 or preferably more observations should be initially considered. Before engaging in rigorous statistical model-building efforts, we also strongly recommend the use of "creative" plotting of the data, such as the simple time series plot and scatter plots of the time series data y_t versus y_{t-1}, y_{t-2}, and so on. For the y_t versus y_{t-1} scatter plot, for example, this can be achieved in a data set of N observations by plotting the first $N - 1$ observations versus the last $N - 1$. Simple time series plots should be used as the preliminary assessment tool for stationarity. The visual inspection of these plots should later be confirmed as described earlier in this chapter. If nonstationarity is suspected, the time series plot of the first (or dth) difference should also be considered. The unit root test by Dickey and Fuller (1979) can also be performed to make sure that the differencing is indeed needed. Once the stationarity of the time series can be presumed, the sample ACF and PACF of the time series of the original time series (or its dth difference if necessary) should be obtained. Depending on the nature of the autocorrelation, the first 20–25 sample autocorrelations and partial autocorrelations should be sufficient. More care should be taken of course if the process exhibits strong autocorrelation and/or seasonality, as we will discuss in the following sections. Table 5.1 together with the $\pm 2/\sqrt{N}$ limits can be used as a guide for identifying AR or MA models. As discussed earlier, the identification of ARMA models would require more care, as both the ACF and PACF will exhibit exponential decay and/or damped sinusoid behavior.

We have already discussed that the differenced series $\{w_t\}$ may have a nonzero mean, say, μ_w. At the identification stage we may obtain an indication of whether or not a nonzero value of μ_w is needed by comparing the sample mean of the differenced series, say, $\bar{w} = \sum_{t=1}^{n-d} [w/(n-d)]$, with its approximate standard error. Box, Jenkins, and Reinsel (2008) give the approximate standard error of $\bar{w}$ for several useful ARIMA(p, d, q) models.

Identification of the appropriate ARIMA model requires skills obtained by experience. Several excellent examples of the identification process are given in Box et al. (2008, Chap. 6), Montgomery et al. (1990), and Bisgaard and Kulahci (2011).

5.7.2 Parameter Estimation

There are several methods such as the methods of moments, maximum likelihood, and least squares that can be employed to estimate the parameters in the tentatively identified model. However, unlike the regression models of Chapter 2, most ARIMA models are **nonlinear** models and require the use of a nonlinear model fitting procedure. This is usually automatically performed by sophisticated software packages such as Minitab JMP, and SAS. In some software packages, the user may have the choice of estimation method and can accordingly choose the most appropriate method based on the problem specifications.

5.7.3 Diagnostic Checking

After a tentative model has been fit to the data, we must examine its adequacy and, if necessary, suggest potential improvements. This is done through residual analysis. The residuals for an ARMA(p, q) process can be obtained from

$$\hat{\varepsilon}_t = y_t - \left(\hat{\delta} + \sum_{i=1}^{p} \hat{\phi}_i y_{t-i} - \sum_{i=1}^{q} \hat{\theta}_i \hat{\varepsilon}_{t-i} \right) \tag{5.82}$$

If the specified model is adequate and hence the appropriate orders p and q are identified, it should transform the observations to a white noise process. Thus the residuals in Eq. (5.82) should behave like white noise.

Let the sample autocorrelation function of the residuals be denoted by $\{r_e(k)\}$. If the model is appropriate, then the residual sample autocorrelation function should have no structure to identify. That is, the autocorrelation should not differ significantly from zero for all lags greater than one. If the form of the model were correct and if we knew the true parameter values, then the standard error of the residual autocorrelations would be $N^{-1/2}$.

Rather than considering the $r_e(k)$ terms individually, we may obtain an indication of whether the first K residual autocorrelations considered together indicate adequacy of the model. This indication may be obtained through an approximate chi-square test of model adequacy. The test statistic is

$$Q = (N - d) \sum_{k=1}^{K} r_e^2(k) \tag{5.83}$$

which is approximately distributed as chi-square with $K - p - q$ degrees of freedom if the model is appropriate. If the model is inadequate, the calculated value of Q will be too large. Thus we should reject the hypothesis of model adequacy if Q exceeds an approximate small upper tail point of the chi-square distribution with $K - p - q$ degrees of freedom. Further details of this test are in Chapter 2 and in the original reference by Box and Pierce (1970). The modification of this test by Ljung and Box (1978) presented in Chapter 2 is also useful in assessing model adequacy.

5.7.4 Examples of Building ARIMA Models

In this section we shall present two examples of the identification, estimation, and diagnostic checking process. One example presents the analysis for a stationary time series, while the other is an example of modeling a nonstationary series.

Example 5.1   Table 5.5 shows the weekly total number of loan applications in a local branch of a national bank for the last 2 years. It is suspected that there should be some relationship (i.e., autocorrelation) between the number of applications in the current week and the number of loan applications in the previous weeks. Modeling that relationship will help the management to proactively plan for the coming weeks through reliable forecasts. As always, we start our analysis with the time series plot of the data, shown in Figure 5.13.

TABLE 5.5   Weekly Total Number of Loan Applications for the Last 2 Years

Week	Applications	Week	Applications	Week	Applications	Week	Applications
1	71	27	62	53	66	79	63
2	57	28	77	54	71	80	61
3	62	29	76	55	59	81	73
4	64	30	88	56	57	82	72
5	65	31	71	57	66	83	65
6	67	32	72	58	51	84	70
7	65	33	66	59	59	85	54
8	82	34	65	60	56	86	63
9	70	35	73	61	57	87	62
10	74	36	76	62	55	88	60
11	75	37	81	63	53	89	67
12	81	38	84	64	74	90	59
13	71	39	68	65	64	91	74
14	75	40	63	66	70	92	61
15	82	41	66	67	74	93	61
16	74	42	71	68	69	94	52
17	78	43	67	69	64	95	55
18	75	44	69	70	68	96	61
19	73	45	63	71	64	97	56
20	76	46	61	72	70	98	61
21	66	47	68	73	73	99	60
22	69	48	75	74	59	100	65
23	63	49	66	75	68	101	55
24	76	50	81	76	59	102	61
25	65	51	72	77	66	103	59
26	73	52	77	78	63	104	63

Figure 5.13 shows that the weekly data tend to have short runs and that the data seem to be indeed autocorrelated. Next, we visually inspect the stationarity. Although there might be a slight drop in the mean for the second year (weeks 53–104), in general, it seems to be safe to assume stationarity.

We now look at the sample ACF and PACF plots in Figure 5.14. Here are possible interpretations of the ACF plot:

1. It cuts off after lag 2 (or maybe even 3), suggesting an MA(2) (or MA(3)) model.

2. It has an (or a mixture of) exponential decay(s) pattern suggesting an AR(p) model.

To resolve the conflict, consider the sample PACF plot. For that, we have only one interpretation; it cuts off after lag 2. Hence we use the second

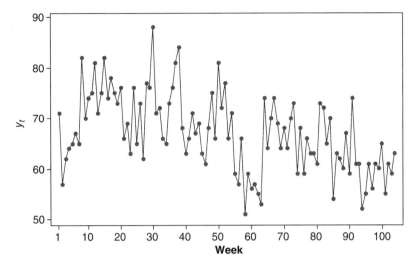

FIGURE 5.13   Time series plot of the weekly total number of loan applications.

interpretation of the sample ACF plot and assume that the appropriate model to fit is the AR(2) model.

Table 5.6 shows the Minitab output for the AR(2) model. The parameter estimates are $\hat{\phi}_1 = 0.27$ and $\hat{\phi}_2 = 0.42$, and they turn out to be significant (see the P-values).

MSE is calculated to be 39.35. The modified Box–Pierce test suggests that there is no autocorrelation left in the residuals. We can also see this in the ACF and PACF plots of the residuals in Figure 5.15.

As the last diagnostic check, we have the 4-in-1 residual plots in Figure 5.16 provided by Minitab: Normal Probability Plot, Residuals versus

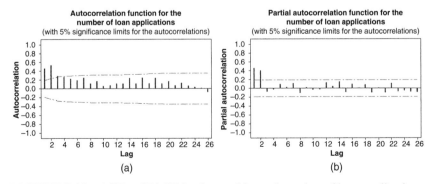

FIGURE 5.14   ACF and PACF for the weekly total number of loan applications.

TABLE 5.6   Minitab Output for the AR(2) Model for the Loan Application Data

```
Final Estimates of Parameters

Type           Coef  SE Coef       T      P
AR    1      0.2682   0.0903    2.97  0.004
AR    2      0.4212   0.0908    4.64  0.000
Constant    20.7642   0.6157   33.73  0.000
Mean         66.844    1.982



Number of observations:   104
Residuals:     SS =  3974.30 (backforecasts excluded)
               MS =    39.35  DF = 101



Modified Box-Pierce (Ljung-Box) Chi-Square statistic

Lag             12     24     36     48
Chi-Square     6.2   16.0   24.9   32.0
DF               9     21     33     45
P-Value      0.718  0.772  0.843  0.927
```

Fitted Value, Histogram of the Residuals, and Time Series Plot of the Residuals. They indicate that the fit is indeed acceptable.

Figure 5.17 shows the actual data and the fitted values. It looks like the fitted values smooth out the highs and lows in the data.

Note that, in this example, we often and deliberately used "vague" words such as "seems" or "looks like." It should be clear by now that

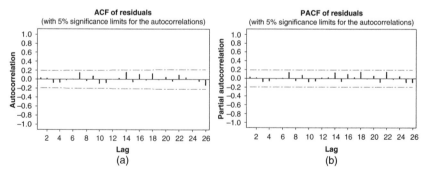

FIGURE 5.15   The sample ACF and PACF of the residuals for the AR(2) model in Table 5.6.

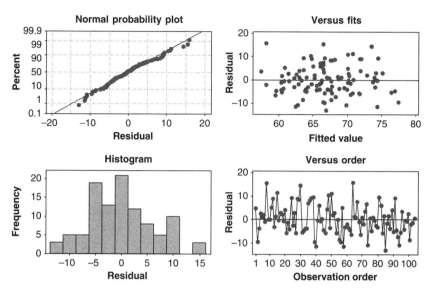

FIGURE 5.16 Residual plots for the AR(2) model in Table 5.6.

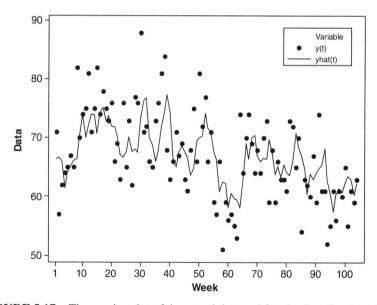

FIGURE 5.17 Time series plot of the actual data and fitted values for the AR(2) model in Table 5.6.

the methodology presented in this chapter has a very sound theoretical foundation. However, as in any modeling effort, we should also keep in mind the subjective component of model identification. In fact, as we mentioned earlier, time series model fitting can be seen as a mixture of science and art and can best be learned by practice and experience. The next example will illustrate this point further.

Example 5.2   Consider the Dow Jones Index data from Chapter 4. A time series plot of the data is given in Figure 5.18. The process shows signs of nonstationarity with changing mean and possibly variance.

Similarly, the slowly decreasing sample ACF and sample PACF with significant value at lag 1, which is close to 1 in Figure 5.19, confirm that indeed the process can be deemed nonstationary. On the other hand, one might argue that the significant sample PACF value at lag 1 suggests that the AR(1) model might also fit the data well. We will consider this interpretation first and fit an AR(1) model to the Dow Jones Index data.

Table 5.7 shows the Minitab output for the AR(1) model. Although it is close to 1, the AR(1) model coefficient estimate $\hat{\phi} = 0.9045$ turns out to be quite significant and the modified Box–Pierce test suggests that there is no autocorrelation left in the residuals. This is also confirmed by the sample ACF and PACF plots of the residuals given in Figure 5.20.

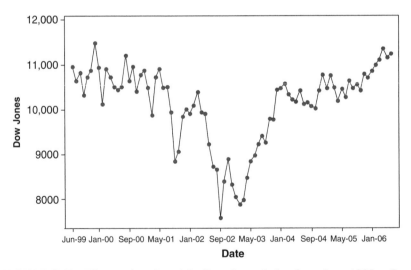

FIGURE 5.18   Time series plot of the Dow Jones Index from June 1999 to June 2006.

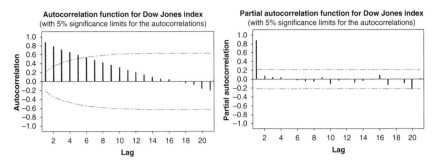

FIGURE 5.19   Sample ACF and PACF of the Dow Jones Index.

TABLE 5.7   Minitab Output for the AR(1) Model for the Dow Jones Index

Final Estimates of Parameters

Type	Coef	SE Coef	T	P
AR 1	0.9045	0.0500	18.10	0.000
Constant	984.94	44.27	22.25	0.000
Mean	10309.9	463.4		

Number of observations: 85
Residuals: SS =   13246015 (backforecasts excluded)
                MS =   159591   DF = 83

Modified Box-Pierce (Ljung-Box) Chi-Square statistic

Lag	12	24	36	48
Chi-Square	2.5	14.8	21.4	29.0
DF	10	22	34	46
P-Value	0.991	0.872	0.954	0.977

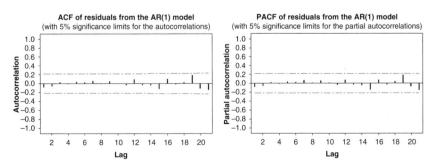

FIGURE 5.20   Sample ACF and PACF of the residuals from the AR(1) model for the Dow Jones Index data.

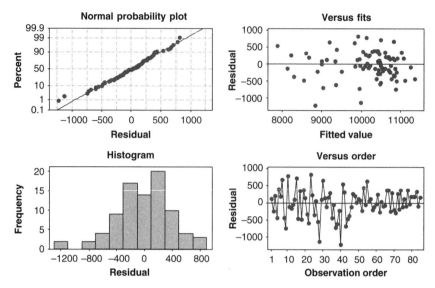

FIGURE 5.21   Residual plots from the AR(1) model for the Dow Jones Index data.

The only concern in the residual plots in Figure 5.21 is in the changing variance observed in the time series plot of the residuals. This is indeed a very important issue since it violates the constant variance assumption. We will discuss this issue further in Section 7.3 but for illustration purposes we will ignore it in this example.

Overall it can be argued that an AR(1) model provides a decent fit to the data. However, we will now consider the earlier interpretation and assume that the Dow Jones Index data comes from a nonstationary process. We then take the first difference of the data as shown in Figure 5.22. While there are once again some serious concerns about changing variance, the level of the first difference remains the same. If we ignore the changing variance and look at the sample ACF and PACF plots given in Figure 5.23, we may conclude that the first difference is in fact white noise. That is, since these plots do not show any sign of significant autocorrelation, a model we may consider for the Dow Jones Index data would be the random walk model, ARIMA(0, 1, 0).

Now the analyst has to decide between the two models: AR(1) and ARIMA(0, 1, 0). One can certainly use some of the criteria we discussed in Section 2.6.2 to choose one of these models. Since these two models are fundamentally quite different, we strongly recommend that the analyst use the subject matter/process knowledge as much as possible. Do we expect

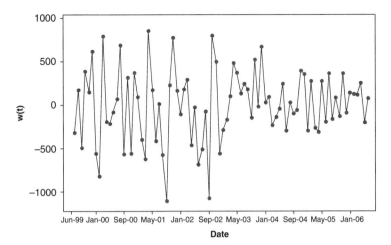

FIGURE 5.22   Time series plot of the first difference $w(t)$ of the Dow Jones Index data.

a financial index such as the Dow Jones Index to wander about a fixed mean as implied by the AR(1)? In most cases involving financial data, the answer would be no. Hence a model such as ARIMA(0, 1, 0) that takes into account the inherent nonstationarity of the process should be preferred. However, we do have a problem with the proposed model. A random walk model means that the price changes are random and cannot be predicted. If we have a higher price today compared to yesterday, that would have no bearing on the forecasts tomorrow. That is, tomorrow's price can be higher or lower than today's and we would have no way to forecast it effectively. This further suggests that the best forecast for tomorrow's price is in fact the price we have today. This is obviously not a reliable and effective forecasting model. This very same issue of the random walk models for

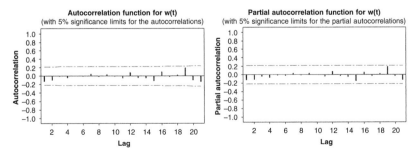

FIGURE 5.23   Sample ACF and PACF plots of the first difference of the Dow Jones Index data.

financial data has been discussed in great detail in the literature. We simply used this data to illustrate that in time series model fitting we can end up with fundamentally different models that will fit the data equally well. At this point, process knowledge can provide the needed guidance in picking the "right" model.

It should be noted that, in this example, we tried to keep the models simple for illustration purposes. Indeed, a more thorough analysis would (and should) pay close attention to the changing variance issue. In fact, this is a very common concern particularly when dealing with financial data. For that, we once again refer the reader to Section 7.3.

5.8 FORECASTING ARIMA PROCESSES

Once an appropriate time series model has been fit, it may be used to generate forecasts of future observations. If we denote the current time by T, the forecast for $y_{T+\tau}$ is called the τ-period-ahead forecast and denoted by $\hat{y}_{T+\tau}(T)$. The standard criterion to use in obtaining the best forecast is the mean squared error for which the expected value of the squared forecast errors, $E[(y_{T+\tau} - \hat{y}_{T+\tau}(T))^2] = E[e_T(\tau)^2]$, is minimized. It can be shown that the best forecast in the mean square sense is the conditional expectation of $y_{T+\tau}$ given current and previous observations, that is, y_T, $y_{T-1}, \ldots$:

$$\hat{y}_{T+\tau}(T) = E\left[y_{T+\tau} \mid y_T, y_{T-1}, \ldots\right] \tag{5.84}$$

Consider, for example, an ARIMA(p, d, q) process at time $T + \tau$ (i.e., τ period in the future):

$$y_{T+\tau} = \delta + \sum_{i=1}^{p+d} \phi_i y_{T+\tau-i} + \varepsilon_{T+\tau} - \sum_{i=1}^{q} \theta_i \varepsilon_{T+\tau-i} \tag{5.85}$$

Further consider its infinite MA representation,

$$y_{T+\tau} = \mu + \sum_{i=0}^{\infty} \psi_i \varepsilon_{T+\tau-i} \tag{5.86}$$

We can partition Eq. (5.86) as

$$y_{T+\tau} = \mu + \sum_{i=0}^{\tau-1} \psi_i \varepsilon_{T+\tau-i} + \sum_{i=\tau}^{\infty} \psi_i \varepsilon_{T+\tau-i} \tag{5.87}$$

In this partition, we can clearly see that the $\sum_{i=0}^{\tau-1} \psi_i \varepsilon_{T+\tau-i}$ component involves the future errors, whereas the $\sum_{i=\tau}^{\infty} \psi_i \varepsilon_{T+\tau-i}$ component involves the present and past errors. From the relationship between the current and past observations and the corresponding random shocks as well as the fact that the random shocks are assumed to have mean zero and to be independent, we can show that the best forecast in the mean square sense is

$$\hat{y}_{T+\tau}(T) = E\left[y_{T+\tau} \middle| y_T, y_{T-1}, \ldots\right] = \mu + \sum_{i=\tau}^{\infty} \psi_i \varepsilon_{T+\tau-i} \qquad (5.88)$$

since

$$E\left[\varepsilon_{T+\tau-i} \middle| y_T, y_{T-1}, \ldots\right] = \begin{cases} 0 & \text{if} \quad i < \tau \\ \varepsilon_{T+\tau-i} & \text{if} \quad i \geq \tau \end{cases}$$

Subsequently, the forecast error is calculated from

$$e_T(\tau) = y_{T+\tau} - \hat{y}_{T+\tau}(T) = \sum_{i=0}^{\tau-1} \psi_i \varepsilon_{T+\tau-i} \qquad (5.89)$$

Since the forecast error in Eq. (5.89) is a linear combination of random shocks, we have

$$E\left[e_T(\tau)\right] = 0 \qquad (5.90)$$

$$\text{Var}\left[e_T(\tau)\right] = \text{Var}\left[\sum_{i=0}^{\tau-1} \psi_i \varepsilon_{T+\tau-i}\right] = \sum_{i=0}^{\tau-1} \psi_i^2 \text{Var}(\varepsilon_{T+\tau-i})$$

$$= \sigma^2 \sum_{i=0}^{\tau-1} \psi_i^2 \qquad (5.91)$$

$$= \sigma^2(\tau), \quad \tau = 1, 2, \ldots$$

It should be noted that the variance of the forecast error gets bigger with increasing forecast lead times τ. This intuitively makes sense as we should expect more uncertainty in our forecasts further into the future. Moreover, if the random shocks are assumed to be normally distributed, $N(0, \sigma^2)$, then the forecast errors will also be normally distributed with $N(0, \sigma^2(\tau))$. We can then obtain the $100(1 - \alpha)$ percent prediction intervals for the future observations from

$$P\left(\hat{y}_{T+\tau}(T) - z_{\alpha/2}\sigma(\tau) < y_{T+\tau} < \hat{y}_{T+\tau}(T) + z_{\alpha/2}\sigma(\tau)\right) = 1 - \alpha \qquad (5.92)$$

where $z_{\alpha/2}$ is the upper $\alpha/2$ percentile of the standard normal distribution, $N(0, 1)$. Hence the $100(1 - \alpha)$ percent prediction interval for $y_{T+\tau}$ is

$$\hat{y}_{T+\tau}(T) \pm z_{\alpha/2}\sigma(\tau) \qquad (5.93)$$

There are two issues with the forecast equation in (5.88). First, it involves infinitely many terms in the past. However, in practice, we will only have a finite amount of data. For a sufficiently large data set, this can be overlooked. Second, Eq. (5.88) requires knowledge of the magnitude of random shocks in the past, which is unrealistic. A solution to this problem is to "estimate" the past random shocks through one-step-ahead forecasts. For the ARIMA model we can calculate

$$\hat{\varepsilon}_t = y_t - \left[\delta + \sum_{i=1}^{p+d}\phi_i y_{t-i} - \sum_{i=1}^{q}\theta_i \hat{\varepsilon}_{t-i}\right] \qquad (5.94)$$

recursively by setting the initial values of the random shocks to zero for $t < p + d + 1$. For more accurate results, these initial values together with the y_t for $t \leq 0$ can also be obtained using back-forecasting. For further details, see Box, Jenkins, and Reinsel (2008).

As an illustration consider forecasting the ARIMA(1, 1, 1) process

$$(1 - \phi B)(1 - B)y_{T+\tau} = (1 - \theta B)\varepsilon_{T+\tau} \qquad (5.95)$$

We will consider two of the most commonly used approaches:

1. As discussed earlier, this approach involves the infinite MA representation of the model in Eq. (5.95), also known as the **random shock** form of the model:

$$\begin{aligned} y_{T+\tau} &= \sum_{i=0}^{\infty}\psi_i \varepsilon_{T+\tau-i} \\ &= \psi_0\varepsilon_{T+\tau} + \psi_1\varepsilon_{T+\tau-1} + \psi_2\varepsilon_{T+\tau-2} + \cdots \end{aligned} \qquad (5.96)$$

Hence the τ-step-ahead forecast can be calculated from

$$\hat{y}_{T+\tau}(T) = \psi_\tau \varepsilon_T + \psi_{\tau+1} \varepsilon_{T-1} + \cdots \qquad (5.97)$$

The weights ψ_i can be calculated from

$$\left(\psi_0 + \psi_1 B + \cdots\right)(1 - \phi B)(1 - B) = (1 - \theta B) \qquad (5.98)$$

and the random shocks can be estimated using the one-step-ahead forecast error; for example, ε_T can be replaced by $e_{T-1}(1) = y_T - \hat{y}_T(T-1)$.

2. Another approach that is often employed in practice is to use **difference equations** as given by

$$y_{T+\tau} = (1 + \phi) y_{T+\tau-1} - \phi y_{T+\tau-2} + \varepsilon_{T+\tau} - \theta \varepsilon_{T+\tau-1} \qquad (5.99)$$

For $\tau = 1$, the best forecast in the mean squared error sense is

$$\hat{y}_{T+1}(T) = E\left[y_{T+1} \big| y_T, y_{T-1}, \ldots\right] = (1 + \phi) y_T - \phi y_{T-1} - \theta e_T(1) \qquad (5.100)$$

We can further show that for lead times $\tau > 2$, the forecast is

$$\hat{y}_{T+\tau}(T) = (1 - \phi)\hat{y}_{T+\tau-1}(T) - \phi \hat{y}_{T+\tau-2}(T) \qquad (5.101)$$

Prediction intervals for forecasts of future observations at time period $T + \tau$ are found using equation 5.87. However, in using Equation 5.87 the ψ weights must be found in order to compute the variance (or standard deviation) of the τ-step ahead forecast error. The ψ weights for the general ARIMA(p, d, q) model may be obtained by equating like powers of B in the expansion of

$$(\psi_0 + \psi_1 B + \psi_2 B^2 + \cdots)(1 - \phi_1 B - \phi_2 B^2 - \cdots - \phi_p B^p)(1 - B)^d$$
$$= (1 - \theta_1 B - \theta_2 B^2 - \cdots - \theta_q B^q)$$

and solving for the ψ weights. We now illustrate this with three examples.

Example 5.3 The ARMA(1, 1) Model   For the ARMA(1, 1) model the product of the required polynomials is

$$(\psi_0 + \psi_1 B + \psi_2 B^2 + \cdots)(1 - \phi B) = (1 - \theta B)$$

Equating like power of B we find that

$$B^0: \psi_0 = 1$$
$$B^1: \psi_1 - \phi = -\theta, \text{ or } \psi_1 = \phi - \theta$$
$$B^2: \psi_2 - \phi\psi_1 = 0, \text{ or } \psi_2 = \phi(\phi - \theta)$$

In general, we can show for the ARMA(1,1) model that $\psi_j = \phi^{j-1}(\phi - \theta)$.

Example 5.4 The AR(2) Model For the AR(2) model the product of the required polynomials is

$$(\psi_0 + \psi_1 B + \psi_2 B^2 + \cdots)(1 - \phi_1 B - \phi_2 B^2) = 1$$

Equating like power of B, we find that

$$B^0: \psi_0 = 1$$
$$B^1: \psi_1 - \phi_1 = 0, \text{ or } \psi_1 = \phi_1$$
$$B^2: \psi_2 - \phi_1\psi_1 - \phi_2 = 0, \text{ or } \psi_2 = \phi_1\psi_1 + \phi_2$$

In general, we can show for the AR(2) model that $\psi_j = \phi_1\psi_{j-1} + \phi_2\psi_{j-2}$.

Example 5.5 The ARIMA(0, 1, 1) or IMA(1,1) Model Now consider a nonstationary model, the IMA(1, 1) model. The product of the required polynomials for this model is

$$(\psi_0 + \psi_1 B + \psi_2 B^2 + \cdots)(1 - B) = (1 - \theta B)$$

It is straightforward to show that the ψ weights for this model are

$$\psi_0 = 1$$
$$\psi_1 = 1 - \theta$$
$$\psi_j = \psi_{j-1}, \, j = 2, 3, \ldots$$

Notice that the prediction intervals will increase in length rapidly as the forecast lead time increases. This is typical of nonstationary ARIMA models. It implies that these models may not be very effective in forecasting more than a few periods ahead.

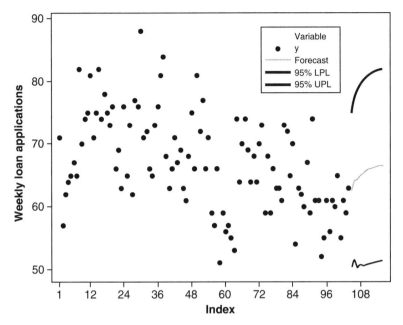

FIGURE 5.24   Time series plot and forecasts for the weekly loan application data.

Example 5.6   Consider the loan applications data given in Table 5.5. Now assume that the manager wants to make forecasts for the next 3 months (12 weeks) using the AR(2) model from Example 5.1. Hence at the 104th week we need to make 1-step, 2-step, ... , 12-step-ahead predictions, which are obtained and plotted using Minitab in Figure 5.24 together with the 95% prediction interval.

Table 5.8 shows the output from JMP for fitting an AR(2) model to the weekly loan application data. In addition to the sample ACF and PACF, JMP provides the model fitting information including the estimates of the model parameters, the forecasts for 10 periods into the future and the associated prediction intervals, and the residual autocorrelation and PACF. The AR(2) model is an excellent fit to the data.

5.9 SEASONAL PROCESSES

Time series data may sometimes exhibit strong periodic patterns. This is often referred to as the time series having a seasonal behavior. This mostly occurs when data is taken in specific intervals—monthly, weekly, and so on. One way to represent such data is through an additive model where the process is assumed to be composed of two parts,

$$y_t = S_t + N_t \tag{5.102}$$

where S_t is the deterministic component with periodicity s and N_t is the stochastic component that may be modeled as an ARMA process. In that, y_t can be seen as a process with predictable periodic behavior with some noise sprinkled on top of it. Since the S_t is deterministic and has periodicity s, we have $S_t = S_{t+s}$ or

$$S_t - S_{t-s} = (1 - B^s)S_t = 0 \tag{5.103}$$

Applying the $(1 - B^s)$ operator to Eq. (5.102), we have

$$
\underbrace{(1 - B^s)y_t}_{\equiv w_t} = \underbrace{(1 - B^s)S_t}_{=0} + (1 - B^s)N_t \\
w_t = (1 - B^s)N_t
\tag{5.104}
$$

TABLE 5.8   JMP AR(2) Output for the Loan Application Data

Time series y(t)

Mean	67.067308
Std	7.663932
N	104
Zero Mean ADF	-0.695158
Single Mean ADF	-6.087814
Trend ADF	-7.396174

TABLE 5.8   *(Continued)*

Time series basic diagnostics

Lag	AutoCorr	plot autocorr	Ljung-Box Q	p-Value
0	1.0000		.	.
1	0.4617		22.8186	<.0001
2	0.5314		53.3428	<.0001
3	0.2915		62.6167	<.0001
4	0.2682		70.5487	<.0001
5	0.2297		76.4252	<.0001
6	0.1918		80.5647	<.0001
7	0.2484		87.5762	<.0001
8	0.1162		89.1255	<.0001
9	0.1701		92.4847	<.0001
10	0.0565		92.8587	<.0001
11	0.0716		93.4667	<.0001
12	0.1169		95.1040	<.0001
13	0.1151		96.7080	<.0001
14	0.2411		103.829	<.0001
15	0.1137		105.430	<.0001
16	0.2540		113.515	<.0001
17	0.1279		115.587	<.0001
18	0.2392		122.922	<.0001
19	0.1138		124.603	<.0001
20	0.1657		128.206	<.0001
21	0.0745		128.944	<.0001
22	0.1320		131.286	<.0001
23	0.0708		131.968	<.0001
24	0.0338		132.125	<.0001
25	0.0057		132.130	<.0001

Lag	Partial	plot partial		
Lag	AutoCorr	plot autocorr		
0	1.0000		Ljung-Box Q	p-Value
1	0.4617			
2	0.4045			
3	−0.0629			
4	−0.0220			
5	0.0976			
6	0.0252			
7	0.1155			
8	−0.1017			
9	0.0145			
10	−0.0330			
11	−0.0250			
12	0.1349			
13	0.0488			
14	0.1489			
15	−0.0842			
16	0.1036			
17	0.0105			
18	0.0830			
19	−0.0938			
20	0.0052			
21	−0.0927			
22	0.1149			
23	−0.0645			
24	−0.0473			
25	−0.0742			

(continued)

TABLE 5.8 *(Continued)*

```
Model Comparison
Model   DF   Variance        AIC        SBC   RSquare     -2LogLH
AR(2)  101  39.458251  680.92398  688.85715     0.343  674.92398
```

Model: AR(2)
Model Summary

DF	101
Sum of Squared Errors	3985.28336
Variance Estimate	39.4582511
Standard Deviation	6.2815803
Akaike's 'A' Information Criterion	680.923978
Schwarz's Bayesian Criterion	688.857151
RSquare	0.34278547
RSquare Adj	0.32977132
MAPE	7.37857799
MAE	4.91939717
-2LogLikelihood	674.923978

Stable	Yes
Invertible	Yes

Parameter Estimates

Term	Lag	Estimate	Std Error	t Ratio	Prob>|t|	Constant Estimate
AR1	1	0.265885	0.089022	2.99	0.0035	21.469383
AR2	2	0.412978	0.090108	4.58	<.0001	
Intercept	0	66.854262	1.833390	36.46	<.0001	

Forecast

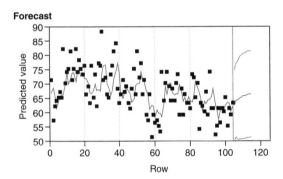

Residuals

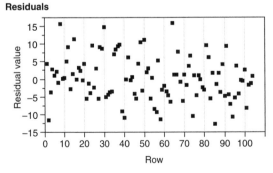

TABLE 5.8   *(Continued)*

Lag	AutoCorr plot autocorr	Ljung-Box Q	p-Value
0	1.0000	.	.
1	0.0320	0.1094	0.7408
2	0.0287	0.1986	0.9055
3	−0.0710	0.7489	0.8617
4	−0.0614	1.1647	0.8839
5	−0.0131	1.1839	0.9464
6	0.0047	1.1864	0.9776
7	0.1465	3.6263	0.8217
8	−0.0309	3.7358	0.8801
9	0.0765	4.4158	0.8820
10	−0.0938	5.4479	0.8593
11	−0.0698	6.0251	0.8717
12	0.0019	6.0255	0.9148
13	0.0223	6.0859	0.9430
14	0.1604	9.2379	0.8155
15	−0.0543	9.6028	0.8440
16	0.1181	11.3501	0.7874
17	−0.0157	11.3812	0.8361
18	0.1299	13.5454	0.7582
19	−0.0059	13.5499	0.8093
20	0.0501	13.8788	0.8366
21	−0.0413	14.1056	0.8650
22	0.0937	15.2870	0.8496
23	0.0409	15.5146	0.8752
24	−0.0035	15.5163	0.9047
25	−0.0335	15.6731	0.9242

Lag	Partial plot partial
0	1.0000
1	0.0320
2	0.0277
3	−0.0729
4	−0.0580
5	−0.0053
6	0.0038
7	0.1399
8	−0.0454
9	0.0715
10	−0.0803

Lag	AutoCorr plot autocorr	Ljung-Box Q	p-Value
11	−0.0586		
12	0.0201		
13	0.0211		
14	0.1306		
15	−0.0669		
16	0.1024		
17	0.0256		
18	0.1477		
19	−0.0027		
20	0.0569		
21	−0.0823		
22	0.1467		
23	−0.0124		
24	0.0448		
25	−0.0869		

The process w_t can be seen as **seasonally stationary**. Since an ARMA process can be used to model N_t, in general, we have

$$\Phi(B)w_t = (1 - B^s)\Theta(B)\varepsilon_t \tag{5.105}$$

where ε_t is white noise.

We can also consider S_t as a stochastic process. We will further assume that after seasonal differencing, $(1 - B^s)$, $(1 - B^s)y_t = w_t$ becomes stationary. This, however, may not eliminate all seasonal features in the process. That is, the seasonally differenced data may still show strong autocorrelation at lags s, $2s$, ... So the seasonal ARMA model is

$$\left(1 - \phi_1^* B^s - \phi_2^* B^{2s} - \cdots - \phi_P^* B^{Ps}\right)w_t = \left(1 - \theta_1^* B^s - \theta_2^* B^{2s} - \cdots - \theta_Q^* B^{Qs}\right)\varepsilon_t \tag{5.106}$$

This representation, however, only takes into account the autocorrelation at seasonal lags s, $2s$, ... Hence a more general seasonal ARIMA model of orders $(p, d, q) \times (P, D, Q)$ with period s is

$$\Phi^*(B^s)\Phi(B)(1 - B)^d(1 - B^s)^D y_t = \delta + \Theta^*(B^s)\Theta(B)\varepsilon_t \tag{5.107}$$

In practice, although it is case specific, it is not expected to have P, D, and Q greater than 1. The results for regular ARIMA processes that we discussed in previous sections apply to the seasonal models given in Eq. (5.107).

As in the nonseasonal ARIMA models, the forecasts for the seasonal ARIMA models can be obtained from the difference equations as illustrated for example in Eq. (5.101) for a nonseasonal ARIMA(1,1,1) process. Similarly the weights in the random shock form given in Eq. (5.96) can be estimated as in Eq. (5.98) to obtain the estimate for the variance of the forecast errors as well as the prediction intervals given in Eqs. (5.91) and (5.92), respectively.

Example 5.7   The ARIMA $(0, 1, 1) \times (0, 1, 1)$ model with $s = 12$ is

$$\underbrace{(1 - B)(1 - B^{12})y_t}_{w_t} = \left(1 - \theta_1 B - \theta_1^* B^{12} + \theta_1 \theta_1^* B^{13}\right)\varepsilon_t$$

For this process, the autocovariances are calculated as

$$\gamma(0) = \mathrm{Var}(w_t) = \sigma^2 \left(1 + \theta_1^2 + \theta_1^{*2} + \left(-\theta_1 \theta_1^* \right)^2 \right)$$
$$= \sigma^2 \left(1 + \theta_1^2 \right) \left(1 + \theta_1^{*2} \right)$$
$$\gamma(1) = \mathrm{Cov}(w_t, w_{t-1}) = \sigma^2 \left(-\theta_1 + \theta_1^* \left(-\theta_1 \theta_1^* \right) \right)$$
$$= -\theta_1 \sigma^2 \left(1 + \theta_1^* \right)$$
$$\gamma(2) = \gamma(3) = \cdots = \gamma(10) = 0$$
$$\gamma(11) = \sigma^2 \theta_1 \theta_1^*$$
$$\gamma(12) = -\sigma^2 \theta_1^* \left(1 + \theta_1^2 \right)$$
$$\gamma(13) = \sigma^2 \theta_1 \theta_1^*$$
$$\gamma(j) = 0, \quad j > 13$$

Example 5.8   Consider the US clothing sales data in Table 4.9. The data obviously exhibit some seasonality and upward linear trend. The sample ACF and PACF plots given in Figure 5.25 indicate a monthly seasonality, $s = 12$, as ACF values at lags 12, 24, 36 are significant and slowly decreasing, and there is a significant PACF value at lag 12 that is close to 1. Moreover, the slowly decreasing ACF in general, also indicates a nonstationarity that can be remedied by taking the first difference. Hence we would now consider $w_t = (1 - B)(1 - B^{12})y_t$.

Figure 5.26 shows that first difference together with seasonal differencing—that is, $w_t = (1 - B)(1 - B^{12})y_t$—helps in terms of stationarity and eliminating the seasonality, which is also confirmed by sample ACF and PACF plots given in Figure 5.27. Moreover, the sample ACF with a significant value at lag 1 and the sample PACF with exponentially

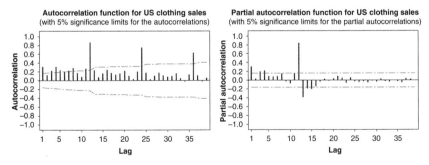

FIGURE 5.25   Sample ACF and PACF plots of the US clothing sales data.

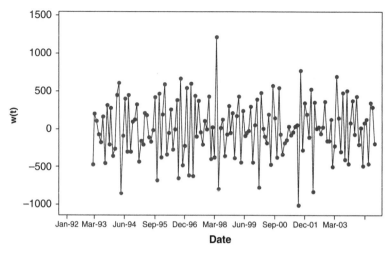

FIGURE 5.26 Time series plot of $w_t = (1 - B)(1 - B^{12})y_t$ for the US clothing sales data.

decaying values at the first 8 lags suggest that a nonseasonal MA(1) model should be used.

The interpretation of the remaining seasonality is a bit more difficult. For that we should focus on the sample ACF and PACF values at lags 12, 24, 36, and so on. The sample ACF at lag 12 seems to be significant and the sample PACF at lags 12, 24, 36 (albeit not significant) seems to be alternating in sign. That suggests that a seasonal MA(1) model can be used as well. Hence an ARIMA$(0, 1, 1) \times (0, 1, 1)_{12}$ model is used to model the data, y_t. The output from Minitab is given in Table 5.9. Both MA(1) and seasonal MA(1) coefficient estimates are significant. As we can see from the sample ACF and PACF plots in Figure 5.28, while there are still some

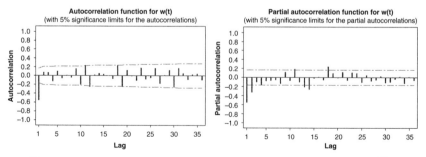

FIGURE 5.27 Sample ACF and PACF plots of $w_t = (1 - B)(1 - B^{12})y_t$.

TABLE 5.9   Minitab Output for the ARIMA(0, 1, 1) × (0, 1, 1)$_{12}$ Model for the US Clothing Sales Data

```
Final Estimates of Parameters

Type        Coef   SE Coef       T       P
MA    1   0.7626   0.0542   14.06   0.000
SMA  12   0.5080   0.0771    6.59   0.000


Differencing: 1 regular, 1 seasonal of order 12
Number of observations:  Original series 155, after
differencing 142
Residuals:     SS =   10033560 (backforecasts excluded)
               MS =   71668   DF = 140


Modified Box-Pierce (Ljung-Box) Chi-Square statistic

Lag               12      24      36      48
Chi-Square      15.8    37.7    68.9    92.6
DF                10      22      34      46
P-Value        0.107   0.020   0.000   0.000
```

small significant values, as indicated by the modified Box pierce statistic most of the autocorrelation is now modeled out.

The residual plots in Figure 5.29 provided by Minitab seem to be acceptable as well.

Finally, the time series plot of the actual and fitted values in Figure 5.30 suggests that the ARIMA(0, 1, 1) × (0, 1, 1)$_{12}$ model provides a reasonable fit to this highly seasonal and nonstationary time series data.

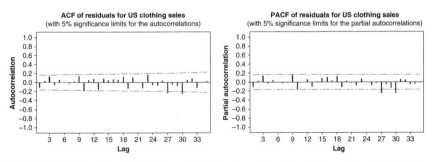

FIGURE 5.28   Sample ACF and PACF plots of residuals from the ARIMA(0, 1, 1) × (0, 1, 1)$_{12}$ model.

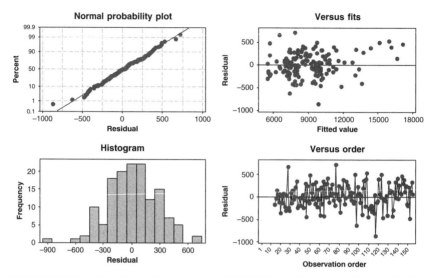

FIGURE 5.29   Residual plots from the ARIMA$(0, 1, 1) \times (0, 1, 1)_{12}$ model for the US clothing sales data.

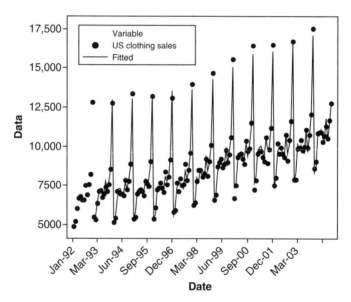

FIGURE 5.30   Time series plot of the actual data and fitted values from the ARIMA$(0, 1, 1) \times (0, 1, 1)_{12}$ model for the US clothing sales data.

5.10 ARIMA MODELING OF BIOSURVEILLANCE DATA

In Section 4.8 we introduced the daily counts of respiratory and gastrointestinal complaints for more than $2\text{-}\frac{1}{2}$ years at several hospitals in a large
metropolitan area from Fricker (2013). Table 4.12 presents the 980 observations from one of these hospitals. Section 4.8 described modeling the
respiratory count data with exponential smoothing. We now present an
ARIMA modeling approach. Figure 5.31 presents the sample ACF, PACF,
and the variogram from JMP for these data. Examination of the original
time series plot in Figure 4.35 and the ACF and variogram indicate that
the daily respiratory syndrome counts may be nonstationary and that the
data should be differenced to obtain a stationary time series for ARIMA
modeling.

The ACF for the differenced series ($d = 1$) shown in Figure 5.32 cuts off
after lag 1 while the PACF appears to be a mixture of exponential decays.
This suggests either an ARIMA(1, 1, 1) or ARIMA(2, 1, 1) model.

The Time Series Modeling platform in JMP allows a group of ARIMA
models to be fit by specifying ranges for the AR, difference, and MA terms.
Table 5.10 summarizes the fits obtained for a constant difference (d = 1),
and both AR (p) and MA (q) parameters ranging from 0 to 2.

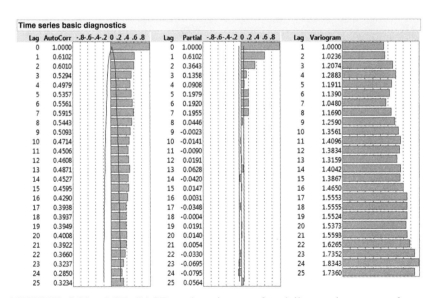

FIGURE 5.31 ACF, PACF, and variogram for daily respiratory syndrome
counts.

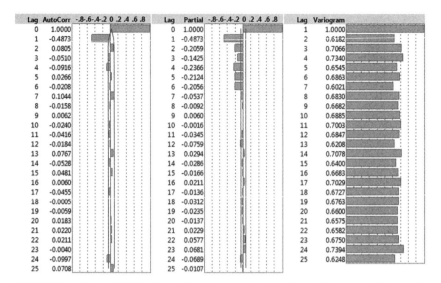

FIGURE 5.32 ACF, PACF, and variogram for the first difference of the daily respiratory syndrome counts.

TABLE 5.10 Summary of Models fit to the Respiratory Syndrome Count Data

Model	Variance	AIC	BIC	RSquare	MAPE	MAE
AR(1)	65.7	6885.5	6895.3	0.4	24.8	6.3
AR(2)	57.1	6748.2	6762.9	0.5	22.9	5.9
MA(1)	81.6	7096.9	7106.6	0.2	28.5	6.9
MA(2)	69.3	6937.6	6952.3	0.3	26.2	6.4
ARMA(1, 1)	52.2	6661.2	6675.9	0.5	21.6	5.6
ARMA(1, 2)	52.1	6661.2	6680.7	0.5	21.6	5.6
ARMA(2, 1)	52.1	6660.7	6680.3	0.5	21.6	5.6
ARMA(2, 2)	52.3	6664.3	6688.7	0.5	21.6	5.6
ARIMA(0, 0, 0)	104.7	7340.4	7345.3	0.0	33.2	8.0
ARIMA(0, 1, 0)*	81.6	7088.2	7093.1	0.2	26.2	7.0
ARIMA(0, 1, 1)*	52.7	6662.8	6672.6	0.5	21.4	5.7
ARIMA(0, 1, 2)	52.6	6662.1	6676.7	0.5	21.4	5.7
ARIMA(1, 1, 0)*	62.2	6824.4	6834.2	0.4	23.2	6.2
ARIMA(1, 1, 1)	52.6	6661.4	6676.1	0.5	21.4	5.7
ARIMA(1, 1, 2)	52.6	6661.9	6681.5	0.5	21.4	5.7
ARIMA(2, 1, 0)	59.6	6783.5	6798.1	0.4	22.7	6.1
ARIMA(2, 1, 1)	52.3	6657.1	6676.6	0.5	21.4	5.6
ARIMA(2, 1, 2)	52.3	6657.8	6682.2	0.5	21.3	5.6

*Indicates that objective function failed during parameter estimation.

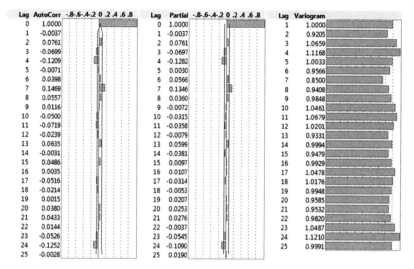

FIGURE 5.33   ACF, PACF, and variogram for the residuals of ARIMA(1, 1, 1) fit to daily respiratory syndrome counts.

In terms of the model summary statistics variance of the errors, AIC and mean absolute prediction error (MAPE) several models look potentially reasonable. For the ARIMA(1, 1, 1) we obtained the following results from JMP:

Parameter estimates

Term	Lag	Estimate	Std error	t Ratio	Prob>|t|	Constant estimate
AR1	1	0.07307009	0.0394408	1.85	0.0642	0.00069557
MA1	1	0.81584055	0.0223680	36.47	<.0001*	
Intercept	0	0.00075040	0.0036018	0.21	0.8350	

Figure 5.33 presents the ACF, PACF, and variogram of the residuals from this model. Other residual plots are in Figure 5.34.

For comparison purposes we also fit the ARIMA(2, 1, 1) model. The parameter estimates obtained from JMP are:

Parameter Estimates

Term	Lag	Estimate	Std Error	t Ratio	Prob>|t|	Constant Estimate
AR1	1	0.09953471	0.0402040	2.48	0.0135*	0.00097496
AR2	2	0.09408008	0.0375486	2.51	0.0124*	
MA1	1	0.84755625	0.0231814	36.56	<.0001*	
Intercept	0	0.00120905	0.0088678	0.14	0.8916	

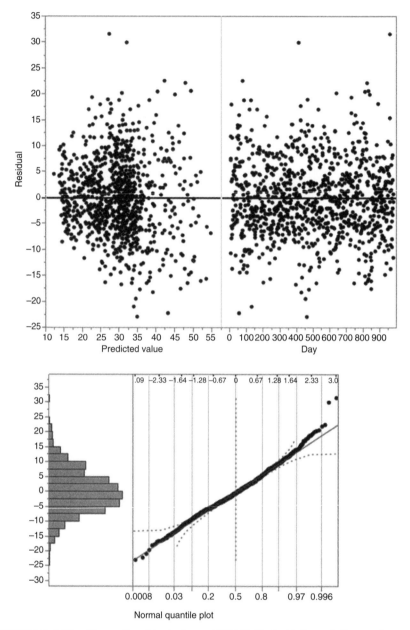

FIGURE 5.34 Plots of residuals from ARIMA(1, 1, 1) fit to daily respiratory syndrome counts.

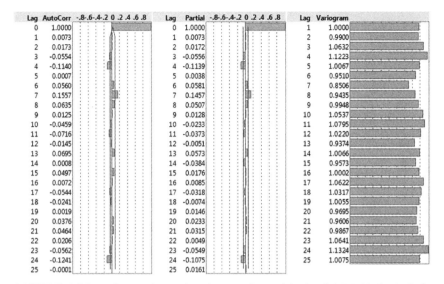

FIGURE 5.35   ACF, PACF, and variogram for residuals of ARIMA(2, 1, 1) fit to daily respiratory syndrome counts.

The lag 2 AR parameter is highly significant. Figure 5.35 presents the plots of the ACF, PACF, and variogram of the residuals from ARIMA(2, 1, 1). Other residual plots are shown in Figure 5.36. Based on the significant lag 2 AR parameter, this model is preferable to the ARIMA(1, 1, 1) model fit previously.

Considering the variation in counts by day of week that was observed previously, a seasonal ARIMA model with a seasonal period of 7 days may be appropriate. The resulting model has an error variance of 50.9, smaller than for the ARIMA(1, 1, 1) and ARIMA(2, 1, 1) models. The AIC is also smaller. Notice that all of the model parameters are highly significant. The residual ACF, PACF, and variogram shown in Figure 5.37 do not suggest any remaining structure. Other residual plots are in Figure 5.38.

Model	Variance	AIC	BIC	RSquare	MAPE	MAE
ARIMA(2, 1, 1)(0, 0, 1)$_7$	50.9	6631.9	6656.4	0.5	21.1	5.6

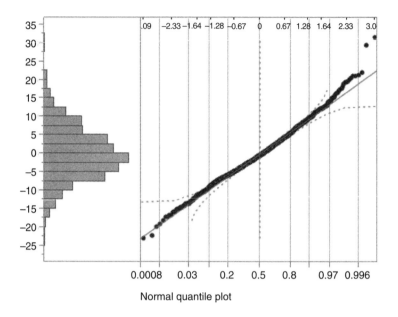

Normal quantile plot

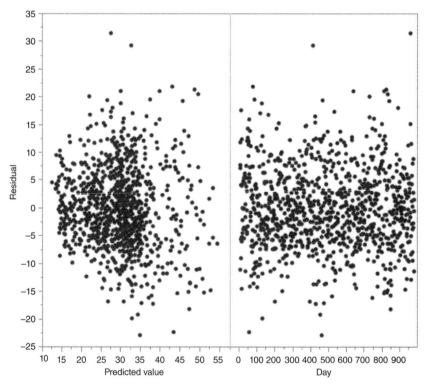

FIGURE 5.36 Plots of residuals from ARIMA(2, 1, 1) fit to daily respiratory syndrome counts.

Parameter estimates

Term	Factor	Lag	Estimate	Std error	t Ratio	Prob>|t|	Constant
AR1,1	1	1	0.1090685	0.0395388	2.76	0.0059*	**estimate**
AR1,2	1	2	0.1186083	0.0376471	3.15	0.0017*	0.00083453
MA1,1	1	1	0.8730535	0.0225127	38.78	<.0001*	
MA2,7	2	7	-0.1744415	0.0328363	-5.31	<.0001*	
Intercept	1	0	0.0010805	0.0051887	0.21	0.8351	

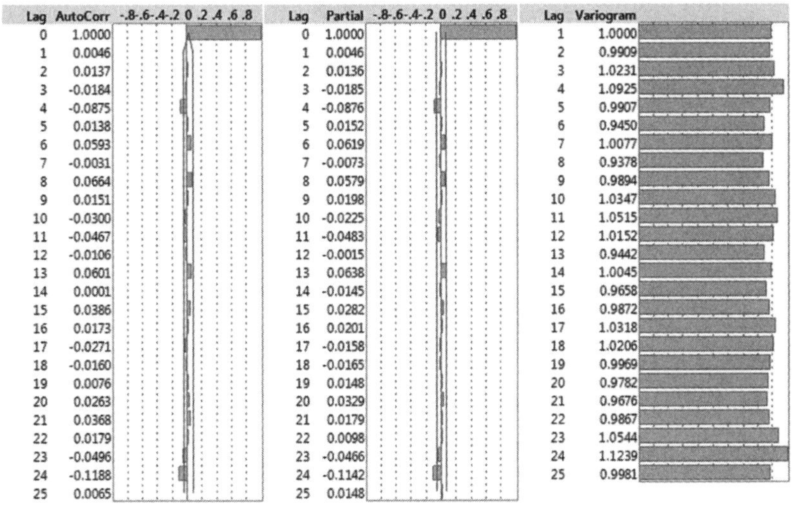

Lag	AutoCorr	-.8 -.6 -.4 -.2 0 .2 .4 .6 .8	Lag	Partial	-.8 -.6 -.4 -.2 0 .2 .4 .6 .8	Lag	Variogram	
0	1.0000		0	1.0000		1	1.0000	
1	0.0046		1	0.0046		2	0.9909	
2	0.0137		2	0.0136		3	1.0231	
3	-0.0184		3	-0.0185		4	1.0925	
4	-0.0875		4	-0.0876		5	0.9907	
5	0.0138		5	0.0152		6	0.9450	
6	0.0593		6	0.0619		7	1.0077	
7	-0.0031		7	-0.0073		8	0.9378	
8	0.0664		8	0.0579		9	0.9894	
9	0.0151		9	0.0198		10	1.0347	
10	-0.0300		10	-0.0225		11	1.0515	
11	-0.0467		11	-0.0483		12	1.0152	
12	-0.0106		12	-0.0015		13	0.9442	
13	0.0601		13	0.0638		14	1.0045	
14	0.0001		14	-0.0145		15	0.9658	
15	0.0386		15	0.0282		16	0.9872	
16	0.0173		16	0.0201		17	1.0318	
17	-0.0271		17	-0.0158		18	1.0206	
18	-0.0160		18	-0.0165		19	0.9969	
19	0.0076		19	0.0148		20	0.9782	
20	0.0263		20	0.0329		21	0.9676	
21	0.0368		21	0.0179		22	0.9867	
22	0.0179		22	0.0098		23	1.0544	
23	-0.0496		23	-0.0466		24	1.1239	
24	-0.1188		24	-0.1142		25	0.9981	
25	0.0065		25	0.0148				

FIGURE 5.37 ACF, PACF, and variogram of residuals from ARIMA$(2, 1, 1) \times (0, 0, 1)_7$ fit to daily respiratory syndrome counts.

5.11 FINAL COMMENTS

ARIMA models (a.k.a. Box–Jenkins models) present a very powerful and flexible class of models for time series analysis and forecasting. Over the years, they have been very successfully applied to many problems in research and practice. However, there might be certain situations where they may fall short on providing the "right" answers. For example, in ARIMA models, forecasting future observations primarily relies on the past data and implicitly assumes that the conditions at which the data is collected will remain the same in the future as well. In many situations this assumption may (and most likely will) not be appropriate. For those cases, the transfer function–noise models, where a set of input variables that may have an effect on the time series are added to the model, provide suitable options. We shall discuss these models in the next chapter. For an excellent discussion of this matter and of time series analysis and forecasting in general, see Jenkins (1979).

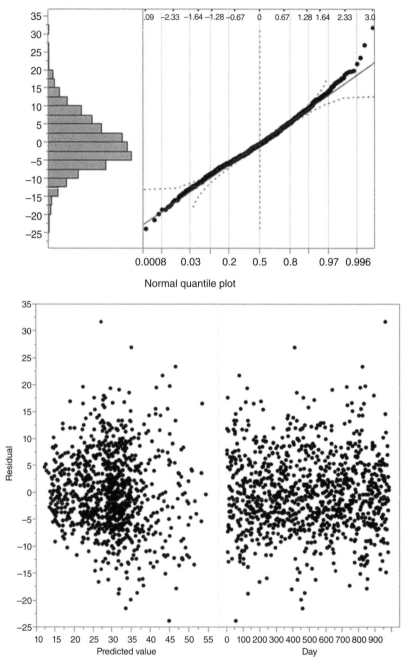

FIGURE 5.38   Plots of residuals from ARIMA(2, 1, 1) × (0, 0, 1)$_7$ fit to daily respiratory syndrome counts.

5.12 R COMMANDS FOR CHAPTER 5

Example 5.1   The loan applications data are in the second column of the array called loan.data in which the first column is the number of weeks. We first plot the data as well as the ACF and PACF.

```
plot(loan.data[,2],type="o",pch=16,cex=.5,xlab='Week',ylab='Loan
Applications')
```

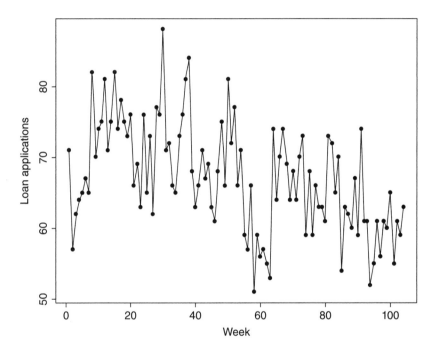

```
par(mfrow=c(1,2),oma=c(0,0,0,0))

acf(loan.data[,2],lag.max=25,type="correlation",main="ACF for the
Number \nof Loan Applications")

acf(loan.data[,2], lag.max=25,type="partial",main="PACF for the
Number \nof Loan Applications")
```

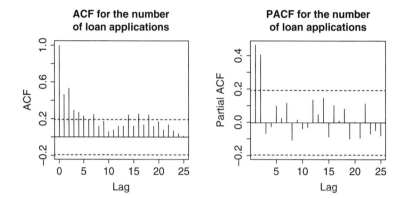

Fit an ARIMA(2,0,0) model to the data using arima function in the stats package.

```
loan.fit.ar2<-arima(loan.data[,2],order=c(2, 0, 0))
loan.fit.ar2

        Call:
        arima(x = loan.data[, 2], order = c(2, 0, 0))

        Coefficients:
                  ar1      ar2   intercept
               0.2659   0.4130     66.8538
        s.e.   0.0890   0.0901      1.8334

        sigma^2 estimated as 38.32:   log likelihood = -337.46,
        aic = 682.92


res.loan.ar2<-as.vector(residuals(loan.fit.ar2))
#to obtain the fitted values we use the function fitted() from
#the forecast package
library(forecast)
fit.loan.ar2<-as.vector(fitted(loan.fit.ar2))

Box.test(res.loan.ar2,lag=48,fitdf=3,type="Ljung")

                Box-Ljung test

        data:  res.loan.ar2
        X-squared = 31.8924, df = 45, p-value = 0.9295


#ACF and PACF of the Residuals
par(mfrow=c(1,2),oma=c(0,0,0,0))
```

```
acf(res.loan.ar2,lag.max=25,type="correlation",main="ACF of the
Residuals \nof AR(2) Model")

acf(res.loan.ar2, lag.max=25,type="partial",main="PACF of the
Residuals \nof AR(2) Model")
```

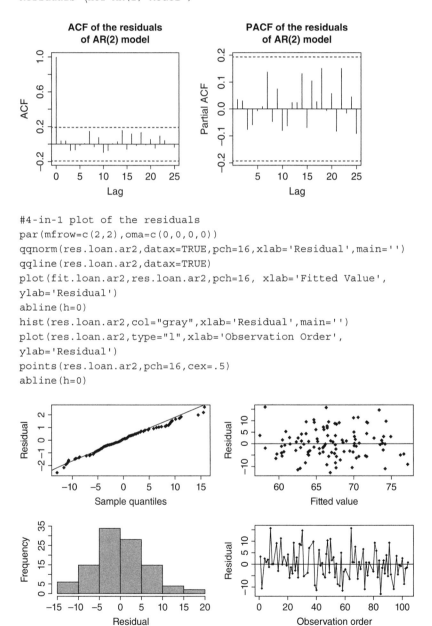

```
#4-in-1 plot of the residuals
par(mfrow=c(2,2),oma=c(0,0,0,0))
qqnorm(res.loan.ar2,datax=TRUE,pch=16,xlab='Residual',main='')
qqline(res.loan.ar2,datax=TRUE)
plot(fit.loan.ar2,res.loan.ar2,pch=16, xlab='Fitted Value',
ylab='Residual')
abline(h=0)
hist(res.loan.ar2,col="gray",xlab='Residual',main='')
plot(res.loan.ar2,type="l",xlab='Observation Order',
ylab='Residual')
points(res.loan.ar2,pch=16,cex=.5)
abline(h=0)
```

Plot fitted values

```
plot(loan.data[,2],type="p",pch=16,cex=.5,xlab='Week',ylab='Loan
Applications')
lines(fit.loan.ar2)
legend(95,88,c("y(t)","yhat(t)"),  pch=c(16,  NA),lwd=c(NA,.5),
cex=.55)
```

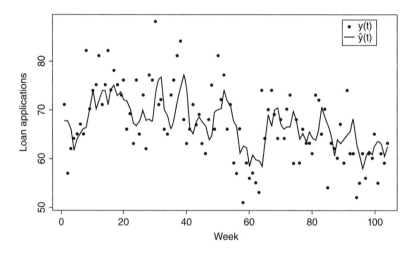

Example 5.2   The Dow Jones index data are in the second column of the array called dji.data in which the first column is the month of the year. We first plot the data as well as the ACF and PACF.

```
plot(dji.data[,2],type="o",pch=16,cex=.5,xlab='Date',ylab='DJI',
xaxt='n')
axis(1,  seq(1,85,12),  dji.data[seq(1,85,12),1])
```

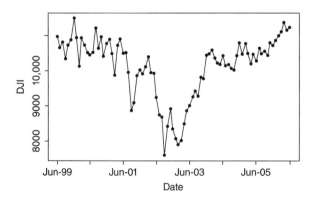

```
par(mfrow=c(1,2),oma=c(0,0,0,0))
acf(dji.data[,2],lag.max=25,type="correlation",main="ACF for the
Number \nof Dow Jones Index")

acf(dji.data[,2], lag.max=25,type="partial",main="PACF for the
Number \nof Dow Jones Index ")
```

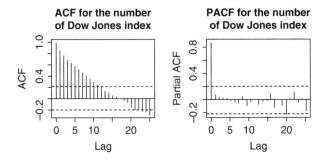

We first fit an ARIMA(1,0,0) model to the data using arima function in the stats package.

```
dji.fit.ar1<-arima(dji.data[,2],order=c(1, 0, 0))
dji.fit.ar1
        Call:
        arima(x = dji.data[, 2], order = c(1, 0, 0))

        Coefficients:
                ar1     intercept
             0.8934   10291.2984
        s.e.   0.0473     373.8723

        sigma^2 estimated as 156691:  log likelihood = -629.8,
        aic = 1265.59


res.dji.ar1<-as.vector(residuals(dji.fit.ar1))
#to obtain the fitted values we use the function fitted() from
#the forecast package
library(forecast)
fit.dji.ar1<-as.vector(fitted(dji.fit.ar1))

Box.test(res.dji.ar1,lag=48,fitdf=3,type="Ljung")

        Box-Ljung test

        data:  res.dji.ar1
        X-squared = 29.9747, df = 45, p-value = 0.9584
```

```
#ACF and PACF of the Residuals
par(mfrow=c(1,2),oma=c(0,0,0,0))
acf(res.dji.ar1,lag.max=25,type="correlation",main="ACF of the
Residuals \nof AR(1) Model")

acf(res.dji.ar1, lag.max=25,type="partial",main="PACF of the
Residuals \nof AR(1) Model")
```

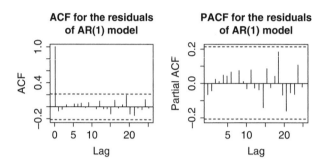

```
#4-in-1 plot of the residuals
par(mfrow=c(2,2),oma=c(0,0,0,0))
qqnorm(res.dji.ar1,datax=TRUE,pch=16,xlab='Residual',main='')
qqline(res.dji.ar1,datax=TRUE)
plot(fit.dji.ar1,res.dji.ar1,pch=16, xlab='Fitted Value',
ylab='Residual')
abline(h=0)
hist(res.dji.ar1,col="gray",xlab='Residual',main='')
plot(res.dji.ar1,type="l",xlab='Observation Order',
ylab='Residual')
points(res.dji.ar1,pch=16,cex=.5)
abline(h=0)
```

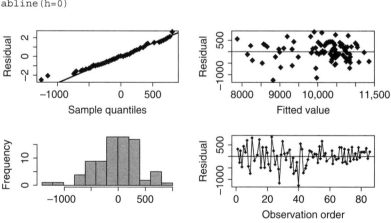

We now consider the first difference of the Dow Jones index.

```
wt.dji<-diff(dji.data[,2])
plot(wt.dji,type="o",pch=16,cex=.5,xlab='Date',ylab='w(t)',
xaxt='n')
axis(1, seq(1,85,12), dji.data[seq(1,85,12),1])
```

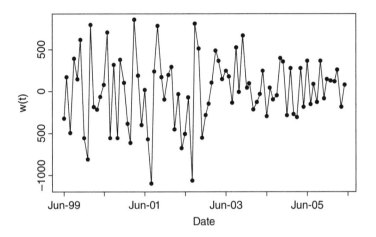

```
par(mfrow=c(1,2),oma=c(0,0,0,0))
acf(wt.dji,lag.max=25,type="correlation",main="ACF for the
Number \nof w(t)")

acf(wt.dji, lag.max=25,type="partial",main="PACF for the
Number \nof w(t)")
```

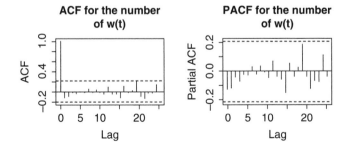

Example 5.6   The loan applications data are in the second column of the array called loan.data in which the first column is the number of weeks. We use the AR(2) model to make the forecasts.

```
loan.fit.ar2<-arima(loan.data[,2],order=c(2, 0, 0))
#to obtain the 1- to 12-step ahead forecasts, we use the
#function forecast() from the forecast package
library(forecast)
loan.ar2.forecast<-as.array(forecast(loan.fit.ar2,h=12))
loan.ar2.forecast
```

Point	Forecast	Lo 80	Hi 80	Lo 95	Hi 95
105	62.58571	54.65250	70.51892	50.45291	74.71851
106	64.12744	55.91858	72.33629	51.57307	76.68180
107	64.36628	55.30492	73.42764	50.50812	78.22444
108	65.06647	55.80983	74.32312	50.90965	79.22330
109	65.35129	55.86218	74.84039	50.83895	79.86362
110	65.71617	56.13346	75.29889	51.06068	80.37167
111	65.93081	56.27109	75.59054	51.15754	80.70409
112	66.13857	56.43926	75.83789	51.30475	80.97240
113	66.28246	56.55529	76.00962	51.40605	81.15887
114	66.40651	56.66341	76.14961	51.50572	81.30730
115	66.49892	56.74534	76.25249	51.58211	81.41572
116	66.57472	56.81486	76.33458	51.64830	81.50114

Note that forecast function provides a list with forecasts as well as 80% and 95% prediction limits. To see the elements of the list, we can do

```
ls(loan.ar2.forecast)
        [1] "fitted"    "level"   "lower"  "mean"   "method"  "model"
        [7] "residuals" "upper"   "x"      "xname"
```

In this list, "mean" stands for the forecasts while "lower" and "upper" provide the 80 and 95% lower and upper prediction limits, respectively. To plot the forecasts and the prediction limits, we have

```
plot(loan.data[,2],type="p",pch=16,cex=.5,xlab='Date',ylab='Loan
Applications',xaxt='n',xlim=c(1,120))
axis(1, seq(1,120,24), dji.data[seq(1,120,24),1])
lines(105:116,loan.ar2.forecast$mean,col="grey40")
lines(105:116,loan.ar2.forecast$lower[,2])
lines(105:116,loan.ar2.forecast$upper[,2])
legend(72,88,c("y","Forecast","95% LPL","95% UPL"), pch=c(16, NA,
NA,NA),lwd=c(NA,.5,.5,.5),cex=.55,col=c("black","grey40","black",
"black"))
```

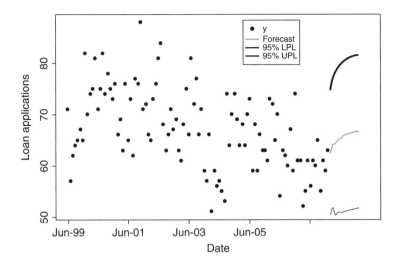

Example 5.8   The clothing sales data are in the second column of the array called closales.data in which the first column is the month of the year. We first plot the data and its ACF and PACF.

```
par(mfrow=c(1,2),oma=c(0,0,0,0))
acf(closales.data[,2],lag.max=50,type="correlation",main="ACF for
the \n Clothing Sales")

acf(closales.data[,2], lag.max=50,type="partial",main="PACF for
the \n Clothing Sales")
```

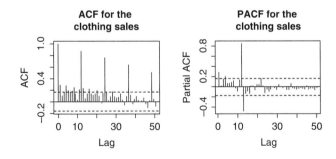

We now take the seasonal and non-seasonal difference of the data.

```
wt.closales<-diff(diff(closales.data[,2],lag=1),lag=12)
#Note that the same result would have been obtained with the
#following command when the order of differencing is reversed
#wt.closales<-diff(diff(closales.data[,2],lag=12),lag=1)
```

```
plot(wt.closales,type="o",pch=16,cex=.5,xlab='Date',ylab='w(t)',
xaxt='n')
axis(1, seq(1,144,24), closales.data[seq(13,144,24),1])
```

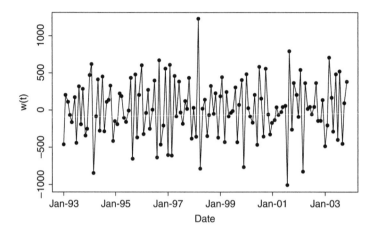

```
par(mfrow=c(1,2),oma=c(0,0,0,0))
acf(wt.closales,lag.max=50,type="correlation",main="ACF for w(t)")
acf(wt.closales, lag.max=50,type="partial",main="PACF for w(t)")
```

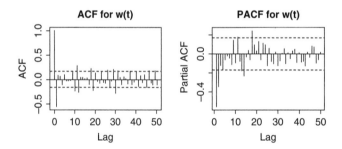

We now fit a seasonal ARIMA$(0,1,1) \times (0,1,1)_{12}$ model to the data. We then plot the residuals plots including ACF and PACF of the residuals. In the end we plot the true and fitted values.

```
closales.fit.sar<-arima(closales.data[,2],order=c(0,1,1),
seasonal=list(order = c(0,1,1),period=12),)

res.closales.sar<-as.vector(residuals(closales.fit.sar))
#to obtain the fitted values we use the function fitted() from
the forecast package
library(forecast)
fit.closales.sar<-as.vector(fitted(closales.fit.sar))
```

```
#ACF and PACF of the Residuals
par(mfrow=c(1,2),oma=c(0,0,0,0))
acf(res.closales.sar,lag.max=50,type="correlation",main="ACF of
the Residuals")

acf(res.closales.sar,lag.max=50,type="partial",main="PACF of the
Residuals")
```

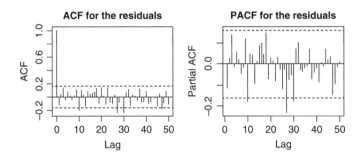

```
#4-in-1 plot of the residuals
par(mfrow=c(2,2),oma=c(0,0,0,0))
qqnorm(res.closales.sar,datax=TRUE,pch=16,xlab='Residual',main='')
qqline(res.closales.sar,datax=TRUE)
plot(fit.closales.sar,res.closales.sar,pch=16, xlab='Fitted
Value',ylab='Residual')
abline(h=0)
hist(res.closales.sar,col="gray",xlab='Residual',main='')
plot(res.closales.sar,type="l",xlab='Observation Order',
ylab='Residual')
points(res.closales.sar,pch=16,cex=.5)
abline(h=0)
```

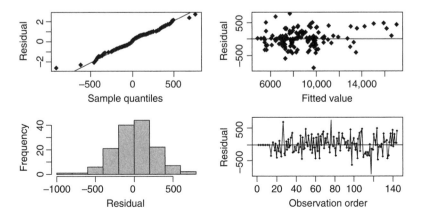

```
plot(closales.data[,2],type="p",pch=16,cex=.5,xlab='Date',
ylab='Clothing Sales',xaxt='n')
axis(1, seq(1,144,24), closales.data[seq(1,144,24),1])
lines(1:144, fit.closales.sar)
legend(2,17500,c("US Clothing Sales","Fitted"), pch=c(16, NA),
lwd=c(NA,.5),cex=.55,col=c("black","black"))
```

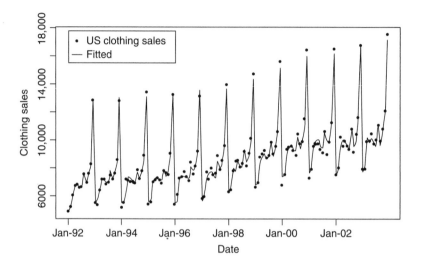

EXERCISES

5.1 Consider the time series data shown in Chapter 4, Table E4.2.

    a. Fit an appropriate ARIMA model to the first 40 observations of this time series.

    b. Make one-step-ahead forecasts of the last 10 observations. Determine the forecast errors.

    c. In Exercise 4.4 you used simple exponential smoothing with $\lambda = 0.2$ to smooth the first 40 time periods of this data and make forecasts of the last 10 observations. Compare the ARIMA forecasts with the exponential smoothing forecasts. How well do both of these techniques work?

5.2 Consider the time series data shown in Table E5.1.

    a. Make a time series plot of the data.

    b. Calculate and plot the sample autocorrelation and PACF. Is there significant autocorrelation in this time series?

TABLE E5.1   Data for Exercise 5.2

Period	y_t	Period	y_t	Period	y_t	Period	y_t	Period	y_t
1	29	11	29	21	31	31	28	41	36
2	20	12	28	22	30	32	30	42	35
3	25	13	28	23	37	33	29	43	33
4	29	14	26	24	30	34	34	44	29
5	31	15	27	25	33	35	30	45	25
6	33	16	26	26	31	36	20	46	27
7	34	17	30	27	27	37	17	47	30
8	27	18	28	28	33	38	23	48	29
9	26	19	26	29	37	39	24	49	28
10	30	20	30	30	29	40	34	50	32

     c. Identify and fit an appropriate ARIMA model to these data. Check for model adequacy.

     d. Make one-step-ahead forecasts of the last 10 observations. Determine the forecast errors.

5.3 Consider the time series data shown in Table E5.2.

     a. Make a time series plot of the data.

     b. Calculate and plot the sample autocorrelation and PA. Is there significant autocorrelation in this time series?

     c. Identify and fit an appropriate ARIMA model to these data. Check for model adequacy.

     d. Make one-step-ahead forecasts of the last 10 observations. Determine the forecast errors.

TABLE E5.2   Data for Exercise 5.3

Period	y_t	Period	y_t	Period	y_t	Period	y_t	Period	y_t
1	500	11	508	21	475	31	639	41	637
2	496	12	510	22	485	32	679	42	606
3	450	13	512	23	495	33	674	43	610
4	448	14	503	24	500	34	677	44	620
5	456	15	505	25	541	35	700	45	613
6	458	16	494	26	555	36	704	46	593
7	472	17	491	27	565	37	727	47	578
8	495	18	487	28	601	38	736	48	581
9	491	19	491	29	610	39	693	49	598
10	488	20	486	30	605	40	65	50	613

5.4   Consider the time series model

$$y_t = 200 + 0.7y_{t-1} + \varepsilon_t$$

a. Is this a stationary time series process?

b. What is the mean of the time series?

c. If the current observation is $y_{100} = 750$, would you expect the next observation to be above or below the mean?

5.5   Consider the time series model

$$y_t = 150 - 0.5y_{t-1} + \varepsilon_t$$

a. Is this a stationary time series process?

b. What is the mean of the time series?

c. If the current observation is $y_{100} = 85$, would you expect the next observation to be above or below the mean?

5.6   Consider the time series model

$$y_t = 50 + 0.8y_{t-1} - 0.15 + \varepsilon_t$$

a. Is this a stationary time series process?

b. What is the mean of the time series?

c. If the current observation is $y_{100} = 160$, would you expect the next observation to be above or below the mean?

5.7   Consider the time series model

$$y_t = 20 + \varepsilon_t + 0.2\varepsilon_{t-1}$$

a. Is this a stationary time series process?

b. Is this an invertible time series?

c. What is the mean of the time series?

d. If the current observation is $y_{100} = 23$, would you expect the next observation to be above or below the mean? Explain your answer.

5.8   Consider the time series model

$$y_t = 50 + 0.8y_{t-1} + \varepsilon_t - 0.2\varepsilon_{t-1}$$

a. Is this a stationary time series process?

b. What is the mean of the time series?

c. If the current observation is $y_{100} = 270$, would you expect the next observation to be above or below the mean?

5.9 The data in Chapter 4, Table E4.4, exhibits a linear trend. Difference the data to remove the trend.

   a. Fit an ARIMA model to the first differences.

   b. Explain how this model would be used for forecasting.

5.10 Table B.1 in Appendix B contains data on the market yield on US Treasury Securities at 10-year constant maturity.

   a. Fit an ARIMA model to this time series, excluding the last 20 observations. Investigate model adequacy. Explain how this model would be used for forecasting.

   b. Forecast the last 20 observations.

   c. In Exercise 4.10, you were asked to use simple exponential smoothing with $\lambda = 0.2$ to smooth the data, and to forecast the last 20 observations. Compare the ARIMA and exponential smoothing forecasts. Which forecasting method do you prefer?

5.11 Table B.2 contains data on pharmaceutical product sales.

   a. Fit an ARIMA model to this time series, excluding the last 10 observations. Investigate model adequacy. Explain how this model would be used for forecasting.

   b. Forecast the last 10 observations.

   c. In Exercise 4.12, you were asked to use simple exponential smoothing with $\lambda = 0.1$ to smooth the data, and to forecast the last 10 observations. Compare the ARIMA and exponential smoothing forecasts. Which forecasting method do you prefer?

   d. How would prediction intervals be obtained for the ARIMA forecasts?

5.12 Table B.3 contains data on chemical process viscosity.

   a. Fit an ARIMA model to this time series, excluding the last 20 observations. Investigate model adequacy. Explain how this model would be used for forecasting.

   b. Forecast the last 20 observations.

   c. Show how to obtain prediction intervals for the forecasts in part b above.

5.13 Table B.4 contains data on the annual US production of blue and gorgonzola cheeses.

    a. Fit an ARIMA model to this time series, excluding the last 10 observations. Investigate model adequacy. Explain how this model would be used for forecasting.

    b. Forecast the last 10 observations.

    c. In Exercise 4.16, you were asked to use exponential smoothing methods to smooth the data, and to forecast the last 10 observations. Compare the ARIMA and exponential smoothing forecasts. Which forecasting method do you prefer?

    d. How would prediction intervals be obtained for the ARIMA forecasts?

5.14 Reconsider the blue and gorgonzola cheese data in Table B.4 and Exercise 5.13. In Exercise 4.17 you were asked to take the first difference of this data and develop a forecasting procedure based on using exponential smoothing on the first differences. Compare this procedure with the ARIMA model of Exercise 5.13.

5.15 Table B.5 shows US beverage manufacturer product shipments. Develop an appropriate ARIMA model and a procedure for forecasting for these data. Explain how prediction intervals would be computed.

5.16 Table B.6 contains data on the global mean surface air temperature anomaly. Develop an appropriate ARIMA model and a procedure for forecasting for these data. Explain how prediction intervals would be computed.

5.17 Reconsider the global mean surface air temperature anomaly data shown in Table B.6 and used in Exercise 5.16. In Exercise 4.20 you were asked to use simple exponential smoothing with the optimum value of λ to smooth the data. Compare the results with those obtained with the ARIMA model in Exercise 5.16.

5.18 Table B.7 contains daily closing stock prices for the Whole Foods Market. Develop an appropriate ARIMA model and a procedure for these data. Explain how prediction intervals would be computed.

5.19 Reconsider the Whole Foods Market data shown in Table B.7 and used in Exercise 5.18. In Exercise 4.22 you used simple exponential smoothing with the optimum value of λ to smooth the data.

Compare the results with those obtained from the ARIMA model in Exercise 5.18.

5.20 Unemployment rate data is given in Table B.8. Develop an appropriate ARIMA model and a procedure for forecasting for these data. Explain how prediction intervals would be computed.

5.21 Reconsider the unemployment rate data shown in Table B.8 and used in Exercise 5.21. In Exercise 4.24 you used simple exponential smoothing with the optimum value of λ to smooth the data. Compare the results with those obtained from the ARIMA model in Exercise 5.20.

5.22 Table B.9 contains yearly data on the international sunspot numbers. Develop an appropriate ARIMA model and a procedure for forecasting for these data. Explain how prediction intervals would be computed.

5.23 Reconsider the sunspot data shown in Table B.9 and used in Exercise 5.22.

   a. In Exercise 4.26 you were asked to use simple exponential smoothing with the optimum value of λ to smooth the data, and to use an exponential smoothing procedure for trends. How do these procedures compare to the ARIMA model from Exercise 5.22? Compare the results with those obtained in Exercise 4.26.

   b. Do you think that using either exponential smoothing procedure would result in better forecasts than those from the ARIMA model?

5.24 Table B.10 contains 7 years of monthly data on the number of airline miles flown in the United Kingdom. This is seasonal data.

   a. Using the first 6 years of data, develop an appropriate ARIMA model and a procedure for these data.

   b. Explain how prediction intervals would be computed.

   c. Make one-step-ahead forecasts of the last 12 months. Determine the forecast errors. How well did your procedure work in forecasting the new data?

5.25 Reconsider the airline mileage data in Table B.10 and used in Exercise 5.24.

   a. In Exercise 4.27 you used Winters' method to develop a forecasting model using the first 6 years of data and you made forecasts for the last 12 months. Compare those forecasts with the ones you made using the ARIMA model from Exercise 5.24.

b. Which forecasting method would you prefer and why?

5.26 Table B.11 contains 8 years of monthly champagne sales data. This is seasonal data.

a. Using the first 7 years of data, develop an appropriate ARIMA model and a procedure for these data.

b. Explain how prediction intervals would be computed.

c. Make one-step-ahead forecasts of the last 12 months. Determine the forecast errors. How well did your procedure work in forecasting the new data?

5.27 Reconsider the monthly champagne sales data in Table B.11 and used in Exercise 5.26.

a. In Exercise 4.29 you used Winters' method to develop a forecasting model using the first 7 years of data and you made forecasts for the last 12 months. Compare those forecasts with the ones you made using the ARIMA model from Exercise 5.26.

b. Which forecasting method would you prefer and why?

5.28 Montgomery et al. (1990) give 4 years of data on monthly demand for a soft drink. These data are given in Chapter 4, Table E4.5.

a. Using the first three years of data, develop an appropriate ARIMA model and a procedure for these data.

b. Explain how prediction intervals would be computed.

c. Make one-step-ahead forecasts of the last 12 months. Determine the forecast errors. How well did your procedure work in forecasting the new data?

5.29 Reconsider the soft drink demand data in Table E4.5 and used in Exercise 5.28.

a. In Exercise 4.31 you used Winters' method to develop a forecasting model using the first 7 years of data and you made forecasts for the last 12 months. Compare those forecasts with the ones you made using the ARIMA model from the previous exercise.

b. Which forecasting method would you prefer and why?

5.30 Table B.12 presents data on the hourly yield from a chemical process and the operating temperature. Consider only the yield data in this exercise. Develop an appropriate ARIMA model and a procedure for forecasting for these data. Explain how prediction intervals would be computed.

5.31 Table B.13 presents data on ice cream and frozen yogurt sales. Develop an appropriate ARIMA model and a procedure for forecasting for these data. Explain how prediction intervals would be computed.

5.32 Table B.14 presents the CO_2 readings from Mauna Loa. Develop an appropriate ARIMA model and a procedure for forecasting for these data. Explain how prediction intervals would be computed.

5.33 Table B.15 presents data on the occurrence of violent crimes. Develop an appropriate ARIMA model and a procedure for forecasting for these data. Explain how prediction intervals would be computed.

5.34 Table B.16 presents data on the US gross domestic product (GDP). Develop an appropriate ARIMA model and a procedure for forecasting for these data. Explain how prediction intervals would be computed.

5.35 Total annual energy consumption is shown in Table B.17. Develop an appropriate ARIMA model and a procedure for forecasting for these data. Explain how prediction intervals would be computed.

5.36 Table B.18 contains data on coal production. Develop an appropriate ARIMA model and a procedure for forecasting for these data. Explain how prediction intervals would be computed.

5.37 Table B.19 contains data on the number of children 0–4 years old who drowned in Arizona. Develop an appropriate ARIMA model and a procedure for forecasting for these data. Explain how prediction intervals would be computed.

5.38 Data on tax refunds and population are shown in Table B.20. Develop an appropriate ARIMA model and a procedure for forecasting for these data. Explain how prediction intervals would be computed.

5.39 Table B.21 contains data from the US Energy Information Administration on monthly average price of electricity for the residential sector in Arizona. This data has a strong seasonal component. Use the data from 2001–2010 to develop an ARIMA model for this data. Use this model to simulate one-month-ahead forecasts for the remaining years. Calculate the forecast errors. Discuss the reasonableness of the forecasts.

5.40 In Exercise 4.44 you were asked to develop a smoothing-type model for the data in Table B.21. Compare the performance of that mode with the performance of the ARIMA model from the previous exercise.

5.41 Table B.22 contains data from the Danish Energy Agency on Danish crude oil production. Develop an appropriate ARIMA model for this data. Compare this model with the smoothing models developed in Exercises 4.46 and 4.47.

5.42 Table B.23 shows Weekly data on positive laboratory test results for influenza are shown in Table B.23. Notice that these data have a number of missing values. In exercise you were asked to develop and implement a scheme to estimate the missing values. This data has a strong seasonal component. Use the data from 1997 to 2010 to develop an appropriate ARIMA model for this data. Use this model to simulate one-week-ahead forecasts for the remaining years. Calculate the forecast errors. Discuss the reasonableness of the forecasts.

5.43 In Exercise 4.48 you were asked to develop a smoothing–type model for the data in Table B.23. Compare the performance of that mode with the performance of the ARIMA model from the previous exercise.

5.44 Data from the Western Regional Climate Center for the monthly mean daily solar radiation (in Langleys) at the Zion Canyon, Utah, station are shown in Table B.24. This data has a strong seasonal component. Use the data from 2003 to 2012 to develop an appropriate ARIMA model for this data. Use this model to simulate one-month-ahead forecasts for the remaining years. Calculate the forecast errors. Discuss the reasonableness of the forecasts.

5.45 In Exercise 4.50 you were asked to develop a smoothing-type model for the data in Table B.24. Compare the performance of that mode with the performance of the ARIMA model from the previous exercise.

5.46 Table B.25 contains data from the National Highway Traffic Safety Administration on motor vehicle fatalities from 1966 to 2012. This data is used by a variety of governmental and industry groups, as well as research organizations. Develop an ARIMA model for forecasting fatalities using the data from 1966 to 2006 to develop the model, and then simulate one-year-ahead forecasts for the remaining years. Compute the forecasts errors. How well does this method seem to work?

5.47 Appendix Table B.26 contains data on monthly single-family residential new home sales from 1963 through 2014. Develop an ARIMA model for forecasting new home sales using the data from 1963 to 2006 to develop the model, and then simulate one-year-ahead forecasts for the remaining years. Compute the forecasts errors. How well does this method seem to work?

5.48 Appendix Table B.27 contains data on the airline best on-time arrival and airport performance. The data is given by month from January 1995 through February 2013. Develop an ARIMA model for forecasting on-time arrivals using the data from 1995 to 2008 to develop the model, and then simulate one-year-ahead forecasts for the remaining years. Compute the forecasts errors. How well does this method seem to work?

5.49 Data from the US Census Bureau on monthly domestic automobile manufacturing shipments (in millions of dollars) are shown in Table B.28. Develop an ARIMA model for forecasting shipments. Note that there is some apparent seasonality in the data. Why does this seasonal behavior occur?

5.50 An ARIMA model has been fit to a time series, resulting in

$$\hat{y}_t = 25 + 0.35y_{t-1} + \varepsilon_t$$

a. Suppose that we are at time period $T = 100$ and $y_{100} = 31$. Determine forecasts for periods 101, 102, 103, ... from this model at origin 100.

b. What is the shape of the forecast function from this model?

c. Suppose that the observation for time period 101 turns out to be $y_{101} = 33$. Revise your forecasts for periods 102, 103, ... using period 101 as the new origin of time.

d. If your estimate $\hat{\sigma}^2 = 2$, find a 95% prediction interval on the forecast of period 101 made at the end of period 100.

5.51 The following ARIMA model has been fit to a time series:

$$\hat{y}_t = 25 + 0.8y_{t-1} - 0.3y_{t-2} + \varepsilon_t$$

   a. Suppose that we are at the end of time period $T = 100$ and we know that $y_{100} = 40$ and $y_{99} = 38$. Determine forecasts for periods 101, 102, 103, ... from this model at origin 100.

   b. What is the shape of the forecast function from this model?

   c. Suppose that the observation for time period 101 turns out to be $y_{101} = 35$. Revise your forecasts for periods 102, 103, ... using period 101 as the new origin of time.

   d. If your estimate $\hat{\sigma}^2 = 1$, find a 95% prediction interval on the forecast of period 101 made at the end of period 100.

5.52 The following ARIMA model has been fit to a time series:

$$\hat{y}_t = 25 + 0.8y_{t-1} - 0.2\varepsilon_{t-1} + \varepsilon_t$$

   a. Suppose that we are at the end of time period $T = 100$ and we know that the forecast for period 100 was 130 and the actual observed value was $y_{100} = 140$. Determine forecasts for periods 101, 102, 103, ... from this model at origin 100.

   b. What is the shape of the forecast function from this model?

   c. Suppose that the observation for time period 101 turns out to be $y_{101} = 132$. Revise your forecasts for periods 102, 103, ... using period 101 as the new origin of time.

   d. If your estimate $\hat{\sigma}^2 = 1.5$, find a 95% prediction interval on the forecast of period 101 made at the end of period 100.

5.53 The following ARIMA model has been fit to a time series:

$$\hat{y}_t = 20 + \varepsilon_t + 0.45\varepsilon_{t-1} - 0.3\varepsilon_{t-2}$$

   a. Suppose that we are at the end of time period $T = 100$ and we know that the observed forecast error for period 100 was 0.5 and for period 99 we know that the observed forecast error was -0.8. Determine forecasts for periods 101, 102, 103, ... from this model at origin 100.

   b. What is the shape of the forecast function that evolves from this model?

   c. Suppose that the observations for the next four time periods turn out to be 17.5, 21.25, 18.75, and 16.75. Revise your forecasts for periods 102, 103, ... using a rolling horizon approach.

   d. If your estimate $\hat{\sigma} = 0.5$, find a 95% prediction interval on the forecast of period 101 made at the end of period 100.

5.54 The following ARIMA model has been fit to a time series:

$$\hat{y}_t = 50 + \varepsilon_t + 0.5\varepsilon_{t-1}$$

a. Suppose that we are at the end of time period $T = 100$ and we know that the observed forecast error for period 100 was 2. Determine forecasts for periods 101, 102, 103, ... from this model at origin 100.

b. What is the shape of the forecast function from this model?

c. Suppose that the observations for the next four time periods turn out to be 53, 55, 46, and 50. Revise your forecasts for periods 102, 103, ... using a rolling horizon approach.

d. If your estimate $\hat{\sigma} = 1$, find a 95% prediction interval on the forecast of period 101 made at the end of period 100.

5.55 For each of the ARIMA models shown below, give the forecasting equation that evolves for lead times $\tau = 1, 2, ..., L$. In each case, explain the shape of the resulting forecast function over the forecast lead time.

a. AR(1)
b. AR(2)
c. MA(1)
d. MA(2)
e. ARMA(1, 1)
f. IMA(1, 1)
g. ARIMA(1, 1, 0)

5.56 Use a random number generator and generate 100 observations from the AR(1) model $y_t = 25 + 0.8y_{t-1} + \varepsilon_t$. Assume that the errors are normally and independently distributed with mean zero and variance $\sigma^2 = 1$.

a. Verify that your time series is AR(1).

b. Generate 100 observations for a $N(0, 1)$ process and add these random numbers to the 100 AR(1) observations in part a to create a new time series that is the sum of AR(1) and "white noise."

c. Find the sample autocorrelation and partial autocorrelation functions for the new time series created in part b. Can you identify the new time series?

d. Does this give you any insight about how the new time series might arise in practical settings?

5.57   Assume that you have fit the following model:

$$\hat{y}_t = y_{t-1} + 0.7\varepsilon_{t-1} + \varepsilon_t$$

a. Suppose that we are at the end of time period $T = 100$. What is the equation for forecasting the time series in period 101?

b. What does the forecast equation look like for future periods 102, 103, ... ?

c. Suppose that we know that the observed value of y_{100} was 250 and forecast error in period 100 was 12. Determine forecasts for periods 101, 102, 103, ... from this model at origin 100.

d. If your estimate $\hat{\sigma} = 1$, find a 95% prediction interval on the forecast of period 101 made at the end of period 100.

e. Show the behavior of this prediction interval for future lead times beyond period 101. Are you surprised at how wide the interval is? Does this tell you something about the reliability of forecasts from this model at long lead times?

5.58   Consider the AR(1) model $y_t = 25 + 0.75y_{t-1} + \varepsilon_t$. Assume that the variance of the white noise process is $\sigma^2 = 1$.

a. Sketch the theoretical ACF and PACF for this model.

b. Generate 50 realizations of this AR(1) process and compute the sample ACF and PACF. Compare the sample ACF and the sample PACF to the theoretical ACF and PACF. How similar to the theoretical values are the sample values?

c. Repeat part b using 200 realizations. How has increasing the sample size impacted the agreement between the sample and theoretical ACF and PACF? Does this give you any insight about the sample sizes required for model building, or the reliability of models built to short time series?

5.59   Consider the AR(1) model $y_t = 25 + 0.75y_{t-1} + \varepsilon_t$. Assume that the variance of the white noise process is $\sigma^2 = 10$.

a. Sketch the theoretical ACF and PACF for this model.

b. Generate 50 realizations of this AR(1) process and compute the sample ACF and PACF. Compare the sample ACF and the sample PACF to the theoretical ACF and PACF. How similar to the theoretical values are the sample values?

c. Compare the results from part b with the results from part b of Exercise 5.47. How much has changing the variance of the white noise process impacted the results?

d. Repeat part b using 200 realizations. How has increasing the sample size impacted the agreement between the sample and theoretical ACF and PACF? Does this give you any insight about the sample sizes required for model building, or the reliability of models built to short time series?

e. Compare the results from part d with the results from part c of Exercise 5.47. How much has changing the variance of the white noise process impacted the results?

5.60 Consider the AR(2) model $y_t = 25 + 0.6y_{t-1} + 0.25y_{t-2} + \varepsilon_t$. Assume that the variance of the white noise process is $\sigma^2 = 1$.

a. Sketch the theoretical ACF and PACF for this model.

b. Generate 50 realizations of this AR(1) process and compute the sample ACF and PACF. Compare the sample ACF and the sample PACF to the theoretical ACF and PACF. How similar to the theoretical values are the sample values?

c. Repeat part b using 200 realizations. How has increasing the sample size impacted the agreement between the sample and theoretical ACF and PACF? Does this give you any insight about the sample sizes required for model building, or the reliability of models built to short time series?

5.61 Consider the MA(1) model $y_t = 40 + 0.4\varepsilon_{t-1} + \varepsilon_t$. Assume that the variance of the white noise process is $\sigma^2 = 2$.

a. Sketch the theoretical ACF and PACF for this model.

b. Generate 50 realizations of this AR(1) process and compute the sample ACF and PACF. Compare the sample ACF and the sample PACF to the theoretical ACF and PACF. How similar to the theoretical values are the sample values?

c. Repeat part b using 200 realizations. How has increasing the sample size impacted the agreement between the sample and theoretical ACF and PACF? Does this give you any insight about the sample sizes required for model building, or the reliability of models built to short time series?

5.62 Consider the ARMA(1, 1) model $y_t = 50 - 0.7y_{t-1} + 0.5\varepsilon_{t-1} + \varepsilon_t$. Assume that the variance of the white noise process is $\sigma^2 = 2$.

a. Sketch the theoretical ACF and PACF for this model.

b. Generate 50 realizations of this AR(1) process and compute the sample ACF and PACF. Compare the sample ACF and the sample

PACF to the theoretical ACF and PACF. How similar to the theoretical values are the sample values?

c. Repeat part b using 200 realizations. How has increasing the sample size impacted the agreement between the sample and theoretical ACF and PACF? Does this give you any insight about the sample sizes required for model building, or the reliability of models built to short time series?

CHAPTER 6

TRANSFER FUNCTIONS AND INTERVENTION MODELS

He uses statistics as a drunken man uses lamp posts – For support rather than illumination

Andrew Lang, Scottish poet

6.1 INTRODUCTION

The ARIMA models discussed in the previous chapter represent a general class of models that can be used very effectively in time series modeling and forecasting problems. An implicit assumption in these models is that the conditions under which the data for the time series process is collected remain the same. If, however, these conditions change over time, ARIMA models can be improved by introducing certain inputs reflecting these changes in the process conditions. This will lead to what is known as **transfer function–noise models**. These models can be seen as regression models in Chapter 3 with serially dependent response, inputs, and the error term. The identification and the estimation of these models can be challenging. Furthermore, not all standard statistical software packages possess the capability to fit such models. So far in this book, we have used

Introduction to Time Series Analysis and Forecasting, Second Edition.
Douglas C. Montgomery, Cheryl L. Jennings and Murat Kulahci.
© 2015 John Wiley & Sons, Inc. Published 2015 by John Wiley & Sons, Inc.

the Minitab and JMP software packages to illustrate time series model fitting. However, Minitab (version 16) lacks the capability of fitting transfer function–noise models. Therefore for Chapters 6 and 7, we will use JMP and R instead.

6.2 TRANSFER FUNCTION MODELS

In Section 5.2, we discussed the **linear filter** and defined it as

$$y_t = L(x_t) = \sum_{i=-\infty}^{+\infty} v_i x_{t-i} \tag{6.1}$$

$$= v(B)x_t,$$

where $v(B) = \sum_{i=-\infty}^{+\infty} v_i B^i$ is called the **transfer function**. Following the definition of a linear filter, Eq. (6.1) is:

1. **Time-invariant** as the coefficients $\{v_i\}$ do not depend on time.
2. **Physically realizable** if $v_i = 0$ for $i < 0$; that is, the output y_t is a linear function of the current and past values of the input:

$$y_t = v_0 x_t + v_1 x_{t-1} + \cdots$$

$$= \sum_{i=0}^{\infty} v_i x_{t-i}. \tag{6.2}$$

3. **Stable if** $\sum_{i=-\infty}^{+\infty} |v_i| < \infty$.

There are two interesting special cases for the input x_t:

Impulse Response Function. If x_t is a unit impulse at time $t = 0$, that is,

$$x_t = \begin{cases} 1, & t = 0 \\ 0, & t \neq 0 \end{cases} \tag{6.3}$$

then the output y_t is

$$y_t = \sum_{i=0}^{\infty} v_i x_{t-i} = v_t \tag{6.4}$$

Therefore the coefficients v_i in Eq. (6.2) are also called the **impulse response function**.

Step Response Function. If x_t is a unit step, that is,

$$x_t = \begin{cases} 0, & t < 0 \\ 1, & t \geq 0 \end{cases} \tag{6.5}$$

then the output y_t is

$$
\begin{aligned}
y_t &= \sum_{i=0}^{\infty} v_i x_{t-i} \\
&= \sum_{i=0}^{t} v_i,
\end{aligned}
\tag{6.6}
$$

which is also called the **step response function**.

A generalization of the step response function is obtained when Eq. (6.5) is modified so that x_t is kept at a certain target value X after $t \geq 0$; that is,

$$x_t = \begin{cases} 0, & t < 0 \\ X, & t \geq 0. \end{cases} \tag{6.7}$$

Hence we have

$$
\begin{aligned}
y_t &= \sum_{i=0}^{\infty} v_i x_{t-i} \\
&= \left(\sum_{i=0}^{t} v_i \right) X \\
&= gX,
\end{aligned}
\tag{6.8}
$$

where g is called the **steady-state gain**.

A more realistic representation of the response is obtained by adding a noise or disturbance term to Eq. (6.2) to account for unanticipated and/or ignored factors that may have an effect on the response as well. Hence the "additive" model representation of the dynamic systems is given as

$$y_t = v(B)x_t + N_t, \tag{6.9}$$

where N_t represents the unobservable noise process. In Eq. (6.9), x_t and N_t are assumed to be independent. The model representation in Eq. (6.9) is also called the **transfer function–noise model**.

Since the noise process is unobservable, the predictions of the response can be made by estimating the impulse response function $\{v_t\}$. Similar to our discussion about the estimation of the coefficients in Wold's decomposition theorem in Chapter 5, attempting to estimate the infinitely many coefficients in $\{v_t\}$ is a futile exercise. Therefore also parallel to the arguments we made in Chapter 5, we will make assumptions about these infinitely many coefficients to be able to represent them with only a handful of parameters. Following the derivations we had for the ARMA models, we will assume that the coefficients in $\{v_t\}$ have a structure and can be represented as

$$
\begin{aligned}
v(B) = \sum_{i=0}^{\infty} v_i B^i &= \frac{w(B)}{\delta(B)} \\
&= \frac{w_0 - w_1 B - \cdots - w_s B^s}{1 - \delta_1 B - \cdots - \delta_r B^r}
\end{aligned}
\tag{6.10}
$$

The interpretation of Eq. (6.10) is quite similar to the one we had for ARMA models; the denominator summarizes the infinitely many coefficients with a certain structure determined by $\{\delta_i\}$ as in the AR part of the ARMA model and the numerator represents the adjustment we may like to make to the strictly structured infinitely many coefficients as in the MA part of the ARMA model.

So the transfer function–noise model in Eq. (6.9) can be rewritten as

$$
y_t = \frac{w(B)}{\delta(B)} x_t + N_t,
$$

where $w(B)/\delta(B) = \delta(B)^{-1} w(B) = \sum_{i=0}^{+\infty} v_i B^i$. For some processes, there may also be a **delay** before a change in the input x_t shows its effect on the response y_t. If we assume that there is b time units of delay between the response and the input, a more general representation for the transfer function–noise models can be obtained as

$$
\begin{aligned}
y_t &= \frac{w(B)}{\delta(B)} x_{t-b} + N_t \\
&= \frac{w(B)}{\delta(B)} B^b x_t + N_t \\
&= v(B) x_t + N_t,
\end{aligned}
\tag{6.11}
$$

Since the denominator $\delta(B)$ in Eq. (6.11) determines the structure of the infinitely many coefficients, the stability of $v(B)$ depends on the coefficients in $\delta(B)$. In fact $v(B)$ is said to be stable if all the roots of $m^r - \delta_1 m^{r-1} - \cdots - \delta_r$ are less than 1 in absolute value.

Once the finite number of parameters in $w(B)$ and $\delta(B)$ are estimated, $v(B)$ can be computed recursively from

$$\delta(B)v(B) = w(B)B^b$$

or

$$v_j - \delta_1 v_{j-1} - \delta_2 v_{j-2} - \cdots - \delta_r v_{j-r} = \begin{cases} -w_{j-b}, & j = b+1, \ldots, b+s \\ 0, & j > b+s \end{cases}$$

$$(6.12)$$

with $v_b = w_0$ and $v_j = 0$ for $j < b$.

The characteristics of the impulse response function are determined by the values of b, r, and s. Recall that in univariate ARIMA modeling, we matched sample autocorrelation and partial autocorrelation functions computed from a time series to theoretical autocorrelation and partial autocorrelation functions of specific ARIMA models to tentatively identify an appropriate model. Thus by knowing the theoretical patterns in the autocorrelation and partial autocorrelation functions for an AR(1) process, for example, we can tentatively identify the AR(1) model when sample autocorrelation and partial autocorrelation functions exhibit the same behavior for an observed time series. The same approach is used in transfer function modeling. However, the primary identification tool is the impulse response function. Consequently, it is necessary that we investigate the nature of the impulse response function and determine what various patterns in the weights imply about the parameters b, r, and s.

Example 6.1   For illustration, we will consider cases for $b = 2$, $r \leq 2$, and $s \leq 2$.

Case 1. $r = 0$ and $s = 2$.

We have

$$y_t = (w_0 - w_1 B - w_2 B^2)x_{t-2}$$

From Eq. (6.12), we have

$$v_0 = v_1 = 0$$
$$v_2 = w_0$$
$$v_3 = -w_1$$
$$v_4 = -w_2$$
$$v_j = 0, \quad j > 4$$

Hence v_t will only be nonzero for $t = 2, 3,$ and 4.

Case 2. $r = 1$ and $s = 2$.

We have

$$y_t = \frac{(w_0 - w_1 B - w_2 B^2)}{1 - \delta_1 B} x_{t-2}$$

As in the AR(1) model, the stability of the transfer function is achieved for $|\delta_1| < 1$. Once again from Eq. (6.12), we have

$$v_0 = v_1 = 0$$
$$v_2 = w_0$$
$$v_3 = \delta_1 w_0 - w_1$$
$$v_4 = \delta_1^2 w_0 - \delta_1 w_1 - w_2$$
$$v_j = \delta_1 v_{j-1}, \quad j > 4$$

Since $|\delta_1| < 1$, the impulse response function will approach zero asymptotically.

Case 3. $r = 2$ and $s = 2$.

We have

$$y_t = \frac{(w_0 - w_1 B - w_2 B^2)}{1 - \delta_1 B - \delta_2 B^2} x_{t-2}$$

The stability of the transfer function depends on the roots of the associated polynomial $m^2 - \delta_1 m^1 - \delta_2$. For stability, the roots obtained by

$$m_1, m_2 = \frac{\delta_1 \pm \sqrt{\delta_1^2 + 4\delta_2}}{2}$$

must satisfy $|m_1|$, $|m_2| < 1$. This also means that

$$\delta_2 - \delta_1 < 1$$
$$\delta_2 + \delta_1 < 1$$
$$-1 < \delta_2 < 1$$

or

$$|\delta_1| < 1 - \delta_2$$
$$-1 < \delta_2 < 1.$$

This set of two equations implies that the stability is achieved with the triangular region given in Figure 6.1. Within that region we might have two real roots or two complex conjugates. For the latter, we need $\delta_1^2 + 4\delta_2 < 0$, which occurs in the area under the curve within the triangle in Figure 6.1. Hence for the values of δ_1 and δ_2 within that curve, the impulse response function would exhibit a damped sinusoid behavior. Everywhere else in the triangle, however, it will have an exponential decay pattern.

Note that when $\delta_2 = 0$ (i.e., $r = 1$), stability is achieved when $|\delta_1| < 1$ as expected.

Table 6.1 summarizes the impulse response functions for the cases we have just discussed with specific values for the parameters.

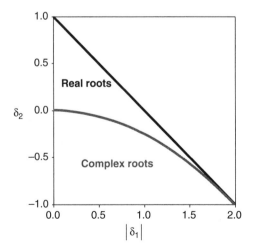

FIGURE 6.1   The stable region for the impulse response function for $r = 2$.

TABLE 6.1 Impulse Response Function with $b = 2$, $r \leq 2$, and $s \leq 2$

$b\ r\ s$	Model		Impulse response function	
2 0 0	$y_t = w_0 x_{t-2}$	$w_0 = 0.5$		$w_0 = -0.5$
2 0 1	$y_t = (w_0 - w_1 B) x_{t-2}$	$w_0 = 0.5$ $w_1 = -0.4$		$w_0 = 0.5$ $w_1 = 0.4$
2 0 2	$y_t = (w_0 - w_1 B - w_2 B^2) x_{t-2}$	$w_0 = 0.5$ $w_1 = -0.4$ $w_2 = -0.6$		$w_0 = 0.5$ $w_1 = 0.4$ $w_2 = -0.6$
2 1 0	$y_t = \dfrac{w_0}{1 - \delta_1 B} x_{t-2}$	$w_0 = 0.5$ $\delta_1 = 0.6$		$w_0 = -0.5$ $\delta_1 = 0.6$
2 1 1	$y_t = \dfrac{w_0 - w_1 B}{1 - \delta_1 B} x_{t-2}$	$w_0 = 0.5$ $w_1 = -0.4$ $\delta_1 = 0.6$		$w_0 = 0.5$ $w_1 = 0.4$ $\delta_1 = 0.6$

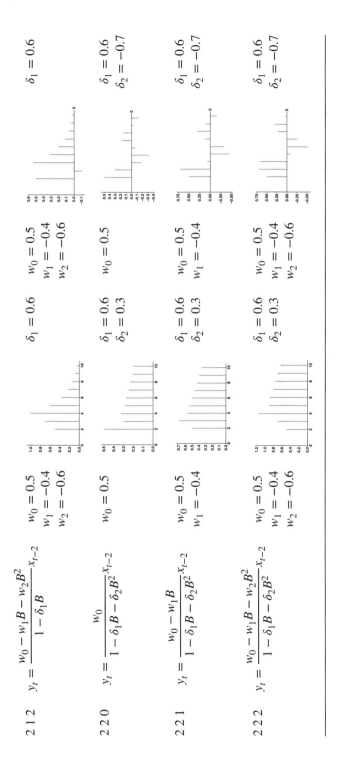

2 1 2 $y_t = \dfrac{w_0 - w_1 B - w_2 B^2}{1 - \delta_1 B} x_{t-2}$ $w_0 = 0.5$ $w_1 = -0.4$ $w_2 = -0.6$ $\delta_1 = 0.6$ $w_0 = 0.5$ $w_1 = -0.4$ $w_2 = -0.6$ $\delta_1 = 0.6$

2 2 0 $y_t = \dfrac{w_0}{1 - \delta_1 B - \delta_2 B^2} x_{t-2}$ $w_0 = 0.5$ $\delta_1 = 0.6$ $\delta_2 = 0.3$ $w_0 = 0.5$ $\delta_1 = 0.6$ $\delta_2 = -0.7$

2 2 1 $y_t = \dfrac{w_0 - w_1 B}{1 - \delta_1 B - \delta_2 B^2} x_{t-2}$ $w_0 = 0.5$ $w_1 = -0.4$ $\delta_1 = 0.6$ $\delta_2 = 0.3$ $w_0 = 0.5$ $w_1 = -0.4$ $\delta_1 = 0.6$ $\delta_2 = -0.7$

2 2 2 $y_t = \dfrac{w_0 - w_1 B - w_2 B^2}{1 - \delta_1 B - \delta_2 B^2} x_{t-2}$ $w_0 = 0.5$ $w_1 = -0.4$ $w_2 = -0.6$ $\delta_1 = 0.6$ $\delta_2 = 0.3$ $w_0 = 0.5$ $w_1 = -0.4$ $w_2 = -0.6$ $\delta_1 = 0.6$ $\delta_2 = -0.7$

435

6.3 TRANSFER FUNCTION–NOISE MODELS

As mentioned in the previous section, in the transfer function–noise model in Eq. (6.11) x_t and N_t are assumed to be independent. Moreover, we will assume that the noise N_t can be represented by an ARIMA(p, d, q) model,

$$\underbrace{\phi(B)\,(1-B)^d}_{=\varphi(B)}\,N_t = \theta(B)\varepsilon_t, \qquad (6.13)$$

where $\{\varepsilon_t\}$ is white noise with $E(\varepsilon_t) = 0$. Hence the transfer function–noise model can be written as

$$\begin{aligned} y_t &= v(B)x_t + \psi(B)\varepsilon_t \\ &= \frac{w(B)}{\delta(B)}x_{t-b} + \frac{\theta(B)}{\varphi(B)}\varepsilon_t \end{aligned} \qquad (6.14)$$

After rearranging Eq. (6.14), we have

$$\underbrace{\delta(B)\varphi(B)}_{=\delta^*(B)}\,y_t = \underbrace{\varphi(B)w(B)}_{=w^*(B)}\,x_{t-b} + \underbrace{\delta(B)\theta(B)}_{=\theta^*(B)}\,\varepsilon_t$$

$$\delta^*(B)y_t = w^*(B)x_{t-b} + \theta^*(B)\varepsilon_t \qquad (6.15)$$

or

$$y_t - \sum_{i=1}^{r^*}\delta_i^* y_{t-i} = w_0^* x_{t-b} - \sum_{i=1}^{s^*} w_i^* x_{t-b-i} + \varepsilon_t - \sum_{i=1}^{q^*}\theta_i^*\varepsilon_{t-i}. \qquad (6.16)$$

Ignoring the terms involving x_t, Eq. (6.16) is the ARMA representation of the response y_t. Due to the addition of x_t, the model in Eq. (6.16) is also called an ARMAX model. Hence the transfer function–noise model as given in Eq. (6.16) can be interpreted as an ARMA model for the response with the additional exogenous factor x_t.

6.4 CROSS-CORRELATION FUNCTION

For the bivariate time series (x_t, y_t), we define the **cross-covariance** function as

$$\gamma_{xy}(t, s) = \mathrm{Cov}(x_t, y_s) \qquad (6.17)$$

Assuming that (x_t, y_t) is (weakly) stationary, we have

$$
\begin{aligned}
E(x_t) &= \mu_x, & \text{constant for all } t \\
E(y_t) &= \mu_y, & \text{constant for all } t \\
\text{Cov}(x_t, x_{t+j}) &= \gamma_x(j), & \text{depends only on } j \\
\text{Cov}(y_t, y_{t+j}) &= \gamma_y(j), & \text{depends only on } j
\end{aligned}
$$

and

$$
\text{Cov}(x_t, y_{t+j}) = \gamma_{xy}(j), \quad \text{depends only on } j \text{ for } j = 0, \pm 1, \pm 2, \dots
$$

Hence the **cross-correlation function** (CCF) is defined as

$$
\rho_{xy}(j) = \text{corr}(x_t, y_{t+j}) = \frac{\gamma_{xy}(j)}{\sqrt{\gamma_x(0)\gamma_y(0)}} \quad \text{for} \quad j = 0, \pm 1, \pm 2, \dots \quad (6.18)
$$

It should be noted that $\rho_{xy}(j) \neq \rho_{xy}(-j)$ but $\rho_{xy}(j) = \rho_{yx}(-j)$.
We then define the correlation matrix at lag j as

$$
\begin{aligned}
\rho(j) &= \begin{bmatrix} \rho_x(j) & \rho_{xy}(j) \\ \rho_{yx}(j) & \rho_y(j) \end{bmatrix} \\
&= \text{corr}\left[\begin{pmatrix} x_t \\ y_t \end{pmatrix}, (x_{t+j} \quad y_{t+j}) \right]
\end{aligned}
\quad (6.19)
$$

For a given sample of N observations, the **sample cross covariance** is estimated from

$$
\hat{\gamma}_{xy}(j) = \frac{1}{N} \sum_{t=1}^{N-j} (x_t - \bar{x})(y_{t+j} - \bar{y}) \quad \text{for} \quad j = 0, 1, 2, \dots \quad (6.20)
$$

and

$$
\hat{\gamma}_{xy}(j) = \frac{1}{N} \sum_{t=1}^{N+j} (x_{t-j} - \bar{x})(y_t - \bar{y}) \quad \text{for} \quad j = -1, -2, \dots \quad (6.21)
$$

Similarly, the **sample cross correlations** are estimated from

$$r_{xy}(j) = \hat{\rho}_{xy}(j) = \frac{\hat{\gamma}_{xy}(j)}{\sqrt{\hat{\gamma}_x(0)\hat{\gamma}_y(0)}} \quad \text{for} \quad j = 0, \pm 1, \pm 2, \dots \quad (6.22)$$

where

$$\hat{\gamma}_x(0) = \frac{1}{N} \sum_{t=1}^{N} (x_t - \bar{x})^2 \quad \text{and} \quad \hat{\gamma}_y(0) = \frac{1}{N} \sum_{t=1}^{N} (y_t - \bar{y})^2$$

Sampling properties such as the mean and variance of the sample CCF are quite complicated. For a few special cases, however, we have the following.

1. For large data sets, $E(r_{xy}(j)) \approx \rho_{xy}(j)$ but the variance is still complicated.
2. If x_t and y_t are autocorrelated but un(cross)correlated at all lags, that is, $\rho_{xy}(j) = 0$, we then have $E(r_{xy}(j)) \approx 0$ and $\mathrm{var}(r_{xy}(j)) \approx (1/N) \sum_{i=-\infty}^{\infty} \rho_x(i)\rho_y(i)$.
3. If $\rho_{xy}(j) = 0$ for all lags j but also x_t is white noise, that is, $\rho_x(j) = 0$ for $j \neq 0$, then we have $\mathrm{var}(r_{xy}(j)) \approx 1/N$ for $j = 0, \pm 1, \pm 2, \dots$.
4. If $\rho_{xy}(j) = 0$ for all lags j but also both x_t and y_t are white noise, then we have $\mathrm{corr}(r_{xy}(i), r_{xy}(j)) \approx 0$ for $i \neq j$.

6.5 MODEL SPECIFICATION

In this section, we will discuss the issues regarding the specification of the model order in a transfer function–noise model. Further discussion can be found in Bisgaard and Kulahci (2006a,b, 2011).

We will first consider the general form of the transfer function–noise model with time delay given as

$$y_t = v(B)x_t + N_t$$
$$= \frac{w(B)}{\delta(B)}x_{t-b} + \frac{\theta(B)}{\varphi(B)}\varepsilon_t. \quad (6.23)$$

The six-step model specification process is outlined next.

Step 1. Obtaining the preliminary estimates of the coefficients in $v(B)$.

One approach is to assume that the coefficients in $v(B)$ are zero except for the first k lags:

$$y_t \cong \sum_{i=0}^{k} v_i x_{t-i} + N_t.$$

We can then attempt to obtain the initial estimates for $v_1, v_2, \ldots, v_k$ through ordinary least squares. However, this approach can lead to highly inaccurate estimates as x_t may have strong autocorrelation. Therefore a method called **prewhitening** of the input is generally preferred.

Method of Prewhitening   For the transfer function–noise model in Eq. (6.23), suppose that x_t follows an ARIMA model as

$$\underbrace{\phi_x(B)(1 - B)^d}_{=\varphi_x(B)} x_t = \theta_x(B)\alpha_t, \qquad (6.24)$$

where α_t is white noise with variance σ_α^2. Equivalently, we have

$$\alpha_t = \theta_x(B)^{-1}\varphi_x(B)x_t. \qquad (6.25)$$

In this notation, $\theta_x(B)^{-1}\varphi_x(B)$ can be seen as a filter that when applied to x_t generates a white noise time series, hence the name "prewhitening."

When we apply this filter to the transfer function–noise model in Eq. (6.23), we obtain

$$\underbrace{\theta_x(B)^{-1}\varphi_x(B)y_t}_{=\beta_t} = \theta_x(B)^{-1}\varphi_x(B)v(B)x_t + \underbrace{\theta_x(B)^{-1}\varphi_x(B)N_t}_{=N_t^*} \qquad (6.26)$$

$$\beta_t = v(B)\alpha_t + N_t^*.$$

The cross covariance between the filtered series α_t and β_t is given by

$$\gamma_{\alpha\beta}(j) = \text{Cov}(\alpha_t, \beta_{t+j}) = \text{Cov}\left(\alpha_t, v(B)\alpha_{t+j} + N^*_{t+j}\right)$$

$$= \text{Cov}\left(\alpha_t, \sum_{i=0}^{\infty} v_i \alpha_{t+j-i} + N^*_{t+j}\right)$$

$$= \text{Cov}\left(\alpha_t, \sum_{i=0}^{\infty} v_i \alpha_{t+j-i}\right) + \underbrace{\text{Cov}\left(\alpha_t, N^*_{t+j}\right)}_{=0} \tag{6.27}$$

$$= \sum_{i=0}^{\infty} v_i \, \text{Cov}(\alpha_t, \alpha_{t+j-i})$$

$$= v_j \, \text{Var}(\alpha_t).$$

Note that $\text{Cov}(\alpha_t, N^*_{t+j}) = 0$ since x_t and N_t are assumed to be independent. From Eq. (6.27), we have $\gamma_{\alpha\beta} = v_j \sigma_\alpha^2$ and hence

$$v_j = \frac{\gamma_{\alpha\beta}(j)}{\sigma_\alpha^2} = \frac{\rho_{\alpha\beta}(j)\sigma_\alpha\sigma_\beta}{\sigma_\alpha^2}$$

$$= \rho_{\alpha\beta}(j)\frac{\sigma_\beta}{\sigma_\alpha}, \tag{6.28}$$

where $\rho_{\alpha\beta}(j) = \text{corr}(\alpha_t, \beta_{t+j})$ is the CCF between α_t and β_t. So through the sample estimates we can obtain the initial estimates for the v_j:

$$\hat{v}_j = r_{\alpha\beta}(j)\frac{\hat{\sigma}_\beta}{\hat{\sigma}_\alpha}. \tag{6.29}$$

Equation (6.29) implies that there is a simple relationship between the impulse response function, $v(B)$, and the cross-correlation function of the prewhitened response and input series. Hence the estimation of the coefficients in $v(B)$ is possible through this relationship as summarized in Eq. (6.29). A similar relationship exists when the response and the input are not prewhitened (see Box et al., 2008). However, the calculations become fairly complicated when the series are not prewhitened. Therefore we strongly recommend the use of prewhitening in model identification and estimation of transfer function–noise models.

Moreover, since α_t is white noise, the variance of $r_{\alpha\beta}(j)$ is relatively easier to obtain than that of $r_{xy}(j)$. In fact, from the special case 3 in the previous section, we have

$$\text{Var}[r_{\alpha\beta}(j)] \approx \frac{1}{N}, \tag{6.30}$$

if $\rho_{\alpha\beta}(j) = 0$ for all lags j. We can then use $\pm 2/\sqrt{N}$ as the *approximate* 95% confidence interval to judge the significance of $r_{\alpha\beta}(j)$.

Step 2. Specifications of the orders r and s.

Once the initial estimates of the v_j from Eq. (6.29) are obtained, we can use them to specify the orders r and s in

$$v(B) = \frac{w(B)}{\delta(B)} B^b$$
$$= \frac{w_0 - w_1 B - \cdots - w_s B^s}{1 - \delta_1 B - \cdots - \delta_r B^r} B^b$$

The specification of the orders r and s can be accomplished by plotting the v_j. In Figure 6.2, we have an example of the plot of the initial estimates for the v_j in which we can see that $\hat{v}_0 \approx 0$, implying that there might be a

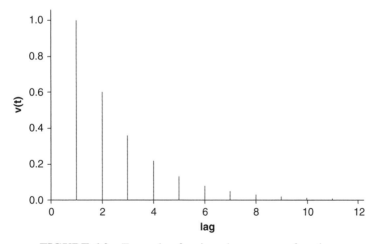

FIGURE 6.2   Example of an impulse response function.

time delay (i.e., $b = 1$). However, for $j > 1$, we have an exponential decay pattern, suggesting that we may have $r = 1$, which implies

$$v_j - \delta v_{j-1} = 0 \quad \text{for} \quad j > 1$$

and

$$s = 0.$$

Hence for this example, our initial attempt in specifying the order of the transfer function noise model will be

$$y_t = \frac{w_0}{1 - \delta B} x_{t-1} + N_t. \tag{6.31}$$

Caution: In model specification, one should be acutely aware of over-parameterization as for an arbitrary η, the model in Eq. (6.31) can also be written as

$$
\begin{aligned}
y_t &= \frac{w_0 (1 - \eta B)}{(1 - \delta B)(1 - \eta B)} x_{t-1} + N_t \\
&= \frac{w_0 - w_1 B}{1 - \delta_1 B - \delta_2 B^2} x_{t-1} + N_t.
\end{aligned}
\tag{6.32}
$$

But the parameters in Eq. (6.32) are not identifiable, since η can arbitrarily take any value.

Step 3. Obtain the estimates of δ_i and w_i.

From $\hat{\delta}(B)\hat{v}(B) = \hat{w}(B)$, we can recursively estimate δ_i and w_i using Eq. (6.12),

$$v_j - \delta_1 v_{j-1} - \delta_2 v_{j-2} - \dots - \delta_r v_{j-r} = \begin{cases} -w_{j-b}, & j = b+1, \dots, b+s \\ 0, & j > b+s \end{cases}$$

with $v_b = w_0$ and $v_j = 0$ for $j < b$. Hence for the example in Step 2, we have

$$\hat{v}_1 = \hat{w}_0$$
$$\hat{v}_2 - \hat{\delta}\hat{v}_1 = 0$$
$$\vdots$$

Step 4. Model the noise.

Once the initial estimates of the model parameters are obtained, the estimated noise can be obtained as

$$\hat{N}_t = y_t - \frac{\hat{w}(B)}{\hat{\delta}(B)}x_{t-\hat{b}}, \tag{6.33}$$

To obtain the estimated noise, we define $\hat{y}_t = (\hat{w}(B)/\hat{\delta}(B))x_{t-\hat{b}}$. We can then calculate $\hat{y}_t$ recursively. To model the estimated noise, we observe its ACF and PACF and determine the orders of the ARIMA model, $\phi(B)(1 - B)^d N_t = \theta(B)\varepsilon_t$.

Step 5. Fitting the overall model.

Steps 1 through 4 provide us with the model specifications and the initial estimates of the parameters in the transfer function–noise model,

$$y_t = \frac{w(B)}{\delta(B)}x_{t-b} + \frac{\theta(B)}{\phi(B)(1 - B)^d}\varepsilon_t.$$

The final estimates of the model parameters are then obtained by a nonlinear model fit. Model selection criteria such as AIC and BIC can be used to pick the "best" model among competing models.

Step 6. Model adequacy checks.

At this step, we check the validity of the two assumptions in the fitted model:

1. The assumption that the noise ε_t is white noise requires the examination of the residuals $\hat{\varepsilon}_t$. We perform the usual checks through analysis of the sample ACF and PACF of the residuals.
2. We should also check the independence between ε_t and x_t. For that, we observe the sample cross-correlation function between $\hat{\varepsilon}_t$ and $\hat{x}_t$. Alternatively, we can examine $r_{\hat{\alpha}\hat{\varepsilon}}(j)$, where $\alpha_t = \hat{\theta}_x(B)^{-1}\hat{\varphi}_x(B)x_t$. Under the assumption the model is adequate, $r_{\hat{\alpha}\hat{\varepsilon}}(j)$ will have 0 mean, $1/\sqrt{N}$ standard deviation, and be independent for different lags j. Hence we can use $\pm 2/\sqrt{N}$ as the limit to check the independence assumption.

TABLE 6.2   The viscosity, y(t) and temperature, x(t)

x(t)	y(t)	x(t)	y(t)	x(t)	y(t)	x(t)	y(t)
0.17	0.30	0.08	0.53	0.00	0.34	−0.04	−0.12
0.13	0.18	0.17	0.54	0.02	0.13	0.11	−0.26
0.19	0.09	0.20	0.42	−0.08	0.21	0.19	0.20
0.09	0.06	0.20	0.37	−0.08	0.06	−0.07	0.18
0.03	0.30	0.27	0.34	−0.26	0.04	−0.10	0.32
0.11	0.44	0.23	0.27	−0.06	−0.06	0.13	0.50
0.15	0.46	0.23	0.34	−0.06	−0.16	0.10	0.40
−0.02	0.44	0.20	0.35	−0.09	−0.47	−0.10	0.41
0.07	0.34	0.08	0.43	−0.14	−0.50	−0.05	0.47
0.00	0.23	−0.16	0.63	−0.10	−0.60	−0.12	0.37
−0.08	0.07	−0.08	0.61	−0.25	−0.49	0.00	0.04
−0.15	0.21	0.14	0.52	−0.23	−0.27	0.03	−0.10
−0.15	0.03	0.17	0.06	−0.11	−0.18	−0.06	−0.34
0.04	−0.20	0.27	−0.11	−0.01	−0.37	0.03	−0.41
0.08	−0.39	0.19	−0.01	−0.17	−0.34	0.04	−0.33
0.10	−0.70	0.10	0.02	−0.23	−0.34	0.09	−0.25
0.07	−0.22	0.13	0.34	−0.28	−0.18	−0.25	−0.18
−0.01	−0.08	−0.05	0.21	−0.26	−0.26	−0.25	−0.06
0.06	0.16	0.13	0.18	−0.19	−0.51	−0.40	0.15
0.07	0.13	−0.02	0.19	−0.26	−0.65	−0.30	−0.32
0.17	0.07	0.04	0.05	−0.20	−0.71	−0.18	−0.32
−0.01	0.23	0.00	0.15	−0.08	−0.82	−0.09	−0.81
0.09	0.33	0.08	0.10	0.03	−0.70	−0.05	−0.87
0.22	0.72	0.08	0.28	−0.08	−0.63	0.09	−0.84
0.09	0.45	0.07	0.20	0.01	−0.29	0.18	−0.73

Example 6.2   In a chemical process it is expected that changes in temperature affect viscosity, a key quality characteristic. It is therefore of great importance to learn more about this relationship. The data are collected every 10 seconds and given in Table 6.2 (Note that for each variable the data are centered by subtracting the respective averages). Figure 6.3 shows the time series plots of the two variables.

Since the data are taken in time and at frequent intervals, we expect the variables to exhibit some autocorrelation and decide to fit a transfer function-noise model following the steps provided earlier.

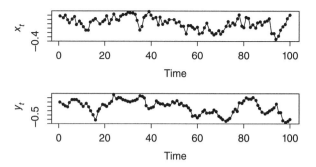

FIGURE 6.3   Time series plots of the viscosity, $y(t)$ and temperature, $x(t)$.

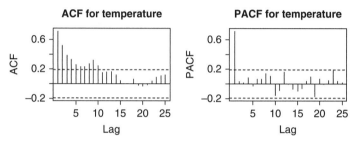

FIGURE 6.4   Sample ACF and PACF of the temperature.

Step 1. Obtaining the preliminary estimates of the coefficients in $v(B)$

In this step we use the prewhitening method. First we fit an ARIMA model to the temperature. Since the time series plot in Figure 6.3 shows that the process is changing around a constant mean and has a constant variance, we will assume that it is stationary.

Sample ACF and PACF plots in Figure 6.4 suggest that an AR(1) model should be used to fit the temperature data. Table 6.3 shows that $\hat{\phi} \cong 0.73$.

TABLE 6.3   AR(1) Model for Temperature, $x(t)$

Parameter Estimate of the AR(1) model for $x(t)$

Term	Coef	SE Coef	T	P-value
AR 1	0.7292	0.0686	10.63	<0.0001

Number of degrees of freedom: 99
MSE = 0.01009
AIC = −171.08

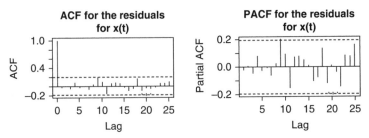

FIGURE 6.5 Sample ACF and PACF of the residuals from the AR(1) model for the Temperature, $x(t)$.

The sample ACF and PACF plots in Figure 6.5 as well as the additional residuals plots in Figure 6.6 reveal that no autocorrelation is left in the data and the model gives a reasonable fit.

Hence we define

$$\alpha_t = (1 - 0.73\,B)x_t$$

and

$$\beta_t = (1 - 0.73\,B)y_t$$

We then compute the sample cross-correlation of α_t and β_t, $r_{\alpha\beta}$ given in Figure 6.7. Since the cross correlation at lags 0, 1 and 2 do not seem to be significant, we conclude that there is a delay of 3 lags (30 seconds) in the system, that is, $b = 3$.

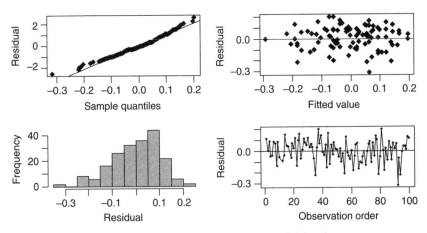

FIGURE 6.6 Residual plots from the AR(1) model for the temperature.

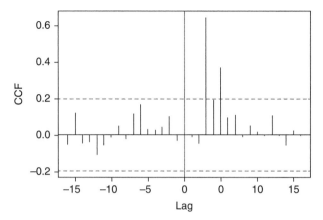

FIGURE 6.7   Sample cross-correlation function between α_t and β_t.

From Eq. (6.34), we have

$$
\begin{aligned}
\hat{v}_j &= r_{\alpha\beta}(j)\frac{\hat{\sigma}_\beta}{\hat{\sigma}_\alpha} \\
&= r_{\alpha\beta}(j)\frac{0.1881}{0.1008},
\end{aligned}
\tag{6.34}
$$

where $\hat{\sigma}_\alpha$ and $\hat{\sigma}_\beta$ are the sample standard deviations of α_t and β_t. The plot of $\hat{v}_j$ is given in Figure 6.8.

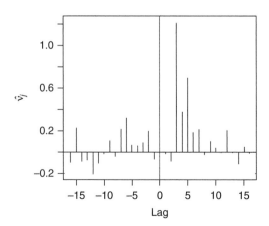

FIGURE 6.8   Plot of the impulse response function for the viscosity data.

Step 2. Specifications of the orders r and s.

To identify the pattern in Figures 6.7 and 6.8, we can refer back to Table 6.1. From the examples of impulse response functions given in that table, we may conclude the denominator of the transfer function is a second order polynomial in B. That is, $r = 2$ so we have $1 - \delta_2 B - \delta_2, B^2$ for the denominator. For the numerator, it seems that $s = 0$ or w_0 would be appropriate. Hence our tentative impulse response function is the following

$$v_t = \frac{w_0}{1 - \delta_1 B - \delta_2 B^2} B^3.$$

Step 3. Obtain the estimates of the δ_i and w_i.

To obtain the estimates of δ_i and w_i, we refer back to Eq. (6.12) which implies that we have

$$\hat{v}_0 \approx 0$$
$$\hat{v}_1 \approx 0$$
$$\hat{v}_2 \approx 0$$
$$\hat{v}_3 = 1.21 = \hat{w}_0$$
$$\hat{v}_4 = 0.37 = 1.21\hat{\delta}_1$$
$$\hat{v}_5 = 0.69 = 0.37\hat{\delta}_1 + 1.21\hat{\delta}_2$$

The parameter estimates are then

$$\hat{w}_0 = 1.21$$
$$\hat{\delta}_1 = 0.31$$
$$\hat{\delta}_2 = 0.48$$

or

$$\hat{v}_t = \frac{1.21}{1 - 0.31 B - 0.48 B^2} B.$$

Step 4. Model the noise.

To model the noise, we first define $\hat{y}_t = \dfrac{\hat{w}(B)}{\hat{\delta}(B)} x_{t-3}$ or

$$\hat{\delta}(B)\hat{y}_t = \hat{w}(B) x_{t-3}$$
$$(1 - 0.31 B - 0.48 B^2)\hat{y}_t = 1.21 x_{t-3}$$
$$\hat{y}_t = 0.31\hat{y}_{t-1} + 0.48\hat{y}_{t-2} + 1.21 x_{t-3}.$$

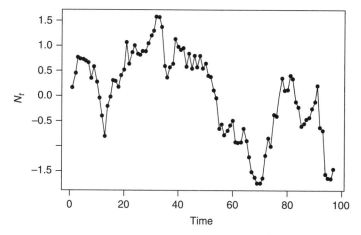

FIGURE 6.9   Time series plot of $\hat{N}_t$.

We then define

$$\hat{N}_t = y_t - \hat{y}_t.$$

Figures 6.9 and 6.10 show the time series plot of $\hat{N}_t$ and its sample ACF/PACF plots respectively which indicate an AR model. Note that partial autocorrelation at lag 3 is borderline significant. However when an AR(3) model is fitted, both ϕ_2 and ϕ_3 are found to be insignificant. Therefore AR(1) model is considered to be the appropriate model.

The parameter estimates for the AR(1) model for $\hat{N}_t$ are given in Table 6.4. Diagnostic checks of the residuals through sample ACF and PACF plots in Figure 6.11 and residuals plots in Figure 6.12 imply that we have a good fit.

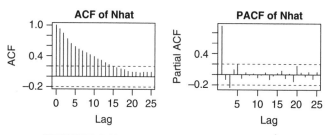

FIGURE 6.10   Sample ACF and PACF of $\hat{N}_t$.

TABLE 6.4 AR(1) Model for N_t

Parameter Estimate of the AR(1) model for $x(t)$

Term	Coef	SE Coef	T	P-value
AR 1	0.9426	0.0300	31.42	<0.0001

Number of degrees of freedom: 96
MSE = 0.0141
AIC = −131.9

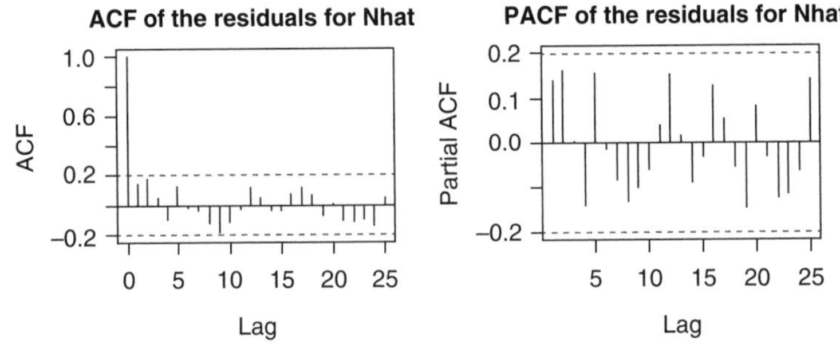

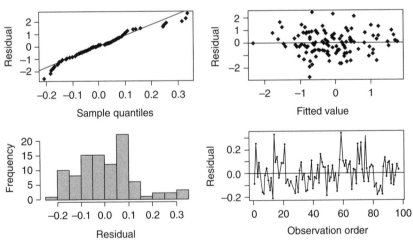

FIGURE 6.11 Sample ACF and PACF of the residuals of the AR(1) model for $\hat{N}_t$.

FIGURE 6.12 Residual plots of the AR(1) model for $\hat{N}_t$.

Note that we do not necessarily need the coefficients estimates as they will be re-estimated in the next step. Thus at this step all we need is a sensible model for $\hat{N}_t$ to put into the overall model.

Step 5. Fitting the overall model.

From Step 4, we have the tentative overall model as

$$y_t = \frac{w_0}{1 - \delta_1 B - \delta_2 B^2} x_{t-3} + \frac{1}{1 - \phi_1 B} \varepsilon_t.$$

The calculations that were made so far could have been performed practically in any statistical package. However as we mentioned at the beginning of the Chapter, unfortunately only a few software packages have the capability to fit the overall transfer function-noise model described above. In the following we provide the output from JMP with which such a model can be fitted. At the end of the chapter, we also provide the R code that can be used to fit the transfer function-noise model.

JMP output for the overall transfer function-noise model is provided in Table 6.5. The estimated coefficients are

$$\hat{w}_0 = 1.3276, \quad \hat{\delta}_1 = 0.3414, \quad \hat{\delta}_2 = 0.2667, \quad \hat{\phi}_1 = 0.8295,$$

and they are all significant.

Step 6. Model adequacy checks

The sample ACF and PACF of the residuals provided in Table 6.5 show no indication of leftover autocorrelation. We further check the cross correlation function between $\alpha_t = (1 - 0.73B)x_t$ and the residuals as given in Figure 6.13. There is a borderline significant cross correlation at lag 5. However we believe that it is at this point safe to claim that the current fitted model is adequate.

Example 6.2 illustrates transfer function modeling with a single input series where both the input and output time series were stationary. It is often necessary to incorporate **multiple input** time series into the model. A simple generalization of the single-input transfer function is to form an additive model for the inputs, say

$$y_t = \sum_{j=1}^{m} \frac{\omega_j(B)}{\delta_j(B)} x_{j,t-b_j} \tag{6.35}$$

TABLE 6.5 JMP Output for the Viscosity-Temperature Transfer Function-Noise Model

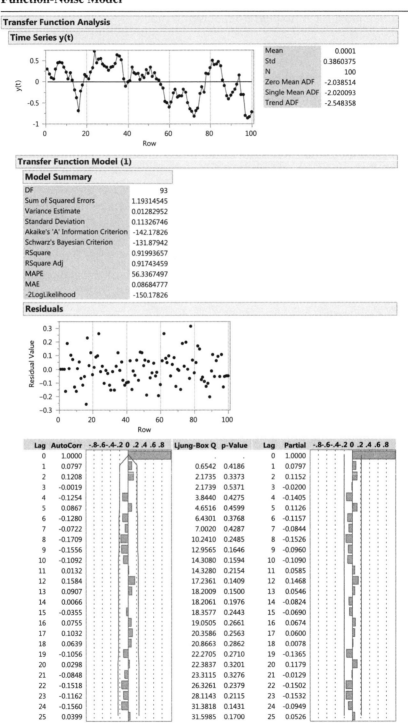

Transfer Function Analysis

Time Series y(t)

Mean	0.0001
Std	0.3860375
N	100
Zero Mean ADF	-2.038514
Single Mean ADF	-2.020093
Trend ADF	-2.548358

Transfer Function Model (1)

Model Summary

DF	93
Sum of Squared Errors	1.19314545
Variance Estimate	0.01282952
Standard Deviation	0.11326746
Akaike's 'A' Information Criterion	-142.17826
Schwarz's Bayesian Criterion	-131.87942
RSquare	0.91993657
RSquare Adj	0.91743459
MAPE	56.3367497
MAE	0.08684777
-2LogLikelihood	-150.17826

Residuals

Lag	AutoCorr	-.8-.6-.4-.2 0 .2 .4 .6 .8	Ljung-Box Q	p-Value	Lag	Partial	-.8-.6-.4-.2 0 .2 .4 .6 .8
0	1.0000		.	.	0	1.0000	
1	0.0797		0.6542	0.4186	1	0.0797	
2	0.1208		2.1735	0.3373	2	0.1152	
3	-0.0019		2.1739	0.5371	3	-0.0200	
4	-0.1254		3.8440	0.4275	4	-0.1405	
5	0.0867		4.6516	0.4599	5	0.1126	
6	-0.1280		6.4301	0.3768	6	-0.1157	
7	-0.0722		7.0020	0.4287	7	-0.0844	
8	-0.1709		10.2410	0.2485	8	-0.1526	
9	-0.1556		12.9565	0.1646	9	-0.0960	
10	-0.1092		14.3080	0.1594	10	-0.1090	
11	0.0132		14.3280	0.2154	11	0.0585	
12	0.1584		17.2361	0.1409	12	0.1468	
13	0.0907		18.2009	0.1500	13	0.0546	
14	0.0066		18.2061	0.1976	14	-0.0824	
15	-0.0355		18.3577	0.2443	15	-0.0690	
16	0.0755		19.0505	0.2661	16	0.0674	
17	0.1032		20.3586	0.2563	17	0.0600	
18	0.0639		20.8663	0.2862	18	0.0078	
19	-0.1056		22.2705	0.2710	19	-0.1365	
20	0.0298		22.3837	0.3201	20	0.1179	
21	-0.0848		23.3115	0.3276	21	-0.0129	
22	-0.1518		26.3261	0.2379	22	-0.1502	
23	-0.1162		28.1143	0.2115	23	-0.1532	
24	-0.1560		31.3818	0.1431	24	-0.0949	
25	0.0399		31.5985	0.1700	25	0.0526	

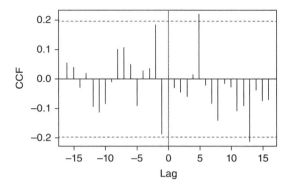

FIGURE 6.13   Sample cross-correlation function between α_1 and the residuals of the transfer function-noise model.

where each input series has a transfer function representation including a potential delay. This is an appropriate approach so long as the input series are uncorrelated with each other. If any of the original series are nonstationary, then differencing may be required. In general, differencing of a higher order than one may be required, and inputs and outputs need not be identically differenced. We now present an example from Montgomery and Weatherby (1980) where two inputs are used in the model.

Example 6.3   Montgomery and Weatherby (1980) present an example of modeling the output viscosity of a chemical process as a function of two inputs, the incoming raw material viscosity $x_{1,t}$ and the reaction temperature $x_{2,t}$. Readings are recorded hourly. Figure 6.14 is a plot of the last 100 readings. All three variables appear to be nonstationary.

Standard univariate ARIMA modeling techniques indicate that the input raw material viscosity can be modeled by an ARIMA(0, 1, 2) or IMA(1,2) process

$$(1 - B)x_{1,t} = (1 + 0.59B + 0.32B^2)\alpha_{1,t},$$

which is then used to prewhiten the output final viscosity. Similarly, an IMA(1,1) model

$$(1 - B)x_{2,t} = (1 + 0.45B)\alpha_{2,t}$$

was used to prewhiten the temperature input. Table 6.6 contains the impulse response functions between the prewhitened inputs and outputs.

Both impulse response functions in Table 6.6 exhibit approximate exponential decay beginning with lag 2 for the initial viscosity input and lag 3

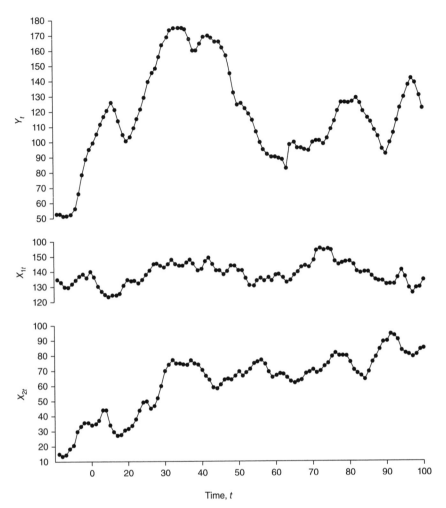

FIGURE 6.14   Hourly readings of final product viscosity y_t, incoming raw material viscosity $x_{r1,t}$, and reaction temperature $x_{2,t}$.

for the temperature input. This is consistent with a tentative model identification of $r = 1, s = 0, b = 2$ for the transfer function relating initial and final viscosities and $r = 1, s = 0, b = 3$ for the transfer function relating temperature and final viscosity. The preliminary parameter estimates for these models are

$$(1 - 0.62B)y_t = 0.34x_{1,t-2}$$
$$(1 - 0.68B)y_t = 0.42x_{2,t-3}.$$

TABLE 6.6   Impulse Response Function for Example 6.3

	Impulse Response Functions	
Lag	Initial and Final Viscosity	Temperature and Final Viscosity
0	−0.422	−0.0358
1	−0.1121	−0.1835
2	0.3446	0.0892
3	0.2127	0.4205
4	0.1327	0.2849
5	0.2418	0.3102
6	0.0851	0.0899
7	0.1491	0.1712
8	0.1402	0.0051

The noise time series is then computed from

$$\hat{N}_t = y_t - \frac{0.34}{1 - 0.62B}x_{1,t-2} - \frac{0.42}{1 - 0.68B}x_{2,t-3}.$$

The sample ACF and PACF of the noise series indicate that it can be modeled as

$$(1 - B)(1 - 0.64B)\hat{N}_t = \varepsilon_t.$$

Therefore, the combined transfer function plus noise model is

$$(1 - B)y_t = \frac{\omega_{10}}{1 - \delta_{11}B}(1 - B)x_{1,t-2} + \frac{\omega_{20}}{1 - \delta_{21}B}(1 - B)x_{2,t-3} + \frac{\varepsilon_t}{1 - \phi_1 B}.$$

The estimates of the parameters in this model are shown in Table 6.7 along with other model summary statistics. The t-test statistics indicate that all model parameters are different from zero.

Residual checks did not reveal any problems with model adequacy. The chi-square statistic for the first 25 residual autocorrelations was 10.412. The first 20 cross-correlations between the residuals and the prewhitened input viscosity produced a chi-square test statistic of 11.028 and the first 20 cross-correlations between the residuals and the prewhitened input temperature produced a chi-square test statistic of 15.109. These chi-square

TABLE 6.7   **Model Summary Statistics for the Two-Input Transfer Function Model in Example 6.3**

Parameter	Estimate	Standard Error	t-statistic for H_0: Parameter $= 0$	95% Confidence Limits	
				Lower	Upper
δ_{11}	0.789	0.110	7.163	0.573	1.005
ω_{10}	0.328	0.103	3.189	0.127	0.530
δ_{21}	0.677	0.186	3.643	0.313	1.041
ω_{20}	0.455	0.124	3.667	0.212	0.698
ϕ_1	0.637	0.082	7.721	0.475	0.799

test statistics are not significant at the 25% level, so we conclude that the model is adequate. The correlation matrix of the parameter estimates is

	δ_{11}	ω_{10}	δ_{21}	ω_{20}	ϕ_1
δ_{11}	1.00	−0.37	−0.06	0.01	0.02
ω_{10}	−0.37	1.00	−0.22	0.00	0.12
δ_{21}	−0.06	−0.22	1.00	−0.14	−0.07
ω_{20}	−0.01	0.00	−0.14	1.00	−0.08
ϕ_1	−0.02	0.07	−0.07	−0.08	1.00

Notice that a complex relationship between two input variables and one output has been modeled with only five parameters and the small off-diagonal elements in the covariance matrix above imply that these parameter estimates are essentially uncorrelated.

6.6 FORECASTING WITH TRANSFER FUNCTION–NOISE MODELS

In this section we discuss making τ-step-ahead forecasts using the transfer function–noise model in Eq. (6.23). We can rearrange Eq. (6.23) and rewrite it in the difference equation form as

$$\delta(B)\varphi(B)y_t = w(B)\varphi(B)x_{t-b} + \theta(B)\delta(B)\varepsilon_t \qquad (6.36)$$

or

$$\delta^*(B)y_t = w^*(B)x_{t-b} + \theta^*(B)\varepsilon_t. \tag{6.37}$$

Then at time $t + \tau$, we will have

$$y_{t+\tau} = \sum_{i=1}^{r+p^*} \delta_i^* y_{t+\tau-i} + w_0^* x_{t+\tau-b} - \sum_{i=1}^{s+p^*} w_i^* x_{t+\tau-b-i} + \varepsilon_{t+\tau} - \sum_{i=1}^{q+r} \theta_i^* \varepsilon_{t+\tau-i},$$

$$\tag{6.38}$$

where r is the order of $\delta(B)$, p^* is the order of $\varphi(B)(= \phi(B)(1 - B)^d)$, and s is the order of $\omega(B)$, and q is the order of $\theta(B)$.

The τ-step ahead MSE forecasts are obtained from

$$\hat{y}_{t+\tau}(\tau) = E[y_{t+\tau}|y_t, y_{t-1}, \dots, x_t, x_{t-1}, \dots]$$

$$= \sum_{i=1}^{r+p^*} \delta_i^* \hat{y}_{t+\tau-i}(t) + w_0^* \hat{x}_{t+\tau-b}(t) \tag{6.39}$$

$$- \sum_{i=1}^{s+p^*} w_i^* \hat{x}_{t+\tau-b-i}(t) - \sum_{i=1}^{q+r} \theta_i^* \varepsilon_{t+\tau-i} \quad \text{for } \tau = 1, 2, \dots, q.$$

Note that the MA terms will vanish for $\tau > q + r$. We obtain Eq. (6.39) using

$$E(\varepsilon_{t+\tau-i}|y_t, y_{t-1}, \dots, x_t, x_{t-1}, \dots) = \begin{cases} \varepsilon_{t+\tau-i}, & i \geq \tau \\ 0, & i < \tau \end{cases}$$

and

$$\hat{x}_t(l) = E(x_{t+l}|y_t, y_{t-1}, \dots, x_t, x_{t-1}, \dots)$$

$$= E(x_{t+l}|x_t, x_{t-1}, \dots). \tag{6.40}$$

Equation (6.40) implies that the relationship between x_t and y_t is unidirectional and that $\hat{x}_t(l)$ is the forecast from the univariate ARIMA model, $\phi_x(B)(1 - B)^d x_t = \theta_x(B)\alpha_t$.

So forecasts $\hat{y}_{t+1}(t), \hat{y}_{t+2}(t), \dots$ can be computed recursively from Eqs. (6.39) and (6.40).

The variance of the forecast errors can be obtained from the infinite MA representations for x_t and N_t given as

$$x_t = \varphi_x(B)^{-1}\theta_x(B)\alpha_t$$
$$= \psi_x(B)\alpha_t \tag{6.41}$$

and

$$N_t = \varphi(B)^{-1}\theta(B)\varepsilon_t$$
$$= \psi(B)\varepsilon_t \tag{6.42}$$
$$= \sum_{i=0}^{\infty} \psi_i \varepsilon_{t-i}.$$

Hence the infinite MA form of the transfer function–noise model is given as

$$y_t = \underbrace{v(B)\psi_x(B)}_{=v^*(B)} \alpha_{t-b} + \psi(B)\varepsilon_t \tag{6.43}$$
$$= \sum_{i=0}^{\infty} v_i^* \alpha_{t-b-i} + \sum_{i=0}^{\infty} \psi_i \varepsilon_{t-i}.$$

Thus the minimum MSE forecast can be represented as

$$\hat{y}_{t+\tau}(t) = \sum_{i=\tau-b}^{\infty} v_i^* \alpha_{t+\tau-b-i} + \sum_{i=\tau}^{\infty} \psi_i \varepsilon_{t+\tau-i} \tag{6.44}$$

and the τ-step-ahead forecast error is

$$e_t(\tau) = y_{t+\tau} - \hat{y}_{t+\tau}(t)$$
$$= \sum_{i=0}^{\tau-b-1} v_i^* \alpha_{t+\tau-b-i} + \sum_{i=0}^{\tau-1} \psi_i \varepsilon_{t+\tau-i}. \tag{6.45}$$

As we can see in Eq. (6.45), the forecast error has two components that are assumed to be independent: forecast errors in forecasting x_t, $\sum_{i=0}^{\tau-b-1} v_i^* \alpha_{t+\tau-b-i}$; and forecast errors in forecasting N_t, $\sum_{i=0}^{\tau-1} \psi_i \varepsilon_{t+\tau-i}$. The forecast variance is simply the sum of the two variances:

$$\sigma^2(\tau) = \text{Var}[e_t(\tau)]$$
$$= \sigma_\alpha^2 \sum_{i=0}^{\tau-b-1} (v_i^*)^2 + \sigma_\varepsilon^2 \sum_{i=0}^{\tau-1} \psi_i^2. \tag{6.46}$$

To check the effect of adding x_t in the model when forecasting, it may be appealing to compare the forecast errors between the transfer function–noise model and the univariate ARIMA model for y_t. Let the forecast error variances for the former and the latter be denoted by $\sigma^2_{\text{TFN}}(\tau)$ and $\sigma^2_{\text{UM}}(\tau)$, respectively. We may then consider

$$R^2(\tau) = 1 - \frac{\sigma^2_{\text{TFN}}(\tau)}{\sigma^2_{\text{UM}}(\tau)}$$

$$= \frac{\sigma^2_{\text{UM}}(\tau) - \sigma^2_{\text{TFN}}(\tau)}{\sigma^2_{\text{UM}}(\tau)}. \tag{6.47}$$

This quantity is expected to go down significantly if the introduction of x_t were indeed appropriate.

Example 6.4 Suppose we need to make forecasts for the next minute (6 observations) for the viscosity data in Example 6.2. We first consider the final model suggested in Example 6.2

$$y_t = \frac{w_0}{1 - \delta_1 B - \delta_2 B^2} x_{t-3} + \frac{1}{1 - \phi_1 B} \varepsilon_t.$$

After some rearrangement, we have

$$y_t = (\delta_1 + \phi_1) y_{t-1} + (\delta_2 - \delta_1 \phi_1) y_{t-2} - \delta_2 \phi_1 y_{t-3} + w_0 x_{t-3}$$
$$- w_0 \phi_1 x_{t-4} + \varepsilon_t - \delta_1 \varepsilon_{t-1} - \delta_2 \varepsilon_{t-2}.$$

From Eq. (6.38), we have the τ-step ahead prediction as

$$\hat{y}_{1+\tau}(t) = (\hat{\delta}_1 + \hat{\phi}_1)[y_{t+\tau-1}] + (\hat{\delta}_2 - \hat{\delta}_1 \hat{\phi}_1)[y_{t+\tau-2}] - \hat{\delta}_2 \hat{\phi}_1 [y_{t+\tau-3}]$$
$$+ \hat{w}_0 [x_{t+\tau-3}] - \hat{w}_0 \hat{\phi}_1 [x_{t+\tau-4}] + [\varepsilon_{t+\tau}] - \hat{\delta}_1 [\varepsilon_{t+\tau-1}] - \hat{\delta}_2 [\varepsilon_{t+\tau-2}],$$

where

$$[y_{t+j}] = \begin{cases} y_{t+j}, & j \leq 0 \\ \hat{y}_{t+j}(t), & j > 0 \end{cases}$$

$$[x_{t+j}] = \begin{cases} x_{t+j}, & j \leq 0 \\ \hat{x}_{t+j}(t), & j > 0 \end{cases}$$

and

$$[\varepsilon_{t+j}] = \begin{cases} \hat{\varepsilon}_{t+j}, & j \leq 0 \\ 0, & j > 0 \end{cases}.$$

Hence for the current and past response and input values, we can use the actual data. For the future response and input values we will instead use their respective forecasts. To forecast the input variable x_t, we will use the AR(1) model, $(1 - 0.73B)x_t = \alpha_t$ from Example 6.2. As for the error estimates, we can use the residuals from the transfer function-noise model or for $b \geq 1$, the one-step-ahead forecast errors for the current and past values of the errors, and set the error estimates equal to zero for future values.

We can obtain the variance of the prediction error from Eq. (6.45). The estimates of σ_α^2 and σ_ε^2 in Eq. (6.45) can be obtained from the univariate AR(1) model for x_t, and the transfer function-noise model from Example 6.2, respectively. Hence for this example we have $\hat{\sigma}_\alpha^2 = 0.0102$ and $\hat{\sigma}_\varepsilon^2 = 0.0128$. The coefficients in $v^*(B)$ and $\psi(B)$ can be calculated from

$$v^*(B) = \sum_{i=0}^{\infty} v_i^* B^i = v(B)\psi_x(B)$$

$$(v_0^* + v_1^* B + v_2^* B^2 + \ldots) = \frac{w_0}{(1 - \delta_1 B - \delta_2 B^2)}(1 - \phi_x B)^{-1}$$

or

$$(v_0^* + v_1^* B + v_2^* B^2 + \ldots)(1 - \delta_1 B - \delta_2 B^2)(1 - \phi_x B) = w_0$$

which means

$$v_0^* = w_0$$
$$v_1^* = (\delta_1 + \phi_x)v_0^* = (\delta_1 + \phi_x)w_0$$
$$v_2^* = (\delta_1 + \phi_x)v_1^* = (\delta_2 - \delta_1\phi_x)v_0^*$$
$$= [(\delta_1 + \phi_x)^2 + (\delta_2 - \delta_1\phi_x)]w_0$$
$$\vdots$$

and

$$\psi_i = \phi_1^i \qquad \text{for } i = 0, 1, 2\ldots$$

Hence the estimates of the coefficients in $v^*(B)$ and $\psi(B)$ can be obtained by using the estimates of the parameters given in Example 6.2. Note that

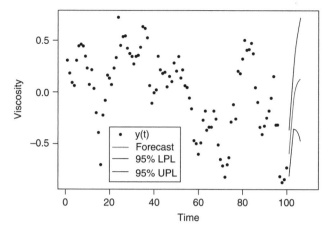

FIGURE 6.15   The time series plots of the actual and 1- to 6-step ahead forecasts for the viscosity data.

for up to 6-step-ahead forecasts, from Eq. (6.45), we will only need to calculate v_0^*, v_1^* and v_2^*.

The time series plot of the forecasts is given in Figure 6.15 together with the approximate 95% prediction limits calculate.by $\pm 2\hat{\sigma}(\tau)$.

For comparison purposes we fit a univariate ARIMA model for y_t. Following the model identification procedure given in Chapter 5, an AR(3) model is deemed a good fit. The estimated standard deviations of the prediction error for the transfer function-noise model and the univariate model are given in Table 6.8. It can be seen that adding the exogenous variable x_t helps to reduce the prediction error standard deviation.

TABLE 6.8   Estimated standard deviations of the prediction error for the transfer function-noise model (TFM) and the univariate model (UM)

	Estimated Standard Deviation of the Prediction Error	
Observation	TFM	UM
101	0.111	0.167
102	0.137	0.234
103	0.149	0.297
104	0.205	0.336
105	0.251	0.362
106	0.296	0.376

6.7 INTERVENTION ANALYSIS

In some cases, the response y_t can be affected by a known event that happens at a specific time such as fiscal policy changes, introduction of new regulatory laws, or switching suppliers. Since these **interventions** do not have to be quantitative variables, we can represent them with indicator variables. Consider, for example, the transfer function–noise model as the following:

$$
\begin{aligned}
y_t &= \frac{w(B)}{\delta(B)} \xi_t^{(T)} + \frac{\theta(B)}{\varphi(B)} \varepsilon_t \\
&= v(B)\xi_t^{(T)} + N_t,
\end{aligned}
\tag{6.48}
$$

where $\xi_t^{(T)}$ is a deterministic indicator variable, taking only the values 0 and 1 to indicate nonoccurrence and occurrence of some event. The model in Eq. (6.48) is called the **intervention model**. Note that this model has only one intervention event. Generalization of this model with several intervention events is also possible.

The most common indicator variables are the pulse and step variables,

$$
P_t^{(T)} = \begin{cases} 0 & \text{if } t \neq T \\ 1 & \text{if } t = T \end{cases}
\tag{6.49}
$$

and

$$
S_t^{(T)} = \begin{cases} 0 & \text{if } t < T \\ 1 & \text{if } t \geq T \end{cases},
\tag{6.50}
$$

where T is a specified occurrence time of the intervention event. The transfer function operator $v(B) = w(B)/\delta(B)$ in Eq. (6.48) usually has a fairly simple and intuitive form.

Examples of Responses to Pulse and Step Inputs

1. We will first consider the pulse indicator variable. We will further assume a simple transfer function–noise model as

$$
y_t = \frac{w_0}{1 - \delta B} P_t^{(T)}.
\tag{6.51}
$$

After rearranging Eq. (6.51), we have

$$(1 - \delta B)\, y_t = w_0 P_t^{(T)} = \begin{cases} 0 & \text{if } t \neq T \\ w_0 & \text{if } t = T \end{cases}$$

or

$$y_t = \delta y_{t-1} + w_0 P_t^{(T)}$$

So we have

$$\begin{aligned}
y_T &= w_0 \\
y_{T+1} &= \delta y_T = \delta w_0 \\
y_{T+2} &= \delta y_{T+1} = \delta^2 y_T = \delta^2 w_0 \\
&\vdots \\
y_{T+k} &= \cdots = \delta^k y_T = \delta^k w_0,
\end{aligned}$$

which means

$$y_t = \begin{cases} 0 & \text{if } t < T \\ w_0 \delta^{t-T} & \text{if } t \geq T \end{cases}.$$

2. For the step indicator variable with the same transfer function–noise model as in the previous case, we have

$$y_t = \frac{w_0}{1 - \delta B} S_t^{(T)}.$$

Solving the difference equation

$$(1 - \delta B)\, y_t = w_0 S_t^{(T)} = \begin{cases} 0 & \text{if } t < T \\ w_0 & \text{if } t \geq T \end{cases}$$

we have

$$\begin{aligned}
y_T &= w_0 \\
y_{T+1} &= \delta y_T + w_0 = w_0(1 + \delta) \\
y_{T+2} &= \delta y_{T+1} + w_0 = w_0(1 + \delta + \delta^2) \\
&\vdots \\
y_{T+k} &= \delta y_{T+k-1} + w_0 = w_0(1 + \delta + \cdots + \delta^k)
\end{aligned}$$

or

$$y_t = w_0(1 + \delta + \cdots + \delta^{t-T}) \quad \text{for } t \geq T$$

In intervention analysis, one of the things we could be interested in may be how permanent the effect of the event will be. Generally, for $y_t = (w(B)/\delta(B))\xi_t^{(T)}$ with stable $\delta(B)$, if the intervention event is a pulse, we will then have a transient (short-lived) effect. On the other hand, if the intervention event is a step, we will have a permanent effect. In general, depending on the form of the transfer function, there are many possible responses to the step and pulse inputs. Table 6.9 displays the output

TABLE 6.9   Output responses to step and pulse inputs.

TABLE 6.10   Weekly Cereal Sales Data

Week	Sales	Week	Sales	Week	Sales	Week	Sales
1	102,450	27	114,980	53	167,170	79	181,560
2	98,930	28	130,250	54	161,200	80	202,130
3	91,550	29	128,070	55	166,710	81	183,740
4	111,940	30	135,970	56	156,430	82	191,880
5	103,380	31	142,370	57	162,440	83	197,950
6	112,120	32	121,300	58	177,260	84	209,040
7	105,780	33	121,380	59	163,920	85	203,990
8	103,000	34	128,790	60	166,040	86	201,220
9	111,920	35	139,290	61	182,790	87	202,370
10	106,170	36	128,530	62	169,510	88	201,100
11	106,350	37	139,260	63	173,940	89	203,210
12	113,920	38	157,960	64	179,350	90	198,770
13	126,860	39	145,310	65	177,980	91	171,570
14	115,680	40	150,340	66	180,180	92	184,320
15	122,040	41	158,980	67	188,070	93	182,460
16	134,350	42	152,690	68	191,930	94	173,430
17	131,200	43	157,440	69	186,070	95	177,680
18	132,990	44	144,500	70	171,860	96	186,460
19	126,020	45	156,340	71	180,240	97	185,140
20	152,220	46	137,440	72	180,910	98	183,970
21	137,350	47	166,750	73	185,420	99	154,630
22	132,240	48	171,640	74	195,470	100	174,720
23	144,550	49	170,830	75	183,680	101	169,580
24	128,730	50	174,250	76	190,200	102	180,310
25	137,040	51	178,480	77	186,970	103	154,080
26	136,830	52	178,560	78	182,330	104	163,560

responses to the unit step and pulse inputs for several transfer function model structures. This table is helpful in model formulation.

Example 6.5   The weekly sales data of a cereal brand for the last two years are given in Table 6.10. As can be seen from Figure 6.16, the sales were showing a steady increase during most of the two-year period. At the end of the summer of the second year (Week 88), the rival company introduced a similar product into the market. Using intervention analysis, we want to study whether that had an effect on the sales. For that, we will first fit an ARIMA model to the preintervention data from Week 1 to Week 87. The sample ACF and PACF of the data for that time period in Figure 6.17

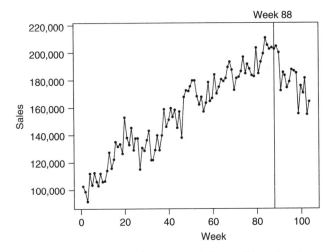

FIGURE 6.16   Time series plot of the weekly sales data.

show that the process is nonstationary. The sample ACF and PACF of the first difference given in Figure 6.18 suggest that an ARIMA(0,1,1) model is appropriate. Then the intervention model has the following form:

$$y_t = w_0 S_t^{(88)} + \frac{1 - \theta B}{1 - B} \varepsilon_t,$$

where

$$S_t^{(88)} = \begin{cases} 0 & \text{if } t < 88 \\ 1 & \text{if } t \geq 88 \end{cases}.$$

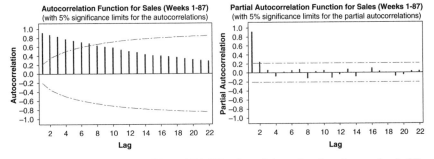

FIGURE 6.17   Sample ACF and PACF pulse of the sales data for weeks 1–87.

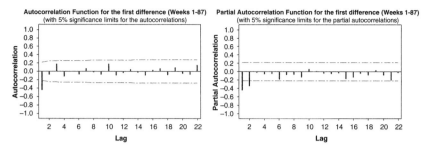

FIGURE 6.18 Sample ACF and PACF plots of the first difference of the sales data for weeks 1–87.

This means that for the intervention analysis we assume that the competition simply slows down (or reverses) the rate of increase in the sales. To fit the model we use the transfer function model option in JMP with $S_t^{(88)}$ as the input. The output in Table 6.11 shows that there was indeed a significant effect on sales due to the introduction of a similar product in the market. The coefficient estimate $\hat{w}_0 = -2369.9$ further suggests that if no appropriate action is taken, the sales will most likely continue to go down.

Example 6.6 Electricity Consumption and the 1973 Arab Oil Embargo
The natural logarithm of monthly electric energy consumption in megawatt hours (MWh) for a regional utility from January 1951 to April 1977 is shown in Figure 6.19. The original data exhibited considerable inequality

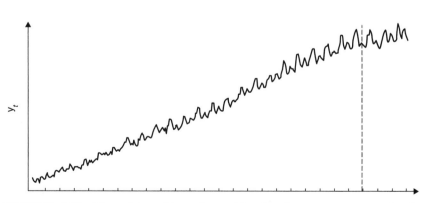

FIGURE 6.19 Natural logarithm of monthly electric energy consumption in megawatt hours (MWh) from January 1951 to April 1977.

TABLE 6.11   JMP Output for the Intervention Analysis in Example 6.5

Transfer Function Analysis
Time Series Sales

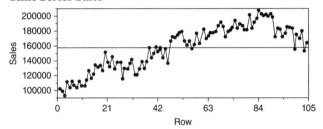

Mean	157540.87
Std	29870.833
N	104
Zero Mean ADF	0.0886817
Single Mean ADF	-2.381355
Trend ADF	-3.770646

Transfer Function Model (1)

Model Summary

DF	101
Sum of Squared Errors	9554575810
Variance Estimate	94599755.5
Standard Deviation	9726.24056
Akaike's 'A' Information Criterion	2186.26524
Schwarz's Bayesian Criterion	2191.5347
RSquare	0.26288323
RSquare Adj	0.25558504
MAPE	4.84506071
MAE	7399.35196
-2LogLikelihood	2182.26524

Parameter Estimates

Variable	Term	Factor	Lag	Estimate	Std Error	t Ratio	Prob>|t|
Step	Num0.0	0	0	-2359.900	1104.753	-2.15	
Sales	MA1,1	1	1	0.557	0.076	7.36	

$(1-B)*Sales_t = - (2369.8995 * Step_t) + (1 - 0.5571*B)*e_t$

Interactive Forecasting

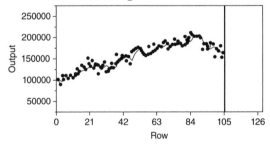

of variance over this time period and the natural logarithm stabilizes the variance. In November of 1973 the Arab oil embargo affected the supply of petroleum products to the United States. Following this event, it is hypothesized that the rate of growth of electricity consumption is smaller than in the pre-embargo period. Montgomery and Weatherby (1980) tested this hypothesis using an intervention model.

Although the embargo took effect in November of 1973, we will assume that first impact of this was not felt until the following month, December, 1973. Therefore, the period from January of 1951 until November of 1973 is assumed to have no intervention effect. These 275 months are analyzed to produce a univariate ARIMA model for the noise model. Both regular and seasonal differencing are required to achieve stationary and a multiplicative $(0, 1, 2) \times (0, 1, 1)_{12}$ was fit to the pre-intervention data. The model is

$$(1 - B)(1 - B^{12}) \ln N_t = (1 - 0.40B - 0.28B^2)(1 - 0.64B^{12})\varepsilon_t.$$

To develop the intervention, it is necessary to hypothesize the effect that the oil embargo may have had on electricity consumption. Perhaps the simplest scenario in which the level of the consumption time series is hypothesized to be permanently changed by a constant amount. Logarithms can be thought of as percent change, so this is equivalent to hypothesizing that the intervention effect is a change in the percent growth of electricity consumption. Thus the input series would be a step function

$$S_t = \begin{cases} 0, t = 1, 2, \ldots, 275 \\ 1, t = 275, 276, 326 \end{cases}.$$

The intervention model is then

$$(1 - B)(1 - B^{12}) \ln y_t = \omega_0 S_t + (1 - \theta_1 B - \theta_2 B^2)(1 - \Theta_{12} B^{12})\varepsilon_t,$$

where Θ_{12} is the seasonal MA parameter at lag 12. Table 6.12 gives the model parameter estimates and the corresponding 95% confidence intervals. Since the 95% confidence intervals do not contain zero, we conclude that all model parameters are statistically significant. The residual standard error for this model is 0.029195. Since the model was built to the natural logarithm of electricity consumption, the standard error has a simple, direct interpretation: namely, the standard deviation of one-step-ahead forecast errors is 2.9195 percent of the level of the time series.

TABLE 6.12   Model Summary Statistics for Example 6.6

Parameter	Point Estimate	95% Confidence Limits	
		Lower	Upper
ω_0	−0.07303	−0.11605	−0.03000
θ_1	0.40170	0.28964	0.51395
θ_2	0.27739	0.16448	0.39030
Θ_{12}	0.64225	0.54836	0.73604

It is also possible to draw conclusions about the intervention effect. This effect is a level change of magnitude $\hat{\omega}_0 = -0.07303$, expressed in the natural logarithm metric. The estimate of the intervention effect in the original MWh metric is $e^{\hat{\omega}_0} = e^{-0.07303} = 0.9296$. That is, the post-intervention level of electricity consumption is 92.96 percent of the pre-intervention level. The effect of the Arab oil embargo has been to reduce the increase in electricity consumption by 7.04 percent. This is a statistically significant effect.

In this example, there are 275 pre-intervention observations and 41 post-intervention observations. Generally, we would like to have as many observations as possible in the post-intervention period to ensure that the power of the test for the intervention effect is high. However, an extremely long post-intervention period may allow other unidentified factors to affect the output, leading to potential confounding of effects. The ability of the procedure to detect an intervention effect is a function of the number of pre- and post-intervention observations, the size of the intervention effect, the form of the noise model, and the parameter values of the process. In many cases, however, an intervention effect can be identified with relatively short record of post-intervention observations.

It is interesting to consider alternative hypotheses regarding the impact of the Arab oil embargo. For example, it may be more reasonable to suspect that the effect of the oil embargo is not to cause an immediate level change in electricity consumption, but a gradual one. This would suggest a model

$$(1 - B)(1 - B^{12}) \ln y_t = \frac{\omega_0}{1 - \delta_1 B} S_t + (1 - \theta_1 B - \theta_2 B^2)(1 - \Theta_{12} B^{12})\varepsilon_t.$$

The results of fitting this model to the data are shown in Table 6.13. Note that the 95% confidence interval for δ_1 includes zero, implying that we can

TABLE 6.13 Model Summary Statistics for the Alternate Intervention Model for Example 6.6

Parameter	Point Estimate	95% Confidence Limits Lower	Upper
ω_0	−0.06553	−0.11918	−0.01187
δ_1	0.18459	−0.51429	0.88347
θ_1	0.40351	0.29064	0.51637
θ_2	0.27634	0.16002	0.38726
Θ_{12}	0.63659	0.54201	0.73117

drop this parameter from the model. This would leave us with the original intervention model that was fit in Example 6.6. Consequently, we conclude that the Arab oil embargo induced an immediate permanent change in the level of electricity consumption.

In some problems there may be multiple intervention effects. Generally, one indicator variable must be used for each intervention effect. For example, suppose that in the electricity consumption example, we think that the oil embargo had two separate effects: the initial impact beginning in month 276, and a second impact beginning three months later. The intervention model to incorporate these effects is

$$(1-B)(1-B^{12})\ln y_t = \omega_{10}S_{1,t} + \omega_{20}S_{2,t} + (1-\theta_1 B - \theta_2 B^2)$$
$$(1-\Theta_{12}B^{12})\varepsilon_t,$$

where

$$S_{1,t} = \begin{cases} 0, & t=1,2,3,\ldots,275 \\ 1, & t=276,277,\ldots,316 \end{cases}$$

and

$$S_{2,t} = \begin{cases} 0, & t=1,2,3,\ldots,278 \\ 1, & t=279,277,\ldots,316 \end{cases}$$

In this model the parameters ω_{10} and ω_{20} represent the initial and secondary effects of the oil embargo and $\omega_{10}+\omega_{20}$ represents the long-term total impact.

There have been many other interesting applications of intervention analysis. For some very good examples, see the following references:

- Box and Tiao (1975) investigate the effects on ozone (O_3) concentration in downtown Los Angeles of a new law that restricted the amount of reactive hydrocarbons in locally sold gasoline, regulations that mandated automobile engine design changes, and the diversion of traffic by opening of the Golden State Freeway. They showed that these interventions did indeed lead to reductions in ozone levels.

- Wichern and Jones (1977) analyzed the impact of the endorsement by the American Dental Association of Crest toothpaste as an effective aid in reducing cavities on the market shares of Crest and Colgate toothpaste. The endorsement led to a significant increase in market share for Crest. See Bisgaard and Kulahci (2011) for a detailed analysis of that example.

- Atkins (1979) used intervention analysis to investigate the effect of compulsory automobile insurance, a company strike, and a change in insurance companies' policies on the number of highway accidents on freeways in British Columbia.

- Izenman and Zabell (1981) study the effect of the 9 November, 1965, blackout in New York City that resulted from a widespread power failure, on the birth rate nine months later. An article in *The New York Times* in August 1966 noted that births were up, but subsequent medical and demographic articles appeared with conflicting statements. Using the weekly birth rate from 1961 to 1966, the authors show that there is no statistically significant increase in the birth rate.

- Ledolter and Chan (1996) used intervention analysis to study the effect of a speed change on rural interstate highways in Iowa on the occurrence of traffic accidents.

Another important application of intervention analysis is in the detection of **time series outliers**. Time series observations are often influenced by external disruptive events, such as strikes, social/political events, economic crises, or wars and civil disturbances. The consequences of these events are observations that are not consistent with the other observations in the time series. These inconsistent observations are called **outliers**. In addition to the external events identified above, outliers can also be caused by more mundane forces, such as data recording or transmission errors. Outliers can have a very disruptive effect on model identification, parameter estimation, and forecasting, so it is important to be able to detect their presence so that they can be removed. Intervention analysis can be useful for this.

There are two kinds of time series outliers: additive outliers and innovation outliers. An additive outlier affects only the level of the t^* observation, while an innovation outlier affects all observations $y_{t^*}, y_{t^*+1}, y_{t^*+2}, \ldots$ beyond time t^* where the original outlier effect occurred. An additive outlier can be modeled as

$$z_t = \frac{\theta(B)}{\phi(B)}\varepsilon_t + \omega I_t^{(t^*)},$$

where $I_t^{(t^*)}$ is an indicator time series defined as

$$I_t^{(t^*)} = \begin{cases} 1 & \text{if } t = t^* \\ 0 & \text{if } t \neq t^* \end{cases}.$$

An innovation outlier is modeled as

$$z_t = \frac{\theta(B)}{\phi(B)}(\varepsilon_t + \omega I_t^{(t^*)}).$$

When the timing of the outlier is known, it is relatively straightforward to fit the intervention model. Then the presence of the outlier can be tested by comparing the estimate of the parameter ω, say, $\hat{\omega}$, to its standard error. When the timing of the outlier is not known, an iterative procedure is required. This procedure is described in Box, Jenkins, and Reinsel (1994) and in Wei (2006). The iterative procedure is capable of identifying multiple outliers in the time series.

6.8 R COMMANDS FOR CHAPTER 6

Example 6.2   The data for this example are in the array called vistemp.data of which the two columns represent the viscosity and the temperature respectively.

Below we first start with the prewhitening step.

```
xt<-vistemp.data[,1]
yt<-vistemp.data[,2]

par(mfrow=c(2,1),oma=c(0,0,0,0))
plot(xt,type="o",pch=16,cex=.5,xlab='Time',ylab=expression
    (italic(x[italic(t)])))
plot(yt,type="o",pch=16,cex=.5,xlab='Time',ylab= expression
    (italic(y[italic(t)])))
```

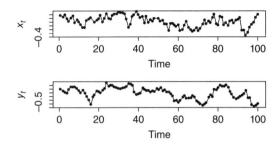

```
#Prewhitening
#Model identification for xt
par(mfrow=c(1,2),oma=c(0,0,0,0))
acf(xt,lag.max=25,type="correlation",main="ACF for Temperature")
acf(xt, lag.max=25,type="partial",main="PACF for Temperature",
    ylab="PACF")
```

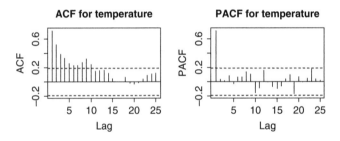

Fit an AR(1) model to xt.

```
xt.ar1<-arima(xt,order=c(1, 0, 0),include.mean=FALSE)
xt.ar1
```

```
        Call:
        arima(x = xt, order = c(1, 0, 0), include.mean = FALSE)

        Coefficients:
                 ar1
             0.7292
        s.e.  0.0686
        sigma^2 estimated as 0.01009:  log likelihood = 87.54,
          aic = -171.08
```

We perform the residual analysis

```
res.xt.ar1(-as.vector(residuals(xt.ar1))
#to obtain the fitted values we use the function fitted() from the
 forecast package
```

```
library(forecast)
fit.xt.ar1<-as.vector(fitted(xt.ar1))
# ACF and PACF of the Residuals
par(mfrow=c(1,2),oma=c(0,0,0,0))
acf(res.xt.ar1,lag.max=25,type="correlation",main="ACF of the
    Residuals for x(t)")
acf(res.xt.ar1, lag.max=25,type="partial",main="PACF of the
    Residuals for x(t)")
```

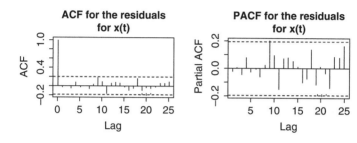

```
# 4-in-1 plot of the residuals
par(mfrow=c(2,2),oma=c(0,0,0,0))
qqnorm(res.xt.ar1,datax=TRUE,pch=16,xlab='Residual',main=")
qqline(res.xt.ar1,datax=TRUE)
plot(fit.xt.ar1,res.xt.ar1,pch=16, xlab='Fitted Value',
    ylab='Residual')
abline(h=0)
hist(res.xt.ar1,col="gray",xlab='Residual',main=")
plot(res.xt.ar1,type="l",xlab='Observation Order',ylab='Residual')
points(res.xt.ar1,pch=16,cex=.5)
abline(h=0)
```

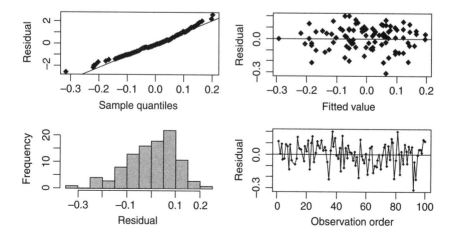

Prewhitening both series using AR(1) coefficients of 0.73.

```
T<-length(xt)
alphat<-xt[2:T]-0.73*xt[1:(T-1)]
betat<- yt[2:T]-0.73*yt[1:(T-1)]
ralbe<-ccf(betat,alphat,main='CCF of alpha(t) and beta(t)',
          ylab='CCF')
abline(v=0,col='blue')
```

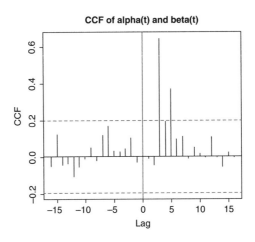

```
#Obtain the estimates of v_t
vhat<-sqrt(var(betat)/var(alphat))*ralbe$acf
nl<-length(vhat)
plot(seq(-(nl-1)/2,(nl-1)/2,1),vhat,
type='h',xlab='Lag',ylab=expression(italic(hat(v)[italic(j)])))
abline(v=0,col='blue')
abline(h=0)
```

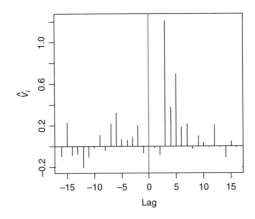

```
#Model the noise using the estimates given in the example

Nhat<-array(0,dim=c(1,T))
for (i in 4:T){
Nhat[i]<-yt[i]+0.31*(Nhat[i-1]-yt[i-1])+0.48*(Nhat[i-2]-yt[i-2])
    +1.21*xt[i-3]
        }
Nhat<-Nhat[4:T]
plot(Nhat,type="o",pch=16,cex=.5,xlab='Time',ylab=expression
    (italic(hat(N)[italic(t)])))
```

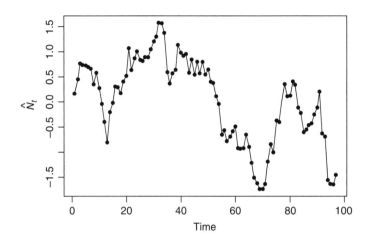

```
par(mfrow=c(1,2),oma=c(0,0,0,0))
acf(Nhat,lag.max=25,type="correlation",main="ACF of Nhat")

acf(Nhat, lag.max=25,type="partial",main="PACF of Nhat")
```

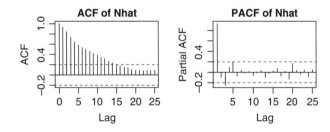

```
#Fit AR(1) and AR(3) models for Nhat

Nhat.ar1<-arima(Nhat,order=c(1, 0, 0),include.mean=FALSE)
Nhat.ar3<-arima(Nhat,order=c(3, 0, 0),include.mean=FALSE)
res.Nhat.ar1<-as.vector(residuals(Nhat.ar1))
```

```
library(forecast)
fit.Nhat.ar1<-as.vector(fitted(Nhat.ar1))
# ACF and PACF of the Residuals
par(mfrow=c(1,2),oma=c(0,0,0,0))
acf(res.Nhat.ar1,lag.max=25,type="correlation",main="ACF of the
    Residuals for Nhat")
acf(res.Nhat.ar1, lag.max=25,type="partial",main="PACF of the
    Residuals for Nhat")
```

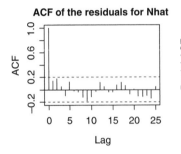
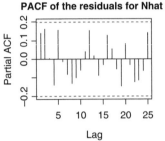

```
# 4-in-1 plot of the residuals
par(mfrow=c(2,2),oma=c(0,0,0,0))
qqnorm(res.Nhat.ar1,datax=TRUE,pch=16,xlab='Residual',main=")
qqline(res.Nhat.ar1,datax=TRUE)
plot(fit.xt.ar1,res.xt.ar1,pch=16, xlab='Fitted Value',
     ylab='Residual')
abline(h=0)
hist(res.Nhat.ar1,col="gray",xlab='Residual',main=")
plot(res.Nhat.ar1,type="l",xlab='Observation Order',ylab='Residual')
points(res.Nhat.ar1,pch=16,cex=.5)
abline(h=0)
```

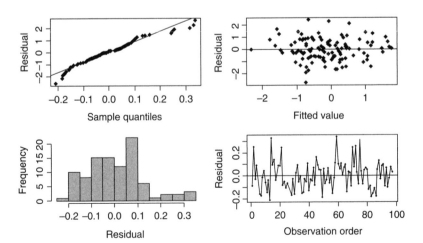

We now fit the following transfer function–noise model

$$y_t = \frac{w_0}{1 - \delta_1 B - \delta_2 B^2} x_{t-3} + \frac{1}{1 - \phi_1 B} \varepsilon_t.$$

For that we will use the "arimax" function in TSA package.

```
library(TSA)
ts.xt<-ts(xt)
lag3.x<-lag(ts.xt,-3)
ts.yt<-ts(yt)
dat3<-cbind(ts.xt,lag3.x,ts.yt)
dimnames(dat3)[[2]]<-c("xt","lag3x","yt")
data2<-na.omit(as.data.frame(dat3))
#Input arguments
#order: determines the model for the error component, i.e. the
#order of the ARIMA model for y(t)
#if there were no x(t)

#xtransf: x(t)
#transfer: the orders (r and s) of the transfer function
visc.tf<-arimax(data2$yt, order=c(1,0,0), xtransf=data.frame(data2$lag3x),
                transfer=list(c(2,0)), include.mean = FALSE)
visc.tf

Call:
arimax(x = data2$yt, order = c(1, 0, 0), include.mean = FALSE, xtransf =
data.frame(data2$lag3x), transfer = list(c(2, 0)))

Coefficients:
         ar1  data2.lag3x-AR1  data2.lag3x-AR2  data2.lag3x-MA0
      0.8295           0.3414           0.2667           1.3276
s.e.  0.0642           0.0979           0.0934           0.1104

sigma^2 estimated as 0.0123:  log likelihood = 75.09,  aic = -142.18


res.visc.tf<-as.vector(residuals(visc.tf))
library(forecast)
fit.visc.tf <-as.vector(fitted(visc.tf))
# ACF and PACF of the Residuals
par(mfrow=c(1,2),oma=c(0,0,0,0))
acf(res.visc.tf,lag.max=25,type="correlation",main="ACF of the
    Residuals \nfor TF-N Model")
acf(res.visc.tf, lag.max=25,type="partial",main="PACF of the
    Residuals \nfor TF-N Model")
```

ACF of the residuals for TF-N model

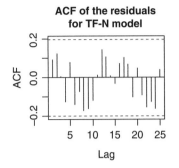

PACF of the residuals for TF-N model

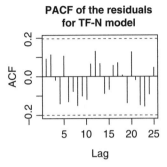

```
# 4-in-1 plot of the residuals
par(mfrow=c(2,2),oma=c(0,0,0,0))
qqnorm(res.visc.tf,datax=TRUE,pch=16,xlab='Residual',main=")
qqline(res.visc.tf,datax=TRUE)
plot(fit.visc.tf,res.visc.tf,pch=16, xlab='Fitted Value',
     ylab='Residual')
abline(h=0)
hist(res.visc.tf,col="gray",xlab='Residual',main=")
plot(res.visc.tf,type="l",xlab='Observation Order',ylab='Residual')
points(res.visc.tf,pch=16,cex=.5)
abline(h=0)
```

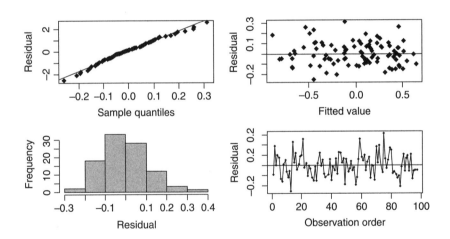

```
T<-length(res.visc.tf)
Ta<-length(alphat)
ccf(res.visc.tf,alphat[(Ta-T+1):Ta],main='CCF of alpha(t) and
    \nResiduals of TF-N Model',ylab='CCF')
abline(v=0,col='blue')
```

**CCF of alpha(t) and
Residuals of TF-N model**

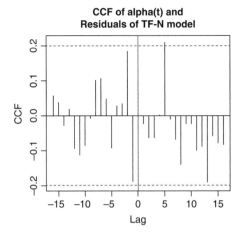

Example 6.4   Note that this is a continuation of Example 6.2. Below we assume that the reader followed the modeling efforts required in Example 6.2. For variable and model names, please refer to R-code for Example 6.2.

For forecasting we use the formula given in the example. Before we proceed, we first make forecasts for x(t) based on the AR(1) model. We will only make 6-step-ahead forecasts for x(t) even if not all of them are needed due to delay.

```
tau<-6
xt.ar1.forecast<-forecast(xt.ar1,h=tau)
```

To make the recursive calculations given in the example simpler, we will simply extend the xt, yt and residuals vectors as the following.

```
xt.new<-c(xt,xt.ar1.forecast$mean)
res.tf.new<-c(rep(0,3),res.visc.tf,rep(0,tau))

yt.new<-c(yt,rep(0,tau))

#Note that 3 0's are added to the beginning to compensate for the
#misalignment between xt and the residuals of transfer function
#noise model due to the delay of 3 lags. Last 6 0's are added
#since the future values of the error are assumed to be 0.
```

We now get the parameter estimates for the transfer function–noise model

```
phi1<-visc.tf[[1]][1]
d1<-visc.tf[[1]][2]
d2<-visc.tf[[1]][3]
w0<-visc.tf[[1]][4]
```

The forecasts are then obtained using:

```
T<-length(yt)
for (i in (T+1):(T+tau)){
        yt.new[i]<-(d1+phi1)*yt.new[i-1]+(d2-d1*phi1)*yt.new[i-2]
                -d2*phi1*yt.new[i-3]+w0*xt.new[i-3]
                -w0*phi1*xt.new[i-4]+res.tf.new[i]-d1*res.tf.new[i-1]-
                d2*res.tf.new[i-1]
}
```

To calculate the prediction limits, we need to first calculate the estimate of forecast error variance given in (6.45). As mentioned in the example, since we only need up to six-step ahead forecasts, we need to calculate only v_0^*, v_1^* and v_2^* and ψ_0 through ψ_5.

```
phix<-xt.ar1[[1]][1]
v0star<-w0
v1star<-(d1+phix)*w0
v2star<-(((d1+phix)^2)+(d2-d1*phix))*w0
vstar<-c(v0star,v1star,v2star)
psi<-phix^(0:(tau-1))
sig2.alpha<-xt.ar1$sigma2
sig2.err<-visc.tf$sigma2
sig2.tfn<-rep(0,6)
b<-3
for (i in 1:6) {
        if ((i-b)<=0) {
                sig2.tfn[i]<- sig2.err*sum(psi[1:i]^2)
        }
        else {
                sig2.tfn[i]<- sig2.alpha*sum(vstar[1:(i-b)]^2)
                + sig2.err*sum(psi[1:i]^2)
        }
}
```

For comparison purposes, we also fit a univariate ARIMA model to y(t). An AR(3) model is considered even though the AR coefficient at third lag is borderline significant. The model is given below.

```
yt.ar3<-arima(yt,order=c(3,0,0),include.mean=FALSE)
> yt.ar3

        Series: x
        ARIMA(3,0,0) with zero mean
```

```
Coefficients:
          ar1      ar2      ar3
       0.9852   0.1298  -0.2700
s.e.   0.0954   0.1367   0.0978

sigma^2 estimated as 0.02779:  log likelihood=36.32
AIC=-66.65    AICc=-66.23   BIC=-56.23
```

To calculate the prediction limits, we need to first calculate the estimate of forecast error variance given in (5.83). To estimate ψ_0 through ψ_5 we use the formula given in (5.46).

```
psi.yt<-vector()
psi.yt[1:4]<-c(0,0,0,1)
sig2.yt<-yt.ar3$sigma2

for (i in 5:(4+tau-1)){
        psi.yt[i]<- yt.ar3[[1]][1]*psi.yt[i-1]+yt.ar3[[1]][2]
*psi.yt[i-2]+yt.ar3[[1]][3]*psi.yt[i-3]
}
psi.yt<-psi.yt[4:(4+tau-1)]
psi.yt
sig2.um<-rep(0,6)
b<-3
for (i in 1:6) {
        sig2.um[i]<- sig2.yt*sum(psi.yt[1:i]^2)
}
```

Thus for the transfer function-noise model and the univariate model we have the following prediction error variances.

```
cbind(sig2.tfn,sig2.um)
            sig2.tfn     sig2.um
     [1,]  0.01230047  0.02778995
     [2,]  0.01884059  0.05476427
     [3,]  0.02231796  0.08841901
     [4,]  0.04195042  0.11308279
     [5,]  0.06331720  0.13109029
     [6,]  0.08793347  0.14171137
```

We can see that adding the exogenous variable x(t) helps to reduce the prediction error variance by half.

To plot the forecasts and the prediction limits, we have

```
plot(yt.new[1:T],type="p",pch=16,cex=.5,xlab='Time',
    ylab='Viscosity',xlim=c(1,110))
lines(101:106,yt.new[101:106],col="grey40")
lines(101:106, yt.new[101:106]+2*sqrt(sig2.tfn))
lines(101:106, yt.new[101:106]-2*sqrt(sig2.tfn))
legend(20,-.4,c("y(t)","Forecast","95% LPL","95% UPL"), pch=c(16,
NA, NA,NA),lwd=c(NA,.5,.5,.5),cex=.55,col=c("black","grey40",
"black","black"))
```

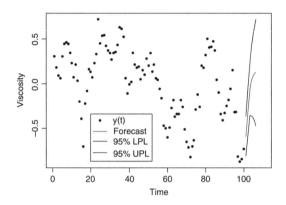

Example 6.5   The data for this example are in the array called cere-alsales.data of which the two columns represent the week and the sales respectively. We first start with the plot of the data

```
yt.sales<-cerealsales.data[,2]
plot(yt.sales,type="o",pch=16,cex=.5,xlab='Week',ylab='Sales')
abline(v=88)
mtext("Week 88", side=3, at=88)
```

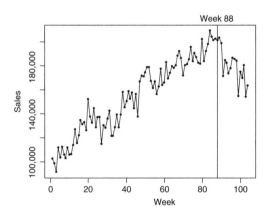

We then try to identify the ARIMA model for the pre-intervention data (up to week 87)

```
par(mfrow=c(1,2),oma=c(0,0,0,0))
acf(yt.sales[1:87],lag.max=25,type="correlation",main="ACF
    for Sales \nWeeks 1-87")
acf(yt.sales[1:87], lag.max=25,type="partial",main="PACF for Sales
\nWeeks 1-87",ylab="PACF")
```

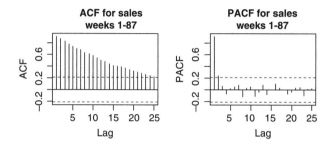

The series appears to be non-stationary. We try the first differences

```
par(mfrow=c(1,2),oma=c(0,0,0,0))
acf(diff(yt.sales[1:87],1),lag.max=25,type="correlation",main="ACF
    for First Differences \nWeeks 1-87")

acf(diff(yt.sales[1:87],1), lag.max=25,type="partial",main="PACF
    for First Differences \nWeeks 1-87",ylab="PACF")
```

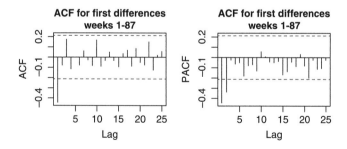

ARIMA(0,0,1) model for the first differences seems to be appropriate. We now fit the transfer function–noise model with the step input. First we define the step indicator variable.

```
library(TSA)
T<-length(yt.sales)
St<-c(rep(0,87),rep(1,(T-87)))
```

```
sales.tf<-arimax(diff(yt.sales), order=c(0,0,1), xtransf= St[2:T],
                    transfer=list(c(0,0)), include.mean = FALSE)
#Note that we adjusted the step function for the differencing we
# did on the yt.sales

sales.tf
        Series: diff(yt.sales)
        ARIMA(0,0,1) with zero mean

        Coefficients:
             ma1      T1-MA0
          -0.5571   -2369.888
        s.e.   0.0757    1104.542

        sigma^2 estimated as 92762871:  log likelihood=-1091.13
        AIC=2186.27   AICc=2186.51   BIC=2194.17
```

EXERCISES

6.1 An input and output time series consists of 300 observations. The prewhitened input series is well modeled by an AR(2) model $y_t = 0.5y_{t-1} + 0.2y_{t-2} + \alpha_t$. We have estimated $\hat{\sigma}_\alpha = 0.2$ and $\hat{\sigma}_\beta = 0.4$. The estimated cross-correlation function between the prewhitened input and output time series is shown below.

Lag, j	0	1	2	3	4	5	6	7	8	9	10
$r_{\alpha\beta}(j)$	0.01	0.03	−0.03	−0.25	−0.35	−0.51	−0.30	−0.15	−0.02	0.09	−0.01

  a. Find the approximate standard error of the cross-correlation function. Which spikes on the cross-correlation function appear to be significant?

  b. Estimate the impulse response function. Tentatively identify the form of the transfer function model.

6.2 Find initial estimates of the parameters of the transfer function model for the situation in Exercise 6.1.

6.3 An input and output time series consists of 200 observations. The prewhitened input series is well modeled by an MA(1) model $y_t = 0.8\alpha_{t-1} + \alpha_t$. We have estimated $\hat{\sigma}_\alpha = 0.4$ and $\hat{\sigma}_\beta = 0.6$. The estimated cross-correlation function between the prewhitened input and output time series is shown below.

Lag, j	0	1	2	3	4	5	6	7	8	9	10
$r_{\alpha\beta}(j)$	0.01	0.55	0.40	0.28	0.20	0.07	0.02	0.01	−0.02	0.01	−0.01

    a. Find the approximate standard error of the cross-correlation function. Which spikes on the cross-correlation function appear to be significant?

    b. Estimate the impulse response function. Tentatively identify the form of the transfer function model.

6.4 Find initial estimates of the parameters of the transfer function model for the situation in Exercise 6.3.

6.5 Write the equations that must be solved in order to obtain initial estimates of the parameters in a transfer function model with $b = 2$, $r = 1$, and $s = 0$.

6.6 Write the equations that must be solved in order to obtain initial estimates of the parameters in a transfer function model with $b = 2$, $r = 2$, and $s = 1$.

6.7 Write the equations that must be solved in order to obtain initial estimates of the parameters in a transfer function model with $b = 2$, $r = 1$, and $s = 1$.

6.8 Consider a transfer function model with $b = 2$, $r = 1$, and $s = 0$. Assume that the noise model is AR(1). Find the forecasts in terms of the transfer function and noise model parameters.

6.9 Consider the transfer function model in Exercise 6.8 with $b = 2$, $r = 1$, and $s = 0$. Now assume that the noise model is AR(2). Find the forecasts in terms of the transfer function and noise model parameters. What difference does this noise model make on the forecasts?

6.10 Consider the transfer function model in the Exercise 6.8 with $b = 2$, $r = 1$, and $s = 0$. Now assume that the noise model is MA(1). Find the forecasts in terms of the transfer function and noise model parameters. What difference does this noise model make on the forecasts?

6.11 Consider the transfer function model

$$y_t = \frac{-0.5 - 0.4B - 0.2B^2}{1 - 0.5B} x_{t-2} + \frac{1}{1 - 0.5B} \varepsilon_t.$$

Find the forecasts that are generated from this model.

6.12 Sketch a graph of the impulse response function for the following transfer function:

$$y_t = \frac{2B}{1 - 0.6B} x_t.$$

6.13 Sketch a graph of the impulse response function for the following transfer function:

$$y_t = \frac{1 - 0.2B}{1 - 0.8B} x_t.$$

6.14 Sketch a graph of the impulse response function for the following transfer function:

$$y_t = \frac{1}{1 - 1.2B + 0.4B^2} x_t.$$

6.15 Box, Jenkins, and Reinsel (1994) fit a transfer function model to data from a gas furnace. The input variable is the volume of methane entering the chamber in cubic feet per minute and the output is the concentration of carbon dioxide emitted. The transfer function model is

$$y_t = \frac{-(0.53 + 0.37B + 0.51B^2)}{1 - 0.57B} x_t + \frac{1}{1 - 0.53B + 0.63B^2} \varepsilon_t,$$

where the input and output variables are measured every nine seconds.

a. What are the values of $b, s,$ and r for this model?

b. What is the form of the ARIMA model for the errors?

c. If the methane input was increased, how long would it take before the carbon dioxide concentration in the output is impacted?

6.16 Consider the global mean surface air temperature anomaly and global CO_2 concentration data in Table B.6 in Appendix B. Fit an appropriate transfer function model to this data, assuming that CO_2 concentration is the input variable.

6.17 Consider the chemical process yield and uncontrolled operating temperature data in Table B.12. Fit an appropriate transfer function model to these data, assuming that temperature is the input variable. Does including the temperature data improve your ability to forecast the yield data?

6.18 Consider the U.S. Internal Revenue tax refunds data in Table B.20. Fit an appropriate transfer function model to these data, assuming that population is the input variable. Does including the population data improve your ability to forecast the tax refund data?

6.19 Find time series data of interest to you where a transfer function–noise model would be appropriate.
  a. Identify and fit the appropriate transfer function–noise model.
  b. Use an ARIMA model to fit only the y_t series.
  c. Compare the forecasting performance of the two models from parts a and b.

6.20 Find a time series of interest to you that you think may be impacted by an outlier. Fit an appropriate ARIMA model to the time series and use either the additive outlier or innovation outlier model to see if the potential outlier is statistically significant.

6.21 Table E6.1 provides 100 observations on a time series.

TABLE E6.1 Time Series Data for Exercise 6.21 (100 observations, read down then across)

86.74	83.79	88.42	84.23	82.20
85.32	84.04	89.65	83.58	82.14
84.74	84.10	97.85	84.13	81.80
85.11	84.85	88.50	82.70	82.32
85.15	87.64	87.06	83.55	81.53
84.48	87.24	85.20	86.47	81.73
84.68	87.52	85.08	86.21	82.54
84.68	86.50	84.44	87.02	82.39
86.32	85.61	84.21	86.65	82.42
88.00	86.83	86.00	85.71	82.21
86.26	84.50	85.57	86.15	82.77
85.83	84.18	83.79	85.80	83.12
83.75	85.46	84.37	85.62	83.22
84.46	86.15	83.38	84.23	84.45
84.65	86.41	85.00	83.57	84.91
84.58	86.05	84.35	84.71	85.76
82.25	86.66	85.34	83.82	85.23
83.38	84.73	86.05	82.42	86.73
83.54	85.95	84.88	83.04	87.00
85.16	86.85	85.42	83.70	85.06

a. Plot the data.

b. There is an apparent outlier in the data. Use intervention analysis to investigate the presence of this outlier.

6.22 Table E6.2 provides 100 observations on a time series. These data represent weekly shipments of a product.

a. Plot the data.

b. Note that there is an apparent increase in the level of the time series at about observation 80. Management suspects that this increase in shipments may be due to a strike at a competitor's plant. Build an appropriate intervention model for these data. Do you think that the impact of this intervention is likely to be permanent?

TABLE E6.2   Time Series Data for Exercise 6.22 (100 observations, read down then across)

1551	1556	1613	1552	1838
1548	1557	1595	1558	1838
1554	1564	1601	1543	1834
1557	1592	1587	1552	1840
1552	1588	1568	1581	1832
1555	1591	1567	1578	1834
1556	1581	1561	1587	1842
1574	1572	1558	1583	1840
1591	1584	1576	1573	1840
1575	1561	1572	1578	1838
1571	1558	1554	1574	1844
1551	1571	1560	1573	1848
1558	1578	1550	1559	1849
1561	1580	1566	1552	1861
1560	1577	1560	1563	1865
1537	1583	1570	1555	1874
1549	1564	1577	1541	1869
1551	1576	1565	1547	1884
1567	1585	1571	1553	1886
1553	1601	1559	1538	1867

6.23 Table B.23 contains data on Danish crude oil production. Historically, oil production increased steadily from 1972 up to about 2000, when the Danish government presented an energy strategy

containing a number of ambitious goals for national energy policy up through 2025. The aim is to reduce Denmark's dependency
on coal, oil and natural gas. The data exhibit a marked downturn
in oil production starting in 2005. Fit and analyze an appropriate
intervention model to these data.

6.24 Table B.25 contains data on annual US motor vehicle fatalities from
1966 through 2012, along with data on several other factors. Fit a
transfer function model to these data using the number of licensed
drivers as the input time series. Compare this transfer function model
to a univariate ARIMA model for the annual fatalities data.

6.25 Table B.25 contains data on annual US motor vehicle fatalities from
1966 through 2012, along with data on several other factors. Fit a
transfer function model to these data using the annual unemployment
rate as the input time series. Compare this transfer function model
to a univariate ARIMA model for the annual fatalities data. Why
do you think that the annual unemployment rate might be a good
predictor of fatalities?

6.26 Table B.25 contains data on annual US motor vehicle fatalities
from 1966 through 2012, along with data on several other factors.
Fit a transfer function model to these data using both the number of licensed drivers and the annual unemployment rate as the
input time series. Compare this two-input transfer function model
to a univariate ARIMA model for the annual fatalities data, and
to the two univariate transfer function models from Exercises 6.24
and 6.25.

CHAPTER 7

SURVEY OF OTHER FORECASTING METHODS

*I always avoid prophesying beforehand, because it is a much better policy
to prophesy after the event has already taken place.*
SIR WINSTON CHURCHILL, *British Prime Minister*

7.1 MULTIVARIATE TIME SERIES MODELS AND FORECASTING

In many forecasting problems, it may be the case that there are more than just one variable to consider. Attempting to model each variable individually may at times work. However, in these situations, it is often the case that these variables are somehow cross-correlated, and that structure can be effectively taken advantage of in forecasting. In the previous chapter we explored this for the "unidirectional" case, where it is assumed that certain inputs have impact on the variable of interest but not the other way around. Multivariate time series models involve several variables that are not only serially but also cross-correlated. As in the univariate case, multivariate or **vector ARIMA** models can often be successfully used in forecasting multivariate time series. Many of the concepts we have seen in Chapter 5

Introduction to Time Series Analysis and Forecasting, Second Edition.
Douglas C. Montgomery, Cheryl L. Jennings and Murat Kulahci.
© 2015 John Wiley & Sons, Inc. Published 2015 by John Wiley & Sons, Inc.

will be directly applicable in the multivariate case as well. We will first start with the property of stationarity.

7.1.1 Multivariate Stationary Process

Suppose that the vector time series $Y_t = (y_{1t}, y_{2t}, \ldots, y_{mt})$ consists of m univariate time series. Then Y_t with finite first and second order moments is said to be **weakly stationary** if

(i) $E(Y_t) = E(Y_{t+s}) = \mu$, constant for all s
(ii) $\text{Cov}(Y_t) = E[(Y_t - \mu)(Y_t - \mu)'] = \Gamma(0)$
(iii) $\text{Cov}(Y_t, Y_{t+s}) = \Gamma(s)$ depends only on s

Note that the diagonal elements of $\Gamma(s)$ give the autocovariance function of the individual time series, $\gamma_{ii}(s)$. Similarly, the autocorrelation matrix is given by

$$\rho(s) = \begin{bmatrix} \rho_{11}(s) & \rho_{12}(s) & \cdots & \rho_{1m}(s) \\ \rho_{21}(s) & \rho_{22}(s) & \cdots & \rho_{2m}(s) \\ \vdots & \vdots & \ddots & \vdots \\ \rho_{m1}(s) & \rho_{m2}(s) & \cdots & \rho_{mm}(s) \end{bmatrix} \tag{7.1}$$

which can also be obtained by defining

$$V = \text{diag}\{\gamma_{11}(0), \gamma_{22}(0), \ldots, \gamma_{mm}(0)\}$$

$$= \begin{bmatrix} \gamma_{11}(0) & 0 & \cdots & 0 \\ 0 & \gamma_{22}(0) & \cdots & 0 \\ \vdots & \vdots & \ddots & \vdots \\ 0 & 0 & \cdots & \gamma_{mm}(0) \end{bmatrix} \tag{7.2}$$

We then have

$$\rho(s) = V^{-1/2}\Gamma(s)V^{-1/2} \tag{7.3}$$

We can further show that $\Gamma(s) = \Gamma(-s)'$ and $\rho(s) = \rho(-s)'$.

7.1.2 Vector ARIMA Models

The stationary vector time series can be represented with a **vector ARMA** model given by

$$\Phi(B)Y_t = \delta + \Theta(B)\varepsilon_t \tag{7.4}$$

where $\mathbf{\Phi}(B) = \mathbf{I} - \mathbf{\Phi}_1 B - \mathbf{\Phi}_2 B^2 - \cdots - \mathbf{\Phi}_p B^p$, $\mathbf{\Theta}(B) = \mathbf{I} - \mathbf{\Theta}_1 B - \mathbf{\Theta}_2 B^2 - \cdots - \mathbf{\Theta}_q B^q$, and $\boldsymbol{\varepsilon}_t$ represents the sequence of independent random vectors with $E(\boldsymbol{\varepsilon}_t) = \mathbf{0}$ and $\text{Cov}(\boldsymbol{\varepsilon}_t) = \mathbf{\Sigma}$. Since the random vectors are independent, we have $\Gamma_\varepsilon(s) = 0$ for all $s \neq 0$.

The process Y_t in Eq. (7.4) is stationary if the roots of

$$\det[\mathbf{\Phi}(B)] = \det[\mathbf{I} - \mathbf{\Phi}_1 B - \mathbf{\Phi}_2 B^2 - \cdots - \mathbf{\Phi}_p B^p] = 0 \quad (7.5)$$

are all greater than one in absolute value. Then the process Y_t is also said to have infinite MA representation given as

$$\begin{aligned} Y_t &= \boldsymbol{\mu} + \mathbf{\Psi}(B)\boldsymbol{\varepsilon}_t \\ &= \boldsymbol{\mu} + \sum_{i=0}^{\infty} \mathbf{\Psi}_i \boldsymbol{\varepsilon}_{t-i} \end{aligned} \quad (7.6)$$

where $\mathbf{\Psi}(B) = \mathbf{\Phi}(B)^{-1}\mathbf{\Theta}(B)$, $\mu = \mathbf{\Phi}(B)^{-1}\delta$, and $\sum_{i=0}^{\infty} \|\mathbf{\Psi}_i\|^2 < \infty$.

Similarly, if the roots of $\det[\mathbf{\Theta}(B)] = \det[\mathbf{I} - \mathbf{\Theta}_1 B - \mathbf{\Theta}_2 B^2 - \cdots - \mathbf{\Theta}_q B^q] = 0$ are greater than unity in absolute value the process Y_t in Eq. (7.4) is invertible.

To illustrate the vector ARMA model given in Eq. (7.4), consider the bivariate ARMA(1,1) model with

$$\begin{aligned} \mathbf{\Phi}(B) &= \mathbf{I} - \mathbf{\Phi}_1 B \\ &= \begin{bmatrix} 1 & 0 \\ 0 & 1 \end{bmatrix} - \begin{bmatrix} \phi_{11} & \phi_{12} \\ \phi_{21} & \phi_{22} \end{bmatrix} B \end{aligned}$$

and

$$\begin{aligned} \mathbf{\Theta}(B) &= \mathbf{I} - \mathbf{\Theta}_1 B \\ &= \begin{bmatrix} 1 & 0 \\ 0 & 1 \end{bmatrix} - \begin{bmatrix} \theta_{11} & \theta_{12} \\ \theta_{21} & \theta_{22} \end{bmatrix} B \end{aligned}$$

Hence the model can be written as

$$\left[\begin{bmatrix} 1 & 0 \\ 0 & 1 \end{bmatrix} - \begin{bmatrix} \phi_{11} & \phi_{12} \\ \phi_{21} & \phi_{22} \end{bmatrix} B \right] Y_t = \begin{bmatrix} \delta_1 \\ \delta_2 \end{bmatrix} + \left[\begin{bmatrix} 1 & 0 \\ 0 & 1 \end{bmatrix} - \begin{bmatrix} \theta_{11} & \theta_{12} \\ \theta_{21} & \theta_{22} \end{bmatrix} B \right] \begin{bmatrix} \varepsilon_{1,t} \\ \varepsilon_{2,t} \end{bmatrix}$$

or

$$y_{1,t} = \delta_1 + \phi_{11} y_{1,t-1} + \phi_{12} y_{2,t-1} + \varepsilon_{1,t} - \theta_{11}\varepsilon_{1,t-1} - \theta_{12}\varepsilon_{2,t-1}$$
$$y_{2,t} = \delta_2 + \phi_{21} y_{1,t-1} + \phi_{22} y_{2,t-1} + \varepsilon_{2,t} - \theta_{21}\varepsilon_{1,t-1} - \theta_{22}\varepsilon_{2,t-1}$$

As in the univariate case, if nonstationarity is present, through an appropriate degree of differencing a stationary vector time series may be achieved. Hence the vector ARIMA model can be represented as

$$\boldsymbol{\Phi}(B)\mathbf{D}(B)Y_t = \boldsymbol{\delta} + \boldsymbol{\Theta}(B)\boldsymbol{\varepsilon}_t$$

where

$$\mathbf{D}(B) = \text{diag}\{(1-B)^{d_1}, (1-B)^{d_2}, \dots, (1-B)^{d_m}\}$$

$$= \begin{bmatrix} (1-B)^{d_1} & 0 & \cdots & 0 \\ 0 & (1-B)^{d_2} & \cdots & 0 \\ \vdots & \vdots & \ddots & \vdots \\ 0 & 0 & \cdots & (1-B)^{d_m} \end{bmatrix}$$

However, the degree of differencing is usually quite complicated and has to be handled with care (Reinsel (1997)).

The identification of the vector ARIMA model can indeed be fairly difficult. Therefore in the next section we will concentrate on the more commonly used and intuitively appealing vector autoregressive models. For a more general discussion see Reinsel (1997), Lütkepohl (2005), Tiao and Box (1981), Tiao and Tsay (1989), Tsay (1989), and Tjostheim and Paulsen (1982).

7.1.3 Vector AR (VAR) Models

The vector AR(p) model is given by

$$\boldsymbol{\Phi}(B)Y_t = \boldsymbol{\delta} + \boldsymbol{\varepsilon}_t \tag{7.7}$$

or

$$Y_t = \boldsymbol{\delta} + \sum_{i=1}^{p} \boldsymbol{\Phi}_i Y_{t-i} + \boldsymbol{\varepsilon}_t$$

For a stationary vector AR process, the infinite MA representation is given as

$$Y_t = \boldsymbol{\mu} + \boldsymbol{\Psi}(B)\boldsymbol{\varepsilon}_t \tag{7.8}$$

where $\boldsymbol{\Psi}(B) = \mathbf{I} + \boldsymbol{\Psi}_1 B + \boldsymbol{\Psi}_2 B^2 + \cdots$ and $\boldsymbol{\mu} = \boldsymbol{\Phi}(B)^{-1}\boldsymbol{\delta}$. Hence we have $E(Y_t) = \boldsymbol{\mu}$ and $\text{Cov}(\boldsymbol{\varepsilon}_t, Y_{t-s}) = 0$ for any $s > 0$ since Y_{t-s} is only concerned with ε_{t-s}, ε_{t-s-1}, ..., which are not correlated with ε_t. Moreover, we also have

$$\text{Cov}(\varepsilon_t, Y_t) = \text{Cov}\left(\varepsilon_t, \varepsilon_t + \Psi_1\varepsilon_{t-1} + \Psi_2\varepsilon_{t-2} + \cdots\right)$$
$$= \text{Cov}(\varepsilon_t, \varepsilon_t)$$
$$= \Sigma$$

and

$$\Gamma(s) = \text{Cov}(Y_{t-s}, Y_t) = \text{Cov}\left(Y_{t-s}, \delta + \sum_{i=1}^{p}\Phi_i Y_{t-i} + \varepsilon_t\right)$$

$$= \text{Cov}\left(Y_{t-s}, \sum_{i=1}^{p}\Phi_i Y_{t-i}\right) + \underbrace{\text{Cov}(Y_{t-s}, \varepsilon_t)}_{=0 \text{ for } s>0}$$

$$= \sum_{i=1}^{p}\text{Cov}(Y_{t-s}, \Phi_i Y_{t-i}) \tag{7.9}$$

$$= \sum_{i=1}^{p}\text{Cov}(Y_{t-s}, Y_{t-i})\Phi_i'$$

Hence we have

$$\Gamma(s) = \sum_{i=1}^{p}\Gamma(s-i)\Phi_i' \tag{7.10}$$

and

$$\Gamma(0) = \sum_{i=1}^{p}\Gamma(-i)\Phi_i' + \Sigma \tag{7.11}$$

As in the univariate case, the Yule–Walker equations can be obtained from the first p equations as

$$\begin{bmatrix} \Gamma(1) \\ \Gamma(2) \\ \vdots \\ \Gamma(p) \end{bmatrix} = \begin{bmatrix} \Gamma(0) & \Gamma(1)' & \cdots & \Gamma(p-1)' \\ \Gamma(1) & \Gamma(0) & \cdots & \Gamma(p-2)' \\ \vdots & \vdots & \ddots & \vdots \\ \Gamma(p-1) & \Gamma(p-2) & \cdots & \Gamma(0) \end{bmatrix} \begin{bmatrix} \Phi_1' \\ \Phi_2' \\ \vdots \\ \Phi_p' \end{bmatrix} \tag{7.12}$$

The model parameters in Φ and Σ can be estimated from Eqs. (7.11) and (7.12).

For the VAR(p), the autocorrelation matrix in Eq. (7.3) will exhibit a decaying behavior following a mixture of exponential decay and damped sinusoid.

Example 7.1 VAR(1) Model   The autocovariance matrix for VAR(1) is given as

$$\boldsymbol{\Gamma}(s) = \boldsymbol{\Gamma}(s-1)\boldsymbol{\Phi}' = (\boldsymbol{\Gamma}(s-2)\boldsymbol{\Phi}')\boldsymbol{\Phi}' = \cdots = \boldsymbol{\Gamma}(0)(\boldsymbol{\Phi}')^s \quad (7.13)$$

and

$$
\begin{aligned}
\boldsymbol{\rho}(s) &= \mathbf{V}^{-1/2}\boldsymbol{\Gamma}(s)\mathbf{V}^{-1/2} \\
&= \mathbf{V}^{-1/2}\boldsymbol{\Gamma}(0)(\boldsymbol{\Phi}')^s\mathbf{V}^{-1/2} \\
&= \mathbf{V}^{-1/2}\boldsymbol{\Gamma}(0)\mathbf{V}^{-1/2}\mathbf{V}^{1/2}(\boldsymbol{\Phi}')^s\mathbf{V}^{-1/2} \\
&= \boldsymbol{\rho}(0)\mathbf{V}^{1/2}(\boldsymbol{\Phi}')^s\mathbf{V}^{-1/2}
\end{aligned}
\quad (7.14)
$$

where $\mathbf{V} = \text{diag}\{\gamma_{11}(0), \gamma_{22}(0), \ldots, \gamma_{mm}(0)\}$. The eigenvalues of $\boldsymbol{\Phi}$ determine the behavior of the autocorrelation matrix. In fact, if the eigenvalues of $\boldsymbol{\Phi}$ are real and/or complex conjugates, the behavior will be a mixture of the exponential decay and damped sinusoid, respectively.

Example 7.2   The pressure readings at two ends of an industrial furnace are taken every 10 minutes and given in Table 7.1. It is expected the individual time series are not only autocorrelated but also cross-correlated. Therefore it is decided to fit a multivariate time series model to this data. The time series plots of the data are given in Figure 7.1. To identify the model we consider the sample ACF plots as well as the cross correlation of the time series given in Figure 7.2. These plots exhibit an exponential decay pattern, suggesting that an autoregressive model may be appropriate. It is further conjectured that a VAR(1) or VAR(2) model may provide a good fit. Another approach to model identification would be to fit ARIMA models to the individual time series and consider the cross correlation of the residuals. For that, we fit an AR(1) model to both time series. The cross-correlation plot of the residuals given in Figure 7.3 further suggests that the VAR(1) model may indeed provide an appropriate fit. Using the SAS ARIMA procedure given in Table 7.2, we fit a VAR(1) model. The SAS output in Table 7.3 confirms that the VAR(1) model provides an appropriate fit for the data. The time series plots of the residuals and the fitted values are given in Figures 7.3, 7.4, and 7.5.

TABLE 7.1 Pressure Readings at Both Ends of the Furnace

Index	Front	Back	Index	Front	Back	Index	Front	Back	Index	Front	Back	Index	Front	Back
	Pressure			Pressure			Pressure			Pressure			Pressure	
1	7.98	20.1	39	10.23	22.2	77	9.23	21.19	115	12.73	19.88	153	5.45	19.46
2	8.64	20.37	40	11.27	18.86	78	10.18	18.52	116	13.86	22.36	154	6.5	18.33
3	10.06	19.99	41	9.57	21.16	79	8.6	21.79	117	12.38	19.04	155	5.23	19.22
4	8.13	19.62	42	10.6	17.89	80	9.66	18.4	118	12.51	23.32	156	4.97	17.7
5	8.84	20.26	43	9.22	20.55	81	8.66	19.17	119	14.32	20.42	157	4.3	18.42
6	10.28	19.46	44	8.91	20.47	82	7.55	18.86	120	13.47	20.88	158	4.47	17.85
7	9.63	20.21	45	10.15	20	83	9.21	19.42	121	12.96	20.25	159	5.48	19.16
8	10.5	19.72	46	11.32	20.07	84	9.45	19.54	122	11.65	20.69	160	5.7	17.91
9	7.91	19.5	47	12.41	20.82	85	10.61	19.79	123	11.99	19.7	161	3.78	19.36
10	9.71	18.97	48	12.41	20.98	86	10.78	18.6	124	9.6	18.51	162	5.86	18.18
11	10.43	22.31	49	10.36	20.91	87	10.68	22.5	125	7.38	18.48	163	6.56	20.82
12	10.99	20.16	50	9.27	20.07	88	14.05	19.1	126	6.98	20.37	164	6.82	18.47
13	10.08	20.73	51	11.77	19.82	89	14.1	23.82	127	8.18	18.21	165	5.18	19.09
14	9.75	20.14	52	11.93	21.81	90	16.11	21.25	128	7.5	20.85	166	6.3	18.91
15	9.37	20.34	53	13.6	20.71	91	13.58	20.46	129	7.04	18.9	167	9.12	20.93
16	11.52	18.83	54	14.26	21.94	92	12.06	22.55	130	9.06	19.84	168	9.3	18.73
17	10.6	24.01	55	14.81	21.75	93	13.76	20.78	131	8.61	19.15	169	10.37	22.17
18	14.31	19.7	56	11.97	18.97	94	13.55	20.94	132	8.93	20.77	170	11.87	19.03
19	13.3	22.53	57	10.99	23.11	95	13.69	21.66	133	9.81	18.95	171	12.36	22.15
20	14.45	20.77	58	10.61	18.92	96	15.07	21.61	134	9.57	20.33	172	14.61	20.67
21	14.8	21.69	59	9.77	20.28	97	15.14	21.69	135	10.31	21.69	173	13.63	22.39

(continued)

TABLE 7.1 (*Continued*)

Index	Pressure Front	Pressure Back	Index	Pressure Front	Pressure Back	Index	Pressure Front	Pressure Back	Index	Pressure Front	Pressure Back	Index	Pressure Front	Pressure Back
22	15.09	20.87	60	11.5	21.18	98	14.01	21.85	136	11.88	18.66	174	13.12	19.75
23	12.96	21.42	61	10.52	19.29	99	12.69	20.87	137	12.36	22.35	175	10.07	18.94
24	11.28	18.95	62	12.58	19.9	100	11.6	20.93	138	12.18	19.34	176	10.14	21.47
25	10.78	22.61	63	12.33	19.87	101	12.15	20.57	139	12.94	22.76	177	11.02	19.79
26	10.42	19.93	64	9.77	19.43	102	12.99	21.17	140	14.25	19.6	178	11.37	21.94
27	9.79	21.88	65	10.71	21.32	103	11.89	19.53	141	12.86	23.74	179	10.98	18.73
28	11.66	18.3	66	10.01	17.85	104	10.85	21.14	142	12.14	18.06	180	10.04	21.41
29	10.81	20.76	67	9.48	21.55	105	11.81	20.09	143	10.06	20.11	181	11.3	19.2
30	9.79	17.66	68	9.39	19.04	106	9.46	18.48	144	10.17	19.56	182	10.59	23
31	10.02	23.09	69	9.05	19.04	107	9.25	20.33	145	7.56	19.27	183	11.69	17.47
32	11.09	17.86	70	9.06	21.39	108	9.26	19.82	146	7.77	18.59	184	10.73	21.59
33	10.28	20.9	71	9.87	17.66	109	8.55	20.07	147	9.03	21.85	185	13.64	21.62
34	9.24	19.5	72	7.84	21.61	110	8.86	19.81	148	10.8	19.21	186	12.92	20.23
35	10.32	22.6	73	7.78	18.05	111	10.32	20.64	149	9.41	19.42			
36	10.65	19	74	6.44	19.07	112	11.39	20.04	150	7.81	19.79			
37	8.51	20.39	75	7.67	19.92	113	11.78	21.52	151	7.99	18.81			
38	11.46	19.23	76	8.48	18.3	114	13.13	20.35	152	5.78	18.46			

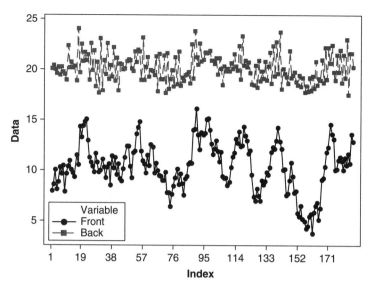

FIGURE 7.1 Time series plots of the pressure readings at both ends of the furnace.

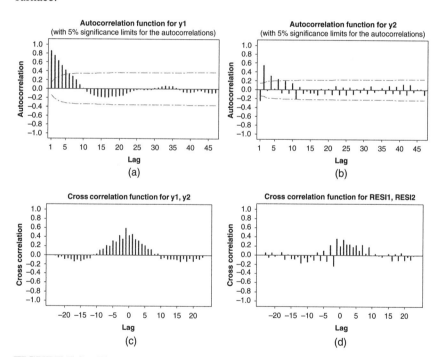

FIGURE 7.2 The sample ACF plot for: (a) the pressure readings at the front end of the furnace, y_1; (b) the pressure readings at the back end of the furnace, y_2; (c) the cross correlation between y_1 and y_2; and (d) the cross correlation between the residuals from the AR(1) model for front pressure and the residuals from the AR(1) model for back pressure.

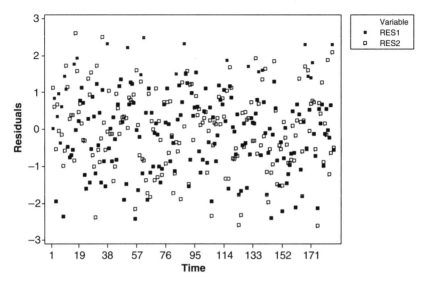

FIGURE 7.3   Time series plots of the residuals from the VAR(1) model.

7.2 STATE SPACE MODELS

In this section we give a brief introduction to an approach to forecasting based on the **state space model**. This is a very general approach that can include regression models and ARIMA models. It can also incorporate a Bayesian approach to forecasting and models with time-varying coefficients. State space models are based on the **Markov property**, which implies the independence of the future of a process from its past, given the present system state. In this type of system, the state of the process at the current time contains all of the past information that is required to predict future process behavior. We will let the system state at time t be

**TABLE 7.2   SAS Commands
to Fit a VAR(1) Model to the
Pressure Data**

```
proc varmax data=simul4;
        model y1 y2 / p=1 ;
        output out=residuals;
run;

proc print data=residuals;
run;
```

TABLE 7.3   SAS Output for the VAR(1) Model for the Pressure Data

```
                        The VARMAX Procedure
        Type of Model                        VAR(1)
        Estimation Method     Least Squares Estimation
                        Constant Estimates
                     Variable        Constant
                        y1             -6.76331
                        y2             27.23208
                     AR Coefficient Estimates
        Lag    Variable             y1               y2
         1       y1              0.73281          0.47405
                 y2              0.41047         -0.56040
                          Schematic
                       Representation of
                       Parameter Estimates
                          Variable/
                     Lag          C     AR1
                     y1           -     ++
                     y2           +     +-
                     + is > 2*std error,
                     - is < -2*std error,
                        . is between,
                           * is N/A
                    Model Parameter Estimates
                             Standard
Equation  Parameter  Estimate   Error    t Value  Pr > |t|   Variable
   y1     CONST1     -6.76331  1.18977    -5.68     0.0001       1
          AR1_1_1     0.73281  0.03772    19.43     0.0001    y1(t-1)
          AR1_1_2     0.47405  0.06463     7.33     0.0001    y2(t-1)
   y2     CONST2     27.23208  1.11083    24.51     0.0001       1
          AR1_2_1     0.41047  0.03522    11.66     0.0001    y1(t-1)
          AR1_2_2    -0.56040  0.06034    -9.29     0.0001    y2(t-1)
                    Covariances of Innovations
             Variable              y1               y2
             y1                1.25114          0.59716
             y2                0.59716          1.09064
                         Information
                          Criteria
                     AICC     0.041153
                     HQC      0.082413
                     AIC      0.040084
                     SBC      0.144528
                     FPEC     1.040904
```

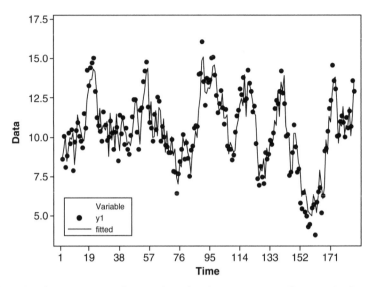

FIGURE 7.4   Actual and fitted values for the pressure readings at the front end of the furnace.

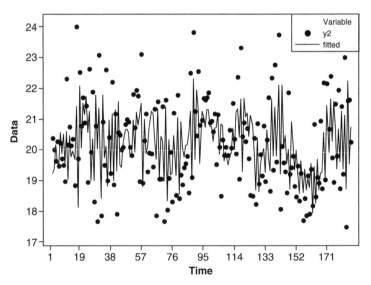

FIGURE 7.5   Actual and fitted values for the pressure readings at the back end of the furnace.

represented by the **state vector** $\mathbf{X}_t$. The elements of this vector are not necessarily observed. A state space model consists of two equations: an **observation** or **measurement equation** that describes how time series observations are produced from the state vector, and a **state** or **system equation** that describes how the state vector evolves through time. We will write these two equations as

$$y_t = \mathbf{h}'_t\mathbf{X}_t + \varepsilon_t \text{ (observation equation)} \tag{7.15}$$

and

$$\mathbf{X}_t = \mathbf{A}\mathbf{X}_{t-1} + \mathbf{G}\mathbf{a}_t \text{ (state equation)} \tag{7.16}$$

respectively. In the observation equation $\mathbf{h}_t$ is a known vector of constants and ε_t is the observation error. If the time series is multivariate then y_t and ε_t become vectors $\mathbf{y}_t$ and ε_t, and the vector $\mathbf{h}_t$ becomes a matrix $\mathbf{H}$. In the state equation $\mathbf{A}$ and $\mathbf{G}$ are known matrices and $\mathbf{a}_t$ is the process noise. Note that the state equation resembles a multivariate AR(1) model, except that it represents the state variables rather than an observed time series, and it has an extra matrix $\mathbf{G}$.

The state space model does not look like any of the time series models we have studied previously. However, we can put many of these models in the state space form. This is illustrated in the following two examples.

Example 7.3   Consider an AR(1) model, which we have previously written as

$$y_t = \phi y_{t-1} + \varepsilon_t$$

In this case we let $X_t = y_t$ and $\mathbf{a}_t = \varepsilon_t$ and write the state equation as

$$\mathbf{X}_t = \mathbf{A}\mathbf{X}_{t-1} + \mathbf{G}\mathbf{a}_t$$
$$[y_t] = [\phi][y_{t-1}] + [1]\varepsilon_t$$

and the observation equation is

$$y_t = \mathbf{h}'_t\mathbf{X}_t + \varepsilon_t$$
$$y_t = [1]\mathbf{X}_t + 0$$
$$y_t = \phi y_{t-1} + \varepsilon_t$$

In the AR(1) model the state vector consists of previous consecutive observations of the time series y_t.

Any ARIMA model can be written in the state space form. Refer to Brockwell and Davis (1991).

Example 7.4   Now let us consider a regression model with one predictor variable and AR(1) errors. We will write this model as

$$y_t = \beta_0 + \beta_1 p_t + \varepsilon_t$$
$$\varepsilon_t = \phi \varepsilon_{t-1} + a_t$$

where p_t is the predictor variable and ε_t is the AR(1) error term. To write this in state space form, define the state vector as

$$\mathbf{X}_t = \begin{bmatrix} \beta_0 \\ \beta_1 \\ p_t - \varepsilon_t \end{bmatrix}$$

The vector $\mathbf{h}_t$ and the matrix $\mathbf{A}$ are

$$\mathbf{h}_t = \begin{bmatrix} 1 \\ p_t \\ 1 \end{bmatrix}, \quad \mathbf{A} = \begin{bmatrix} 1 & 0 & 0 \\ 0 & 1 & 0 \\ 0 & 0 & \phi \end{bmatrix}$$

and the state space representation of this model becomes

$$y_t = [1, p_t, 1]\mathbf{X}_t + \varepsilon_t$$

$$\mathbf{X}_t = \begin{bmatrix} 1 & 0 & 0 \\ 0 & 1 & 0 \\ 0 & 0 & \phi \end{bmatrix} \begin{bmatrix} \beta_0 \\ \beta_1 \\ p_{t-1} - \varepsilon_{t-1} \end{bmatrix} + \begin{bmatrix} 0 \\ 0 \\ \phi \varepsilon_{t-1} \end{bmatrix}$$

Multiplying these equations out will produce the time series regression model with one predictor and AR(1) errors.

The state space formulation does not admit any new forecasting techniques. Consequently, it does not produce better forecasts than any of the other methods. The state space approach does admit a Bayesian formulation of the problem, in which the model parameters have a prior distribution that represents our degree of belief concerning the values of these coefficients. Then after some history of the process (observation) becomes available, this prior distribution is updated into a posterior distribution. Another formulation allows the coefficients in the regression model to vary through time.

The state space formulation does allow a common mathematical framework to be used for model development. It also permits relatively easy generalization of many models. This has some advantages for researchers. It also would allow common computer software to be employed for making forecasts from a variety of techniques. This could have some practical appeal to forecasters.

7.3   ARCH AND GARCH MODELS

In the standard regression and time series models we have covered so far, many diagnostic checks were based on the assumptions that we imposed on the errors: independent, identically distributed with zero mean, and constant variance. Our main concern has mostly been about the **independence** of the errors. The constant variance assumption is often taken as a given. In many practical cases and particularly in finance, it is fairly common to observe the violation of this assumption. Figure 7.6, for example, shows the S&P500 Index (weekly close) from 1995 to 1998. Most of the 1990s enjoyed a bull market up until toward the end when the dotcom bubble burst. The worrisome market resulted in high volatility (i.e., increasing variance). A linear trend model, an exponential smoother, or even an ARIMA model would have failed to capture this phenomenon, as all assume constant variance of the errors. This will in turn result in

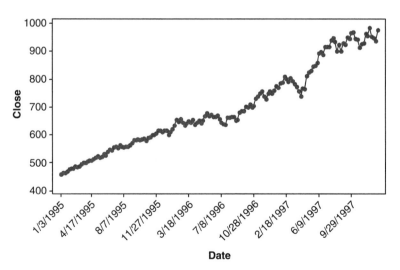

FIGURE 7.6   Time series plot of S&P500 Index weekly close from 1995 to 1998.

the underestimation of the standard errors calculated using OLS and will lead to erroneous conclusions. There are different ways of dealing with this situation. For example, if the changes in the variance at certain time intervals are known, weighted regression can be employed. However, it is often the case that these changes are unknown to the analyst. Moreover, it is usually of great value to the analyst to know why, when, and how these changes in the variance occur. Hence, if possible, modeling these changes (i.e., the variance) can be quite beneficial.

Consider, for example, the simple AR(p) model from Chapter 5 given as

$$y_t = \delta + \phi_1 y_{t-1} + \phi_2 y_{t-2} + \cdots + \phi_p y_{t-p} + e_t \qquad (7.17)$$

where e_t is the uncorrelated, zero mean noise with changing variance. Please note that we used e_t to distinguish it from our general white noise error ε_t. Since we let the variance of e_t change in time, one approach is to model e_t^2 as an AR(l) model as

$$e_t^2 = \xi_0 + \xi_1 e_{t-1}^2 + \xi_2 e_{t-2}^2 + \cdots + \xi_l e_{t-l}^2 + a_t \qquad (7.18)$$

where a_t is a white noise sequence with zero mean and constant variance σ_a^2. In this notation e_t is said to follow an **autoregressive conditional heteroskedastic** process of order l, ARCH(l).

To check for a need for an ARCH model, once the ARIMA or regression model is fitted, not only the standard residual analysis and diagnostics checks have to be performed but also some serial dependence checks for e_t^2 should be made.

To further generalize the ARCH model, we will consider the alternative representation originally proposed by Engle (1982). Let us assume that the error can be represented as

$$e_t = \sqrt{v_t} w_t \qquad (7.19)$$

where w_t is independent and identically distributed with mean 0 and variance 1, and

$$v_t = \zeta_0 + \zeta_1 e_{t-1}^2 + \zeta_2 e_{t-2}^2 + \cdots + \zeta e_{t-l}^2 \qquad (7.20)$$

Hence the conditional variance of e_t is

$$\begin{aligned}
\mathrm{Var}(e_t | e_{t-1}, \ldots) &= E\left(e_t^2 | e_{t-1}^2, \ldots\right) \\
&= v_t \\
&= \zeta_0 + \zeta_1 e_{t-1}^2 + \zeta_2 e_{t-2}^2 + \cdots + l e_{t-l}^2 \qquad (7.21)
\end{aligned}$$

We can also argue that the current conditional variance should also depend on the previous conditional variances as

$$v_t = \zeta_0 + \varsigma_1 v_{t-1} + \varsigma_2 v_{t-2} + \cdots + \varsigma_k v_{t-k} + \zeta_1 e_{t-1}^2 + \zeta_2 e_{t-2}^2 + \cdots + \zeta_l e_{t-l}^2$$

$$(7.22)$$

In this notation, the error term e_t is said to follow a **generalized autoregressive conditional heteroskedastic** process of orders k and l, GARCH(k, l), proposed by Bollerslev (1986). In Eq. (7.22) the model for conditional variance resembles an ARMA model. However, it should be noted that the model in Eq. (7.22) is not a proper ARMA model, as this would have required a white noise error term with a constant variance for the MA part. But none of the terms on the right-hand side of the equation possess this property. For further details, see Hamilton (1994), Bollerslev et al. (1992), and Degiannakis and Xekalaki (2004). Further extensions of ARCH models also exist for various specifications of v_t in Eq. (7.22); for example, Integrated GARCH (I-GARCH) by Engle and Bollerslev (1986), Exponential GARCH (E-GARCH) by Nelson (1991), Nonlinear GARCH by Glosten et al. (1993), and GARCH for multivariate data by Engle and Kroner (1993). But they are beyond the scope of this book. For a brief overview of these models, see Hamilton (1994).

Example 7.5   Consider the weekly closing values for the S&P500 Index from 1995 to 1998 given in Table 7.4. Figure 7.6 shows that the data exhibits nonstationarity. But before taking the first difference of the data, we decided to take the log transformation of the data first. As observed in Chapters 2 and 3, the log transformation is sometimes used for financial data when we are interested, for example, in the rate of change or percentage changes in the price of a stock. For further details, see Granger and Newbold (1986). The time series plot of the first differences of the log of the S&P500 Index is given in Figure 7.7, which shows that while the mean seems to be stable around 0, the changes in the variance are worrisome. The ACF and PACF plots of the first difference given in Figure 7.8 suggest that, except for some borderline significant ACF values at seemingly arbitrary lags, there is no autocorrelation left in the data. As in the case of the Dow Jones Index in Chapter 5, this suggests that the S&P500 Index follows a random walk process. However, the time series plot of the differences does not exhibit a constant variance behavior. For that, we consider the ACF and PACF of the squared differences given in Figure 7.9, which suggests that an AR(3) model can be used. Thus we fit the ARCH(3) model for the variance using the AUTOREG procedure in SAS given in Table 7.5. The SAS output in

TABLE 7.4 Weekly Closing Values for the S&P500 Index from 1995 to 1998

Date	Close	Date	Close	Date	Close	Date	Close	Date	Close
1/3/1995	460.68	8/14/1995	559.21	3/25/1996	645.5	11/4/1996	730.82	6/16/1997	898.7
1/9/1995	465.97	8/21/1995	560.1	4/1/1996	655.86	11/11/1996	737.62	6/23/1997	887.3
1/16/1995	464.78	8/28/1995	563.84	4/8/1996	636.71	11/18/1996	748.73	6/30/1997	916.92
1/23/1995	470.39	9/5/1995	572.68	4/15/1996	645.07	11/25/1996	757.02	7/7/1997	916.68
1/30/1995	478.65	9/11/1995	583.35	4/22/1996	653.46	12/2/1996	739.6	7/14/1997	915.3
2/6/1995	481.46	9/18/1995	581.73	4/29/1996	641.63	12/9/1996	728.64	7/21/1997	938.79
2/13/1995	481.97	9/25/1995	584.41	5/6/1996	652.09	12/16/1996	748.87	7/28/1997	947.14
2/21/1995	488.11	10/2/1995	582.49	5/13/1996	668.91	12/23/1996	756.79	8/4/1997	933.54
2/27/1995	485.42	10/9/1995	584.5	5/20/1996	678.51	12/30/1996	748.03	8/11/1997	900.81
3/6/1995	489.57	10/16/1995	587.46	5/28/1996	669.12	1/6/1997	759.5	8/18/1997	923.54
3/13/1995	495.52	10/23/1995	579.7	6/3/1996	673.31	1/13/1997	776.17	8/25/1997	899.47
3/20/1995	500.97	10/30/1995	590.57	6/10/1996	665.85	1/20/1997	770.52	9/2/1997	929.05
3/27/1995	500.71	11/6/1995	592.72	6/17/1996	666.84	1/27/1997	786.16	9/8/1997	923.91
4/3/1995	506.42	11/13/1995	600.07	6/24/1996	670.63	2/3/1997	789.56	9/15/1997	950.51
4/10/1995	509.23	11/20/1995	599.97	7/1/1996	657.44	2/10/1997	808.48	9/22/1997	945.22
4/17/1995	508.49	11/27/1995	606.98	7/8/1996	646.19	2/18/1997	801.77	9/29/1997	965.03
4/24/1995	514.71	12/4/1995	617.48	7/15/1996	638.73	2/24/1997	790.82	10/6/1997	966.98
5/1/1995	520.12	12/11/1995	616.34	7/22/1996	635.9	3/3/1997	804.97	10/13/1997	944.16
5/8/1995	525.55	12/18/1995	611.95	7/29/1996	662.49	3/10/1997	793.17	10/20/1997	941.64
5/15/1995	519.19	12/26/1995	615.93	8/5/1996	662.1	3/17/1997	784.1	10/27/1997	914.62
5/22/1995	523.65	1/2/1996	616.71	8/12/1996	665.21	3/24/1997	773.88	11/3/1997	927.51
5/30/1995	532.51	1/8/1996	601.81	8/19/1996	667.03	3/31/1997	757.9	11/10/1997	928.35
6/5/1995	527.94	1/15/1996	611.83	8/26/1996	651.99	4/7/1997	737.65	11/17/1997	963.09
6/12/1995	539.83	1/22/1996	621.62	9/3/1996	655.68	4/14/1997	766.34	11/24/1997	955.4
6/19/1995	549.71	1/29/1996	635.84	9/9/1996	680.54	4/21/1997	765.37	12/1/1997	983.79
6/26/1995	544.75	2/5/1996	656.37	9/16/1996	687.03	4/28/1997	812.97	12/8/1997	953.39
7/3/1995	556.37	2/12/1996	647.98	9/23/1996	686.19	5/5/1997	824.78	12/15/1997	946.78
7/10/1995	559.89	2/20/1996	659.08	9/30/1996	701.46	5/12/1997	829.75	12/22/1997	936.46
7/17/1995	553.62	2/26/1996	644.37	10/7/1996	700.66	5/19/1997	847.03	12/29/1997	975.04
7/24/1995	562.93	3/4/1996	633.5	10/14/1996	710.82	5/27/1997	848.28		
7/31/1995	558.94	3/11/1996	641.43	10/21/1996	700.92	6/2/1997	858.01		
8/7/1995	555.11	3/18/1996	650.62	10/28/1996	703.77	6/9/1997	893.27		

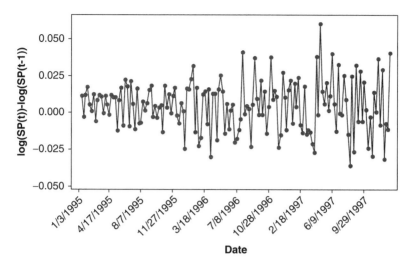

FIGURE 7.7   Time series plot of the first difference of the log transformation of the weekly close for S&P500 Index from 1995 to 1998.

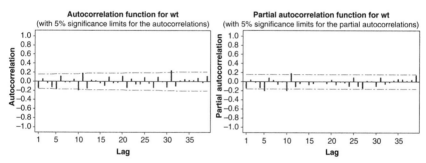

FIGURE 7.8   ACF and PACF plots of the first difference of the log transformation of the weekly close for the S&P500 Index from 1995 to 1998.

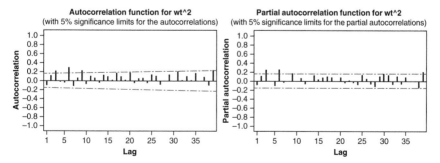

FIGURE 7.9   ACF and PACF plots of the square of the first difference of the log transformation of the weekly close for S&P500 Index from 1995 to 1998.

TABLE 7.5 SAS Commands to Fit the ARCH(3) Model[a]

```
proc autoreg data=sp5003;
    model dlogc = /   garch=( q=3);
run;
```

[a] dlogc is the first difference of the log trans-formed data.

TABLE 7.6 SAS output for the ARCH(3) model

GARCH Estimates

SSE	0.04463228	Observations		156
MSE	0.0002861	Uncond Var		0.00030888
Log Likelihood	422.53308	Total R-Square		.
SBC	-824.86674	AIC		-837.06616
Normality Test	1.6976	Pr > ChiSq		0.4279

The AUTOREG Procedure

Variable	DF	Standard Estimate	Approx Error	t Value	Pr > |t|
Intercept	1	0.004342	0.001254	3.46	0.0005
ARCH0	1	0.000132	0.0000385	3.42	0.0006
ARCH1	1	4.595E-10	3.849E-11	11.94	<.0001
ARCH2	1	0.2377	0.1485	1.60	0.1096
ARCH3	1	0.3361	0.1684	2.00	0.0460

Table 7.6 gives the coefficient estimates for the ARCH(3) model for the variance.

There are other studies on financial indices also yielding the ARCH(3) model for the variance, for example, Bodurtha and Mark (1991) and Attanasio (1991). In fact, successful implementations of reasonably simple, low-order ARCH/GARCH models have been reported in various research studies; see, for example, French et al. (1987).

7.4 DIRECT FORECASTING OF PERCENTILES

Throughout this book we have stressed the concept that a forecast should almost always be more than a point estimate of the value of some future event. A prediction interval should accompany most point forecasts, because the PI will give the decision maker some idea about the inherent variability of the forecast and the likely forecast error that could be experienced. Most of the forecasting techniques in this book have been presented showing how both point forecasts and PIs are obtained.

A PI can be thought of as an estimate of the percentiles of the distribution of the forecast variable. Typically, a PI is obtained by forecasting the mean and then adding appropriate multiples of the standard deviation of forecast error to the estimate of the mean. In this section we present and illustrate a different method that directly smoothes the percentiles of the distribution of the forecast variable.

Suppose that the forecast variable y_t has a probability distribution $f(y)$. We will assume that the variable y_t is either stationary or is changing slowly with time. Therefore a model for y_t that is correct at least locally is

$$y_t = \mu + \varepsilon_t$$

Let the observations on y_t be classified into a finite number of bins, where the bins are defined with limits

$$B_0 < B_1 < \cdots < B_n$$

The n bins should be defined so that they do not overlap; that is, each observation can be classified into one and only one bin. The bins do not have to be of equal width. In fact, there may be situations where bins may be defined with unequal width to obtain more information about specific percentiles that are of interest. Typically, $10 \le n \le 20$ bins are used.

Let p_k be the probability that the variable y_t falls in the bin defined by the limits B_{k-1} and B_k. That is,

$$p_k = P(B_{k-1} < y_t \le B_k), \quad k = 1, 2, \ldots, n$$

Assume that $\sum_{k=1}^{n} p_k = 1$. Also, note that $P(y_t \le B_k) = \sum_{j=1}^{k} p_j$. Now let us consider estimating the probabilities. Write the probabilities as an $n \times 1$ vector $\mathbf{p}$ defined as

$$\mathbf{p} = \begin{bmatrix} p_1 \\ p_2 \\ \vdots \\ p_n \end{bmatrix}$$

Let the estimate of the vector $\mathbf{p}$ at time period T be

$$\hat{\mathbf{p}}(T) = \begin{bmatrix} \hat{p}_1(T) \\ \hat{p}_2(T) \\ \vdots \\ \hat{p}_n(T) \end{bmatrix}$$

Note that if we wanted to estimate the percentile of the distribution of y_t corresponding to B_k at time period T we could do this by calculating $\sum_{j=1}^{k} \hat{p}_j(T)$.

We will use an exponential smoothing procedure to compute the estimated probabilities in the vector $\hat{\mathbf{p}}(T)$. Suppose that we are at the end of time period t and the current observation y_T is known. Let $u_k(T)$ be an indicator variable defined as follows:

$$u_k(T) = \begin{cases} 1 & \text{if } B_{k-1} < y_T \leq B_k \\ 0 & \text{otherwise} \end{cases}$$

So the indicator variable $u_k(T)$ is equal to unity if the observation y_T in period T falls in the kth bin. Note that $\sum_{t=1}^{T} u_k(t)$ is the total number of observations that fell in the kth bin during the time periods $t = 1, 2, \dots, T$. Define the $n \times 1$ observation vector $\mathbf{u}(T)$ as

$$\mathbf{u}(T) = \begin{bmatrix} u_1(T) \\ u_2(T) \\ \vdots \\ u_n(T) \end{bmatrix}$$

This vector will have $n - 1$ elements equal to zero and one element equal to unity. The exponential smoothing procedure for revising the probabilities $\hat{p}_k(T - 1)$ given that we have a new observation y_T is

$$\hat{p}_k(T) = \lambda u_k(T) + (1 - \lambda)\hat{p}_k(T - 1), \quad k = 1, 2, \dots, n \quad (7.23)$$

where $0 < \lambda < 1$ is the smoothing constant. In vector form, Eq. (7.23) for updating the probabilities is

$$\hat{\mathbf{p}}_k(T) = \lambda \mathbf{u}_k(T) + (1 - \lambda)\hat{\mathbf{p}}_k(T - 1)$$

This smoothing procedure produces an unbiased estimate of the probabilities p_k. Furthermore, because $u_k(T)$ is a Bernoulli random variable with parameter p_k, the variance of $\hat{p}_k(T)$ is

$$V[\hat{p}_k(T)] = \frac{\lambda}{2 - \lambda} p_k(1 - p_k)$$

Starting estimates or initial values of the probabilities at time $T = 0$ are required. These starting values $\hat{p}_k(0)$, $k = 1, 2, \dots, n$ could be subjective estimates or they could be obtained from an analysis of historical data.

The estimated probabilities can be used to obtain estimates of specific percentiles of the distribution of the variable y_t. One way to do this would be to estimate the cumulative probability distribution of y_t at time T as follows:

$$F(y) = \begin{cases} 0, & \text{if } y \leq B_0 \\ \sum_{j=1}^{k} \hat{p}_j(T), & \text{if } y = B_k, k = 1, 2, \ldots, n \\ 1, & \text{if } y \geq B_n \end{cases}$$

The values of the cumulative distribution could be plotted on a graph with $F(y)$ on the vertical axis and y on the horizontal axis and the points connected by a smooth curve. Then to obtain an estimate of any specific percentile, say, $\hat{F}_{1-\gamma} = 1 - \gamma$, all you would need to do is determine the value of y on the horizontal axis corresponding to the desired percentile $1 - \gamma$ on the vertical axis. For example, to find the 95th percentile of the distribution of y, find the value of y on the horizontal axis that corresponds to 0.95 on the vertical axis. This can also be done mathematically. If the desired percentile $1 - \gamma$ exactly matches one of the bin limits so that $F(B_k) = 1 - \gamma$, then the solution is easy and the desired percentile estimate is $\hat{F}_{1-\gamma} = B_k$. However, if the desired percentile $1 - \gamma$ is between two of the bin limits, say, $F(B_{k-1}) < 1 - \gamma < F(B_k)$, then interpolation is required. A linear interpolation formula is

$$\hat{F}_{1-\gamma} = \frac{[F(B_k) - (1-\gamma)]B_{k-1} + [(1-\gamma) - F(B_{k-1})]B_k}{F(B_k) - F(B_{k-1})} \quad (7.24)$$

In the extreme tails of the distribution or in cases where the bins are very wide, it may be desirable to use a nonlinear interpolation scheme.

Example 7.6   A financial institution is interested in forecasting the number of new automobile loan applications generated each week by a particular business channel. The information in Table 7.7 is known at the end of week $T - 1$. The next-to-last column of this table is the cumulative distribution of loan applications at the end of week $T - 1$. This cumulative distribution is shown in Figure 7.10.

Suppose that 74 loan applications are received during the current week, T. This number of loan applications fall into the eighth bin ($k = 7$ in Table 7.7). Therefore we can construct the observation vector $\mathbf{u}(T)$

as follows:

$$\mathbf{u}(T) = \begin{bmatrix} 0 \\ 0 \\ 0 \\ 0 \\ 0 \\ 0 \\ 0 \\ 1 \\ 0 \\ 0 \\ 0 \end{bmatrix}$$

TABLE 7.7   Distribution of New Automobile Loan Applications

k	B_{k-1}	B_k	$\hat{p}_k(T-1)$	$F(B_k)$, at the end of week $T-1$	$F(B_k)$, at the end of week $T-1$
0	0	10	0.02	0.02	0.018
1	10	20	0.04	0.06	0.054
2	20	30	0.05	0.11	0.099
3	30	40	0.05	0.16	0.144
4	40	50	0.09	0.25	0.225
5	50	60	0.10	0.35	0.315
6	60	70	0.13	0.48	0.432
7	70	80	0.16	0.64	0.676
8	80	90	0.20	0.84	0.856
9	90	100	0.10	0.94	0.946
10	100	110	0.06	1.00	1.000

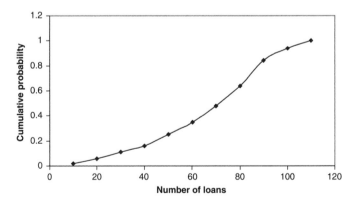

FIGURE 7.10   Cumulative distribution of the number of loan applications, week $T-1$.

Equation (7.23) is now used with $\lambda = 0.10$ to update the probabilities:

$$\hat{\mathbf{p}}_k(T) = \lambda \mathbf{u}_k(T) + (1 - \lambda)\hat{\mathbf{p}}_k(T - 1)$$

$$= 0.1 \begin{bmatrix} 0 \\ 0 \\ 0 \\ 0 \\ 0 \\ 0 \\ 0 \\ 1 \\ 0 \\ 0 \\ 0 \end{bmatrix} + 0.9 \begin{bmatrix} 0.02 \\ 0.04 \\ 0.05 \\ 0.05 \\ 0.09 \\ 0.10 \\ 0.13 \\ 0.16 \\ 0.20 \\ 0.10 \\ 0.06 \end{bmatrix} = \begin{bmatrix} 0.018 \\ 0.036 \\ 0.045 \\ 0.045 \\ 0.081 \\ 0.090 \\ 0.117 \\ 0.244 \\ 0.180 \\ 0.090 \\ 0.054 \end{bmatrix}$$

Therefore the new cumulative distribution of loan applications is found by summing the cumulative probabilities in $\hat{\mathbf{p}}_k(T - 1)$:

$$F(B_k) = \begin{bmatrix} 0.018 \\ 0.054 \\ 0.099 \\ 0.144 \\ 0.225 \\ 0.315 \\ 0.432 \\ 0.676 \\ 0.856 \\ 0.946 \\ 1.000 \end{bmatrix}$$

These cumulative probabilities are also listed in the last column of Table 7.7. The graph of the updated cumulative distribution is shown in Figure 7.11.

Now suppose that we want to find the number of loan applications that corresponds to a particular percentile of this distribution. If this percentile corresponds exactly to one of the cumulative probabilities, such as the 67.6 th percentile, the problem is easy. From the last column of Table 7.7 we would find that

$$\hat{F}_{0.676} = 80$$

That is, in about two of every three weeks we would expect to have 80 or fewer loan applications from this particular channel. However, if the desired

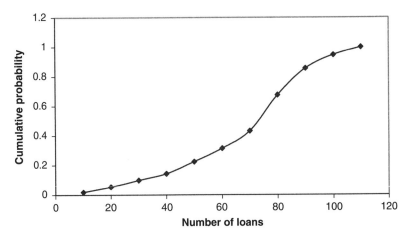

FIGURE 7.11   Cumulative distribution of the number of loan applications, week T.

percentile does not correspond to one of the cumulative probabilities in the last column of Table 7.7, we will need to interpolate using Eq. (7.24). For instance, if we want the 75th percentile, we would use Eq. (7.24) as follows:

$$\hat{F}_{0.75} = \frac{[F(B_k) - (0.75)]B_{k-1} + [(0.75) - [F(B_{k-1})]B_k}{F(B_k) - F(B_{k-1})}$$

$$= \frac{(0.856 - 0.75)90 + (0.75 - 0.676)80}{0.856 - 0.676}$$

$$= 85.89 \approx 86$$

Therefore, in about three of every four weeks, we would expect to have approximately 86 or fewer loan applications from this loan channel.

7.5   COMBINING FORECASTS TO IMPROVE PREDICTION PERFORMANCE

Readers have been sure to notice that any time series can be modeled and forecast using several methods. For example, it is not at all unusual to find that the time series y_t, $t = 1, 2, \ldots$, which contains a trend (say), can be forecast by both an exponential smoothing approach and an ARIMA model. In such situations, it seems inefficient to use one forecast and ignore all of the information in the other. It turns out that the forecasts from the two methods can be combined to produce a forecast that is superior in terms of forecast error than either forecast alone. For a review paper on the combination of forecasts, see Clemen (1989).

Bates and Granger (1969) suggested using a linear combination of the two forecasts. Let $\hat{y}_{1,T+\tau}(T)$ and $\hat{y}_{2,T+\tau}(T)$ be the forecasts from two different methods at the end of time period T for some future period $T + \tau$ for the time series y_t. The combined forecast is

$$\hat{y}^c_{T+\tau} = k_1\hat{y}_{1,T+\tau}(T) + k_2\hat{y}_{2,T+\tau}(T) \tag{7.25}$$

where k_1 and k_2 are weights. If these weights are chosen properly, the combined forecast $\hat{y}^c_{T+\tau}$ can have some nice properties. Let the two individual forecasts be unbiased. Then we should choose $k_2 = 1 - k_1$ so that the combined forecast will also be unbiased. Let $k = k_1$ so that the combined forecast is

$$\hat{y}^c_{T+\tau} = k\hat{y}_{1,T+\tau}(T) + (1 - k)\hat{y}_{2,T+\tau}(T) \tag{7.26}$$

Let the error from the combined forecast be $e^c_{T+\tau}(T) = y_{T+\tau} - \hat{y}^c_{T+\tau}(T)$. The variance of this forecast error is

$$\begin{aligned}
\text{Var}\left[e^c_{T+\tau}(T)\right] &= \text{Var}\left[y_{T+\tau} - \hat{y}^c_{T+\tau}(T)\right] \\
&= \text{Var}[ke_{1,T+\tau}(T) + (1 - k)e_{2,T+\tau}(T)] \\
&= k^2\sigma_1^2 + (1 - k)^2\sigma_2^2 + 2k(1 - k)\rho\sigma_1\sigma_2
\end{aligned}$$

where $e_{1,T+\tau}(T)$ and $e_{2s,T+\tau}(T)$ are the forecast errors in period $T + \tau$ for the two individual forecasting methods, σ_1^2 and σ_2^2 are the variances of the individual forecast errors for the two forecasting methods, and ρ is the correlation between the two individual forecast errors. A good combined forecast would be one that minimizes the variance of the combined forecast error. If we choose the weight k equal to

$$k^* = \frac{\sigma_2^2 - \rho\sigma_1\sigma_2}{\sigma_1^2 + \sigma_2^2 - 2\rho\sigma_1\sigma_2} \tag{7.27}$$

this will minimize the variance of the combined forecast error. By choosing this value for the weight, the minimum variance of the combined forecast error is equal to

$$\text{Min Var}\left[e^c_{T+\tau}(T)\right] = \frac{\sigma_1^2\sigma_2^2(1 - \rho^2)}{\sigma_1^2 + \sigma_2^2 - 2\rho\sigma_1\sigma_2} \tag{7.28}$$

and this minimum variance of the combined forecast error is less than or equal to the minimum of the variance of the forecast errors of the two individual forecasting methods. That is,

$$\text{Min Var} \left[e^c_{T+\tau}(T) \right] \leq \min \left(\sigma^2_1, \sigma^2_2 \right)$$

It turns out that the variance of the combined forecast error depends on the correlation coefficient. Let σ^2_1 be the smaller of the two individual forecast error variances. Then we have the following:

1. If $\rho = \sigma_1 / \sigma_2$, then Min Var $[e^c_{T+\tau}(T)] = \sigma^2_1$.
2. If $\rho = 0$, then Var $[e^c_{T+\tau}(T)] = \sigma^2_1 \sigma^2_2 / (\sigma^2_1 + \sigma^2_2)$.
3. If $\rho \to -1$, then Var $[e^c_{T+\tau}(T)] \to 0$.
4. If $\rho \to 1$, then Var $[e^c_{T+\tau}(T)] \to 0$ if $\sigma^2_1 \neq \sigma^2_2$.

Clearly, we would be happiest if the two forecasting methods have forecast errors with large negative correlation. The best possible case is when the two individual forecasting methods produce forecast errors that are perfectly negatively correlated. However, even if the two individual forecasting methods have forecast errors that are positively correlated, the combined forecast will still be superior to the individual forecasts, provided that $\rho \neq \sigma_1 / \sigma_2$.

Example 7.7   Suppose that two forecasting methods can be used for a time series, and that the two variances of the forecast errors are $\sigma^2_1 = 20$ and $\sigma^2_2 = 40$. If the correlation coefficient $\rho = -0.6$, then we can calculate the optimum value of the weight from Eq. (7.27) as follows:

$$
\begin{aligned}
k^* &= \frac{\sigma^2_2 - \rho \sigma_1 \sigma_2}{\sigma^2_1 + \sigma^2_2 - 2\rho \sigma_1 \sigma_2} \\
&= \frac{40 - (-0.6)\sqrt{(40)(20)}}{40 + 20 - 2(-0.6)\sqrt{(40)(20)}} \\
&= \frac{56.9706}{93.9411} \\
&= 0.6065
\end{aligned}
$$

So the combined forecasting equation is

$$\hat{y}^c_{T+\tau} = 0.6065 \hat{y}_{1,T+\tau}(T) + 0.3935 \hat{y}_{2,T+\tau}(T)$$

Forecasting method one, which has the smallest individual forecast error variance, receives about 1.5 times the weight of forecasting method two.

The variance of the forecast error for the combined forecast is computed from Eq. (7.28):

$$
\begin{aligned}
\text{Min Var}\left[e^c_{T+\tau}(T)\right] &= \frac{\sigma_1^2\sigma_2^2(1-\rho^2)}{\sigma_1^2+\sigma_2^2-2\rho\sigma_1\sigma_2} \\
&= \frac{(20)(40)[1-(-0.6)^2]}{20+40-2(-0.6)\sqrt{(20)(40)}} \\
&= \frac{512}{93.9411} \\
&= 5.45
\end{aligned}
$$

This is a considerable reduction in the variance of forecast error. If the correlation had been positive instead of negative, then

$$
\begin{aligned}
k^* &= \frac{\sigma_2^2-\rho\sigma_1\sigma_2}{\sigma_1^2+\sigma_2^2-2\rho\sigma_1\sigma_2} \\
&= \frac{40-(0.6)\sqrt{(40)(20)}}{40+20-2(0.6)\sqrt{(40)(20)}} \\
&= \frac{23.0294}{26.0589} \\
&= 0.8837
\end{aligned}
$$

Now forecasting method one, which has the smallest variance of forecast error, receives much more weight. The variance of the forecast error for the combined forecast is

$$
\begin{aligned}
\text{Min Var}\left[e^c_{T+\tau}(T)\right] &= \frac{\sigma_1^2\sigma_2^2(1-\rho^2)}{\sigma_1^2+\sigma_2^2-2\rho\sigma_1\sigma_2} \\
&= \frac{(20)(40)[1-(0.6)^2]}{20+40-2(0.6)\sqrt{(20)(40)}} \\
&= \frac{512}{26.0589} \\
&= 19.6478
\end{aligned}
$$

In this situation, there is very little improvement in the forecast error resulting from the combination of forecasts.

Newbold and Granger (1974) have extended this technique to the combination of n forecasts. Let $\hat{y}_{i,T+\tau}(T)$, $i=1,2,\ldots,n$ be the n unbiased forecasts at the end of period T for some future period $T + \tau$ for the time series y_t. The combined forecast is

$$\hat{y}^c_{T+\tau}(T) = \sum_{i=1}^{n} k_i \hat{y}_{T+\tau}(T)$$
$$= \mathbf{k}' \hat{\mathbf{y}}_{T+\tau}(T)$$

where $\mathbf{k}' = [k_1, k_2, \dots, k_n]$ is the vector of weights, and $\hat{\mathbf{y}}^c_{T+\tau}(T)$ is a vector of the individual forecasts. We require that all of the weights $0 \leq k_i \leq 1$ and $\sum_{i=1}^{n} k_i = 1$. The variance of the forecast error is minimized if the weights are chosen as

$$\mathbf{k} = \frac{\Sigma^{-1}_{T+\tau}(T)\mathbf{1}}{\mathbf{1}' \Sigma^{-1}_{T+\tau}(T)\mathbf{1}}$$

where $\Sigma_{T+\tau}(T)$ is the covariance matrix of the lead τ forecast errors given by

$$\Sigma_{T+\tau}(T) = E[\mathbf{e}_{T+\tau}(T)\mathbf{e}'_{T+\tau}(T)]$$

$\mathbf{1}' = [1, 1, \dots, 1]$ is a vector of ones, and $\mathbf{e}_{T+\tau}(T) = y_{T+\tau}\mathbf{1} - \hat{\mathbf{y}}_{T+\tau}(T)$ is a vector of the individual forecast errors.

The elements of the covariance matrix are usually unknown and will need to be estimated. This can be done by straightforward methods for estimating variances and covariances (refer to Chapter 2). It may also be desirable to regularly update the estimates of the covariance matrix so that these quantities reflect current forecasting performance. Newbold and Granger (1974) suggested several methods for doing this, and Montgomery, Johnson, and Gardiner (1990) investigate several of these methods. They report that a smoothing approach for updating the elements of the covariance matrix seems to work well in practice.

7.6 AGGREGATION AND DISAGGREGATION OF FORECASTS

Suppose that you wish to forecast the unemployment level of the state in which you live. One way to do this would be to forecast this quantity directly, using the time series of current and previous unemployment data, plus any other predictors that you think are relevant. Another way to do this would be to forecast unemployment at a substate level (say, by county and/or metropolitan area), and then to obtain the state level forecast by summing up the forecasts for each substate region. Thus individual forecasts of a collection of subseries are aggregated to form the forecasts of the quantity of interest. If the substate level forecasts are useful in their own right (as they probably are), this second approach seems very useful. However, there is another way to do this. First, forecast the state level unemployment and then disaggregate this forecast into the individual substate level regional forecasts. This disaggregation could be accomplished by multiplying the state level forecasts by a series of indices that reflect the proportion of total statewide unemployment that is accounted for by each region at the substate level. These indices also evolve with time, so it will be necessary to forecast them as well as part of a complete system.

This problem is sometimes referred to as the **top–down** versus **bottom–up** forecasting problem. In many, if not most, of these problems, we are interested in both forecasts for the top level quantity (the aggregate time series) and forecasts for the bottom level time series that are the components of the aggregate.

This leads to an obvious question: is it better to forecast the aggregate or top level quantity directly and then disaggregate, or to forecast the individual components directly and then aggregate them to form the forecast of the total? In other words, is it better to forecast from the top down or from the bottom up? The literature in statistical forecasting, business forecasting and econometrics, and time series analysis suggests that this question is far from settled at either the theoretical or empirical levels. Sometimes the aggregate quantity is more accurate than the disaggregated components, and sometimes the aggregate quantity is subject to less measurement error. It may be more complete and timely as well, and these aspects of the problem should encourage those who consider forecasting the aggregate quantity and then disaggregating. On the other hand, sometimes the bottom level data is easier to obtain and is at least thought to be more timely and accurate, and this would suggest that a bottom–up approach would be superior to the top–down approach.

In any specific practical application it will be difficult to argue on theoretical grounds what the correct approach should be. Therefore, in most situations, this question will have to be settled empirically by trying both approaches. With modern computer software for time series analysis and forecasting, this is not difficult. However, in conducting such a study it is a good idea to have an adequate amount of data for identifying and fitting the time series models for both the top level series and the bottom level series, and a reasonable amount of data for testing the two approaches. Obviously, **data splitting** should be done here, and the data used for model building should not be used for investigating forecasting model performance. Once an approach is determined, the forecasts should be carefully monitored over time to make sure that the dynamics of the problem have not changed, and that the top–down approach that was found to be optimal in testing (say) is now no longer as effective as the bottom–up approach. The methods for monitoring forecasting model performance presented in Chapter 2 are useful in this regard.

There are some results available about the effect of adding time series together. This is a special case of a more general problem called **temporal aggregation**, in which several time series may be combined as, for instance, when monthly data are aggregated to form quarterly data. For example, suppose that we have a top level time series Y_t that is the sum of two independent time series $y_{1,t}$ and $y_{2,t}$, and let us assume that both of the bottom level time series are moving average (MA) processes of orders q_1 and q_2, respectively. So, using the notation for ARIMA models introduced in Chapter 5, we have

$$Y_t = \theta_1(B)a_t + \theta_2(B)b_t$$

where a_t and b_t are independent white noise processes. Now let q be the maximum of q_1 and q_2. The autocorrelation function for the top level time series Y_t must be zero for all of the lags beyond q. This means that there is a representation of the top level time series as an MA process

$$Y_t = \theta_3(B)u_t$$

where u_t is white noise. This moving average process has the same order as the higher order bottom level time series.

Now consider the general ARIMA(p_1, d, q_1) model

$$\phi_1(B)\nabla^d y_t = \theta_1(B)a_t$$

and suppose that we are interested in the sum of two time series $z_t = y_t + w_t$. A practical situation where this occurs, in addition to the top–down versus bottom–up problem, is when the time series y_t we are interested in cannot be observed directly and w_t represents added noise due to measurement error. We want to know something about the nature of the sum of the two series, z_t. The sum can be written as

$$\phi_1(B)\nabla^d z_t = \theta_1(B)a_t + \phi_1(B)\nabla^d w_t$$

Assume that the time series w_t can be represented as a stationary ARMA$(p_2, 0, q_2)$ model

$$\phi_2(B)w_t = \theta_2(B)b_t$$

where b_t is white noise independent of a_t. Then the top level time series is

$$\phi_1(B)\phi_2(B)\nabla^d z_t = \phi_2(B)\theta_1(B)a_t + \phi_1(B)\theta_2(B)\nabla^d b_t$$

The term on the left-hand side is a polynomial of order $P = p_1 + p_2$, the first term on the right-hand side is a polynomial of order $q_1 + p_2$, and the second term on the right-hand side is a polynomial of order $p_1 + q_2 + d$. Let Q be the maximum of $q_1 + p_2$ and $p_1 + q_2 + d$. Then the top level time series is an ARIMA(P, d, Q) model, say,

$$\phi_3(B)\nabla^d z_t = \theta_3(B)u_t$$

where u_t is white noise.

Example 7.8 Suppose that we have a time series that is represented by an IMA(1, 1) model, and to this time series is added white noise. This could be a situation where measurements on a periodic sample of some characteristic in the output of a chemical process are made with a laboratory procedure, and the laboratory procedure has some built-in measurement error, represented by the white noise. Suppose that the underlying IMA(1, 1) model is

$$y_t = y_{t-1} - 0.6a_{t-1} + a_t$$

Let D_t be the first difference of the observed time series $z_t = y_t + w_t$, where w_t is white noise:

$$D_t = z_t - z_{t-1}$$
$$= (1 - \theta B)a_t + (1 - B)w_t$$

The autocovariances of the differenced series are

$$\gamma_0 = \sigma_a^2(1 + \theta^2) + 2\sigma_w^2$$
$$\gamma_1 = -\sigma_a^2\theta - \sigma_w^2$$
$$\gamma_j = 0, \quad j \geq 2$$

Because the autocovariances at and beyond lag 2 are zero, we know that the observed time series will be IMA(1, 1). In general, we could write this as

$$z_t = z_{t-1} - \theta^* u_{t-1} + u_t$$

where the parameter θ^* is unknown. However, we can find θ^* easily. The autocovariances of the first differences of this observed time series are

$$\gamma_0 = \sigma_u^2(1 + \theta^{*2})$$
$$\gamma_1 = -\sigma_u^2\theta^*$$
$$\gamma_j = 0, \quad j \geq 2$$

Now all we have to do is to equate the autocovariances for this observed series in terms of the parameter θ^* with the autocovariances of the time series D_t and we can solve for θ^* and σ_u^2. This gives the following:

$$\frac{\theta^*}{1 - \theta^*} = \frac{0.6}{1 - 0.6 + \sigma_w^2/\sigma_a^2}$$
$$\sigma_u^2 = \sigma_a^2 \frac{(0.6)^2}{\theta^{*2}}$$

Suppose that $\sigma_a^2 = 2$ and $\sigma_w^2 = 1$. Then it turns out that the solution is $\theta^* = 0.4$ and $\sigma_u^2 = 4.50$. Adding the measurement error from the laboratory procedure to the original sample property has inflated the variability of the observed value rather considerably over the original variability that was present in the sample property.

7.7 NEURAL NETWORKS AND FORECASTING

Neural networks, or more accurately **artificial neural networks**, have been motivated by the recognition that the human brain processes information in a way that is fundamentally different from the typical digital computer. The neuron is the basic structural element and information-processing module of the brain. A typical human brain has an enormous number of them (approximately 10 billion neurons in the cortex and 60 trillion synapses or connections between them) arranged in a highly complex, nonlinear, and parallel structure. Consequently, the human brain is a very efficient structure for information processing, learning, and reasoning.

An artificial neural network is a structure that is designed to solve certain types of problems by attempting to emulate the way the human brain would solve the problem. The general form of a neural network is a "black-box" type of model that is often used to model high-dimensional, nonlinear data. In the forecasting environment, neural networks are sometimes used to solve prediction problems instead of using a formal model building approach or development of the underlying knowledge of the system that would be required to develop an analytical forecasting procedure. If it was a successful approach that might be satisfactory. For example, a company might want to forecast demand for its products. If a neural network procedure can do this quickly and accurately, the company may have little interest in developing a specific analytical forecasting model to do it. Hill et al. (1994) is a basic reference on artificial neural networks and forecasting.

Multilayer feedforward artificial neural networks are multivariate statistical models used to relate p predictor variables $x_1, x_2, \ldots, x_p$ to one or more output or response variables y. In a forecasting application, the inputs could be explanatory variables such as would be used in a regression model, and they could be previous values of the outcome or response variable (lagged variables). The model has several **layers**, each consisting of either the original or some constructed variables. The most common structure involves three layers: the **inputs**, which are the original predictors; the **hidden layer**, comprised of a set of constructed variables; and the output layer, made up of the responses. Each variable in a layer is called a **node**. Figure 7.12 shows a typical three-layer artificial neural network for forecasting the output variable y in terms of several predictors.

A node takes as its input a transformed linear combination of the outputs from the nodes in the layer below it. Then it sends as an output a transformation of itself that becomes one of the inputs to one or more nodes on the next layer. The transformation functions are usually either sigmoidal (S shaped) or linear and are usually called **activation functions** or **transfer**

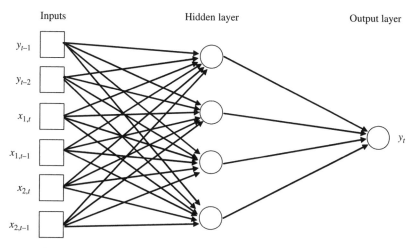

FIGURE 7.12   Artificial neural network with one hidden layer.

functions. Let each of the k hidden layer nodes a_u be a linear combination of the input variables:

$$a_u = \sum_{j=1}^{p} w_{1ju} x_j + \theta_u$$

where the w_{1ju} are unknown parameters that must be estimated (called weights) and θ_u is a parameter that plays the role of an intercept in linear regression (this parameter is sometimes called the bias node).

Each node is transformed by the activation function $g(\ )$. Much of the neural networks literature refers to these activation functions notationally as σ_u because of their S shape (the use of σ is an unfortunate choice of notation so far as statisticians are concerned). Let the output of node a_u be denoted by $z_u = g(a_u)$. Now we form a linear combination of these outputs, say, $b = \sum_{u=1}^{k} w_{uev} z_u$. Finally, the output response or the predicted value for y is a transformation of the b, say, $y = \tilde{g}(b)$, where $\tilde{g}(b)$ is the activation function for the response.

The response variable y is a transformed linear combination of the original predictors. For the hidden layer, the activation function is often chosen to be either a logistic function or a hyperbolic tangent function. The choice of activation function for the output layer often depends on the nature of the response variable. If the response is bounded or dichotomous, the output activation function is usually taken to be sigmoidal, while if it is continuous, an identity function is often used.

The neural network model is a very flexible form containing many parameters, and it is this feature that gives a neural network a nearly universal approximation property. That is, it will fit many historical data sets very well. However, the parameters in the underlying model must be estimated (parameter estimation is called "training" in the neural network literature), and there are a lot of them. The usual approach is to estimate the parameters by minimizing the overall residual sum of squares taken over all responses and all observations. This is a nonlinear least squares problem, and a variety of algorithms can be used to solve it. Often a procedure called **backpropagation** (which is a variation of steepest descent) is used, although derivative-based gradient methods have also been employed. As in any nonlinear estimation procedure, starting values for the parameters must be specified in order to use these algorithms. It is customary to standardize all the input variables, so small essentially random values are chosen for the starting values.

With so many parameters involved in a complex nonlinear function, there is considerable danger of **overfitting**. That is, a neural network will provide a nearly perfect fit to a set of historical or "training" data, but it will often predict new data very poorly. Overfitting is a familiar problem to statisticians trained in empirical model building. The neural network community has developed various methods for dealing with this problem, such as reducing the number of unknown parameters (this is called "optimal brain surgery"), stopping the parameter estimation process before complete convergence and using cross-validation to determine the number of iterations to use, and adding a penalty function to the residual sum of squares that increases as a function of the sum of the squares of the parameter estimates.

There are also many different strategies for choosing the number of layers and number of neurons and the form of the activation functions. This is usually referred to as choosing the **network architecture**. Cross-validation can be used to select the number of nodes in the hidden layer.

Artificial neural networks are an active area of research and application in many fields, particularly for the analysis of large, complex, highly nonlinear problems. The overfitting issue is frequently overlooked by many users and even the advocates of neural networks, and because many members of the neural network community do not have sound training in empirical model building, they often do not appreciate the difficulties overfitting may cause. Furthermore, many computer programs for implementing neural networks do not handle the overfitting problem particularly well. Studies of the ability of neural networks to predict future values of a time series that were not used in parameter estimation (fitting) have been, in many cases, disappointing. Our view is that neural networks are a complement to the familiar statistical tools of forecasting, and they might be one of the approaches you should consider, but they are not a replacement for them.

7.8 SPECTRAL ANALYSIS

This book has been focused on the analysis and modeling of time series in the **time domain**. This is a natural way to develop models, since time series all are observed as a function of time. However, there is another approach to describing and analyzing time series that uses a **frequency domain** approach. This approach consists of using the Fourier representation of a time series, given by

$$y_t = \sum_{k=1}^{T} a_k \sin(2\pi f_k t) + b_k \cos(2\pi f_k t) \tag{7.29}$$

where $f_k = k/T$. This model is named after J.B.J Fourier, an 18th century French mathematician, who claimed that any periodic function could be represented as a series of harmonically related sinusoids. Other contributors to Fourier analysis include Euler, D. Bernoulli, Laplace, Lagrange, and Dirichlet. The original work of Fourier was focused on phenomena in continuous time, such as vibrating strings, and there are still many such applications today from such diverse fields as geophysics, oceanography, atmospheric science, astronomy, and many disciplines of engineering. However, the key ideas carry over to discrete time series. We confine our discussion to **stationary** discrete time series.

Computing the constants a_k and b_k turns out to be quite simple:

$$a_k = \frac{2}{T} \sum_{k=1}^{T} \cos(2\pi f_k t) \tag{7.30}$$

and

$$b_k = \frac{2}{T} \sum_{k=1}^{T} \sin(2\pi f_k t) \tag{7.31}$$

These coefficients are combined to form a periodogram

$$I(f_k) = \frac{T}{2} \left(a_k^2 + b_k^2 \right) \tag{7.32}$$

The periodogram is then usually smoothed and scaled to produce the **spectrum** or a **spectral density function**. The spectral density function is just the Fourier transform of the autocorrelation function, so it conveys similar information as is found in the autocorrelations. However, sometimes the spectral density is easier to interpret than the autocorrelation function because adjacent sample autocorrelations can be highly correlated while estimates of the spectrum at adjacent frequencies are approximately independent. Generally, if the frequency k/T is important then $I(f_k)$ will be large, and if the frequency k/T is not important then $I(f_k)$ will be small.

It can be shown that

$$\sum_{k=1}^{T} I(f_k) = \sigma^2 \tag{7.33}$$

where σ^2 is the variance of the time series. Thus the spectrum decomposes the variance of the time series into individual components, each of which is associated with a particular frequency. So we can think of spectral analysis

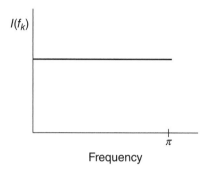

Frequency

FIGURE 7.13 The spectrum of a white noise process.

as an analysis of variance technique. It decomposes the variability in the time series by frequency.

It is helpful to know what the spectrum looks like for some simple ARMA models. If the time series white noise (uncorrelated observations with constant variance σ^2), it can be shown that the spectrum is a horizontal straight line as shown in Figure 7.13. This means that the contribution to the variance at all frequencies is equal. A logical use for the spectrum is to calculate it for the residuals from a time series model and see if the spectrum is reasonably flat.

Now consider the AR(1) process. The shape of the spectrum depends on the value of the AR(1) parameter ϕ. When $\phi > 0$, which results in a positively autocorrelated time series, the spectrum is dominated by low-frequency components. These low-frequency or long-period components result in a relative smooth time series. When $\phi < 0$, the time series is negatively autocorrelated, and the time series has a more ragged or volatile appearance. This produces a spectrum dominated by high-frequency or short-period components. Examples are shown in Figure 7.14.

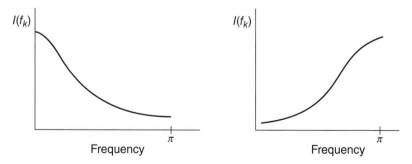

Frequency   Frequency

FIGURE 7.14 The spectrum of AR(1) processes.

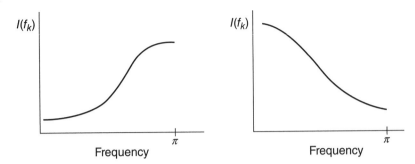

FIGURE 7.15   The spectrum of MA(1) processes.

The spectrum of the MA(1) process is shown in Figure 7.15. When the MA(1) parameter is positive, that is, when $\theta > 0$, the time series is negatively autocorrelated and has a more volatile appearance. Thus the spectrum is dominated by higher frequencies. When the MA(1) parameter is negative ($\theta > 0$), the time series is negatively autocorrelated and has a smoother appearance. This results in a spectrum that is dominated by low frequencies.

The spectrum of seasonal processes will exhibit peaks at the harmonically related seasonal frequencies. For example, consider the simple seasonal model with period 12, as might be used to represent monthly data with an annual seasonal cycle:

$$(1 - \phi^* B^{12})y_t = \varepsilon_t$$

If ϕ^* is positive, the spectrum will exhibit peaks at frequencies 0 and $2\pi kt/12$, $k = 1, 2, 3, 4, 5, 6$. Figure 7.16 shows the spectrum.

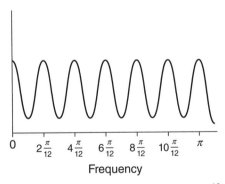

FIGURE 7.16   The spectrum of the seasonal $(1 - \phi^* B^{12})y_t = \varepsilon_t$ process.

Fisher's Kappa statistic tests the null hypothesis that the values in the series are drawn from a normal distribution with variance 1 against the alternative hypothesis that the series has some periodic component. The test statistic kappa (κ) is the ratio of the maximum value of the periodogram, $I(f_k)$, to its average value. The probability of observing a larger Kappa if the null hypothesis is true is given by

$$P(\kappa > k) = 1 - \sum_{j=0}^{q} (-1)^j \binom{q}{j} \left[\max\left(1 - \frac{jk}{q}, 0\right) \right]^{q-1}$$

where k is the observed value of the kappa statistic, $q = T/2$ if T is even and $q = (T - 1)/2$ if T is odd. The null hypothesis is rejected if the computed probability is less that the desired significance level. There is also a Kolmogorov–Smirnov test due to Bartlett that compares the normalized cumulative periodogram to the cumulative distribution function of the uniform distribution on the interval (0, 1). The test statistic equals the maximum absolute difference of the cumulative periodogram and the uniform CDF. If this quantity exceeds $a/\sqrt{q}$, then we should reject the hypothesis that the series comes from a normal distribution. The values $a = 1.36$ and $a = 1.63$ correspond to significance levels of 0.05 and 0.01, respectively.

In general, we have found it difficult to determine the exact form of an ARIMA model purely from examination of the spectrum. The autocorrelation and partial autocorrelation functions are almost always more useful and easier to interpret. However, the spectrum is a complimentary tool and should always be considered as a useful supplement to the ACF and PACF. For further reading on spectral analysis and its many applications, see Jenkins and Watts (1969), Percival and Walden (1992), and Priestley (1991).

Example 7.9   JMP can be used to compute and display the spectrum for time series. We will illustrate the JMP output using the monthly US beverage product shipments. These data are shown originally in Figure 1.5 and are in Appendix Table B. These data were also analyzed in Chapter 2 to illustrate decomposition techniques. Figure 7.17 presents the JMP output, including a time series plot, the sample ACF, PACF, and variogram, and the spectral density function. Notice that there is a prominent peak in the spectral density at frequency 0.0833 that corresponds to a seasonal period of 12 months. The JMP output also provides the Fisher kappa statistic and the P-value indicates that there is at least one periodic component. The Bartlett Kolmogorov–Smirnov test statistic is also significant at the 0.01 level indicating that the data do not come from a normal distribution.

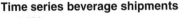

Time series beverage shipments

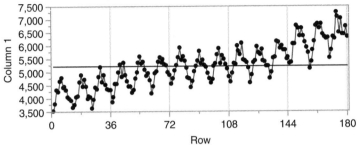

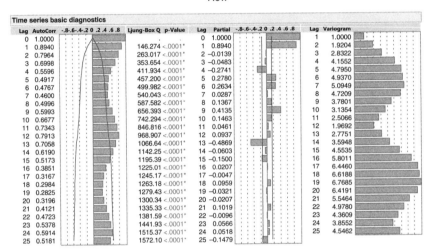

Time series basic diagnostics

Lag	AutoCorr	-.8-.6-.4-.2 0 .2 .4 .6 .8	Ljung-Box Q	p-Value	Lag	Partial	-.8-.6-.4-.2 0 .2 .4 .6 .8	Lag	Variogram
0	1.0000		.		0	1.0000		1	1.0000
1	0.8940		146.274	<.0001*	1	0.8940		2	1.9204
2	0.7964		263.017	<.0001*	2	-0.0139		3	2.8322
3	0.6998		353.654	<.0001*	3	-0.0483		4	4.1552
4	0.5596		411.934	<.0001*	4	-0.2741		5	4.7950
5	0.4917		457.200	<.0001*	5	0.2780		6	4.9370
6	0.4767		499.982	<.0001*	6	0.2634		7	5.0949
7	0.4600		540.043	<.0001*	7	0.0287		8	4.7209
8	0.4996		587.582	<.0001*	8	0.1367		9	3.7801
9	0.5993		656.393	<.0001*	9	0.4135		10	3.1354
10	0.6677		742.294	<.0001*	10	0.1463		11	2.5066
11	0.7343		846.816	<.0001*	11	0.0461		12	1.9692
12	0.7913		968.907	<.0001*	12	0.0937		13	2.7751
13	0.7058		1066.64	<.0001*	13	-0.4869		14	3.5948
14	0.6190		1142.25	<.0001*	14	-0.0603		15	4.5535
15	0.5173		1195.39	<.0001*	15	-0.1500		16	5.8011
16	0.3851		1225.01	<.0001*	16	0.0207		17	6.4460
17	0.3167		1245.17	<.0001*	17	-0.0047		18	6.6188
18	0.2984		1263.18	<.0001*	18	0.0959		19	6.7685
19	0.2825		1279.43	<.0001*	19	-0.0321		20	6.4191
20	0.3196		1300.34	<.0001*	20	-0.0207		21	5.5464
21	0.4121		1335.33	<.0001*	21	0.1019		22	4.9780
22	0.4723		1381.59	<.0001*	22	-0.0096		23	4.3609
23	0.5378		1441.93	<.0001*	23	0.0566		24	3.8552
24	0.5914		1515.37	<.0001*	24	0.0518		25	5.4462
25	0.5181		1572.10	<.0001*	25	-0.1479			

Spectral density

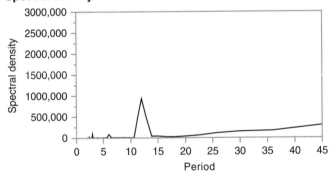

FIGURE 7.17   JMP output showing the spectrum, ACF, PACF, and variogram for the beverage shipment data.

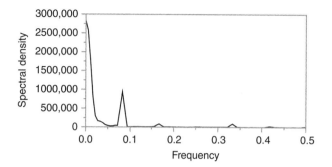

White noise test

Fisher's Kappa	28.784553
Prob > Kappa	1.041e-13
Bartlett's Kolmogorov-Smirnov	0.7545515

FIGURE 7.17 (*Continued*)

7.9 BAYESIAN METHODS IN FORECASTING

In many forecasting problems there is little or no historical information available at the time initial forecasts are required. Consequently, the initial forecasts must be based on subjective considerations. As information becomes available, this subjective information can be modified in light of actual data. An example of this is forecasting demand for seasonal clothing, which, because of style obsolescence, has a relatively short life. In this industry a common practice is to, at the start of the season, make a forecast of total sales for a product during the season and then as the season progresses the original forecast can be modified taking into account actual sales.

Bayesian methods can be useful in problems of this general type. The original subjective estimates of the forecast are translated into subjective estimates of the forecasting model parameters. Then Bayesian methods are used to update these parameter estimates when information in the form of time series data becomes available. This section gives a brief overview of the Bayesian approach to parameter estimation and demonstrates the methodology with a simple time series model.

The method of parameter estimation makes use of the Bayes' theorem. Let y be a random variable with probability density function that is characterized by an unknown parameter θ. We write this density as $f(y|\theta)$ to show that the distribution depends on θ. Assume that θ is a random variable with probability distribution $h_0(\theta)$ which is called the **prior distribution** for θ.

The prior distribution summarizes the subjective information that we have about θ, and the treatment of θ as a random variable is the major difference between Bayesian and classical methods of estimation. If we are relatively confident about the value of θ we should choose prior distribution with a small variance and if we are relatively uncertain about the value of θ we should choose prior distribution with a large variance.

In a time series or forecasting scenario, the random variable y is a sequence of observations, say $y_1, y_2, \ldots, y_T$. The new estimate of the parameter θ will be in the form of a revised distribution, $h_1(\theta|y)$, called the **posterior distribution** for θ. Using Bayes' theorem the posterior distribution is

$$h_1(\theta|y) = \frac{h_0(\theta)f(y|\theta)}{\int_\theta h_0(\theta)f(y|\theta)d\theta} = \frac{h_0(\theta)f(y|\theta)}{g(y)} \tag{7.34}$$

where $f(y|\theta)$ is usually called the likelihood of y, given the value of θ, and $g(y)$ is the unconditional distribution of y averaged over all θ. If the parameter θ is a discrete random variable then the integral in Eq. (7.34) should be replaced by a summation sign. This equation basically blends the observed information with the prior information to obtain a revised description of the uncertainty in the value of θ in the form of a posterior probability distribution. The **Bayes' estimator of** θ, which we will denote by θ^*, is defined as the expected value of the posterior distribution:

$$\theta^* = \int_\theta \theta h_1(\theta|y)d\theta \tag{7.35}$$

Typically we would use θ^* as the estimate of θ in the forecasting model. In some cases, it turns out that θ^* is optimal in the sense of minimizing the variance of forecast error.

We will illustrate the procedure with a relatively simple example. Suppose that y is normally distributed with mean μ and variance σ_y^2; that is,

$$f(y|\mu) = N\left(\mu, \sigma_y^2\right) = \left(2\pi\sigma_y^2\right)^{-1/2} \exp\left[-\frac{1}{2}\left(\frac{y-\mu}{\sigma_y}\right)^2\right]$$

We will assume that σ_y^2 is known. The prior distribution for μ is assumed to be normal with mean μ' and variance $\sigma_\mu^{2'}$:

$$h_0(\mu|y) = N\left(\mu', \sigma_\mu^{2'}\right) = \left(2\pi\sigma_\mu^{2'}\right)^{-1/2} \exp\left[-\frac{1}{2}\left(\frac{\mu-\mu'}{\sigma_\mu'}\right)^2\right]$$

The posterior distribution of μ given the observation y is

$$h_1(\theta|y) = \frac{2\pi\left(\sigma_\mu^{2'}\sigma_y^2\right)^{-1/2}\exp\left[\frac{1}{2}(\mu-\mu')/\sigma_\mu^{2'}+(y-\mu)/\sigma_y^2\right]}{\int_{-\infty}^{\infty}2\pi\left(\sigma_\mu^{2'}\sigma_y^2\right)^{-1/2}\exp\left[\frac{1}{2}(\mu-\mu')/\sigma_\mu^{2'}+(y-\mu)/\sigma_y^2\right]d\mu}$$

$$= \left(2\pi\frac{\sigma_\mu^{2'}\sigma_y^2}{\sigma_\mu^{2'}+\sigma_y^2}\right)^{-1/2}\exp\left[-\frac{1}{2}\frac{\left[(\mu-\left(y\sigma_\mu^{2'}+\mu'\sigma_y^2\right)/\left(\sigma_\mu^{2'}+\sigma_y^2\right)\right]}{\sigma_\mu^{2'}\sigma_y^2/\left(\sigma_\mu^{2'}+\sigma_y^2\right)}\right]$$

which is a normal distribution with mean and variance

$$\mu'' = E(\mu|y) = \frac{y\sigma_\mu^{2'}+\mu'\sigma_y^2}{\sigma_\mu^{2'}+\sigma_y^2}$$

and

$$\sigma_\mu^{2''} = V(\mu|y) = \frac{\sigma_\mu^{2'}\sigma_y^2}{\sigma_\mu^{2'}+\sigma_y^2}$$

respectively. Refer to Winkler (2003), Raiffa and Schlaifer (1961), Berger (1985), and West and Harrison (1997) for more details of Bayesian statistical inference and decision making and additional examples.

Now let us consider a simple time series model, the constant process, defined in Eq. (4.1) as

$$y_t = \mu + \varepsilon_t$$

where μ is the unknown mean and the random component is ε_t, which we will assume to have a normal distribution with mean zero and known variance σ_y^2 Consequently, we are assuming that the observation in any period t has a normal distribution, say

$$f(y_t|\mu) = N\left(\mu, \sigma_y^2\right)$$

Since the variance σ_y^2 is known, the problem is to estimate μ.

Suppose that at the start of the forecasting process, time $t = 0$, we estimate the mean demand rate to be μ' and the variance $\sigma_\mu^{2'}$ captures the uncertainty in this estimate. So the prior distribution for μ is the normal prior

$$h_0(\mu) = N\left(\mu', \sigma_\mu^{2'}\right)$$

After one period, the observation y_1 is known. The estimate μ' and the variance $\sigma_\mu^{2'}$ can now be updated using the results obtained above for a normal sampling process and a normal prior:

$$h_1(\mu|y_1) = N[\mu''(1), \sigma^{2''}(1)]$$

where

$$\mu''(1) = E(\mu|y_1) = \frac{y\sigma_\mu^{2'} + \mu'\sigma_y^2}{\sigma_\mu^{2'} + \sigma_y^2}$$

and

$$\sigma_\mu^{2''}(1) = V(\mu|y_1) = \frac{\sigma_\mu^{2'}\sigma_y^2}{\sigma_\mu^{2'} + \sigma_y^2}$$

At the end of period 2, when the next observation y_2 becomes available, the Bayesian updating process transforms $h_1(\mu|y_1)$ into $h_2(\mu|y_1,y_2)$ in the following way:

$$h_2(\mu|y_1,y_2) = \frac{h_1(\mu|y_1)f(y_2|\mu)}{\int_\mu h_1(\mu|y_1)f(y_2|\mu)d\mu}$$

Here the old posterior h_1 is now used as a prior and combined with the likelihood of y_2 to obtain the new posterior distribution of μ at the end of period 2. Using our previous results, we now have

$$h_2(\mu|y_1,y_2) = N[\mu''(2), \sigma^{2''}(2)]$$

and

$$\mu''(2) = E(\mu|y_1,y_2) = \frac{\bar{y}\sigma_\mu^{2'} + \mu'\left(\sigma_y^2/2\right)}{\sigma_\mu^{2'} + \left(\sigma_y^2/2\right)}$$

$$\sigma_\mu^{2''}(2) = V(\mu|y_1,y_2) = \frac{\sigma_\mu^{2'}\sigma_y^2}{2\sigma_\mu^{2'} + \sigma_y^2}$$

where $\bar{y} = (y_1 + y_2)/2$. It is easy to show that $h_2(\mu|y_1,y_2) = h_2(\mu|\bar{y})$; that is, the same posterior distribution is obtained using the sample average $\bar{y}$ as from using y_1 and y_2 sequentially because the sample average is a sufficient statistic for estimating the mean μ.

In general, we can show that after observing y_T, we can calculate the posterior distribution as

$$h_2(\mu|y_1, y_2, \ldots, y_T) = N[\mu''(T), \sigma^{2''}(T)]$$

where

$$\mu''(T) = \frac{\bar{y}\sigma_\mu^{2'} + \mu'\left(\sigma_y^2/T\right)}{\sigma_\mu^{2'} + \left(\sigma_y^2/T\right)}$$

$$\sigma_\mu^{2''}(T) = \frac{\sigma_\mu^{2'}\sigma_y^2}{T\sigma_\mu^{2'} + \sigma_y^2}$$

where $\bar{y} = (y_1 + y_2 + \cdots + y_T)/T$. The Bayes estimator of μ after T periods is $\mu^*(T) = \mu''(T)$. We can write this as

$$\mu^*(T) = \frac{T}{r + T}\bar{y} + \frac{r}{r + T}\mu' \tag{7.36}$$

where $r = \sigma_y^2/\sigma_\mu^{2'}$. Consequently, the Bayes estimator of μ is just a weighted average of the sample mean and the initial subjective estimate μ'. The Bayes estimator can be written in a recursive form as

$$\mu^*(T) = \lambda(T)y_T + [1 - \lambda(T)]\mu^*(T - 1) \tag{7.37}$$

where

$$\lambda(T) = \frac{1}{r + T} = \frac{\sigma_\mu^{2'}}{T\sigma_\mu^{2'} + \sigma_y^2}$$

Equation (7.36) shows that the estimate of the mean in period T is updated at each period by a form that is similar to first-order exponential smoothing. However, notice that the smoothing factor $\lambda(T)$ is a function of T, and it becomes smaller as T increases. Furthermore, since $\sigma_\mu^{2''}(T) = \lambda(T)\sigma_y^2$, the uncertainty in the estimate of the mean decreases to zero as time T becomes large. Also, the weight given to the prior estimate of the mean decreases as T becomes large. Eventually, as more data becomes available, a permanent forecasting procedure could be adopted, perhaps involving exponential smoothing. This estimator is optimal in the sense that it minimizes the variance of forecast error even if the process is not normally distributed.

We assumed that the variance of the demand process was known, or at least a reasonable estimate of it was available. Uncertainty in the value of this parameter could be handled by also treating it as a random variable. Then the prior distribution would be a joint distribution that would reflect the uncertainty in both the mean and the variance. The Bayesian updating process in this case is considerably more complicated than in the known-variance case. Details of the procedure and some useful advice on choosing a prior are in Raiffa and Schlaifer (1961).

Once the prior has been determined, the forecasting process is relatively straightforward. For a constant process, the forecasting equation is

$$\hat{y}_{T+\tau}(T) = \hat{\mu}(T) = \mu^*(T)$$

using the Bayes estimate as the current estimate of the mean. Our uncertainty in the estimate of the mean is just the posterior variance. So the variance of the forecast is

$$V[\hat{y}_{T+\tau}(T)] = \sigma_\mu^{2''}(T)$$

and the variance of forecast error is

$$V[y_{T+\tau} - \hat{y}_{T+\tau}(T)] = V[e_\tau(T+\tau)] = \sigma_y^2 + \sigma_\mu^{2''}(T) \qquad (7.38)$$

The variance of forecast error is independent of the lead time in the Bayesian case for a constant process. If we assume that y and μ are normally distributed, then we can use Eq. (3.38) to find a $100(1-\alpha)\%$ prediction interval on the forecast $V[\hat{y}_{T+\tau}(T)]$ as follows:

$$\mu^*(T) \pm Z_{\alpha/2} \sqrt{\sigma_y^2 + \sigma_\mu^{2''}(T)} \qquad (7.39)$$

where $Z_{\alpha/2}$ is the usual $\alpha/2$ percentage point of the standard normal distribution.

Example 7.10   Suppose that we are forecasting weekly demand for a new product. We think that demand is normally distributed, and that at least in the short run that a constant model is appropriate. There is no useful historical information, but a reasonable prior distribution for μ is $N(100, 25)$ and σ_y^2 is estimated to be 150. At time period $T = 0$ the forecast for period 1 is

$$\hat{y}_1(0) = 100$$

The variance of forecast error is $150 + 25 = 175$, so a 96% prediction interval for y_1 is

$$100 \pm 1.96\sqrt{175} \quad \text{or} \quad [74.1, 125.9]$$

Suppose the actual demand experienced in period 1 is $y_1 = 86$. We can use Eq. (7.37) to update the estimate of the mean. First, calculate $r = \sigma_y^2/\sigma_\mu^{2'} = 150/25 = 6$ and $\lambda(1) = 1/(6+1) = 0.143$, then

$$\mu^*(1) = \mu''(1) = \lambda(1)y_1 + [1 - \lambda(1)]\mu^*(0)$$
$$= 0.143(86) + (1 - 0.143)100$$
$$= 98.0$$

and

$$\sigma_\mu^{2''}(1) = \lambda(1)\sigma_y^2$$
$$= 0.143(150)$$
$$= 21.4$$

The forecast for period 2 is now

$$\hat{y}_2(1) = 98.0$$

The corresponding 95% prediction interval for y_2 is

$$98.0 \pm 1.96\sqrt{150 + 21.4} \quad \text{or} \quad [72.3, 123.7]$$

In time period the actual demand experienced is 94. Now $\lambda(2) = 1/(6 + 2) = 0.125$, and

$$\mu^*(2) = \mu''(2) = \lambda(2)y_2 + [1 - \lambda(2)]\mu^*(1)$$
$$= 0.125(94) + (1 - 0.125)98.0$$
$$= 97.5$$

So the forecast for period 3 is

$$\hat{y}_3(3) = 97.5$$

and the updated variance estimate is

$$\sigma_\mu^{2''}(2) = \lambda(2)\sigma_y^2$$
$$= 0.125(150)$$
$$= 18.8$$

Therefore the 96% prediction interval for y_3 is

$$97.5 \pm 1.96\sqrt{150 + 18.8} \quad \text{or} \quad [72.0, 123.0]$$

This procedure would be continued until it seems appropriate to change to a more permanent forecasting procedure. For example, a change to first-order exponential smoothing could be made when $\lambda(T)$ drop to a target level, say $0.05 < \lambda(T) < 0.1$. Then after sufficient data has been observed, an appropriate time series model could be fit to the data.

7.10 SOME COMMENTS ON PRACTICAL IMPLEMENTATION AND USE OF STATISTICAL FORECASTING PROCEDURES

Over the last 35 years there has been considerable information accumulated about forecasting techniques and how these methods are applied in a wide variety of settings. Despite the development of excellent analytical techniques, many business organizations still rely on judgment forecasts by their marketing, sales, and managerial/executive teams. The empirical evidence regarding judgment forecasts is that they are not as successful as statistically based forecasts. There are some fields, such as financial investments, where there is considerable strong evidence that this is so. There are a number of reasons why we would expect judgment forecasts to be inferior to statistical methods.

Inconsistency, or changing one's mind for no compelling or obvious reason, is a significant source of judgment forecast errors. Formalizing the forecasting process through the use of analytical methods is one approach to eliminating inconsistency as a source of error. Formal decision rules that predict the variables of interest using relatively few inputs invariably predict better than humans, because humans are inconsistent over time in their choice of input factors to consider, and how to weight them.

Letting more **recent events** dominate one's thinking, instead of weighting current and previous experience more evenly, is another source of judgment forecast errors. If these recent events are essentially random in nature, they can have undue impact on current forecasts. A good forecasting system will certainly monitor and evaluate recent events and experiences, but will only incorporate them into the forecasts if there is sufficient evidence to indicate that they represent real effects.

Mistaking **correlation** for **causality** can also be a problem. This is the belief that two (or more) variables are related in a causal manner and taking action on this, when the variables exhibit only a correlation between them. It is not difficult to find correlative relationships; any two variables that are monotonically related will exhibit strong correlation. So company sales may appear to be related to some factor that over a short time period is moving synchronously with sales, but relying on this as a causal relationship will lead to problems. The statistical significance of patterns and relationships does not necessarily imply a cause-and-effect relationship.

Judgment forecasts are often dominated by **optimistic thinking**. Most humans are naturally optimistic. An executive wants sales for the product line to increase because his/her bonus may depend on the results. A product manager wants his/her product to be successful. Sometimes bonus payouts are made for exceeding sales goals, and this can lead to unrealistically low forecasts, which in turn are used to set the goals. However, unrealistic forecasts, whether too high or too low, always result in problems downstream in the organization where forecast errors have meaningful impact on efficiency, effectiveness, and bottom-line results.

Humans are notorious for **underestimating variability**. Judgment forecasts rarely incorporate uncertainty in any formal way and, as a result, often underestimate its magnitude and impact. A judgment forecaster often completely fails to express any uncertainty in his/her forecast. Because all forecasts are wrong, one must have some understanding of the magnitude of forecast errors. Furthermore, planning for appropriate actions in the face of likely forecast error should be part of the decision-making process that is driven by the forecast. Statistical forecasting methods can be accompanied by prediction intervals. In our view, every forecast should be accompanied by a PI that adequately expresses for the decision maker how much uncertainty is associated with the point forecast.

In general, both the users of forecasts (decision makers) and the preparers (forecasters) have reasonably good awareness of many of the basic analytical forecasting techniques, such as exponential smoothing and regression-based methods. They are less familiar with time series models such as the ARIMA model, transfer function models, and other more sophisticated methods. Decision makers are often unsatisfied with subjective and judgment methods and want better forecasts. They often feel that analytical methods can be helpful in this regard.

This leads to a discussion of **expectations**. What kind of results can one reasonably expect to obtain from analytical forecasting methods? By results, we mean forecast errors. Obviously, the results that a specific forecaster obtains are going to depend on the specific situation: what variables are being forecast, the availability and quality of data, the methods that can be applied to the problem, and the tools and expertise that are available. However, because there have been many surveys of both forecasters and users of forecasts, as well as forecast competitions (e.g., see Makridakis et al. (1993)) where many different techniques have been applied in head-to-head challenges, some broad conclusions can be drawn.

In general, exponential smoothing type methods, including Winters' method, typically experience mean absolute prediction errors ranging from 10% to 15% for lead-one forecasts. As the lead time increases, the prediction error increases, with mean absolute prediction errors typically in the 17–25% range at lead times of six periods. At 12 period lead times, the mean absolute prediction error can range from 18% to 45%. More sophisticated time series models such as ARIMA models are not usually much better, with the mean absolute prediction error ranging from about 10% for lead-one forecasts, to about 17% for lead-six forecasts, and up to 25% for 12 period lead times. This probably accounts for some of the dissatisfaction that forecasters often express with the more sophisticated techniques; they can be much harder to use, but they do not have substantial payback in terms of reducing forecasting errors. Regression methods often produce mean absolute prediction errors ranging from 12% to 18% for lead-one forecasts. As the lead time increases, the prediction error increases, with mean absolute prediction errors typically in the 17–20% range for six period lead times. At 12 period lead times, the mean absolute prediction error can range from 20% to 25%. Seasonal time series are often easier to predict than nonseasonal ones, because seasonal patterns are relatively stable through time, and relatively simple methods such as Winters' method and seasonal adjustment procedures typically work very well as forecasting techniques. Interestingly, seasonal adjustment techniques are not used nearly as widely as we would expect, given their relatively good performance.

When forecasting is done well in an organization, it is typically done by a group of individuals who have some training and experience in the techniques, have access to the right information, and have an opportunity to see how the forecasts are used. If higher levels of management routinely intervene in the process and use their judgment to modify the forecasts, it is highly desirable if the forecast preparers can interact with these managers to learn why the original forecasts require modification. Unfortunately, in many organizations, forecasting is done in an informal way, and the forecasters are often marketing or sales personnel, or market researchers, for whom forecasting is only a (sometimes small) part of their responsibilities. There is often a great deal of turnover in these positions, and so no long-term experience base or continuity builds up. The lack of a formal, organized process is often a big part of the reason why forecasting is not as successful as it should be.

Any evaluation of a forecasting effort in an organization should consider at least the following questions:

1. What methods are being used? Are the methods appropriate to organizational needs, when planning horizons and other business issues are taken into account? Is there an opportunity to use more than

one forecasting procedure? Could forecasts be combined to improve results?

2. Are the forecasting methods being used correctly?

3. Is an appropriate set of data being used in preparing the forecasts? Is data quality an issue? Are the underlying assumptions of the methods employed satisfied at least well enough for the methods to be successful?

4. Is uncertainty being addressed adequately? Are prediction intervals used as part of the forecast report? Do forecast users understand the PIs?

5. Does the forecasting system take economic/market forces into account? Is there an ability to capitalize on current events, natural forces, and swings in customer preferences and tastes?

6. Is forecasting separate from planning? Very often the forecast is really just a plan or schedule. For example, it may reflect a production plan, not a forecast of what we could realistically expect to sell (i.e., demand). Many individuals do not understand the difference between a forecast and a plan.

In the short-to-medium term, most businesses can benefit by taking advantage of the relative stability of seasonal patterns and the inertia present in most time series of interest. These are the methods we have focused on in this book.

7.11 R COMMANDS FOR CHAPTER 7

Example 7.11   The data for this example are in the array called pressure .data of which the two columns represent the viscosity and the temperature, respectively. To model the multivariate data we use the "VAR" function in R package "vars." But we first start with time series, acf, pacf, ccf plots as suggested in the example

```
library(vars)

pf<-pressure.data[,1]
pb<-pressure.data[,2]
plot(pf,type="o",pch=16,cex=.5,xlab='Time', ylab='Pressure',ylim=
  c(4,25))
lines(pb, type="o",pch=15,cex=.5, col="grey40")
legend(1,7,c("Variable","Front","Back"),
```

```
pch=c(NA,16,15),lwd=c(NA,.5,.5),cex=.55,col=c("black","black",
  "grey40"))
```

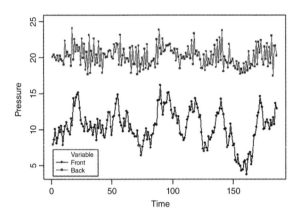

```
res.pf<- as.vector(residuals(arima(pf,order=c(1,0,0))))
res.pb<- as.vector(residuals(arima(pb,order=c(1,0,0))))
par(mfrow=c(2,2),oma=c(0,0,0,0))
acf(pf,lag.max=25,type="correlation",main="ACF for Front
  Pressure")
acf(pb, lag.max=25, type="correlation",main="PACF for Back
  Pressure",ylab="PACF")
ccf(pb,pf,main='CCF of \nFront and Back Pressures',ylab='CCF')
ccf(res.pb,res.pf,main='CCF of Residuals for \nFront and Back
  Pressures',ylab='CCF')
```

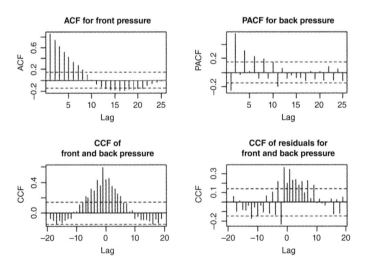

We now fit a VAR(1) model to the data using VAR function:

```
> pressure.var1<-VAR(pressure.data)
> pressure.var1

    VAR Estimation Results:
    =======================

    Estimated coefficients for equation pfront:
    ===========================================
    Call:
    pfront = pfront.l1 + pback.l1 + const

     pfront.l1    pback.l1       const
     0.7329529   0.4735983  -6.7555089

    Estimated coefficients for equation pback:
    ==========================================
    Call:
    pback = pfront.l1 + pback.l1 + const

     pfront.l1    pback.l1       const
     0.4104251  -0.5606308  27.2369791
```

Note that there is also a VARselect function that will automatically select the best p order of the VAR(p) model. In this case we tried p upto 5.

```
> VARselect(pressure.data,lag.max=5)

    $selection
    AIC(n)   HQ(n)   SC(n)  FPE(n)
         1       1       1       1

    $criteria
                       1          2          3          4          5
    AIC(n)  0.05227715 0.09275642 0.1241682 0.1639090 0.1882266
    HQ(n)   0.09526298 0.16439946 0.2244684 0.2928665 0.3458413
    SC(n)   0.15830468 0.26946896 0.3715657 0.4819916 0.5769942
    FPE(n)  1.05367413 1.09722530 1.1322937 1.1783005 1.2074691
```

The output shows that VAR(1) was indeed the right choice. We now plot the residuals.

```
plot(residuals(pressure.var1)[,1],type="p",pch=15,cex=.5,xlab=
  'Time', ylab='Residuals',ylim=c(-3,3))
points(residuals(pressure.var1)[,2],pch=1,cex=.5, col="grey40")
```

```
legend(100,3,c("Residuals","Front","Back"),
pch=c(NA,15,1),lwd=c(NA,.5,.5),cex=.55,col=c("black","black",
  "grey40"))
```

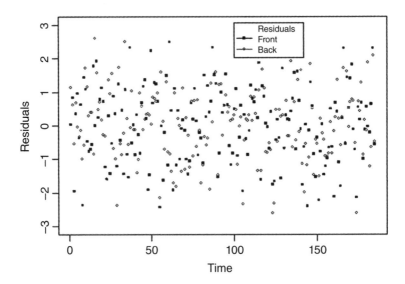

Example 7.12   The data for this example are in the array called sp500.data of which the two columns represent the date and the S&P 500 closing values respectively. To model the multivariate data we use the "garch" function in R package "tseries." But we first start with time series, acf, pacf plots as suggested in the example

```
library(tseries)

sp<-ts(sp500.data[,2])
logsp.d1<-diff(log(sp))
T<-length(logsp.d1)
plot(logsp.d1,type="o",pch=16,cex=.5,xlab='Date', ylab='log(SP(t))
  -log(SP(t-1))',xaxt='n')
lablist<-as.vector(sp500.data[seq(1,T+1,40),1])
axis(1, seq(1,T+1,40), labels=FALSE)
text(seq(1,T+1,40),par("usr")[3]-.01, ,labels = lablist, srt = 45,
  pos = 1, xpd = TRUE)
```

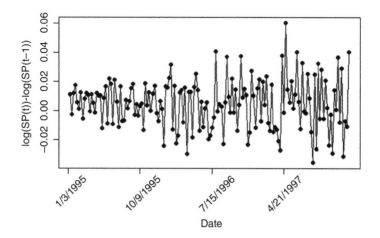

```
# ACF and PACF of the first difference of the log transformation
par(mfrow=c(1,2),oma=c(0,0,0,0))
acf(logsp.d1,lag.max=25,type="correlation",main="ACF of the wt")
acf(logsp.d1, lag.max=25,type="partial",main="PACF of the wt")
```

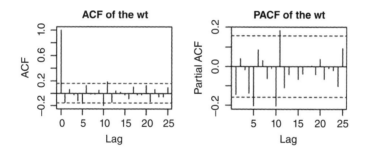

```
# ACF and PACF of the square of the first difference of the log
# transformation
par(mfrow=c(1,2),oma=c(0,0,0,0))
acf(logsp.d1^2,lag.max=25,type="correlation",main="ACF of the wt^2")
acf(logsp.d1^2, lag.max=25,type="partial",main="PACF of the wt^2")
```

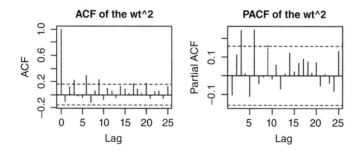

```
# Fit a GARCH(0,3) model
sp.arch3<-garch(logsp.d1, order = c(0,3), trace = FALSE)
summary(sp.arch3)
      Call:
      garch(x = logsp.d1, order = c(0, 3), trace = FALSE)

      Model:
      GARCH(0,3)

      Residuals:
         Min       1Q  Median       3Q      Max
      -2.0351 -0.4486  0.3501  0.8869  2.9320

      Coefficient(s):
         Estimate  Std. Error  t value Pr(>|t|)
      a0 2.376e-04  6.292e-05    3.776 0.000159 ***
      a1 2.869e-15  1.124e-01    0.000 1.000000
      a2 7.756e-02  8.042e-02    0.964 0.334876
      a3 9.941e-02  1.124e-01    0.884 0.376587
      --
      Signif. codes: 0 '***' 0.001 '**' 0.01 '*' 0.05 '.' 0.1 ' ' 1

      Diagnostic Tests:
              Jarque Bera Test

      data:  Residuals
      X-squared = 0.4316, df = 2, p-value = 0.8059


              Box-Ljung test

      data:  Squared.Residuals
      X-squared = 2.2235, df = 1, p-value = 0.1359
```

EXERCISES

7.1 Show that an AR(2) model can be represented in state space form.

7.2 Show that an MA(1) model can be written in state space form.

7.3 Consider the information on weekly spare part demand shown in Table E7.1. Suppose that 74 requests for 55 parts are received during the current week, T. Find the new cumulative distribution of demand. Use $\lambda = 0.1$. What is your forecast of the 70th percentile of the demand distribution?

7.4 Consider the information on weekly luxury car rentals shown in Table E7.2. Suppose that 37 requests for rentals are received during

TABLE E7.1 Spare Part Demand Information for Exercise 7.3

k	B_{k-1}	B_k	$\hat{p}_k(T-1)$	$F(B_k)$, at the end of week $T-1$
0	0	5	0.02	0.02
1	5	10	0.03	0.05
2	10	15	0.04	0.09
3	15	20	0.05	0.14
4	20	25	0.08	0.22
5	25	30	0.09	0.31
6	30	35	0.12	0.43
7	35	40	0.17	0.60
8	45	50	0.21	0.81
9	50	55	0.11	0.92
10	55	60	0.08	1.00

the current week, T. Find the new cumulative distribution of demand. Use $\lambda = 0.1$. What is your forecast of the 90th percentile of the demand distribution?

7.5 Rework Exercise 7.3 using $\lambda = 0.4$. How much difference does changing the value of the smoothing parameter make in your estimate of the 70th percentile of the demand distribution?

7.6 Rework Exercise 7.4 using $\lambda = 0.4$. How much difference does changing the value of the smoothing parameter make in your estimate of the 70th percentile of the demand distribution?

TABLE E7.2 Luxury Car Rental Demand Information for Exercise 7.4

k	B_{k-1}	B_k	$\hat{p}_k(T-1)$	$F(B_k)$, at the end of week $T-1$
0	0	5	0.06	0.06
1	5	10	0.07	0.13
2	10	15	0.08	0.21
3	15	20	0.09	0.30
4	20	25	0.15	0.45
5	25	30	0.22	0.67
6	30	35	0.24	0.91
7	35	40	0.05	0.96
8	45	50	0.04	1.00

7.7 Suppose that two forecasting methods can be used for a time series, and that the two variances of the forecast errors are $\sigma_1^2 = 10$ and $\sigma_2^2 = 25$. If the correlation coefficient $\rho = -0.75$, calculate the optimum value of the weight used to optimally combine the two individual forecasts. What is the variance of the combined forecast?

7.8 Suppose that two forecasting methods can be used for a time series, and that the two variances of the forecast errors are $\sigma_1^2 = 15$ and $\sigma_2^2 = 20$. If the correlation coefficient $\rho = -0.4$, calculate the optimum value of the weight used to optimally combine the two individual forecasts. What is the variance of the combined forecast?

7.9 Suppose that two forecasting methods can be used for a time series, and that the two variances of the forecast errors are $\sigma_1^2 = 8$ and $\sigma_2^2 = 16$. If the correlation coefficient $\rho = -0.3$, calculate the optimum value of the weight used to optimally combine the two individual forecasts. What is the variance of the combined forecast?

7.10 Suppose that two forecasting methods can be used for a time series, and that the two variances of the forecast errors are $\sigma_1^2 = 1$ and $\sigma_2^2 = 8$. If the correlation coefficient $\rho = -0.65$, calculate the optimum value of the weight used to optimally combine the two individual forecasts. What is the variance of the combined forecast?

7.11 Rework Exercise 7.8 assuming that $\rho = 0.4$. What effect does changing the sign of the correlation coefficient have on the weight used to optimally combine the two forecasts? What is the variance of the combined forecast?

7.12 Rework Exercise 7.9 assuming that $\rho = 0.3$. What effect does changing the sign of the correlation coefficient have on the weight used to optimally combine the two forecasts? What is the variance of the combined forecast?

7.13 Suppose that there are three lead-one forecasts available for a time series, and the covariance matrix of the three forecasts is as follows:

$$\Sigma_{T+1}(T) = \begin{bmatrix} 10 & -4 & -2 \\ -4 & 6 & -3 \\ -2 & -3 & 15 \end{bmatrix}$$

Find the optimum weights for combining these three forecasts. What is the variance of the combined forecast?

7.14 Suppose that there are three lead-one forecasts available for a time series, and the covariance matrix of the three forecasts is as follows:

$$\mathbf{\Sigma}_{T+1}(T) = \begin{bmatrix} 8 & -2 & -1 \\ -2 & 3 & -2 \\ -1 & -2 & 10 \end{bmatrix}$$

Find the optimum weights for combining these three forecasts. What is the variance of the combined forecast?

7.15 Table E7.3 presents 25 forecast errors for two different forecasting techniques applied to the same time series. Is it possible to combine the two forecasts to improve the forecast errors? What is the optimum weight for combining the forecasts? What is the variance of the combined forecast?

TABLE E7.3   Forecast Errors for Exercise 7.15

Time period	Forecast errors, method 1	Forecast errors, method 2
1	−0.78434	6.9668
2	−0.31111	4.5512
3	2.15622	−1.2681
4	−1.81293	6.8967
5	−0.77498	1.6574
6	2.31673	−8.7601
7	−0.94866	0.7472
8	0.81314	−0.7457
9	−2.95718	−0.5355
10	0.08175	−1.3458
11	1.08915	−5.8232
12	−0.20637	1.2722
13	0.57157	−2.4561
14	0.41435	4.3111
15	0.47138	5.9894
16	1.23274	−6.8757
17	−0.66288	1.5996
18	1.71193	10.5031
19	−2.00317	9.8664
20	−2.87901	3.0399
21	−2.87901	14.1992
22	−0.16103	9.0080
23	2.12427	−0.4551
24	0.60598	0.7123
25	0.18259	1.7346

7.16 Show that when combining two forecasts, if the correlation between the two sets of forecast errors is $\rho = \sigma_1/\sigma_2$, then Min Var $[e^c_{T+\tau}(T)] = \sigma_1^2$, where σ_1^2 is the smaller of the two forecast error variances.

7.17 Show that when combining two forecasts, if the correlation between the two sets of forecast errors is $\rho = 0$, then Var $[e^c_{T+\tau}(T)] = \sigma_1^2\sigma_2^2/(\sigma_1^2 + \sigma_2^2)$.

7.18 Let y_t be an IMA(1, 1) time series with parameter $\theta = 0.4$. Suppose that this time series is observed with an additive white noise error.

   a. What is the model form of the observed error?

   b. Find the parameters of the observed time series, assuming that the variances of the errors in the original time series and the white noise are equal.

7.19 Show that an AR(1) time series that is observed with an additive white noise error is an ARMA(1, 1) process.

7.20 Generate 100 observations of an ARIMA(1, 1, 0) time series. Add 100 observations of white noise to this time series. Calculate the sample ACF and sample PACF of the new time series. Identify the model form and estimate the parameters.

7.21 Generate 100 observations of an ARIMA(1, 1, 0) time series. Generate another 100 observations of an AR(1) time series and add these observations to the original time series. Calculate the sample ACF and sample PACF of the new time series. Identify the model form and estimate the parameters.

7.22 Generate 100 observations of an AR(2) time series. Generate another 100 observations of an AR(1) time series and add these observations to the original time series. Calculate the sample ACF and sample PACF of the new time series. Identify the model form and estimate the parameters.

7.23 Generate 100 observations of an MA(2) time series. Generate another 100 observations of an MA(1) time series and add these observations to the original time series. Calculate the sample ACF and sample PACF of the new time series. Identify the model form and estimate the parameters.

7.24 Table E7.4 presents data on the type of heating fuel used in new single-family houses built in the United States from 1971 through

TABLE E7.4 Data for Exercise 7.24

	Number of Houses (in thousands)			
Year	Total	Gas	Electricity	Oil	Other types or none
1971	1014	605	313	83	15
1972	1143	621	416	93	13
1973	1197	560	497	125	16
1974	940	385	458	85	11
1975	875	347	429	82	18
1976	1034	407	499	110	19
1977	1258	476	635	120	28
1978	1369	511	710	109	40
1979	1301	512	662	86	41
1980	957	394	482	29	52
1981	819	339	407	16	57
1982	632	252	315	17	48
1983	924	400	448	22	53
1984	1025	460	492	24	49
1985	1072	466	528	36	42
1986	1120	527	497	52	45
1987	1123	583	445	58	38
1988	1085	587	402	60	36
1989	1026	596	352	50	28
1990	966	573	318	48	27
1991	838	505	267	37	29
1992	964	623	283	36	22
1993	1039	682	303	34	20
1994	1160	772	333	39	16
1995	1066	708	305	37	16
1996	1129	781	299	37	11
1997	1116	771	296	38	11
1998	1160	809	307	34	10
1999	1270	884	343	35	9
2000	1242	868	329	37	8
2001	1256	875	336	35	9
2002	1325	907	371	38	10
2003	1386	967	377	31	12
2004	1532	1052	440	29	10
2005	1636	1082	514	31	9

2005. Develop an appropriate multivariate time series model for the gas, electricity, and oil time series.

7.25 Reconsider the data on heating fuel in Table E7.4. Suppose that you are interested in forecasting the aggregate series (the Total column in Table E7.4). One way to do this is to forecast the total directly. Another way is to forecast the individual component series and sum the forecasts of the components to obtain a forecast for the total. Investigate these approaches for this data and report on your conclusions.

7.26 Reconsider the data on heating fuel in Table E7.4. Suppose that you are interested in forecasting the four individual components series (the Gas, Electricity, Oil, and Other Types columns in Table E7.4). One way to do this is to forecast the individual time series directly. Another way is to forecast the total and obtain forecasts of the individual component series by decomposing the forecast for the totals into component parts. Investigate these approaches for this data and report on your conclusions.

7.27 Table E7.5 contains data on property crimes reported to the police in the United States. Both the number of property crimes and the crime rate per 100,000 individuals are shown. Using the data on the number of crimes reported, develop an appropriate multivariate time series model for the burglary, larceny-theft, and motor vehicle theft time series.

7.28 Repeat Exercise 7.27 using the property crime rate data. Compare the models obtained using the number of crimes reported versus the crime rate.

7.29 Reconsider the data on property crimes in Table E7.5. Suppose that you are interested in forecasting the aggregate crime rate series. One way to do this is to forecast the total directly. Another way is to forecast the individual component series and sum the forecasts of the components to obtain a forecast for the total. Investigate these approaches for this data and report on your conclusions.

7.30 Reconsider the data on property crimes in Table E7.5. Suppose that you are interested in forecasting the four individual component series (the Burglary, Larceny-Theft, and Motor Vehicle Theft columns in Table E7.5). One way to do this is to forecast the individual time series directly. Another way is to forecast the total

TABLE E7.5   Property Crime Data for Exercise 7.27

	Property Crime (in thousands)			
Year	Total	Burglary	Larceny-theft	Motor vehicle theft
1960	3096	912	1855	328
1961	3199	950	1913	336
1962	3451	994	2090	367
1963	3793	1086	2298	408
1964	4200	1213	2514	473
1965	4352	1283	2573	497
1966	4793	1410	2822	561
1967	5404	1632	3112	660
1968	6125	1859	3483	784
1969	6749	1982	3889	879
1970	7359	2205	4226	928
1971	7772	2399	4424	948
1972	7414	2376	4151	887
1973	7842	2566	4348	929
1974	9279	3039	5263	977
1975	10,253	3265	5978	1010
1976	10,346	3109	6271	966
1977	9955	3072	5906	978
1978	10,123	3128	5991	1004
1979	11,042	3328	6601	1113
1980	12,064	3795	7137	1132
1981	12,062	3780	7194	1088
1982	11,652	3447	7143	1062
1983	10,851	3130	6713	1008
1984	10,608	2984	6592	1032
1985	11,103	3073	6926	1103
1986	11,723	3241	7257	1224
1987	12,025	3236	7500	1289
1988	12,357	3218	7706	1433
1989	12,605	3168	7872	1565
1990	12,655	3074	7946	1636
1991	12,961	3157	8142	1662
1992	12,506	2980	7915	1611
1993	12,219	2835	7821	1563
1994	12,132	2713	7880	1539
1995	12,064	2594	7998	1472
1996	11,805	2506	7905	1394
1997	11,558	2461	7744	1354
1998	10,952	2333	7376	1243
1999	10,208	2102	6956	1152
2000	10,183	2051	6972	1160
2001	10,437	2117	7092	1228
2002	10,451	2152	7053	1246

TABLE E7.5 (*Continued*)

	Crime Rate (per 100,000 population)		
Year	Total	Burglary	Larceny-theft	Motor vehicle theft
1960	1726.3	508.6	1034.7	183.0
1961	1747.9	518.9	1045.4	183.6
1962	1857.5	535.2	1124.8	197.4
1963	2012.1	576.4	1219.1	216.6
1964	2197.5	634.7	1315.5	247.4
1965	2248.8	662.7	1329.3	256.8
1966	2450.9	721.0	1442.9	286.9
1967	2736.5	826.6	1575.8	334.1
1968	3071.8	932.3	1746.6	393.0
1969	3351.3	984.1	1930.9	436.2
1970	3621.0	1084.9	2079.3	456.8
1971	3768.8	1163.5	2145.5	459.8
1972	3560.4	1140.8	1993.6	426.1
1973	3737.0	1222.5	2071.9	442.6
1974	4389.3	1437.7	2489.5	462.2
1975	4810.7	1532.1	2804.8	473.7
1976	4819.5	1448.2	2921.3	450.0
1977	4601.7	1419.8	2729.9	451.9
1978	4642.5	1434.6	2747.4	460.5
1979	5016.6	1511.9	2999.1	505.6
1980	5353.3	1684.1	3167.0	502.2
1981	5263.9	1647.2	3135.3	474.1
1982	5032.5	1488.0	3083.1	458.6
1983	4637.4	1338.7	2871.3	431.1
1984	4492.1	1265.5	2795.2	437.7
1985	4666.4	1291.7	2911.2	463.5
1986	4881.8	1349.8	3022.1	509.8
1987	4963.0	1335.7	3095.4	531.9
1988	5054.0	1316.2	3151.7	586.1
1989	5107.1	1283.6	3189.6	634.0
1990	5073.1	1232.2	3185.1	655.8
1991	5140.2	1252.1	3229.1	659.0
1992	4903.7	1168.4	303.6	631.6
1993	4740.0	1099.7	3033.9	606.3
1994	4660.2	1042.1	3026.9	591.3
1995	4590.5	987.0	3043.2	560.3
1996	4451.0	945.0	2980.3	525.7
1997	4316.3	918.8	2891.8	505.7
1998	4052.5	863.2	2729.5	459.9
1999	3743.6	770.4	2550.7	422.5
2000	3618.3	728.8	2477.3	412.2
2001	3658.1	741.8	2485.7	430.5
2002	3624.1	746.2	2445.8	432.1

and obtain forecasts of the individual component series by decomposing the forecast for the totals into component parts. Investigate these approaches using the crime rate data, and report on your conclusions.

7.31 Table B.1 contains data on market yield of US treasury securities at 10-year constant maturity. Compute the spectrum for this time series. What features of the time series are apparent from examination of the spectrum?

7.32 Table B.3 contains on the viscosity of a chemical product. Compute the spectrum for this time series. What features of the time series are apparent from examination of the spectrum?

7.33 Table B.6 contains the global mean surface air temperature anomaly and the global CO_2 concentration data. Compute the spectrum for both of these time series. What features of the two time series are apparent from examination of the spectrum?

7.34 Table B.11 contains sales data on sales of Champagne. Compute the spectrum for this time series. Is the seasonal nature of the time series apparent from examination of the spectrum?

7.35 Table B.21 contains data on the average monthly retail price of electricity in the residential sector for Arizona from 2001 through 2014. Take the first difference of this timer series and compute the spectrum. Is the seasonal nature of the time series apparent from examination of the spectrum?

7.36 Table B.22 contains data on Danish crude oil production. Compute the spectrum for this time series. What features of the time series are apparent from examination of the spectrum?

7.37 Table B.23 contains data on positive test results for influenza in the US. Compute the spectrum for this time series. Is the seasonal nature of the time series apparent from examination of the spectrum?

7.38 Table B.24 contains data on monthly mean daily solar radiation. Compute the spectrum for this time series. Is the seasonal nature of the time series apparent from examination of the spectrum?

7.39 Table B.27 contains data on airline on-time arrival performance. Compute the spectrum for this time series. Is there any evidence of seasonality in the time series that is apparent from examination of the spectrum?

7.40 Table B.28 contains data on US automobile manufacturing shipments. Compute the spectrum for this time series. Is there any evidence of seasonality in the time series that is apparent from examination of the spectrum?

7.41 Weekly demand for a spare part is assumed to follow a Poisson distribution:

$$f(y|\lambda) = \frac{e^{-\lambda}\lambda^y}{y!}, \quad y = 0, 1, \ldots$$

The mean λ of the demand distribution is assumed to be a random variable with a gamma distribution

$$h(\lambda) = \frac{b^a}{(a-1)!}\lambda^{a-1}e^{-b\lambda}, \quad \lambda > 0$$

where a and b are parameters having subjectively determined values. In the week following the establishment of this prior distribution d parts were demanded. What is the posterior distribution of λ?

APPENDIX A

STATISTICAL TABLES

Introduction to Time Series Analysis and Forecasting, Second Edition.
Douglas C. Montgomery, Cheryl L. Jennings and Murat Kulahci.
© 2015 John Wiley & Sons, Inc. Published 2015 by John Wiley & Sons, Inc.

TABLE A.1 Cumulative Standard Normal Distribution

$$\Phi(z) = P(Z \le z) = \int_{-\infty}^{z} \frac{1}{\sqrt{2\pi}} e^{-u^2/2}\, du$$

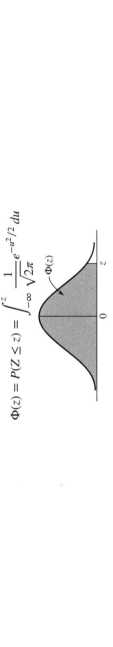

z	−0.09	−0.08	−0.07	−0.06	−0.05	−0.04	−0.03	−0.02	−0.01	−0.00	z
−3.9	0.000033	0.000034	0.000036	0.000037	0.000039	0.000041	0.000042	0.000044	0.000046	0.000048	−3.9
−3.8	0.000050	0.000052	0.000054	0.000057	0.000059	0.000062	0.000064	0.000067	0.000069	0.000072	−3.8
−3.7	0.000075	0.000078	0.000082	0.000085	0.000088	0.000092	0.000096	0.000100	0.000104	0.000108	−3.7
−3.6	0.000112	0.000117	0.000121	0.000126	0.000131	0.000136	0.000142	0.000147	0.000153	0.000159	−3.6
−3.5	0.000165	0.000172	0.000179	0.000185	0.000193	0.000200	0.000208	0.000216	0.000224	0.000233	−3.5
−3.4	0.000242	0.000251	0.000260	0.000270	0.000280	0.000291	0.000302	0.000313	0.000325	0.000337	−3.4
−3.3	0.000350	0.000362	0.000376	0.000390	0.000404	0.000419	0.000434	0.000450	0.000467	0.000483	−3.3
−3.2	0.000501	0.000519	0.000538	0.000557	0.000577	0.000598	0.000619	0.000641	0.000664	0.000687	−3.2
−3.1	0.000711	0.000736	0.000762	0.000789	0.000816	0.000845	0.000874	0.000904	0.000935	0.000968	−3.1
−3.0	0.001001	0.001035	0.001070	0.001107	0.001144	0.001183	0.001223	0.001264	0.001306	0.001350	−3.0
−2.9	0.001395	0.001441	0.001489	0.001538	0.001589	0.001641	0.001695	0.001750	0.001807	0.001866	−2.9
−2.8	0.001926	0.001988	0.002052	0.002118	0.002186	0.002256	0.002327	0.002401	0.002477	0.002555	−2.8
−2.7	0.002635	0.002718	0.002803	0.002890	0.002980	0.003072	0.003167	0.003264	0.003364	0.003467	−2.7
−2.6	0.003573	0.003681	0.003793	0.003907	0.004025	0.004145	0.004269	0.004396	0.004527	0.004661	−2.6
−2.5	0.004799	0.004940	0.005085	0.005234	0.005386	0.005543	0.005703	0.005868	0.006037	0.006210	−2.5

z	.00	.01	.02	.03	.04	.05	.06	.07	.08	.09	z
-2.4	0.008198	0.007976	0.007760	0.007549	0.007344	0.007143	0.006947	0.006756	0.006569	0.006387	-2.4
-2.3	0.010724	0.010444	0.010170	0.009903	0.009642	0.009387	0.009137	0.008894	0.008656	0.008424	-2.3
-2.2	0.013903	0.013553	0.013209	0.012874	0.012545	0.012224	0.011911	0.011604	0.011304	0.011011	-2.2
-2.1	0.017864	0.017429	0.017003	0.016586	0.016177	0.015778	0.015386	0.015003	0.014629	0.014262	-2.1
-2.0	0.022750	0.022216	0.021692	0.021178	0.020675	0.020182	0.019699	0.019226	0.018763	0.018309	-2.0
-1.9	0.028717	0.028067	0.027429	0.026803	0.026190	0.025588	0.024998	0.024419	0.023852	0.023295	-1.9
-1.8	0.035930	0.035148	0.034379	0.033625	0.032884	0.032157	0.031443	0.030742	0.030054	0.029379	-1.8
-1.7	0.044565	0.043633	0.042716	0.041815	0.040929	0.040059	0.039204	0.038364	0.037538	0.036727	-1.7
-1.6	0.054799	0.053699	0.052616	0.051551	0.050503	0.049471	0.048457	0.047460	0.046479	0.045514	-1.6
-1.5	0.066807	0.065522	0.064256	0.063008	0.061780	0.060571	0.059380	0.058208	0.057053	0.055917	-1.5
-1.4	0.080757	0.079270	0.077804	0.076359	0.074934	0.073529	0.072145	0.070781	0.069437	0.068112	-1.4
-1.3	0.096801	0.095098	0.093418	0.091759	0.090123	0.088508	0.086915	0.085343	0.083793	0.082264	-1.3
-1.2	0.115070	0.113140	0.111233	0.109349	0.107488	0.105650	0.103835	0.102042	0.100273	0.098525	-1.2
-1.1	0.135666	0.133500	0.131357	0.129238	0.127143	0.125072	0.123024	0.121001	0.119000	0.117023	-1.1
-1.0	0.158655	0.156248	0.153864	0.151505	0.149170	0.146859	0.144572	0.142310	0.140071	0.137857	-1.0
-0.9	0.184060	0.181411	0.178786	0.176185	0.173609	0.171056	0.168528	0.166023	0.163543	0.161087	-0.9
-0.8	0.211855	0.208970	0.206108	0.203269	0.200454	0.197662	0.194894	0.192150	0.189430	0.186733	-0.8
-0.7	0.241964	0.238852	0.235762	0.232695	0.229650	0.226627	0.223627	0.220650	0.217695	0.214764	-0.7
-0.6	0.274253	0.270931	0.267629	0.264347	0.261086	0.257846	0.254627	0.251429	0.248252	0.245097	-0.6
-0.5	0.308538	0.305026	0.301532	0.298056	0.294599	0.291160	0.287740	0.284339	0.280957	0.277595	-0.5
-0.4	0.344578	0.340903	0.337243	0.333598	0.329969	0.326355	0.322758	0.319178	0.315614	0.312067	-0.4
-0.3	0.382089	0.378281	0.374484	0.370700	0.366928	0.363169	0.359424	0.355691	0.351973	0.348268	-0.3
-0.2	0.420740	0.416834	0.412936	0.409046	0.405165	0.401294	0.397432	0.393580	0.389739	0.385908	-0.2
-0.1	0.460172	0.456205	0.452242	0.448283	0.444330	0.440382	0.436441	0.432505	0.428576	0.424655	-0.1

(continued)

TABLE A.1 *(Continued)*

z	−0.09	−0.08	−0.07	−0.06	−0.05	−0.04	−0.03	−0.02	−0.01	−0.00	z
0.0	0.464144	0.468119	0.472097	0.476078	0.480061	0.484047	0.488033	0.492022	0.496011	0.500000	0.0
0.0	0.500000	0.503989	0.507978	0.511967	0.515953	0.519939	0.523922	0.527903	0.531881	0.535856	0.0
0.1	0.539828	0.543795	0.547758	0.551717	0.555760	0.559618	0.563559	0.567495	0.571424	0.575345	0.1
0.2	0.579260	0.583166	0.587064	0.590954	0.594835	0.598706	0.602568	0.606420	0.610261	0.614092	0.2
0.3	0.617911	0.621719	0.625516	0.629300	0.633072	0.636831	0.640576	0.644309	0.648027	0.651732	0.3
0.4	0.655422	0.659097	0.662757	0.666402	0.670031	0.673645	0.677242	0.680822	0.684386	0.687933	0.4
0.5	0.691462	0.694974	0.698468	0.701944	0.705401	0.708840	0.712260	0.715661	0.719043	0.722405	0.5
0.6	0.725747	0.729069	0.732371	0.735653	0.738914	0.742154	0.745373	0.748571	0.751748	0.754903	0.6
0.7	0758036	0.761148	0.764238	0.767305	0.770350	0.773373	0.776373	0.779350	0.782305	0.785236	0.7
0.8	0.788145	0.791030	0.793892	0.796731	0.799546	0.802338	0.805106	0.807850	0.810570	0.813267	0.8
0.9	0.815940	0.818589	0.821214	0.823815	0.826391	0.828944	0.831472	0.833977	0.836457	0.838913	0.9
1.0	0.841345	0.843752	O.846136	0.848495	0.850830	0.853141	0.855428	0.857690	0.859929	0.862143	1.0
1.1	0.864334	0.866500	0.868643	0.870762	0.872857	0.874928	0.876976	0.878999	0.881000	0.882977	1.1
1.2	0.884930	0.886860	0.888767	0.890651	0.892512	0.894350	0.896165	0.897958	0.899727	0.901475	1.2
1.3	0.903199	0.904902	0.906582	0.908241	0.909877	0.911492	0.913085	0.914657	0.916207	0.917736	1.3
1.4	0.919243	0.920730	0.922196	0.923641	0.925066	0.926471	0.927855	0.929219	0.930563	0.931888	1.4
1.5	0.933193	0.934478	0.935744	0936992	0.938220	0.939429	0.940620	0.941792	0.942947	0.944083	1.5
1.6	0.945201	0.946301	0.947384	0.948449	0.949497	0.950529	0.951543	0.952540	0.953521	0.954486	1.6
1.7	0.955435	0.956367	0.957284	0.958185	0.959071	0.959941	0.960796	0.961636	0.962462	0.963273	1.7
1.8	0.964070	0.964852	0.965621	0.966375	0.967116	0.967843	0.968557	0.969258	0.969946	0.970621	1.8
1.9	0.971283	0.971933	0.972571	0.973197	0.973810	0.974412	0.975002	0.975581	0.976148	0.976705	1.9

2.0	0.977250	0.977784	0.978308	0.978822	0.979325	0.979818	0.980301	0.980774	0.981237	0.981691	2.0
2.1	0.982136	0.982571	0.982997	0.983414	0.983823	0.984222	0.984614	0.984997	0.985371	0.985738	2.1
2.2	0.986097	0.986447	0.986791	0.987126	0.987455	0.987776	0.988089	0.988396	0.988696	0.988989	2.2
2.3	0.989276	0.989556	0.989830	0.990097	0.990358	0.990613	0.990863	0.991106	0.991344	0.991576	2.3
2.4	0.991802	0.992024	0.992240	0.992451	0.992656	0.992857	0.993053	0.993244	0.993431	0.993613	2.4
2.5	0.993790	0.993963	0.994132	0.994297	0.994457	0.994614	0.994766	0.994915	0.995060	0.995201	2.5
2.6	0.995339	0.995473	0.995604	0.995731	0.995855	0.995975	0.996093	0.996207	0.996319	0.996427	2.6
2.7	0.996533	0.996636	0.996736	0.996833	0.996928	0.997020	0.997110	0.997197	0.997282	0.997365	2.7
2.8	0.997445	0.997523	0.997599	0.997673	0.997744	0.997814	0.997882	0.997948	0.998012	0.998074	2.8
2.9	0.998134	0.998193	0.998250	0.998305	0.998359	0.998411	0.998462	0.998511	0.998559	0.998605	2.9
3.0	0.998650	0.998694	0.998736	0.998777	0.998817	0.998856	0.998893	0.998930	0.998965	0.998999	3.0
3.1	0.999032	0.999065	0.999096	0.999126	0.999155	0.999184	0.999211	0.999238	0.999264	0.999289	3.1
3.2	0.999313	0.999336	0.999359	0.999381	0.999402	0.999423	0.999443	0.999462	0.999481	0.999499	3.2
3.3	0.999517	0.999533	0.999550	0.999566	0.999581	0.999596	0.999610	0.999624	0.999638	0.999650	3.3
3.4	0.999663	0.999675	0.999687	0.999698	0.999709	0.999720	0.999730	0.999740	0.999749	0.999758	3.4
3.5	0.999767	0.999776	0.999784	0.999792	0.999800	0.999807	0.999815	0.999821	0.999828	0.999835	3.5
3.6	0.999841	0.999847	0.999853	0.999858	0.999864	0.999869	0.999874	0.999879	0.999883	0.999888	3.6
3.7	0.999892	0.999896	0.999900	0.999904	0.999908	0.999912	0.999915	0.999918	0.999922	0.999925	3.7
3.8	0.999928	0.999931	0.999933	0.999936	0.999938	0.999941	0.999943	0.999946	0.999948	0.999950	3.8
3.9	0.999952	0.999954	0.999956	0.999958	0.999959	0.999961	0.999963	0.999964	0.999966	0.999967	3.9

TABLE A.2   Percentage Points $t_{\alpha,v}$ of the t Distribution

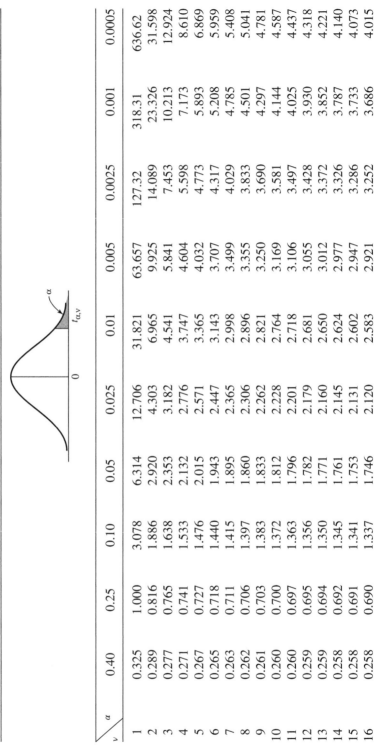

v \ α	0.40	0.25	0.10	0.05	0.025	0.01	0.005	0.0025	0.001	0.0005
1	0.325	1.000	3.078	6.314	12.706	31.821	63.657	127.32	318.31	636.62
2	0.289	0.816	1.886	2.920	4.303	6.965	9.925	14.089	23.326	31.598
3	0.277	0.765	1.638	2.353	3.182	4.541	5.841	7.453	10.213	12.924
4	0.271	0.741	1.533	2.132	2.776	3.747	4.604	5.598	7.173	8.610
5	0.267	0.727	1.476	2.015	2.571	3.365	4.032	4.773	5.893	6.869
6	0.265	0.718	1.440	1.943	2.447	3.143	3.707	4.317	5.208	5.959
7	0.263	0.711	1.415	1.895	2.365	2.998	3.499	4.029	4.785	5.408
8	0.262	0.706	1.397	1.860	2.306	2.896	3.355	3.833	4.501	5.041
9	0.261	0.703	1.383	1.833	2.262	2.821	3.250	3.690	4.297	4.781
10	0.260	0.700	1.372	1.812	2.228	2.764	3.169	3.581	4.144	4.587
11	0.260	0.697	1.363	1.796	2.201	2.718	3.106	3.497	4.025	4.437
12	0.259	0.695	1.356	1.782	2.179	2.681	3.055	3.428	3.930	4.318
13	0.259	0.694	1.350	1.771	2.160	2.650	3.012	3.372	3.852	4.221
14	0.258	0.692	1.345	1.761	2.145	2.624	2.977	3.326	3.787	4.140
15	0.258	0.691	1.341	1.753	2.131	2.602	2.947	3.286	3.733	4.073
16	0.258	0.690	1.337	1.746	2.120	2.583	2.921	3.252	3.686	4.015

ν										
17	0.257	0.689	1.333	1.740	2.110	2.567	2.898	3.222	3.646	3.965
18	0.257	0.688	1.330	1.734	2.101	2.552	2.878	3.197	3.610	3.922
19	0.257	0.688	1.328	1.729	2.093	2.539	2.861	3.174	3.579	3.883
20	0.257	0.687	1.325	1.725	2.086	2.528	2.845	3.153	3.552	3.850
21	0.257	0.686	1.323	1.721	2.080	2.518	2.831	3.135	3.527	3.819
22	0.256	0.686	1.321	1.717	2.074	2.508	2.819	3.119	3.505	3.792
23	0.256	0.685	1.319	1.714	2.069	2.500	2.807	3.104	3.485	3.767
24	0.256	0.685	1.318	1.711	2.064	2.492	2.797	3.091	3.467	3.745
25	0.256	0.684	1.316	1.708	2.060	2.485	2.787	3.078	3.450	3.725
26	0.256	0.684	1.315	1.706	2.056	2.479	2.779	3.067	3.435	3.707
27	0.256	0.684	1.314	1.703	2.052	2.473	2.771	3.057	3.421	3.690
28	0.256	0.683	1.313	1.701	2.048	2.467	2.763	3.047	3.408	3.674
29	0.256	0.683	1.311	1.699	2.045	2.462	2.756	3.038	3.396	3.659
30	0.256	0.683	1.310	1.697	2.042	2.457	2.750	3.030	3.385	3.646
40	0.255	0.681	1.303	1.684	2.001	2.423	2.704	2.971	3.307	3.551
60	0.254	0.679	1.296	1.671	2.000	2.390	2.660	2.915	3.232	3.460
120	0.254	0.677	1.289	1.658	1.980	2.358	2.617	2.860	3.160	3.373
∞	0.253	0.674	1.282	1.645	1.960	2.326	2.576	2.807	3.090	3.291

$ν$ = degrees of freedom.

TABLE A.3 Percentage Points $\chi^2_{\alpha,\nu}$ of the Chi-Square Distribution

ν \ α	0.995	0.990	0.975	0.950	0.900	0.500	0.100	0.050	0.025	0.010	0.005
1	0.00+	0.00+	0.00+	0.00+	0.02	0.45	2.71	3.84	5.02	6.63	7.88
2	0.01	0.02	0.05	0.10	0.21	1.39	4.61	5.99	7.38	9.21	10.60
3	0.07	0.11	0.22	0.35	0.58	2.37	6.25	7.81	9.35	11.34	12.84
4	0.21	0.30	0.48	0.71	1.06	3.36	7.78	9.49	11.14	13.28	14.86
5	0.41	0.55	0.83	1.15	1.61	4.35	9.24	11.07	12.83	15.09	16.75
6	0.68	0.87	1.24	1.64	2.20	5.35	10.65	12.59	14.45	16.81	18.55
7	0.99	1.24	1.69	2.17	2.83	6.35	12.03	14.07	16.01	18.48	20.28
8	1.34	1.65	2.18	2.73	3.49	7.34	13.36	15.51	17.53	20.09	21.96
9	1.73	2.09	2.70	3.33	4.17	8.34	14.68	16.92	19.02	21.67	23.59
10	2.16	2.56	3.25	3.94	4.87	9.34	15.99	18.31	20.48	23.21	25.19
11	2.60	3.05	3.82	4.57	5.58	10.34	17.28	19.68	21.92	24.72	26.76
12	3.07	3.57	4.40	5.23	6.30	11.34	18.55	21.03	23.34	26.22	28.30
13	3.57	4.11	5.01	5.89	7.04	12.34	19.81	22.36	24.74	27.69	29.82
14	4.07	4.66	5.63	6.57	7.79	13.34	21.06	23.68	26.12	29.14	31.32
15	4.60	5.23	6.27	7.26	8.55	14.34	22.31	25.00	27.49	30.58	32.80

ν											
16	5.14	5.81	6.91	7.96	9.31	15.34	23.54	26.30	28.85	32.00	34.27
17	5.70	6.41	7.56	8.67	10.09	16.34	24.77	27.59	30.19	33.41	35.72
18	6.26	7.01	8.23	9.39	10.87	17.34	25.99	28.87	31.53	34.81	37.16
19	6.84	7.63	8.91	10.12	11.65	18.34	27.20	30.14	32.85	36.19	38.58
20	7.43	8.26	9.59	10.85	12.44	19.34	28.41	31.41	34.17	37.57	40.00
21	8.03	8.90	10.28	11.59	13.24	20.34	29.62	32.67	35.48	38.93	41.40
22	8.64	9.54	10.98	12.34	14.04	21.34	30.81	33.92	36.78	40.29	42.80
23	9.26	10.20	11.69	13.09	14.85	22.34	32.01	35.17	38.08	41.64	44.18
24	9.89	10.86	12.40	13.85	15.66	23.34	33.20	36.42	39.36	42.98	45.56
25	10.52	11.52	13.12	14.61	16.47	24.34	34.28	37.65	40.65	44.31	46.93
26	11.16	12.20	13.84	15.38	17.29	25.34	35.56	38.89	41.92	45.64	48.29
27	11.81	12.88	14.57	16.15	18.11	26.34	36.74	40.11	43.19	46.96	49.65
28	12.46	13.57	15.31	16.93	18.94	27.34	37.92	41.34	44.46	48.28	50.99
29	13.12	14.26	16.05	17.71	19.77	28.34	39.09	42.56	45.72	49.59	52.34
30	13.79	14.95	16.79	18.49	20.60	29.34	40.26	43.77	46.98	50.89	53.67
40	20.71	22.16	24.43	26.51	29.05	39.34	51.81	55.76	59.34	63.69	66.77
50	27.99	29.71	32.36	34.76	37.69	49.33	63.17	67.50	71.42	76.15	79.49
60	35.53	37.48	40.48	43.19	46.46	59.33	74.40	79.08	83.30	88.38	91.95
70	43.28	45.44	48.76	51.74	55.33	69.33	85.53	90.53	95.02	100.42	104.22
80	51.17	53.54	57.15	60.39	64.28	79.33	96.58	101.88	106.63	112.33	116.32
90	59.20	61.75	65.65	69.13	73.29	89.33	107.57	113.14	118.14	124.12	128.30
100	67.33	70.06	74.22	77.93	82.36	99.33	118.50	124.34	129.56	135.81	140.17

ν = degrees of freedom.

TABLE A.4 Percentage Points $f_{\alpha,\mu,\nu}$ of the F Distribution

Degrees of Freedom for the Numerator (μ)

ν	1	2	3	4	5	6	7	8	9	10	12	15	20	24	30	40	60	120	∞
1	5.83	7.50	8.20	8.58	8.82	8.98	9.10	9.19	9.26	9.32	9.41	9.49	9.58	9.63	9.67	9.71	9.76	9.80	9.85
2	2.57	3.00	3.15	3.23	3.28	3.31	3.34	3.35	3.37	3.38	3.39	3.41	3.43	3.43	3.44	3.45	3.46	3.47	3.48
3	2.02	2.28	2.36	2.39	2.41	2.42	2.43	2.44	2.44	2.44	2.45	2.46	2.46	2.46	2.47	2.47	2.47	2.47	2.47
4	1.81	2.00	2.05	2.06	2.07	2.08	2.08	2.08	2.08	2.08	2.08	2.08	2.08	2.08	2.08	2.08	2.08	2.08	2.08
5	1.69	1.85	1.88	1.89	1.89	1.89	1.89	1.89	1.89	1.89	1.89	1.89	1.88	1.88	1.88	1.88	1.87	1.87	1.87
6	1.62	1.76	1.78	1.79	1.79	1.78	1.78	1.78	1.77	1.77	1.77	1.76	1.76	1.75	1.75	1.75	1.74	1.74	1.74
7	1.57	1.70	1.72	1.72	1.71	1.71	1.70	1.70	1.70	1.69	1.68	1.68	1.67	1.67	1.66	1.66	1.65	1.65	1.65
8	1.54	1.66	1.67	1.66	1.66	1.65	1.64	1.64	1.63	1.63	1.62	1.62	1.61	1.60	1.60	1.59	1.59	1.58	1.58
9	1.51	1.62	1.63	1.63	1.62	1.61	1.60	1.60	1.59	1.59	1.58	1.57	1.56	1.56	1.55	1.54	1.54	1.53	1.53
10	1.49	1.60	1.60	1.59	1.59	1.58	1.57	1.56	1.56	1.55	1.54	1.53	1.52	1.52	1.51	1.51	1.50	1.49	1.48
11	1.47	1.58	1.58	1.57	1.56	1.55	1.54	1.53	1.53	1.52	1.51	1.50	1.49	1.49	1.48	1.47	1.47	1.46	1.45
12	1.46	1.56	1.56	1.55	1.54	1.53	1.52	1.51	1.51	1.50	1.49	1.48	1.47	1.46	1.45	1.45	1.44	1.43	1.42
13	1.45	1.55	1.55	1.53	1.52	1.51	1.50	1.49	1.49	1.48	1.47	1.46	1.45	1.44	1.43	1.42	1.42	1.41	1.40
14	1.44	1.53	1.53	1.52	1.51	1.50	1.49	1.48	1.47	1.46	1.45	1.44	1.43	1.42	1.41	1.41	1.40	1.39	1.38

Degrees of Freedom for the Denominator (ν)

Degrees of Freedom for the Denominator (ν)																			
15	1.52	1.52	1.51	1.49	1.48	1.47	1.46	1.46	1.45	1.44	1.43	1.41	1.41	1.40	1.39	1.38	1.37	1.36	1.43
16	1.51	1.51	1.50	1.48	1.47	1.46	1.45	1.44	1.44	1.43	1.41	1.40	1.39	1.38	1.37	1.36	1.35	1.34	1.42
17	1.51	1.50	1.49	1.47	1.46	1.45	1.44	1.43	1.43	1.41	1.40	1.39	1.38	1.37	1.36	1.35	1.34	1.33	1.42
18	1.50	1.49	1.48	1.46	1.45	1.44	1.43	1.42	1.42	1.40	1.39	1.38	1.37	1.36	1.35	1.34	1.33	1.32	1.41
19	1.49	1.49	1.47	1.46	1.44	1.43	1.42	1.41	1.41	1.40	1.38	1.37	1.36	1.35	1.34	1.33	1.32	1.30	1.41
20	1.49	1.48	1.47	1.45	1.44	1.43	1.42	1.41	1.40	1.39	1.37	1.36	1.35	1.34	1.33	1.32	1.31	1.29	1.40
21	1.48	1.48	1.46	1.44	1.43	1.42	1.41	1.40	1.39	1.38	1.37	1.35	1.34	1.33	1.32	1.31	1.30	1.28	1.40
22	1.48	1.47	1.45	1.44	1.42	1.41	1.40	1.39	1.39	1.37	1.36	1.34	1.33	1.32	1.31	1.30	1.29	1.28	1.40
23	1.47	1.47	1.45	1.43	1.42	1.41	1.40	1.39	1.38	1.37	1.35	1.34	1.33	1.32	1.31	1.30	1.28	1.27	1.39
24	1.47	1.46	1.44	1.43	1.41	1.40	1.39	1.38	1.38	1.36	1.35	1.33	1.32	1.31	1.30	1.29	1.28	1.26	1.39
25	1.47	1.46	1.44	1.42	1.41	1.40	1.39	1.38	1.37	1.36	1.34	1.33	1.32	1.31	1.29	1.28	1.27	1.25	1.39
26	1.46	1.45	1.44	1.42	1.41	1.39	1.38	1.37	1.37	1.35	1.34	1.32	1.31	1.30	1.29	1.28	1.26	1.25	1.38
27	1.46	1.45	1.43	1.42	1.40	1.39	1.38	1.37	1.36	1.35	1.33	1.32	1.31	1.30	1.28	1.27	1.26	1.24	1.38
28	1.46	1.45	1.43	1.41	1.40	1.39	1.38	1.37	1.36	1.34	1.33	1.31	1.30	1.29	1.28	1.27	1.25	1.24	1.38
29	1.45	1.45	1.43	1.41	1.40	1.38	1.37	1.36	1.35	1.34	1.32	1.31	1.30	1.29	1.27	1.26	1.25	1.23	1.38
30	1.45	1.44	1.42	1.41	1.39	1.38	1.37	1.36	1.35	1.34	1.32	1.30	1.29	1.28	1.27	1.26	1.24	1.23	1.38
40	1.44	1.42	1.40	1.39	1.37	1.35	1.34	1.33	1.31	1.30	1.28	1.26	1.25	1.24	1.22	1.21	1.19	1.19	1.36
60	1.42	1.41	1.38	1.37	1.35	1.33	1.32	1.31	1.29	1.27	1.25	1.24	1.22	1.21	1.19	1.17	1.15	1.15	1.35
120	1.40	1.39	1.37	1.35	1.33	1.31	1.30	1.29	1.26	1.24	1.22	1.21	1.19	1.18	1.16	1.13	1.10	1.10	1.34
∞	1.39	1.37	1.35	1.33	1.31	1.28	1.27	1.25	1.24	1.22	1.19	1.18	1.16	1.14	1.12	1.08	1.00	—	1.32

(continued)

TABLE A.4 (*Continued*)

$$f_{0.10,\mu,\nu}$$

ν	Degrees of Freedom for the Numerator (μ)																		
	1	2	3	4	5	6	7	8	9	10	12	15	20	24	30	40	60	120	∞
1	39.86	49.50	53.59	55.83	57.24	58.20	58.91	59.44	59.86	60.19	60.71	61.22	61.74	62.00	62.26	62.53	62.79	63.06	63.33
2	8.53	9.00	9.16	9.24	9.29	9.33	9.35	9.37	9.38	9.39	9.41	9.42	9.44	9.45	9.46	9.47	9.47	9.48	9.49
3	5.54	5.46	5.39	5.34	5.31	5.28	5.27	5.25	5.24	5.23	5.22	5.20	5.18	5.18	5.17	5.16	5.15	5.14	5.13
4	4.54	4.32	4.19	4.11	4.05	4.01	3.98	3.95	3.94	3.92	3.90	3.87	3.84	3.83	3.82	3.80	3.79	3.78	3.76
5	4.06	3.78	3.62	3.52	3.45	3.40	3.37	3.34	3.32	3.30	3.27	3.24	3.21	3.19	3.17	3.16	3.14	3.12	3.10
6	3.78	3.46	3.29	3.18	3.11	3.05	3.01	2.98	2.96	2.94	2.90	2.87	2.84	2.82	2.80	2.78	2.76	2.74	2.72
7	3.59	3.26	3.07	2.96	2.88	2.83	2.78	2.75	2.72	2.70	2.67	2.63	2.59	2.58	2.56	2.54	2.51	2.49	2.47
8	3.46	3.11	2.92	2.81	2.73	2.67	2.62	2.59	2.56	2.54	2.50	2.46	2.42	2.40	2.38	2.36	2.34	2.32	2.29
9	3.36	3.01	2.81	2.69	2.61	2.55	2.51	2.47	2.44	2.42	2.38	2.34	2.30	2.28	2.25	2.23	2.21	2.18	2.16
10	3.29	2.92	2.73	2.61	2.52	2.46	2.41	2.38	2.35	2.32	2.28	2.24	2.20	2.18	2.16	2.13	2.11	2.08	2.06
11	3.23	2.86	2.66	2.54	2.45	2.39	2.34	2.30	2.27	2.25	2.21	2.17	2.12	2.10	2.08	2.05	2.03	2.00	1.97
12	3.18	2.81	2.61	2.48	2.39	2.33	2.28	2.24	2.21	2.19	2.15	2.10	2.06	2.04	2.01	1.99	1.96	1.93	1.90
13	3.14	2.76	2.56	2.43	2.35	2.28	2.23	2.20	2.16	2.14	2.10	2.05	2.01	1.98	1.96	1.93	1.90	1.88	1.85
14	3.10	2.73	2.52	2.39	2.31	2.24	2.19	2.15	2.12	2.10	2.05	2.01	1.96	1.94	1.91	1.89	1.86	1.83	1.80
15	3.07	2.70	2.49	2.36	2.27	2.21	2.16	2.12	2.09	2.06	2.02	1.97	1.92	1.90	1.87	1.85	1.82	1.79	1.76

Degrees of Freedom for the Denominator (ν)

(continued)

Degrees of Freedom for the Denominator (ν)																			
16	3.05	2.67	2.46	2.33	2.24	2.18	2.13	2.09	2.06	2.03	1.99	1.94	1.89	1.87	1.84	1.81	1.78	1.75	1.72
17	3.03	2.64	2.44	2.31	2.22	2.15	2.10	2.06	2.03	2.00	1.96	1.91	1.86	1.84	1.81	1.78	1.75	1.72	1.69
18	3.01	2.62	2.42	2.29	2.20	2.13	2.08	2.04	2.00	1.98	1.93	1.89	1.84	1.81	1.78	1.75	1.72	1.69	1.66
19	2.99	2.61	2.40	2.27	2.18	2.11	2.06	2.02	1.98	1.96	1.91	1.86	1.81	1.79	1.76	1.73	1.70	1.67	1.63
20	2.97	2.59	2.38	2.25	2.16	2.09	2.04	2.00	1.96	1.94	1.89	1.84	1.79	1.77	1.74	1.71	1.68	1.64	1.61
21	2.96	2.57	2.36	2.23	2.14	2.08	2.02	1.98	1.95	1.92	1.87	1.83	1.78	1.75	1.72	1.69	1.66	1.62	1.59
22	2.95	2.56	2.35	2.22	2.13	2.06	2.01	1.97	1.93	1.90	1.86	1.81	1.76	1.73	1.70	1.67	1.64	1.60	1.57
23	2.94	2.55	2.34	2.21	2.11	2.05	1.99	1.95	1.92	1.89	1.84	1.80	1.74	1.72	1.69	1.66	1.62	1.59	1.55
24	2.93	2.54	2.33	2.19	2.10	2.04	1.98	1.94	1.91	1.88	1.83	1.78	1.73	1.70	1.67	1.64	1.61	1.57	1.53
25	2.92	2.53	2.32	2.18	2.09	2.02	1.97	1.93	1.89	1.87	1.82	1.77	1.72	1.69	1.66	1.63	1.59	1.56	1.52
26	2.91	2.52	2.31	2.17	2.08	2.01	1.96	1.92	1.88	1.86	1.81	1.76	1.71	1.68	1.65	1.61	1.58	1.54	1.50
27	2.90	2.51	2.30	2.17	2.07	2.00	1.95	1.91	1.87	1.85	1.80	1.75	1.70	1.67	1.64	1.60	1.57	1.53	1.49
28	2.89	2.50	2.29	2.16	2.06	2.00	1.94	1.90	1.87	1.84	1.79	1.74	1.69	1.66	1.63	1.59	1.56	1.52	1.48
29	2.89	2.50	2.28	2.15	2.06	1.99	1.93	1.89	1.86	1.83	1.78	1.73	1.68	1.65	1.62	1.58	1.55	1.51	1.47
30	2.88	2.49	2.28	2.14	2.03	1.98	1.93	1.88	1.85	1.82	1.77	1.72	1.67	1.64	1.61	1.57	1.54	1.50	1.46
40	2.84	2.44	2.23	2.09	2.00	1.93	1.87	1.83	1.79	1.76	1.71	1.66	1.61	1.57	1.54	1.51	1.47	1.42	1.38
60	2.79	2.39	2.18	2.04	1.95	1.87	1.82	1.77	1.74	1.71	1.66	1.60	1.54	1.51	1.48	1.44	1.40	1.35	1.29
120	2.75	2.35	2.13	1.99	1.90	1.82	1.77	1.72	1.68	1.65	1.60	1.55	1.48	1.45	1.41	1.37	1.32	1.26	1.19
∞	2.71	2.30	2.08	1.94	1.85	1.77	1.72	1.67	1.63	1.60	1.55	1.49	1.42	1.38	1.34	1.30	1.24	1.17	1.00

TABLE A.4 (*Continued*)

$$f_{0.05,\mu,\nu}$$

Degrees of Freedom for the Numerator (μ)

ν	1	2	3	4	5	6	7	8	9	10	12	15	20	24	30	40	60	120	∞
1	161.4	199.5	215.7	224.6	230.2	234.0	236.8	238.9	240.5	241.9	243.9	245.9	248.0	249.1	250.1	251.1	252.2	253.3	254.3
2	18.51	19.00	19.16	19.25	19.30	19.33	19.35	19.37	19.38	19.40	19.41	19.43	19.45	19.45	19.46	19.47	19.48	19.49	19.50
3	10.13	9.55	9.28	9.12	9.01	8.94	8.89	8.85	8.81	8.79	8.74	8.70	8.66	8.64	8.62	8.59	8.57	8.55	8.53
4	7.71	6.94	6.59	6.39	6.26	6.16	6.09	6.04	6.00	5.96	5.91	5.86	5.80	5.77	5.75	5.72	5.69	5.66	5.63
5	6.61	5.79	5.41	5.19	5.05	4.95	4.88	4.82	4.77	4.74	4.68	4.62	4.56	4.53	4.50	4.46	4.43	4.40	4.36
6	5.99	5.14	4.76	4.53	4.39	4.28	4.21	4.15	4.10	4.06	4.00	3.94	3.87	3.84	3.81	3.77	3.74	3.70	3.67
7	5.59	4.74	4.35	4.12	3.97	3.87	3.79	3.73	3.68	3.64	3.57	3.51	3.44	3.41	3.38	3.34	3.30	3.27	3.23
8	5.32	4.46	4.07	3.84	3.69	3.58	3.50	3.44	3.39	3.35	3.28	3.22	3.15	3.12	3.08	3.04	3.01	2.97	2.93
9	5.12	4.26	3.86	3.63	3.48	3.37	3.29	3.23	3.18	3.14	3.07	3.01	2.94	2.90	2.86	2.83	2.79	2.75	2.71
10	4.96	4.10	3.71	3.48	3.33	3.22	3.14	3.07	3.02	2.98	2.91	2.85	2.77	2.74	2.70	2.66	2.62	2.58	2.54
11	4.84	3.98	3.59	3.36	3.20	3.09	3.01	2.95	2.90	2.85	2.79	2.72	2.65	2.61	2.57	2.53	2.49	2.45	2.40
12	4.75	3.89	3.49	3.26	3.11	3.00	2.91	2.85	2.80	2.75	2.69	2.62	2.54	2.51	2.47	2.43	2.38	2.34	2.30
13	4.67	3.81	3.41	3.18	3.03	2.92	2.83	2.77	2.71	2.67	2.60	2.53	2.46	2.42	2.38	2.34	2.30	2.25	2.21
14	4.60	3.74	3.34	3.11	2.96	2.85	2.76	2.70	2.65	2.60	2.53	2.46	2.39	2.35	2.31	2.27	2.22	2.18	2.13
15	4.54	3.68	3.29	3.06	2.90	2.79	2.71	2.64	2.59	2.54	2.48	2.40	2.33	2.29	2.25	2.20	2.16	2.11	2.07

Degrees of Freedom for the Denominator (ν)

(continued)

Degrees of Freedom for the Denominator (ν)

ν																			
16	4.49	3.63	3.24	3.01	2.85	2.74	2.66	2.59	2.54	2.49	2.42	2.35	2.28	2.24	2.19	2.15	2.11	2.06	2.01
17	4.45	3.59	3.20	2.96	2.81	2.70	2.61	2.55	2.49	2.45	2.38	2.31	2.23	2.19	2.15	2.10	2.06	2.01	1.96
18	4.41	3.55	3.16	2.93	2.77	2.66	2.58	2.51	2.46	2.41	2.34	2.27	2.19	2.15	2.11	2.06	2.02	1.97	1.92
19	4.38	3.52	3.13	2.90	2.74	2.63	2.54	2.48	2.42	2.38	2.31	2.23	2.16	2.11	2.07	2.03	1.98	1.93	1.88
20	4.35	3.49	3.10	2.87	2.71	2.60	2.51	2.45	2.39	2.35	2.28	2.20	2.12	2.08	2.04	1.99	1.95	1.90	1.84
21	4.32	3.47	3.07	2.84	2.68	2.57	2.49	2.42	2.37	2.32	2.25	2.18	2.10	2.05	2.01	1.96	1.92	1.87	1.81
22	4.30	3.44	3.05	2.82	2.66	2.55	2.46	2.40	2.34	2.30	2.23	2.15	2.07	2.03	1.98	1.94	1.89	1.84	1.78
23	4.28	3.42	3.03	2.80	2.64	2.53	2.44	2.37	2.32	2.27	2.20	2.13	2.05	2.01	1.96	1.91	1.86	1.81	1.76
24	4.26	3.40	3.01	2.78	2.62	2.51	2.42	2.36	2.30	2.25	2.18	2.11	2.03	1.98	1.94	1.89	1.84	1.79	1.73
25	4.24	3.39	2.99	2.76	2.60	2.49	2.40	2.34	2.28	2.24	2.16	2.09	2.01	1.96	1.92	1.87	1.82	1.77	1.71
26	4.23	3.37	2.98	2.74	2.59	2.47	2.39	2.32	2.27	2.22	2.15	2.07	1.99	1.95	1.90	1.85	1.80	1.75	1.69
27	4.21	3.35	2.96	2.73	2.57	2.46	2.37	2.31	2.25	2.20	2.13	2.06	1.97	1.93	1.88	1.84	1.79	1.73	1.67
28	4.20	3.34	2.95	2.71	2.56	2.45	2.36	2.29	2.24	2.19	2.12	2.04	1.96	1.91	1.87	1.82	1.77	1.71	1.65
29	4.18	3.33	2.93	2.70	2.55	2.43	2.35	2.28	2.22	2.18	2.10	2.03	1.94	1.90	1.85	1.81	1.75	1.70	1.64
30	4.17	3.32	2.92	2.69	2.53	2.42	2.33	2.27	2.21	2.16	2.09	2.01	1.93	1.89	1.84	1.79	1.74	1.68	1.62
40	4.08	3.23	2.84	2.61	2.45	2.34	2.25	2.18	2.12	2.08	2.00	1.92	1.84	1.79	1.74	1.69	1.64	1.58	1.51
60	4.00	3.15	2.76	2.53	2.37	2.25	2.17	2.10	2.04	1.99	1.92	1.84	1.75	1.70	1.65	1.59	1.53	1.47	1.39
120	3.92	3.07	2.68	2.45	2.29	2.17	2.09	2.02	1.96	1.91	1.83	1.75	1.66	1.61	1.55	1.50	1.43	1.35	1.25
∞	3.84	3.00	2.60	2.37	2.21	2.10	2.01	1.94	1.88	1.83	1.75	1.67	1.57	1.52	1.46	1.39	1.32	1.22	1.00

TABLE A.4 (Continued)

$$f_{0.25,\mu,\nu}$$

ν	Degrees of Freedom for the Numerator (μ)																		
	1	2	3	4	5	6	7	8	9	10	12	15	20	24	30	40	60	120	∞
1	647.8	799.5	864.2	899.6	921.8	937.1	948.2	956.7	963.3	968.6	976.7	984.9	993.1	997.2	1001	1006	1010	1014	1018
2	38.51	39.00	39.17	39.25	39.30	39.33	39.36	39.37	39.39	39.40	39.41	39.43	39.45	39.46	39.46	39.47	39.48	39.49	39.50
3	17.44	16.04	15.44	15.10	14.88	14.73	14.62	14.54	14.47	14.42	14.34	14.25	14.17	14.12	14.08	14.04	13.99	13.95	13.90
4	12.22	10.65	9.98	9.60	9.36	9.20	9.07	8.98	8.90	8.84	8.75	8.66	8.56	8.51	8.46	8.41	8.36	8.31	8.26
5	10.01	8.43	7.76	7.39	7.15	6.98	6.85	6.76	6.68	6.62	6.52	6.43	6.33	6.28	6.23	6.18	6.12	6.07	6.02
6	8.81	7.26	6.60	6.23	5.99	5.82	5.70	5.60	5.52	5.46	5.37	5.27	5.17	5.12	5.07	5.01	4.96	4.90	4.85
7	8.07	6.54	5.89	5.52	5.29	5.12	4.99	4.90	4.82	4.76	4.67	4.57	4.47	4.42	4.36	4.31	4.25	4.20	4.14
8	7.57	6.06	5.42	5.05	4.82	4.65	4.53	4.43	4.36	4.30	4.20	4.10	4.00	3.95	3.89	3.84	3.78	3.73	3.67
9	7.21	5.71	5.08	4.72	4.48	4.32	4.20	4.10	4.03	3.96	3.87	3.77	3.67	3.61	3.56	3.51	3.45	3.39	3.33
10	6.94	5.46	4.83	4.47	4.24	4.07	3.95	3.85	3.78	3.72	3.62	3.52	3.42	3.37	3.31	3.26	3.20	3.14	3.08
11	6.72	5.26	4.63	4.28	4.04	3.88	3.76	3.66	3.59	3.53	3.43	3.33	3.23	3.17	3.12	3.06	3.00	2.94	2.88
12	6.55	5.10	4.47	4.12	3.89	3.73	3.61	3.51	3.44	3.37	3.28	3.18	3.07	3.02	2.96	2.91	2.85	2.79	2.72
13	6.41	4.97	4.35	4.00	3.77	3.60	3.48	3.39	3.31	3.25	3.15	3.05	2.95	2.89	2.84	2.78	2.72	2.66	2.60
14	6.30	4.86	4.24	3.89	3.66	3.50	3.38	3.29	3.21	3.15	3.05	2.95	2.84	2.79	2.73	2.67	2.61	2.55	2.49
15	6.20	4.77	4.15	3.80	3.58	3.41	3.29	3.20	3.12	3.06	2.96	2.86	2.76	2.70	2.64	2.59	2.52	2.46	2.40

Degrees of Freedom for the Denominator (ν)

Degrees of Freedom for the Denominator (v)																			
16	6.12	4.69	4.08	3.73	3.50	3.34	3.22	3.12	3.05	2.99	2.89	2.79	2.68	2.63	2.57	2.51	2.45	2.38	2.32
17	6.04	4.62	4.01	3.66	3.44	3.28	3.16	3.06	2.98	2.92	2.82	2.72	2.62	2.56	2.50	2.44	2.38	2.32	2.25
18	5.98	4.56	3.95	3.61	3.38	3.22	3.10	3.01	2.93	2.87	2.77	2.67	2.56	2.50	2.44	2.38	2.32	2.26	2.19
19	5.92	4.51	3.90	3.56	3.33	3.17	3.05	2.96	2.88	2.82	2.72	2.62	2.51	2.45	2.39	2.33	2.27	2.20	2.13
20	5.87	4.46	3.86	3.51	3.29	3.13	3.01	2.91	2.84	2.77	2.68	2.57	2.46	2.41	2.35	2.29	2.22	2.16	2.09
21	5.83	4.42	3.82	3.48	3.25	3.09	2.97	2.87	2.80	2.73	2.64	2.53	2.42	2.37	2.31	2.25	2.18	2.11	2.04
22	5.79	4.38	3.78	3.44	3.22	3.05	2.93	2.84	2.76	2.70	2.60	2.50	2.39	2.33	2.27	2.21	2.14	2.08	2.00
23	5.75	4.35	3.75	3.41	3.18	3.02	2.90	2.81	2.73	2.67	2.57	2.47	2.36	2.30	2.24	2.18	2.11	2.04	1.97
24	5.72	4.32	3.72	3.38	3.15	2.99	2.87	2.78	2.70	2.64	2.54	2.44	2.33	2.27	2.21	2.15	2.08	2.01	1.94
25	5.69	4.29	3.69	3.35	3.13	2.97	2.85	2.75	2.68	2.61	2.51	2.41	2.30	2.24	2.18	2.12	2.05	1.98	1.91
26	5.66	4.27	3.67	3.33	3.10	2.94	2.82	2.73	2.65	2.59	2.49	2.39	2.28	2.22	2.16	2.09	2.03	1.95	1.88
27	5.63	4.24	3.65	3.31	3.08	2.92	2.80	2.71	2.63	2.57	2.47	2.36	2.25	2.19	2.13	2.07	2.00	1.93	1.85
28	5.61	4.22	3.63	3.29	3.06	2.90	2.78	2.69	2.61	2.55	2.45	2.34	2.23	2.17	2.11	2.05	1.98	1.91	1.83
29	5.59	4.20	3.61	3.27	3.04	2.88	2.76	2.67	2.59	2.53	2.43	2.32	2.21	2.15	2.09	2.03	1.96	1.89	1.81
30	5.57	4.18	3.59	3.25	3.03	2.87	2.75	2.65	2.57	2.51	2.41	2.31	2.20	2.14	2.07	2.01	1.94	1.87	1.79
40	5.42	4.05	3.46	3.13	2.90	2.74	2.62	2.53	2.45	2.39	2.29	2.18	2.07	2.01	1.94	1.88	1.80	1.72	1.64
60	5.29	3.93	3.34	3.01	2.79	2.63	2.51	2.41	2.33	2.27	2.17	2.06	1.94	1.88	1.82	1.74	1.67	1.58	1.48
120	5.15	3.80	3.23	2.89	2.67	2.52	2.39	2.30	2.22	2.16	2.05	1.94	1.82	1.76	1.69	1.61	1.53	1.43	1.31
∞	5.02	3.69	3.12	2.79	2.57	2.41	2.29	2.19	2.11	2.05	1.94	1.83	1.71	1.64	1.57	1.48	1.39	1.27	1.00

(continued)

TABLE A.4 (*Continued*)

$f_{0.01,\mu,v}$

v		Degrees of Freedom for the Numerator (μ)																	
	1	2	3	4	5	6	7	8	9	10	12	15	20	24	30	40	60	120	∞
1	4052	4999.5	5403	5625	5764	5859	5928	5982	6022	6056	6106	6157	6209	6235	6261	6287	6313	6339	6366
2	98.50	99.00	99.17	99.25	99.30	99.33	99.36	99.37	99.39	99.40	99.42	99.43	99.45	99.46	99.47	99.47	99.48	99.49	99.50
3	34.12	30.82	29.46	28.71	28.24	27.91	27.67	27.49	27.35	27.23	27.05	26.87	26.69	26.00	26.50	26.41	26.32	26.22	26.13
4	21.20	18.00	16.69	15.98	15.52	15.21	14.98	14.80	14.66	14.55	14.37	14.20	14.02	13.93	13.84	13.75	13.65	13.56	13.46
5	16.26	13.27	12.06	11.39	10.97	10.67	10.46	10.29	10.16	10.05	9.89	9.72	9.55	9.47	9.38	9.29	9.20	9.11	9.02
6	13.75	10.92	9.78	9.15	8.75	8.47	8.26	8.10	7.98	7.87	7.72	7.56	7.40	7.31	7.23	7.14	7.06	6.97	6.88
7	12.25	9.55	8.45	7.85	7.46	7.19	6.99	6.84	6.72	6.62	6.47	6.31	6.16	6.07	5.99	5.91	5.82	5.74	5.65
8	11.26	8.65	7.59	7.01	6.63	6.37	6.18	6.03	5.91	5.81	5.67	5.52	5.36	5.28	5.20	5.12	5.03	4.95	4.46
9	10.56	8.02	6.99	6.42	6.06	5.80	5.61	5.47	5.35	5.26	5.11	4.96	4.81	4.73	4.65	4.57	4.48	4.40	4.31
10	10.04	7.56	6.55	5.99	5.64	5.39	5.20	5.06	4.94	4.85	4.71	4.56	4.41	4.33	4.25	4.17	4.08	4.00	3.91
11	9.65	7.21	6.22	5.67	5.32	5.07	4.89	4.74	4.63	4.54	4.40	4.25	4.10	4.02	3.94	3.86	3.78	3.69	3.60
12	9.33	6.93	5.95	5.41	5.06	4.82	4.64	4.50	4.39	4.30	4.16	4.01	3.86	3.78	3.70	3.62	3.54	3.45	3.36
13	9.07	6.70	5.74	5.21	4.86	4.62	4.44	4.30	4.19	4.10	3.96	3.82	3.66	3.59	3.51	3.43	3.34	3.25	3.17
14	8.86	6.51	5.56	5.04	4.69	4.46	4.28	4.14	4.03	3.94	3.80	3.66	3.51	3.43	3.35	3.27	3.18	3.09	3.00
15	8.68	6.36	5.42	4.89	4.36	4.32	4.14	4.00	3.89	3.80	3.67	3.52	3.37	3.29	3.21	3.13	3.05	2.96	2.87
16	8.53	6.23	5.29	4.77	4.44	4.20	4.03	3.89	3.78	3.69	3.55	3.41	3.26	3.18	3.10	3.02	2.93	2.84	2.75
17	8.40	6.11	5.18	4.67	4.34	4.10	3.93	3.79	3.68	3.59	3.46	3.31	3.16	3.08	3.00	2.92	2.83	2.75	2.65

Degrees of Freedom for the Denominator (v)

Degrees of Freedom for the Denominator (v)																			
18	8.29	6.01	5.09	4.58	4.25	4.01	3.84	3.71	3.60	3.51	3.37	3.23	3.08	3.00	2.92	2.84	2.75	2.66	2.57
19	8.18	5.93	5.01	4.50	4.17	3.94	3.77	3.63	3.52	3.43	3.30	3.15	3.00	2.92	2.84	2.76	2.67	2.58	2.59
20	8.10	5.85	4.94	4.43	4.10	3.87	3.70	3.56	3.46	3.37	3.23	3.09	2.94	2.86	2.78	2.69	2.61	2.52	2.42
21	8.02	5.78	4.87	4.37	4.04	3.81	3.64	3.51	3.40	3.31	3.17	3.03	2.88	2.80	2.72	2.64	2.55	2.46	2.36
22	7.95	5.72	4.82	4.31	3.99	3.76	3.59	3.45	3.35	3.26	3.12	2.98	2.83	2.75	2.67	2.58	2.50	2.40	2.31
23	7.88	5.66	4.76	4.26	3.94	3.71	3.54	3.41	3.30	3.21	3.07	2.93	2.78	2.70	2.62	2.54	2.45	2.35	2.26
24	7.82	5.61	4.72	4.22	3.90	3.67	3.50	3.36	3.26	3.17	3.03	2.89	2.74	2.66	2.58	2.49	2.40	2.31	2.21
25	7.77	5.57	4.68	4.18	3.85	3.63	3.46	3.32	3.22	3.13	2.99	2.85	2.70	2.62	2.54	2.45	2.36	2.27	2.17
26	7.72	5.53	4.64	4.14	3.82	3.59	3.42	3.29	3.18	3.09	2.96	2.81	2.66	2.58	2.50	2.42	2.33	2.23	2.13
27	7.68	5.49	4.60	4.11	3.78	3.56	3.39	3.26	3.15	3.06	2.93	2.78	2.63	2.55	2.47	2.38	2.29	2.20	2.10
28	7.64	5.45	4.57	4.07	3.75	3.53	3.36	3.23	3.12	3.03	2.90	2.75	2.60	2.52	2.44	2.35	2.26	2.17	2.06
29	7.60	5.42	4.54	4.04	3.73	3.50	3.33	3.20	3.09	3.00	2.87	2.73	2.57	2.49	2.41	2.33	2.23	2.14	2.03
30	7.56	5.39	4.51	4.02	3.70	3.47	3.30	3.17	3.07	2.98	2.84	2.70	2.55	2.47	2.39	2.30	2.21	2.11	2.01
40	7.31	5.18	4.31	3.83	3.51	3.29	3.12	2.99	2.89	2.80	2.66	2.52	2.37	2.29	2.20	2.11	2.02	1.92	1.80
60	7.08	4.98	4.13	3.65	3.34	3.12	2.95	2.82	2.72	2.63	2.50	2.35	2.20	2.12	2.03	1.94	1.84	1.73	1.60
120	6.85	4.79	3.95	3.48	3.17	2.96	2.79	2.66	2.56	2.47	2.34	2.19	2.03	1.95	1.86	1.76	1.66	1.53	1.38
∞	6.63	4.61	3.78	3.32	3.02	2.80	2.64	2.51	2.41	2.32	2.18	2.04	1.88	1.79	1.70	1.59	1.47	1.32	1.00

TABLE A.5 Critical Values of the Durbin–Watson Statistic

Sample Size	Probability in Lower Tail (Significance Level = α)	k = Number of Regressors (Excluding the Intercept)									
		1		2		3		4		5	
		d_L	d_U	d_L	d_U	d_L	d_U	d_L	d_U	d_L	d_U
	0.01	0.81	1.07	0.70	1.25	0.59	1.46	0.49	1.70	0.39	1.96
15	0.025	0.95	1.23	0.83	1.40	0.71	1.61	0.59	1.84	0.48	2.09
	0.05	1.08	1.36	0.95	1.54	0.82	1.75	0.69	1.97	0.56	2.21
	0.01	0.95	1.15	0.86	1.27	0.77	1.41	0.63	1.57	0.60	1.74
20	0.025	1.08	1.28	0.99	1.41	0.89	1.55	0.79	1.70	0.70	1.87
	0.05	1.20	1.41	1.10	1.54	1.00	1.68	0.90	1.83	0.79	1.99
	0.01	1.05	1.21	0.98	1.30	0.90	1.41	0.83	1.52	0.75	1.65
25	0.025	1.13	1.34	1.10	1.43	1.02	1.54	0.94	1.65	0.86	1.77
	0.05	1.20	1.45	1.21	1.55	1.12	1.66	1.04	1.77	0.95	1.89
	0.01	1.13	1.26	1.07	1.34	1.01	1.42	0.94	1.51	0.88	1.61
30	0.025	1.25	1.38	1.18	1.46	1.12	1.54	1.05	1.63	0.98	1.73
	0.05	1.35	1.49	1.28	1.57	1.21	1.65	1.14	1.74	1.07	1.83
	0.01	1.25	1.34	1.20	1.40	1.15	1.46	1.10	1.52	1.05	1.58
40	0.025	1.35	1.45	1.30	1.51	1.25	1.57	1.20	1.63	1.15	1.69
	0.05	1.44	1.54	1.39	1.60	1.34	1.66	1.29	1.72	1.23	1.79
	0.01	1.32	1.40	1.28	1.45	1.24	1.49	1.20	1.54	1.16	1.59
50	0.025	1.42	1.50	1.38	1.54	1.34	1.59	1.30	1.64	1.26	1.69
	0.05	1.50	1.59	1.46	1.63	1.42	1.67	1.38	1.72	1.34	1.77
	0.01	1.38	1.45	1.35	1.48	1.32	1.52	1.28	1.56	1.25	1.60
60	0.025	1.47	1.54	1.44	1.57	1.40	1.61	1.37	1.65	1.33	1.69
	0.05	1.55	1.62	1.51	1.65	1.48	1.69	1.44	1.73	1.41	1.77
	0.01	1.47	1.52	1.44	1.54	1.42	1.57	1.39	1.60	1.36	1.62
80	0.025	1.54	1.59	1.52	1.62	1.49	1.65	1.47	1.67	1.44	1.70
	0.05	1.61	1.66	1.59	1.69	1.56	1.72	1.53	1.74	1.51	1.77
	0.01	1.52	1.56	1.50	1.58	1.48	1.60	1.45	1.63	1.44	1.65
100	0.025	1.59	1.63	1.57	1.65	1.55	1.67	1.53	1.70	1.51	1.72
	0.05	1.65	1.69	1.63	1.72	1.61	1.74	1.59	1.76	1.57	1.78

Source: Adapted from J. Durbin and G. S. Watson [1951]. Testing for serial correlation in least squares regression II.*Biometrika* **38**, with permission of the publisher.

APPENDIX B

DATA SETS FOR EXERCISES

Introduction to Time Series Analysis and Forecasting, Second Edition.
Douglas C. Montgomery, Cheryl L. Jennings and Murat Kulahci.
© 2015 John Wiley & Sons, Inc. Published 2015 by John Wiley & Sons, Inc.

TABLE B.1   Market Yield on US Treasury Securities at 10-Year Constant Maturity

Month	Rate (%)	Month	Rate (%)	Month	Rate (%)	Month	Rate (%)
Apr-1953	2.83	Oct-1966	5.01	Apr-1980	11.47	Oct-1993	5.33
May-1953	3.05	Nov-1966	5.16	May-1980	10.18	Nov-1993	5.72
Jun-1953	3.11	Dec-1966	4.84	Jun-1980	9.78	Dec-1993	5.77
Jul-1953	2.93	Jan-1967	4.58	Jul-1980	10.25	Jan-1994	5.75
Aug-1953	2.95	Feb-1967	4.63	Aug-1980	11.10	Feb-1994	5.97
Sep-1953	2.87	Mar-1967	4.54	Sep-1980	11.51	Mar-1994	6.48
Oct-1953	2.66	Apr-1967	4.59	Oct-1980	11.75	Apr-1994	6.97
Nov-1953	2.68	May-1967	4.85	Nov-1980	12.68	May-1994	7.18
Dec-1953	2.59	Jun-1967	5.02	Dec-1980	12.84	Jun-1994	7.10
Jan-1954	2.48	Jul-1967	5.16	Jan-1981	12.57	Jul-1994	7.30
Feb-1954	2.47	Aug-1967	5.28	Feb-1981	13.19	Aug-1994	7.24
Mar-1954	2.37	Sep-1967	5.30	Mar-1981	13.12	Sep-1994	7.46
Apr-1954	2.29	Oct-1967	5.48	Apr-1981	13.68	Oct-1994	7.74
May-1954	2.37	Nov-1967	5.75	May-1981	14.10	Nov-1994	7.96
Jun-1954	2.38	Dec-1967	5.70	Jun-1981	13.47	Dec-1994	7.81
Jul-1954	230	Jan-1968	5.53	Jul-1981	14.28	Jan-1995	7.78
Aug-1954	2.36	Feb-1968	5.56	Aug-1981	14.94	Feb-1995	7.47
Sep-1954	2.38	Mar-1968	5.74	Sep-1981	15.32	Mar-1995	7.20
Oct-1954	2.43	Apr-1968	564	Oct-1981	15.15	Apr-1995	7.06
Nov-1954	2.48	May-1968	5.87	Nov-1981	13.39	May-1995	6.63
Dec-1954	2.51	Jun-1968	5.72	Dec-1981	13.72	Jun-1995	6.17
Jan-1955	2.61	Jul-1968	5.50	Jan-1982	14.59	Jul-1995	6.28
Feb-1955	2.65	Aug-1968	5.42	Feb-1982	14.43	Aug-1995	6.49
Mar-1955	2.68	Sep-1968	5.46	Mar-1982	13.86	Sep-1995	6.20
Apr-1955	2.75	Oct-1968	5.58	Apr-1982	13.87	Oct-1995	6.04
May-1955	2.76	Nov-1968	5.70	May-1982	13.62	Nov-1995	5.93
Jun-1955	2.78	Dec-1968	6.03	Jun-1982	14.30	Dec-1995	5.71
Jul-1955	2.90	Jan-1969	6.04	Jul-1982	13.95	Jan-1996	5.65
Aug-1955	2.97	Feb-1969	6.19	Aug-1982	13.06	Feb-1996	5.81
Sep-1955	2.97	Mar-1969	6.30	Sep-1982	12.34	Mar-1996	6.27
Oct-1955	2.88	Apr-1969	6.17	Oct-1982	10.91	Apr-1996	6.51
Nov-1955	2.89	May-1969	6.32	Nov-1982	10.55	May-1996	6.74
Dec-1955	2.96	Jun-1969	6.57	Dec-1982	10.54	Jun-1996	6.91
Jan-1956	2.90	Jul-1969	6.72	Jan-1983	10.46	Jul-1996	6.87
Feb-1956	2.84	Aug-1969	6.69	Feb-1983	10.72	Aug-1996	6.64
Mar-1956	2.96	Sep-1969	7.16	Mar-1983	10.51	Sep-1996	6.83
Apr-1956	3.18	Oct-1969	7.10	Apr-1983	10.40	Oct-1996	6.53
May-1956	3.07	Nov-1969	7.14	May-1983	10.38	Nov-1996	6.20
Jun-1956	3.00	Dec-1969	7.65	Jun-1983	10.85	Dec-1996	6.30

(*continued*)

TABLE B.1 (*Continued*)

Month	Rate (%)	Month	Rate (%)	Month	Rate (%)	Month	Rate (%)
Jul-1956	3.11	Jan-1970	7.79	Jul-1983	11.38	Jan-1997	6.58
Aug-1956	3.33	Feb-1970	7.24	Aug-1983	11.85	Feb-1997	6.42
Sep-1956	3.38	Mar-1970	7.07	Sep-1983	11.65	Mar-1997	6.69
Oct-1956	3.34	Apr-1970	7.39	Oct-1983	11.54	Apr-1997	6.89
Nov-1956	3.49	May-1970	7.91	Nov-1983	11.69	May-1997	6.71
Dec-1956	3.59	Jun-1970	7.84	Dec-1983	11.83	Jun-1997	6.49
Jan-1957	3.46	Jul-1970	7.46	Jan-1984	11.67	Jul-1997	6.22
Feb-1957	3.34	Aug-1970	7.53	Feb-1984	11.84	Aug-1997	6.30
Mar-1957	3.41	Sep-1970	7.39	Mar-1984	12.32	Sep-1997	6.21
Apr-1957	3.48	Oct-1970	7.33	Apr-1984	12.63	Oct-1997	6.03
May-1957	3.60	Nov-1970	6.84	May-1984	13.41	Nov-1997	5.88
Jun-1957	3.80	Dec-1970	6.39	Jun-1984	13.56	Dec-1997	5.81
Jul-1957	3.93	Jan-1971	6.24	Jul-1984	13.36	Jan-1998	5.54
Aug-1957	3.93	Feb-1971	6.11	Aug-1984	12.72	Feb-1998	5.57
Sep-1957	3.92	Mar-1971	5.70	Sep-1984	12.52	Mar-1998	5.65
Oct-1957	3.97	Apr-1971	5.83	Oct-1984	12.16	Apr-1998	5.64
Nov-1957	3.72	May-1971	6.39	Nov-1984	11.57	May-1998	5.65
Dec-1957	3.21	Jun-1971	6.52	Dec-1984	11.50	Jun-1998	5.50
Jan-1958	3.09	Jul-1971	6.73	Jan-1985	11.38	Jul-1998	5.46
Feb-1958	3.05	Aug-1971	6.58	Feb-1985	11.51	Aug-1998	5.34
Mar-1958	2.98	Sep-1971	6.14	Mar-1985	11.86	Sep-1998	4.81
Apr-1958	2.88	Oct-1971	5.93	Apr-1985	11.43	Oct-1998	4.53
May-1958	2.92	Nov-1971	5.81	May-1985	10.85	Nov-1998	4.83
Jun-1958	2.97	Dec-1971	5.93	Jun-1985	10.16	Dec-1998	4.65
Jul-1958	3.20	Jan-1972	5.95	Jul-1985	10.31	Jan-1999	4.72
Aug-1958	3.54	Feb-1972	6.08	Aug-1985	10.33	Feb-1999	5.00
Sep-1958	3.76	Mar-1972	6.07	Sep-1985	10.37	Mar-1999	5.23
Oct-1958	3.80	Apr-1972	6.19	Oct-1985	10.24	Apr-1999	5.18
Nov-1958	3.74	May-1972	6.13	Nov-1985	9.78	May-1999	5.54
Dec-1958	3.86	Jun-1972	6.11	Dec-1985	9.26	Jun-1999	5.90
Jan-1959	4.02	Jul-1972	6.11	Jan-1986	9.19	Jul-1999	5.79
Feb-1959	3.96	Aug-1972	6.21	Feb-1986	8.70	Aug-1999	5.94
Mar-1959	3.99	Sep-1972	6.55	Mar-1986	7.78	Sep-1999	5.92
Apr-1959	4.12	Oct-1972	6.48	Apr-1986	7.30	Oct-1999	6.11
May-1959	4.31	Nov-1972	6.28	May-1986	7.71	Nov-1999	6.03
Jun-1959	4.34	Dec-1972	6.36	Jun-1986	7.80	Dec-1999	6.28
Jul-1959	4.40	Jan-1973	6.46	Jul-1986	7.30	Jan-2000	6.66
Aug-1959	4.43	Feb-1973	6.64	Aug-1986	7.17	Feb-2000	6.52
Sep-1959	4.68	Mar-1973	6.71	Sep-1986	7.45	Mar-2000	6.26
Oct-1959	4.53	Apr-1973	6.67	Oct-1986	7.43	Apr-2000	5.99
Nov-1959	4.53	May-1973	6.85	Nov-1986	7.25	May-2000	6.44

TABLE B.1   (*Continued*)

Month	Rate (%)	Month	Rate (%)	Month	Rate (%)	Month	Rate (%)
Dec-1959	4.69	Jun-1973	6.90	Dec-1986	7.11	Jun-2000	6.10
Jan-1960	4.72	Jul-1973	7.13	Jan-1987	7.08	Jul-2000	6.05
Feb-1960	4.49	Aug-1973	7.40	Feb-1987	7.25	Aug-2000	5.83
Mar-1960	4.25	Sep-1973	7.09	Mar-1987	7.25	Sep-2000	5.80
Apr-1960	4.28	Oct-1973	6.79	Apr-1987	8.02	Oct-2000	5.74
May-1960	4.35	Nov-1973	6.73	May-1987	8.61	Nov-2000	5.72
Jun-1960	4.15	Dec-1973	6.74	Jun-1987	8.40	Dec-2000	5.24
Jul-1960	3.90	Jan-1974	6.99	Jul-1987	8.45	Jan-2001	5.16
Aug-1960	3.80	Feb-1974	6.96	Aug-1987	8.76	Feb-2001	5.10
Sep-1960	3.80	Mar-1974	7.21	Sep-1987	9.42	Mar-2001	4.89
Oct-1960	3.89	Apr-1974	7.51	Oct-1987	9.52	Apr-2001	5.14
Nov-1960	3.93	May-1974	7.58	Nov-1987	8.86	May-2001	5.39
Dec-1960	3.84	Jun-1974	7.54	Dec-1987	8.99	Jun-2001	5.28
Jan-1961	3.84	Jul-1974	7.81	Jan-1988	8.67	Jul-2001	5.24
Feb-1961	3.78	Aug-1974	8.04	Feb-1988	8.21	Aug-2001	4.97
Mar-1961	3.74	Sep-1974	8.04	Mar-1988	8.37	Sep-2001	4.73
Apr-1961	3.78	Oct-1974	7.90	Apr-1988	8.72	Oct-2001	4.57
May-1961	3.71	Nov-1974	7.68	May-1988	9.09	Nov-2001	4.65
Jun-1961	3.88	Dec-1974	7.43	Jun-1988	8.92	Dec-2001	5.09
Jul-1961	3.92	Jan-1975	7.50	Jul-1988	9.06	Jan-2002	5.04
Aug-1961	4.04	Feb-1975	7.39	Aug-1988	9.26	Feb-2002	4.91
Sep-1961	3.98	Mar-1975	7.73	Sep-1988	8.98	Mar-2002	5.28
Oct-1961	3.92	Apr-1975	8.23	Oct-1988	8.80	Apr-2002	5.21
Nov-1961	3.94	May-1975	8.06	Nov-1988	8.96	May-2002	5.16
Dec-1961	4.06	Jun-1975	7.86	Dec-1988	9.11	Jun-2002	4.93
Jan-1962	4.08	Jul-1975	8.06	Jan-1989	9.09	Jul-2002	4.65
Feb-1962	4.04	Aug-1975	8.40	Feb-1989	9.17	Aug-2002	4.26
Mar-1962	3.93	Sep-1975	8.43	Mar-1989	9.36	Sep-2002	3.87
Apr-1962	3.84	Oct-1975	8.14	Apr-1989	9.18	Oct-2002	3.94
May-1962	3.87	Nov-1975	8.05	May-1989	8.86	Nov-2002	4.05
Jun-1962	3.91	Dec-1975	8.00	Jun-1989	8.28	Dec-2002	4.03
Jul-1962	4.01	Jan-1976	7.74	Jul-1989	8.02	Jan-2003	4.05
Aug-1962	3.98	Feb-1976	7.79	Aug-1989	8.11	Feb-2003	3.90
Sep-1962	3.98	Mar-1976	7.73	Sep-1989	8.19	Mar-2003	3.81
Oct-1962	3.93	Apr-1976	7.56	Oct-1989	8.01	Apr-2003	3.96
Nov-1962	3.92	May-1976	7.90	Nov-1989	7.87	May-2003	3.57
Dec-1962	3.86	Jun-1976	7.86	Dec-1989	7.84	Jun-2003	3.33
Jan-1963	3.83	Jul-1976	7.83	Jan-1990	8.21	Jul-2003	3.98
Feb-1963	3.92	Aug-1976	7.77	Feb-1990	8.47	Aug-2003	4.45
Mar-1963	3.93	Sep-1976	7.59	Mar-1990	8.59	Sep-2003	4.27
Apr-1963	3.97	Oct-1976	7.41	Apr-1990	8.79	Oct-2003	4.29

(*continued*)

TABLE B.1 (*Continued*)

Month	Rate (%)	Month	Rate (%)	Month	Rate (%)	Month	Rate (%)
May-1963	3.93	Nov-1976	7.29	May-1990	8.76	Nov-2003	4.30
Jun-1963	3.99	Dec-1976	6.87	Jun-1990	8.48	Dec-2003	4.27
Jul-1963	4.02	Jan-1977	7.21	Jul-1990	8.47	Jan-2004	4.15
Aug-1963	4.00	Feb-1977	7.39	Aug-1990	8.75	Feb-2004	4.08
Sep-1963	4.08	Mar-1977	7.46	Sep-1990	8.89	Mar-2004	3.83
Ocl-1963	4.11	Apr-1977	7.37	Oct-1990	8.72	Apr-2004	4.35
Nov-1963	4.12	May-1977	7.46	Nov-1990	8.39	May-2004	4.72
Dec-1963	4.13	Jun-1977	7.28	Dec-1990	8.08	Jun-2004	4.73
Jan-1964	4.17	Jul-1977	7.33	Jan-1991	8.09	Jul-2004	4.50
Feb-1964	4.15	Aug-1977	7.40	Feb-1991	7.85	Aug-2004	4.28
Mar-1964	4.22	Sep-1977	7.34	Mar-1991	8.11	Sep-2004	4.13
Apr-1964	4.23	Oct-1977	7.52	Apr-1991	8.04	Oct-2004	4.10
May-1964	4.20	Nov-1977	7.58	May-1991	8.07	Nov-2004	4.19
Jun-1964	4.17	Dec-1977	7.69	Jun-1991	8.28	Dec-2004	4.23
Jul-1964	4.19	Jan-1978	7.96	Jul-1991	8.27	Jan-2005	4.22
Aug-1964	4.19	Feb-1978	8.03	Aug-1991	7.90	Feb-2005	4.17
Sep-1964	4.20	Mar-1978	8.04	Sep-1991	7.65	Mar-2005	4.50
Oct-1964	4.19	Apr-1978	8.15	Oct-1991	7.53	Apr-2005	4.34
Nov-1964	4.15	May-1978	8.35	Nov-1991	7.42	May-2005	4.14
Dec-1964	4.18	Jun-1978	8.46	Dec-1991	7.09	Jun-2005	4.00
Jan-1965	4.19	Jul-1978	8.64	Jan-1992	7.03	Jul-2005	4.18
Feb-1965	4.21	Aug-1978	8.41	Feb-1992	7.34	Aug-2005	4.26
Mar-1965	4.21	Sep-1978	8.42	Mar-1992	7.54	Sep-2005	4.20
Apr-1965	4.20	Oct-1978	8.64	Apr-1992	7.48	Oct-2005	4.46
May-1965	4.21	Nov-1978	8.81	May-1992	7.39	Nov-2005	4.54
Jun-1965	4.21	Dec-1978	9.01	Jun-1992	7.26	Dec-2005	4.47
Jul-1965	4.20	Jan-1979	9.10	Jul-1992	6.84	Jan-2006	4.42
Aug-1965	4.25	Feb-1979	9.10	Aug-1992	6.59	Feb-2006	4.57
Sep-1965	4.29	Mar-1979	9.12	Sep-1992	6.42	Mar-2006	4.72
Oct-1965	4.35	Apr-1979	9.18	Oct-1992	6.59	Apr-2006	4.99
Nov-1965	4.45	May-1979	9.25	Nov-1992	6.87	May-2006	5.11
Dec-1965	4.62	Jun-1979	8.91	Dec-1992	6.77	Jun-2006	5.11
Jan-1966	4.61	Jul-1979	8.95	Jan-1993	6.60	Jul-2006	5.09
Feb-1966	4.83	Aug-1979	9.03	Feb-1993	6.26	Aug-2006	4.88
Mar-1966	4.87	Sep-1979	9.33	Mar-1993	5.98	Sep-2006	4.72
Apr-1966	4.75	Oct-1979	10.30	Apr-1993	5.97	Oct-2006	4.73
May-1966	4.78	Nov-1979	10.65	May-1993	6.04	Nov-2006	4.60
Jun-1966	4.81	Dec-1979	10.39	Jun-1993	5.96	Dec-2006	4.56
Jul-1966	5.02	Jan-1980	10.80	Jul-1993	5.81	Jan-2007	4.76
Aug-1966	5.22	Feb-1980	12.41	Aug-1993	5.68	Feb-2007	4.72
Sep-1966	5.18	Mar-1980	12.75	Sep-1993	5.36		

TABLE B.2 Pharmaceutical Product Sales

Week	Sales (In Thousands)	Week	Sales (In Thousands)	Week	Sales (In Thousands)	Week	Sales (In Thousands)
1	10618.1	31	10334.5	61	10538.2	91	10375.4
2	10537.9	32	10480.1	62	10286.2	92	10123.4
3	10209.3	33	10387.6	63	10171.3	93	10462.7
4	10553.0	34	10202.6	64	10393.1	94	10205.5
5	9934.9	35	10219.3	65	10162.3	95	10522.7
6	10534.5	36	10382.7	66	10164.5	96	10253.2
7	10196.5	37	10820.5	67	10327.0	97	10428.7
8	10511.8	38	10358.7	68	10365.1	98	10615.8
9	10089.6	39	10494.6	69	10755.9	99	10417.3
10	10371.2	40	10497.6	70	10463.6	100	10445.4
11	10239.4	41	10431.5	71	10080.5	101	10690.6
12	10472.4	42	10447.8	72	10479.6	102	10271.8
13	10827.2	43	10684.4	73	9980.9	103	10524.8
14	10640.8	44	10176.5	74	10039.2	104	9815.0
15	10517.8	45	10616.0	75	10246.1	105	10398.5
16	10154.2	46	10627.7	76	10368.0	106	10553.1
17	9969.2	47	10684.0	77	10446.3	107	10655.8
18	10260.4	48	10246.7	78	10535.3	108	10199.1
19	10737.0	49	10265.0	79	10786.9	109	10416.6
20	10430.0	50	10090.4	80	9975.8	110	10391.3
21	10689.0	51	9881.1	81	10160.9	111	10210.1
22	10430.4	52	10449.7	82	10422.1	112	10352.5
23	10002.4	53	10276.3	83	10757.2	113	10423.8
24	10135.7	54	10175.2	84	10463.8	114	10519.3
25	10096.2	55	10212.5	85	10307.0	115	10596.7
26	10288.7	56	10395.5	86	10134.7	116	10650.0
27	10289.1	57	10545.9	87	10207.7	117	10741.6
28	10589.9	58	10635.7	88	10488.0	118	10246.0
29	10551.9	59	10265.2	89	10262.3	119	10354.4
30	10208.3	60	10551.6	90	10785.9	120	10155.4

TABLE B.3 Chemical Process Viscosity

Time Period	Reading	Time Period	Reading	Time Period	Reading	Time Period	Reading
1	86.7418	26	87.2397	51	85.5722	76	84.7052
2	85.3195	27	87.5219	52	83.7935	77	83.8168
3	84.7355	28	86.4992	53	84.3706	78	82.4171
4	85.1113	29	85.6050	54	83.3762	79	83.0420
5	85.1487	30	86.8293	55	84.9975	80	83.6993
6	84.4775	31	84.5004	56	84.3495	81	82.2033
7	84.6827	32	84.1844	57	85.3395	82	82.1413
8	84.6757	33	85.4563	58	86.0503	83	81.7961
9	86.3169	34	86.1511	59	84.8839	84	82.3241
10	88.0006	35	86.4142	60	85.4176	85	81.5316
11	86.2597	36	86.0498	61	84.2309	86	81.7280
12	85.8286	37	86.6642	62	83.5761	87	82.5375
13	83.7500	38	84.7289	63	84.1343	88	82.3877
14	84.4628	39	85.9523	64	82.6974	89	82.4159
15	84.6476	40	86.8473	65	83.5454	90	82.2102
16	84.5751	41	88.4250	66	86.4714	91	82.7673
17	82.2473	42	89.6481	67	86.2143	92	83.1234
18	83.3774	43	87.8566	68	87.0215	93	83.2203
19	83.5385	44	88.4997	69	86.6504	94	84.4510
20	85.1620	45	87.0622	70	85.7082	95	84.9145
21	83.7881	46	85.1973	71	86.1504	96	85.7609
22	84.0421	47	85.0767	72	85.8032	97	85.2302
23	84.1023	48	84.4362	73	85.6197	98	86.7312
24	84.8495	49	84.2112	74	84.2339	99	87.0048
25	87.6416	50	85.9952	75	83.5737	100	85.0572

TABLE B.4   US Production of Blue and Gorgonzola Cheeses

Year	Production (10^3 lb)	Year	Production (10^3 lb)
1950	7,657	1974	28,262
1951	5,451	1975	28,506
1952	10,883	1976	33,885
1953	9,554	1977	34,776
1954	9,519	1978	35,347
1955	10,047	1979	34,628
1956	10,663	1980	33,043
1957	10,864	1981	30,214
1958	11,447	1982	31,013
1959	12,710	1983	31,496
1960	15,169	1984	34,115
1961	16,205	1985	33,433
1962	14,507	1986	34,198
1963	15,400	1987	35,863
1964	16,800	1988	37,789
1965	19,000	1989	34,561
1966	20,198	1990	36,434
1967	18,573	1991	34,371
1968	19,375	1992	33,307
1969	21,032	1993	33,295
1970	23,250	1994	36,514
1971	25,219	1995	36,593
1972	28,549	1996	38,311
1973	29,759	1997	42,773

Source: http://www.nass.usda.gov/QuickStats/.

TABLE B.5 US Beverage Manufacturer Product Shipments, Unadjusted

Month	Dollars (In Millions)	Month	Dollars (In Millions)	Month	Dollars (In Millions)	Month	Dollars (In Millions)
Jan-1992	3519	Oct-1995	4681	Jul-1999	5339	Apr-2003	5576
Feb-1992	3803	Nov-1995	4466	Aug-1999	5474	May-2003	6160
Mar-1992	4332	Dec-1995	4463	Sep-1999	5278	Jun-2003	6121
Apr-1992	4251	Jan-1996	4217	Oct-1999	5184	Jul-2003	5900
May-1992	4661	Feb-1996	4322	Nov-1999	4975	Aug-2003	5994
Jun-1992	4811	Mar-1996	4779	Dec-1999	4751	Sep-2003	5841
Jul-1992	4448	Apr-1996	4988	Jan-2000	4600	Oct-2003	5832
Aug-1992	4451	May-1996	5383	Feb-2000	4718	Nov-2003	5505
Sep-1992	4343	Jun-1996	5591	Mar-2000	5218	Dec-2003	5573
Oct-1992	4067	Jul-1996	5322	Apr-2000	5336	Jan-2004	5331
Nov-1992	4001	Aug-1996	5404	May-2000	5665	Feb-2004	5355
Dec-1992	3934	Sep-1996	5106	Jun-2000	5900	Mar-2004	6057
Jan-1993	3652	Oct-1996	4871	Jul-2000	5330	Apr-2004	6055
Feb-1993	3768	Nov-1996	4977	Aug-2000	5626	May-2004	6771
Mar-1993	4082	Dec-1996	4706	Sep-2000	5512	Jun-2004	6669
Apr-1993	4101	Jan-1997	4193	Oct-2000	5293	Jul-2004	6375
May-1993	4628	Feb-1997	4460	Nov-2000	5143	Aug-2004	6666
Jun-1993	4898	Mar-1997	4956	Dec-2000	4842	Sep-2004	6383
Jul-1993	4476	Apr-1997	5022	Jan-2001	4627	Oct-2004	6118
Aug-1993	4728	May-1997	5408	Feb-2001	4881	Nov-2004	5927
Sep-1993	4458	Jun-1997	5565	Mar-2001	5321	Dec-2004	5750
Oct-1993	4004	Jul-1997	5360	Apr-2001	5290	Jan-2005	5122

Month	Value	Month	Value	Month	Value	Month	Value
Nov-1993	4095	Aug-1997	5490	May-2001	6002	Feb-2005	5398
Dec-1993	4056	Sep-1997	5286	Jun-2001	5811	Mar-2005	5817
Jan-1994	3641	Oct-1997	5257	Jul-2001	5671	Apr-2005	6163
Feb-1994	3966	Nov-1997	5002	Aug-2001	6102	May-2005	6763
Mar-1994	4417	Dec-1997	4897	Sep-2001	5482	Jun-2005	6835
Apr-1994	4367	Jan-1998	4577	Oct-2001	5429	Jul-2005	6678
May-1994	4821	Feb-1998	4764	Nov-2001	5356	Aug-2005	6821
Jun-1994	5190	Mar-1998	5052	Dec-2001	5167	Sep-2005	6421
Jul-1994	4638	Apr-1998	5251	Jan-2002	4608	Oct-2005	6338
Aug-1994	4904	May-1998	5558	Feb-2002	4889	Nov-2005	6265
Sep-1994	4528	Jun-1998	5931	Mar-2002	5352	Dec-2005	6291
Oct-1994	4383	Jul-1998	5476	Apr-2002	5441	Jan-2006	5540
Nov-1994	4339	Aug-1998	5603	May-2002	5970	Feb-2006	5822
Dec-1994	4327	Sep-1998	5425	Jun-2002	5750	Mar-2006	6318
Jan-1995	3856	Oct-1998	5177	Jul-2002	5670	Apr-2006	6268
Feb-1995	4072	Nov-1998	4792	Aug-2002	5860	May-2006	7270
Mar-1995	4563	Dec-1998	4776	Sep-2002	5449	Jun-2006	7096
Apr-1995	4561	Jan-1999	4450	Oct-2002	5401	Jul-2006	6505
May-1995	4984	Feb-1999	4659	Nov-2002	5240	Aug-2006	7039
Jun-1995	5316	Mar-1999	5043	Dec-2002	5229	Sep-2006	6440
Jul-1995	4843	Apr-1999	5233	Jan-2003	4770	Oct-2006	6446
Aug-1995	5383	May-1999	5423	Feb-2003	5006	Nov-2006	6717
Sep-1995	4889	Jun-1999	5814	Mar-2003	5518	Dec-2006	6320

Source: http://www.census.gov/indicator/www/m3/nist/nalcshist2.htm.

TABLE B.6 Global Mean Surface Air Temperature Anomaly and Global CO_2 Concentration

Year	Anomaly (°C)	CO_2 (ppmv)	Year	Anomaly (°C)	CO_2 (ppmv)	Year	Anomaly (°C)	CO_2 (ppmv)
1880	-0.11	290.7	1922	-0.09	303.8	1964	-0.25	319.2
1881	-0.13	291.2	1923	-0.16	304.1	1965	-0.15	320.0
1882	-0.01	291.7	1924	-0.11	304.5	1966	-0.07	321.1
1883	-0.04	292.1	1925	-0.15	305.0	1967	-0.02	322.0
1884	-0.42	292.6	1926	0.04	305.4	1968	-0.09	322.9
1885	-0.23	293.0	1927	-0.05	305.8	1969	0.00	324.2
1886	-0.25	293.3	1928	0.01	306.3	1970	0.04	325.2
1887	-0.45	293.6	1929	-0.22	306.8	1971	-0.10	326.1
1888	-0.23	293.8	1930	-0.03	307.2	1972	-0.05	327.2
1889	0.04	294.0	1931	0.03	307.7	1973	0.18	328.8
1890	-0.22	294.2	1932	0.04	308.2	1974	-0.06	329.7
1891	-0.55	294.3	1933	-0.11	308.6	1975	-0.02	330.7
1892	-0.40	294.5	1934	0.05	309.0	1976	-0.21	331.8
1893	-0.39	294.6	1935	-0.08	309.4	1977	0.16	333.3
1894	-0.32	294.7	1936	0.01	309.8	1978	0.07	334.6
1895	-0.32	294.8	1937	0.12	310.0	1979	0.13	336.9
1896	-0.27	294.9	1938	0.15	310.2	1980	0.27	338.7
1897	-0.15	295.0	1939	-0.02	310.3	1981	0.40	339.9
1898	-0.21	295.2	1940	0.14	310.4	1982	0.10	341.1
1899	-0.25	295.5	1941	0.11	310.4	1983	0.34	342.8

Year	Anomaly	CO2	Year	Anomaly	CO2	Year	Anomaly	CO2
1900	−0.05	295.8	1942	0.10	310.3	1984	0.16	344.4
1901	−0.05	296.1	1943	0.06	310.2	1985	0.13	345.9
1902	−0.30	296.5	1944	0.10	310.1	1986	0.19	347.2
1903	−0.35	296.8	1945	−0.01	310.1	1987	0.35	348.9
1904	−0.42	297.2	1946	0.01	310.1	1988	0.42	351.5
1905	−0.25	297.6	1947	0.12	310.2	1989	0.28	352.9
1906	−0.15	298.1	1948	−0.03	310.3	1990	0.49	354.2
1907	−0.41	298.5	1949	−0.09	310.5	1991	0.44	355.6
1908	−0.30	298.9	1950	−0.17	310.7	1992	0.16	356.4
1909	−0.31	299.3	1951	−0.02	311.1	1993	0.18	357.0
1910	−0.21	299.7	1952	0.03	311.5	1994	0.31	358.9
1911	−0.25	300.1	1953	0.12	311.9	1995	0.47	360.9
1912	−0.33	300.4	1954	−0.09	312.4	1996	0.36	362.6
1913	−0.28	300.8	1955	−0.09	313.0	1997	0.40	363.8
1914	−0.02	301.1	1956	−0.18	313.6	1998	0.71	366.6
1915	0.06	301.4	1957	0.08	314.2	1999	0.43	368.3
1916	−0.20	301.7	1958	0.10	314.9	2000	0.41	369.5
1917	−0.46	302.1	1959	0.05	315.8	2001	0.56	371.0
1918	−0.33	302.4	1960	−0.02	316.6	2002	0.70	373.1
1919	−0.09	302.7	1961	0.10	317.3	2003	0.66	375.6
1920	−0.15	303.0	1962	0.05	318.1	2004	0.60	377.4
1921	−0.04	303.4	1963	0.03	318.7			

Source: http://data.giss.nasa.gov.gistemp/.

593

TABLE B.7 Whole Foods Market Stock Price, Daily Closing Adjusted for Splits

Date[a]	Dollars	Date	Dollars	Date	Dollars	Date	Dollars	Date	Dollars
1/2/01	28.05	3/15/01	22.01	5/25/01	27.88	8/7/01	32.24	10/23/01	35.20
1/3/01	28.23	3/16/01	22.26	5/29/01	27.78	8/8/01	31.60	10/24/01	35.30
1/4/01	26.25	3/19/01	22.35	5/30/01	28.03	8/9/01	31.78	10/25/01	35.65
1/5/01	25.41	3/20/01	23.06	5/31/01	28.36	8/10/01	32.99	10/26/01	35.96
1/8/01	26.25	3/21/01	22.78	6/1/01	28.31	8/13/01	32.69	10/29/01	35.86
1/9/01	26.03	3/22/01	22.19	6/4/01	27.58	8/14/01	33.31	10/30/01	35.61
1/10/01	26.09	3/23/01	22.19	6/5/01	27.43	8/15/01	32.78	10/31/01	34.42
1/11/01	26.28	3/26/01	22.66	6/6/01	27.16	8/16/01	32.78	11/1/01	34.55
1/12/01	26.00	3/27/01	22.50	6/7/01	27.92	8/17/01	32.82	11/2/01	35.43
1/16/01	25.63	3/28/01	21.36	6/8/01	27.36	8/20/01	33.04	11/5/01	34.92
1/17/01	25.57	3/29/01	20.71	6/11/01	27.17	8/21/01	33.79	11/6/01	35.56
1/18/01	25.57	3/30/01	20.86	6/12/01	27.39	8/22/01	32.69	11/7/01	35.85
1/19/01	25.16	4/2/01	20.95	6/13/01	27.58	8/23/01	32.40	11/8/01	36.89
1/22/01	26.52	4/3/01	20.12	6/14/01	27.55	8/24/01	32.91	11/9/01	37.24
1/23/01	27.18	4/4/01	19.50	6/15/01	27.49	8/27/01	33.38	11/12/01	37.01
1/24/01	26.93	4/5/01	20.30	6/18/01	27.70	8/28/01	34.72	11/13/01	37.52
1/25/01	26.50	4/6/01	20.09	6/19/01	27.19	8/29/01	35.22	11/14/01	37.24
1/26/01	26.50	4/9/01	20.38	6/20/01	26.76	8/30/01	34.77	11/15/01	40.36
1/29/01	27.27	4/10/01	21.13	6/21/01	26.53	8/31/01	34.85	11/16/01	39.42
1/30/01	27.70	4/11/01	20.63	6/22/01	26.45	9/4/01	33.91	11/19/01	40.16
1/31/01	28.17	4/12/01	20.35	6/25/01	25.97	9/5/01	34.39	11/20/01	42.64
2/1/01	28.26	4/16/01	20.39	6/26/01	26.11	9/6/01	34.49	11/21/01	41.86
2/2/01	28.29	4/17/01	20.95	6/27/01	26.50	9/7/01	34.37	11/23/01	42.58
2/5/01	28.23	4/18/01	21.94	6/28/01	26.98	9/10/01	33.44	11/26/01	42.63

Date	Value	Date	Value	Date	Value	Date	Value	Date	Value
2/6/01	28.54	4/19/01	21.43	6/29/01	26.84	9/17/01	33.24	11/27/01	42.14
2/7/01	28.94	4/20/01	21.37	7/2/01	28.03	9/18/01	33.18	11/28/01	41.62
2/8/01	28.51	4/23/01	21.24	7/3/01	28.00	9/19/01	31.26	11/29/01	42.59
2/9/01	27.55	4/24/01	21.13	7/5/01	28.01	9/20/01	31.04	11/30/01	42.50
2/12/01	28.05	4/25/01	22.36	7/6/01	27.20	9/21/01	30.33	12/3/01	42.38
2/13/01	27.98	4/26/01	22.93	7/9/01	27.92	9/24/01	30.69	12/4/01	42.77
2/14/01	23.55	4/27/01	23.26	7/10/01	27.10	9/25/01	30.84	12/5/01	43.80
2/15/01	24.21	4/30/01	24.07	7/11/01	27.15	9/26/01	29.95	12/6/01	45.13
2/16/01	23.92	5/1/01	23.79	7/12/01	27.19	9/27/01	29.22	12/7/01	45.40
2/20/01	23.77	5/2/01	24.56	7/13/01	26.69	9/28/01	31.11	12/10/01	43.81
2/21/01	23.74	5/3/01	24.43	7/16/01	26.79	10/1/01	30.93	12/11/01	42.16
2/22/01	23.55	5/4/01	24.29	7/17/01	27.17	10/2/01	30.98	12/12/01	41.24
2/23/01	23.34	5/7/01	23.33	7/18/01	26.72	10/3/01	32.59	12/13/01	40.91
2/26/01	23.22	5/8/01	25.20	7/19/01	26.33	10/4/01	32.50	12/14/01	41.05
2/27/01	22.87	5/9/01	24.94	7/20/01	26.23	10/5/01	32.12	12/17/01	41.13
2/28/01	21.36	5/10/01	24.95	7/23/01	26.59	10/8/01	32.09	12/18/01	41.55
3/1/01	21.30	5/11/01	25.25	7/24/01	26.82	10/9/01	32.85	12/19/01	41.35
3/2/01	21.51	5/14/01	25.70	7/25/01	27.24	10/10/01	33.44	12/20/01	41.27
3/5/01	21.32	5/15/01	26.33	7/26/01	28.49	10/11/01	32.68	12/21/01	42.46
3/6/01	21.67	5/16/01	27.81	7/27/01	31.65	10/12/01	32.54	12/24/01	42.96
3/7/01	21.48	5/17/01	28.04	7/30/01	34.47	10/15/01	32.07	12/26/01	43.63
3/8/01	21.85	5/18/01	28.75	7/31/01	33.63	10/16/01	33.18	12/27/01	43.63
3/9/01	21.49	5/21/01	28.72	8/1/01	32.58	10/17/01	33.45	12/28/01	43.59
3/12/01	21.48	5/22/01	28.33	8/2/01	32.62	10/18/01	34.35	12/31/01	43.14
3/13/01	22.10	5/23/01	27.61	8/3/01	32.09	10/19/01	33.95		
3/14/01	21.79	5/24/01	27.98	8/6/01	32.41	10/22/01	34.42		

[a]Date: Month/Day/Year.

TABLE B.8 Unemployment Rate—Full-Time Labor Force, Not Seasonally Adjusted

Month	Rate (%)	Month	Rate (%)	Month	Rate (%)	Month	Rate (%)	Month	Rate (%)	Month	Rate (%)
Jan-1963	6.8	Jan-1970	3.8	Jan-1977	7.9	Jan-1984	8.9	Jan-1991	7.0	Jan-1998	5.0
Feb-1963	6.8	Feb-1970	4.3	Feb-1977	8.2	Feb-1984	8.5	Feb-1991	7.4	Feb-1998	4.8
Mar-1963	6.2	Mar-1970	4.2	Mar-1977	7.6	Mar-1984	8.3	Mar-1991	7.1	Mar-1998	4.8
Apr-1963	5.6	Apr-1970	4.1	Apr-1977	6.7	Apr-1984	7.8	Apr-1991	6.6	Apr-1998	4.0
May-1963	5.4	May-1970	4.2	May-1977	6.4	May-1984	7.4	May-1991	6.7	May-1998	4.2
Jun-1963	6.0	Jun-1970	5.5	Jun-1977	7.3	Jun-1984	7.4	Jun-1991	7.0	Jun-1998	4.6
Jul-1963	5.4	Jul-1970	5.1	Jul-1977	6.9	Jul-1984	7.6	Jul-1991	6.9	Jul-1998	4.6
Aug-1963	4.9	Aug-1970	4.7	Aug-1977	6.6	Aug-1984	7.2	Aug-1991	6.5	Aug-1998	4.3
Sep-1963	4.3	Sep-1970	4.5	Sep-1977	6.0	Sep-1984	6.8	Sep-1991	6.3	Sep-1998	4.1
Oct-1963	4.4	Oct-1970	4.5	Oct-1977	5.9	Oct-1994	6.9	Oct-1991	6.3	Oct-1998	3.9
Nov-1963	4.9	Nov-1970	4.9	Nov-1977	5.9	Nov-1984	6.8	Nov-1991	6.6	Nov-1998	3.8
Dec-1963	5.1	Dec-1970	5.2	Dec-1977	5.7	Dec-1984	7.1	Dec-1991	7.0	Dec-1998	3.9
Jan-1964	6.2	Jan-1971	6.1	Jan-1978	6.7	Jan-1985	8.0	Jan-1992	8.1	Jan-1999	4.6
Feb-1964	6.1	Feb-1971	6.2	Feb-1978	6.6	Feb-1985	7.9	Feb-1992	8.3	Feb-1999	4.6
Mar-1964	5.7	Mar-1971	5.9	Mar-1978	6.2	Mar-1985	7.5	Mar-1992	7.9	Mar-1999	4.3
Apr-1964	5.1	Apr-1971	5.4	Apr-1978	5.5	Apr-1985	7.1	Apr-1992	7.4	Apr-1999	4.0
May-1964	4.7	May-1971	5.2	May-1978	5.5	May-1985	7.0	May-1992	7.4	May-1999	3.9
Jun-1964	5.7	Jun-1971	6.4	Jun-1978	6.1	Jun-1985	7.5	Jun-1992	8.0	Jun-1999	4.3
Jul-1964	4.7	Jul-1971	6.0	Jul-1978	6.1	Jul-1985	7.4	Jul-1992	7.8	Jul-1999	4.4
Aug-1964	4.5	Aug-1971	5.6	Aug-1978	5.5	Aug-1985	6.8	Aug-1992	7.3	Aug-1999	4.1
Sep-1964	4.0	Sep-1971	5.1	Sep-1978	5.1	Sep-1985	6.6	Sep-1992	7.0	Sep-1999	3.8
Oct-1964	4.0	Oct-1971	4.8	Oct-1978	4.8	Oct-1985	6.5	Oct-1992	6.7	Oct-1999	3.7
Nov-1964	4.0	Nov-1971	5.1	Nov-1978	5.0	Nov-1986	6.6	Nov-1992	7.0	Nov-1999	3.6

Date	Value	Date	Value	Date	Value	Date	Value	Date	Value	Date	Value
Dec-1964	4.3	Dec-1971	5.2	Dec-1978	5.2	Dec-1985	6.6	Dec-1992	7.1	Dec-1999	3.7
Jan-1965	5.3	Jan-1972	6.1	Jan-1979	5.9	Jan-1986	7.3	Jan-1993	7.9	Jan-2000	4.3
Feb-1965	5.5	Feb-1972	6.0	Feb-1979	6.1	Feb-1986	7.8	Feb-1993	7.9	Feb-2000	4.2
Mar-1965	4.9	Mar-1972	5.8	Mar-1979	5.7	Mar-1986	7.5	Mar-1993	7.5	Mar-2000	4.1
Apr-1965	4.5	Apr-1972	5.2	Apr-1979	5.3	Apr-1986	7.0	Apr-1993	6.9	Apr-2000	3.5
May-1965	4.1	May-1972	5.1	May-1979	5.0	May-1986	7.1	May-1993	6.9	May-2000	3.7
Jun-1965	5.1	Jun-1972	6.0	Jun-1979	5.9	Jun-1986	7.3	Jun-1993	7.2	Jun-2000	4.0
Jul-1965	4.2	Jul-1972	5.7	Jul-1979	5.7	Jul-1986	7.0	Jul-1993	7.1	Jul-2000	4.0
Aug-1965	3.9	Aug-1972	5.2	Aug-1979	5.5	Aug-1986	6.4	Aug-1993	6.5	Aug-2000	3.9
Sep-1965	3.4	Sep-1972	4.6	Sep-1979	5.1	Sep-1986	6.5	Sep-1993	6.2	Sep-2000	3.5
Oct-1965	3.2	Oct-1972	4.5	Oct-1979	5.1	Oct-1986	6.3	Oct-1993	6.1	Oct-2000	3.5
Nov-1965	3.3	Nov-1972	4.2	Nov-1979	5.2	Nov-1986	6.4	Nov-1993	6.0	Nov-2000	3.5
Dec-1965	3.4	Dec-1972	4.2	Dec-1979	5.3	Dec-1986	6.3	Dec-1993	6.2	Dec-2000	3.6
Jan-1966	4.1	Jan-1973	5.1	Jan-1980	6.5	Jan-1987	7.2	Jan-1994	7.5	Jan-2001	4.5
Feb-1966	4.0	Feb-1973	5.2	Feb-1980	6.5	Feb-1987	7.1	Feb-1994	7.4	Feb-2001	4.4
Mar-1966	3.8	Mar-1973	4.9	Mar-1980	6.4	Mar-1987	6.7	Mar-1994	7.0	Mar-2001	4.4
Apr-1966	3.5	Apr-1973	4.4	Apr-1980	6.6	Apr-1987	6.1	Apr-1994	6.3	Apr-2001	4.0
May-1966	3.4	May-1973	4.2	May-1980	7.2	May-1987	6.1	May-1994	5.9	May-2001	4.1
Jun-1966	4.3	Jun-1973	5.1	Jun-1980	8.0	Jun-1987	6.4	Jun-1994	6.3	Jun-2001	4.6
Jul-1966	3.7	Jul-1973	4.7	Jul-1980	8.1	Jul-1987	6.1	Jul-1994	6.4	Jul-2001	4.6
Aug-1966	3.2	Aug-1973	4.3	Aug-1980	7.6	Aug-1987	5.6	Aug-1994	5.8	Aug-2001	4.7
Sep-1966	2.9	Sep-1973	3.9	Sep-1980	6.9	Sep-1987	5.3	Sep-1994	5.5	Sep-2001	4.7
Oct-1966	2.8	Oct-1973	3.6	Oct-1980	6.8	Oct-1987	5.3	Oct-1994	5.3	Oct-2001	4.9
Nov-1966	3.0	Nov-1973	4.0	Nov-1980	7.0	Nov-1987	5.4	Nov-1994	5.2	Nov-2001	5.2
Dec-1966	3.1	Dec-1973	4.1	Dec-1980	7.0	Dec-1987	5.3	Dec-1994	5.0	Dec-2001	5.5
Jan-1967	3.8	Jan-1974	5.2	Jan-1981	8.0	Jan-1988	6.1	Jan-1995	6.1	Jan-2002	6.5

(*continued*)

TABLE B.8 (*Continued*)

Month	Rate (%)	Month	Rate (%)	Month	Rate (%)	Month	Rate (%)	Month	Rate (%)	Month	Rate (%)
Feb-1967	3.6	Feb-1974	5.3	Feb-1981	8.0	Feb-1988	6.1	Feb-1995	5.8	Feb-2002	6.3
Mar-1967	3.5	Mar-1974	5.0	Mar-1981	7.6	Mar-1988	5.8	Mar-1995	5.7	Mar-2002	6.2
Apr-1967	3.2	Apr-1974	4.6	Apr-1981	7.0	Apr-1988	5.2	Apr-1995	5.5	Apr-2002	5.9
May-1967	3.0	May-1974	4.5	May-1981	7.2	May-1988	5.4	May-1995	5.4	May-2002	5.7
Jun-1967	4.3	Jurv1974	5.6	Jun-1981	7.8	Jun-1988	5.4	Jun-1995	5.7	Jun-2002	5.2
Jul-1967	3.7	Jul-1974	5.4	Jul-1981	7.4	Jul-1988	5.4	Jul-1995	5.8	Jul-2002	6.0
Aug-1967	3.4	Aug-1974	4.9	Aug-1981	7.0	Aug-1988	5.2	Aug-1995	5.5	Aug-2002	5.6
Sep-1967	3.1	Sep-1974	4.9	Sep-1981	6.9	Sep-1988	4.8	Sep-1995	5.2	Sep-2002	5.3
Oct-1967	3.1	Oct-1974	5.0	Oct-1981	7.3	Oct-1988	4.7	Oct-1995	5.0	Oct-2002	5.4
Nov-1967	3.0	Nov-1974	5.7	Nov-1981	7.8	Nov-1988	4.9	Nov-1995	5.1	Nov-2002	5.7
Dec-1967	3.0	Dec-1974	6.3	Dec-1981	8.5	Dec-1988	5.0	Dec-1995	5.2	Dec-2002	5.9
Jan-1968	3.7	Jan-1975	8.7	Jan-1982	9.5	Jan-1989	5.7	Jan-1996	6.2	Jan-2003	6.6
Feb-1968	3.8	Feb-1975	9.0	Feb-1982	9.6	Feb-1989	5.5	Feb-1996	5.9	Feb-2003	6.5
Mar-1968	3.4	Mar-1975	9.1	Mar-1982	9.7	Mar-1989	5.2	Mar-1996	5.8	Mar-2003	6.3
Apr-1968	2.9	Apr-1975	8.7	Apr-1982	9.4	Apr-1989	5.0	Apr-1996	5.3	Apr-2003	6.0
May-1968	2.7	May-1975	8.6	May-1982	9.5	May-1989	5.0	May-1996	5.3	May-2003	6.0
Jun-1968	4.2	Jun-1975	9.3	Jun-1982	10.3	Jun-1989	5.3	Jun-1996	5.4	Jun-2003	6.6

Date	Value	Date	Value	Date	Value	Date	Value	Date	Value	Date	Value
Jul-1968	3.7	Jul-1975	8.7	Jul-1982	10.1	Jul-1989	5.3	Jul-1996	5.6	Jul-2003	6.4
Aug-1966	3.1	Aug-1975	7.9	Aug-1982	9.8	Aug-1989	4.9	Aug-1996	4.9	Aug-2003	6.1
Sep-1968	2.7	Sep-1975	7.6	Sep-1982	9.7	Sep-1969	4.7	Sep-1996	4.8	Sep-2003	5.7
Oct-1968	2.7	Oct-1975	7.4	Oct-1982	10.1	Oct-1969	4.6	Oct-1996	4.7	Oct-2003	5.6
Nov-1968	2.6	Nov-1975	7.5	Nov-1962	10.6	Nov-1989	4.9	Nov-1996	4.9	Nov-2003	5.7
Dec-1968	2.5	Dec-1975	7.5	Dec-1962	11.0	Dec-1989	4.9	Dec-1996	4.9	Dec-2003	5.6
Jan-1969	3.3	Jan-1976	8.6	Jan-1983	11.9	Jan-1990	5.8	Jan-1997	5.8	Jan-2004	6.3
Feb-1969	3.3	Feb-1976	8.4	Feb-1983	11.9	Feb-1990	5.6	Feb-1997	5.5	Feb-2004	6.1
Mar-1969	3.1	Mar-1976	8.0	Mar-1983	11.4	Mar-1990	5.4	Mar-1997	5.4	Mar-2004	6.2
Apr-1969	2.9	Apr-1976	7.2	Apr-1983	10.6	Apr-1990	5.2	Apr-1997	4.7	Apr-2004	5.4
May-1969	2.7	May-1976	6.8	May-1983	10.3	May-1990	5.1	May-1997	4.7	May-2004	5.5
Jun-1969	4.0	Jun-1976	8.1	Jun-1983	10.6	Jun-1990	5.3	Jun-1997	5.1	Jun-2004	5.8
Jul-1969	3.6	Jul-1976	7.6	Jul-1983	9.8	Jul-1990	5.4	JuL-1997	5.0	Jul-2004	5.8
Aug-1969	3.1	Aug-1976	7.3	Aug-1983	9.4	Aug-1990	5.2	Aug-1997	4.6	Aug-2004	5.3
Sep-1969	3.0	Sep-1976	6.9	Sep-1983	8.8	Sep-1990	5.3	Sep-1997	4.5	Sep-2004	5.1
Oct-1969	2.8	Oct-1976	6.7	Oct-1983	8.4	Oct-1990	5.2	Oct-1997	4.2	Oct-2004	5.0
Nov-1969	2.7	Nov-1976	7.0	Nov-1983	8.2	Nov-1990	5.7	Nov-1997	4.1	Nov-2004	5.1
Dec-1969	2.8	Dec-1976	7.2	Dec-1983	6.2	Dec-1990	6.0	Dec-1997	4.3	Dec-2004	5.2

Source: http://data.bls.gov/cgi-bin/srgate.

TABLE B.9   International Sunspot Numbers

Year	Sunspot Number	Year	Sunspot Number	Year	Sunspot Number	Year	Sunspot Number	Year	Sunspot Number
1700	5.1	1761	86	1622	4.1	1883	63.8	1944	9.7
1701	11.1	1762	61.3	1623	1.9	1884	63.6	1945	33.3
1702	16.1	1763	45.2	1824	8.6	1885	52.3	1946	92.7
1703	23.1	1764	36.5	1825	16.7	1886	25.5	1947	151.7
1704	36.1	1765	21	1826	36.4	1887	13.2	1948	136.4
1705	58.1	1766	11.5	1827	49.7	1888	6.9	1949	134.8
1706	29.1	1767	37.9	1828	64.3	1889	64	1950	84
1707	20.1	1768	69.9	1829	67.1	1890	7.2	1951	69.5
1708	10.1	1769	106.2	1830	71	1891	35.7	1952	31.6
1709	8.1	1770	100.9	1831	47.9	1892	73.1	1953	14
1710	3.1	1771	81.7	1832	27.6	1893	85.2	1954	4.5
1711	0.1	1772	66.6	1833	8.6	1894	78.1	1955	38.1
1712	0.1	1773	34.9	J834	13.3	1895	64.1	1956	141 .8
1713	2.1	1774	30.7	1835	57	1896	41 .9	1957	190.3
1714	11.1	1775	7.1	1836	121.6	1897	26.3	1958	184.9
1715	27.1	1776	19.9	1837	138.4	1898	26.8	1959	159.1
1716	47.1	1777	92.6	1838	103.3	1899	12.2	1960	112.4
1717	63.1	1778	154. 5	1839	85.8	1900	9.6	1961	54
1718	60.1	1779	126	1840	64.7	1901	2.8	1962	37.7
1719	39.1	1780	84.9	1841	36.8	1902	5.1	1963	28
1720	28.1	1781	68.2	1842	24.3	1903	24.5	1964	10.3
1721	26.1	1782	38.6	1843	10.8	1904	42.1	1965	152
1722	22.1	1783	22.9	1844	15.1	1905	63.6	1966	47.1
1723	11.1	1784	10.3	1845	40.2	1906	53.9	1967	93.8
1724	21.1	1785	24.2	1846	61.6	1907	62.1	1966	106
1725	40.1	1786	83	1847	98.6	1908	48.6	1969	105.6
1726	78.1	1787	132.1	1848	124. 8	1909	44	1970	104.6
1727	122.1	1788	131	1849	96.4	1910	18.7	1971	66.7
1728	103.1	1789	118.2	1850	66.7	1911	5.8	1972	69
1729	73.1	1790	90	1851	64.6	1912	3.7	1973	38.1
1730	47.1	1791	66.7	1852	54.2	1913	1.5	1974	34.6
1731	35.1	1792	60.1	1853	39.1	1914	9.7	1975	15.6
1732	11.1	1793	47	1854	20.7	1915	47.5	1976	12.7
1733	5.1	1794	41.1	1855	6.8	1916	57.2	1977	27.6
1734	16.1	1795	21.4	1856	4.4	1917	104	1978	92.6
1735	34.1	1796	16.1	1857	22.8	1918	80.7	1979	155.5
1736	70.1	1797	6.5	1858	54.9	1919	63.7	1980	154.7
1737	81.1	1798	4.2	1859	93.9	1920	37.7	1981	140.6
1738	111.1	1799	6.9	1860	95.9	1921	26.2	1982	116

TABLE B.9   (*Continued*)

Year	Sunspot Number	Year	Sunspot Number	Year	Sunspot Number	Year	Sunspot Number	Year	Sunspot Number
1739	101.1	1800	14. 6	1861	77.3	1922	14.3	1983	66.7
1740	73.1	1801	34.1	1862	59.2	1923	59	1984	46
1741	40.1	1802	45.1	1863	44.1	1924	16.8	1985	18
1742	20.1	1803	43.2	1864	47.1	1925	44.4	1986	13.5
1743	16.1	1804	47.6	1865	30.6	1926	64	1987	29.3
1744	5.1	1805	42.3	1866	16.4	1927	69.1	1988	100.3
1745	11.1	1606	28.2	1867	7.4	1928	77.9	1989	157.7
1746	22.1	1807	10.2	1868	37.7	1929	65	1990	142.7
1747	40.1	1808	8.2	1869	74.1	1930	35.8	1991	145.8
1748	60.1	1809	2.6	1870	139.1	1931	21.3	1992	94.4
1749	81	1810	0.1	1871	111.3	1932	11.2	1993	54.7
1750	83.5	1811	1.5	1872	101.7	1933	5.8	1994	30
1751	47.8	1812	5.1	1873	66.3	1934	8.8	1995	17.6
1752	47.9	1813	12.3	1874	44.8	1935	36.2	1996	8.7
1753	30.8	1814	14	1875	17.1	1936	79.8	1997	21.6
1754	12.3	1815	35.5	1876	11.4	1937	114.5	1998	64.4
1755	9.7	1816	45.9	1877	12.5	1938	109.7	1999	93.4
1756	10.3	1817	41.1	1878	3.5	1939	88.9	2000	119.7
1757	32.5	1818	30.2	1879	6.1	1940	67.9	2001	111.1
1758	47.7	1819	24	1880	32.4	1941	47.6	2002	104.1
1759	54.1	1620	15.7	1881	54.4	1942	30.7	2003	63.8
1760	63	1821	6.7	1882	59.8	1943	16.4	2004	40.5

Source: http://sidc.oma.be/html/sunspot.html (yearly sunspot number).

TABLE B.10 United Kingdom Airline Miles Flown

Month	Miles (In Millions)	Month	Miles (In Millions)
Jan-1964	7.269	Jul-1967	12.222
Feb-1964	6.775	Aug-1967	12.246
Mar-1964	7.819	Sep-1967	13.281
Apr-1964	8.371	Oct-1967	10.366
May-1964	9.069	Nov-1967	8.730
Jun-1964	10.248	Dec-1967	9.614
Jul-1964	11.030	Jan-1968	8.639
Aug-1964	10.882	Feb-1968	8.772
Sep-1964	10.333	Mar-1968	10.894
Oct-1964	9.109	Apr-1968	10.455
Nov-1964	7.685	May-1968	11.179
Dec-1964	7.682	Jun-1968	10.588
Jan-1965	8.350	Jul-1968	10.794
Feb-1965	7.829	Aug-1968	12.770
Mar-1965	8.829	Sep-1968	13.812
Apr-1965	9.948	Oct-1968	10.857
May-1965	10.638	Nov-1968	9.290
Jun-1965	11.253	Dec-1968	10.925
Jul-1965	11.424	Jan-1969	9.491
Aug-1965	11.391	Feb-1969	8.919
Sep-1965	10.665	Mar-1969	11.607
Oct-1965	9.396	Apr-1969	8.852
Nov-1965	7.775	May-1969	12.537
Dec-1965	7.933	Jun-1969	14.759
Jan-1966	8.186	Jul-1969	13.667
Feb-1966	7.444	Aug-1969	13.731
Mar-1966	8.484	Sep-1969	15.110
Apr-1966	9.864	Oct-1969	12.185
May-1966	10.252	Nov-1969	10.645
Jun-1966	12.282	Dec-1969	12.161
Jul-1966	11.637	Jan-1970	10.840
Aug-1966	11.577	Feb-1970	10.436
Sep-1966	12.417	Mar-1970	13.589
Oct-1966	9.637	Apr-1970	13.402
Nov-1966	8.094	May-1970	13.103
Dec-1966	9.280	Jun-1970	14.933
Jan-1967	8.334	Jul-1970	14.147
Feb-1967	7.899	Aug-1970	14.057
Mar-1967	9.994	Sep-1970	16.234
Apr-1967	10.078	Oct-1970	12.389
May-1967	10.801	Nov-1970	11.594
Jun-1967	12.953	Dec-1970	12.772

Source: Adapted from Montgomery, Johnson, and Gardner (1990), with permission of the publisher.

TABLE B.11   Champagne Sales

Month	Sales (In Thousands of Bottles)	Month	Sales (In Thousands Bottles)	Month	Sales (In Thousands Bottles)
Jan-1962	2.851	Sep-1964	3.528	May-1967	4.968
Feb-1962	2.672	Oct-1964	5.211	Jun-1967	4.677
Mar-1962	2.755	Nov-1964	7.614	Jul-1967	3.523
Apr-1962	2.721	Dec-1964	9.254	Aug-1967	1.821
May-1962	2.946	Jan-1965	5.375	Sep-1967	5.222
Jun-1962	3.036	Feb-1965	3.088	Oct-1967	6.873
Jul-1962	2.282	Mar-1965	3.718	Nov-1967	10.803
Aug-1962	2.212	Apr-1965	4.514	Dec-1967	13.916
Sep-1962	2.922	May-1965	4.520	Jan-1968	2.639
Oct-1962	4.301	Jun-1965	4.539	Feb-1968	2.899
Nov-1962	5.764	Jul-1965	3.663	Mar-1968	3.370
Dec-1962	7.132	Aug-1965	1.643	Apr-1968	3.740
Jan-1963	2.541	Sep-1965	4.739	May-1968	2.927
Feb-1963	2.475	Oct-1965	5.428	Jun-1968	3.986
Mar-1963	3.031	Nov-1965	8.314	Jul-1968	4.217
Apr-1963	3.266	Dec-1965	10.651	Aug-1968	1.738
May-1963	3.776	Jan-1966	3.633	Sep-1968	5.221
Jun-1963	3.230	Feb-1966	4.292	Oct-1968	6.424
Jul-1963	3.028	Mar-1966	4.154	Nov-1968	9.842
Aug-1963	1.759	Apr-1966	4.121	Dec-1968	13.076
Sep-1963	3.595	May-1966	4.647	Jan-1969	3.934
Oct-1963	4.474	Jun-1966	4.753	Feb-1969	3.162
Nov-1963	6.838	Jul-1966	3.965	Mar-1969	4.286
Dec-1963	8.357	Aug-1966	1.723	Apr-1969	4.676
Jan-1964	3.113	Sep-1966	5.048	May-1969	5.010
Feb-1964	3.006	Oct-1966	6.922	Jun-1969	4.874
Mar-1964	4.047	Nov-1966	9.858	Jul-1969	4.633
Apr-1964	3.523	Dec-1966	11.331	Aug-1969	1.659
May-1964	3.937	Jan-1967	4.016	Sep-1969	5.951
Jun-1964	3.986	Feb-1967	3.957	Oct-1969	6.981
Jul-1964	3.260	Mar-1967	4.510	Nov-1969	9.851
Aug-1964	1.573	Apr-1967	4.276	Dec-1969	12.670

TABLE B.12 **Chemical Process Yield, with Operating Temperature (Uncontrolled)**

Hour	Yield (%)	Temperature (°F)	Hour	Yield (%)	Temperature (°F)
1	89.0	153	26	99.4	152
2	90.5	152	27	99.6	153
3	91.5	153	28	99.8	153
4	93.2	153	29	98.8	154
5	93.9	154	30	99.9	154
6	94.6	151	31	98.2	153
7	94.7	153	32	98.7	153
8	93.5	152	33	97.5	154
9	91.2	151	34	97.9	152
10	89.3	150	35	98.3	152
11	85.6	150	36	98.8	151
12	80.3	149	37	99.1	150
13	75.9	149	38	99.2	149
14	75.3	147	39	98.6	148
15	78.3	146	40	95.3	147
16	89.1	143	41	94.2	146
17	88.3	148	42	91.3	148
18	89.2	151	43	90.6	145
19	90.1	152	44	91.2	143
20	94.3	153	45	88.3	145
21	97.7	154	46	84.1	150
22	98.6	152	47	86.5	147
23	98.7	153	48	88.2	150
24	98.9	152	49	89.5	151
25	99.2	152	50	89.5	152

TABLE B.13   US Production of Ice Cream and Frozen Yogurt

Year	Ice Cream (10^3 gal)	Frozen Yogurt (10^3 gal)	Year	Ice Cream (10^3 gal)	Frozen Yogurt (10^3 gal)
1950	554,351	–	1975	836,552	–
1951	568,849	–	1976	818,241	–
1952	592,705	–	1977	809,849	–
1953	605,051	–	1978	815,360	–
1954	596,821	–	1979	811,079	–
1955	628,525	–	1980	829,798	–
1956	641,333	–	1981	832,450	–
1957	650,583	–	1982	852,072	–
1958	657,175	–	1983	881,543	–
1959	698,931	–	1984	894,468	–
1960	697,552	–	1985	901,449	–
1961	697,151	–	1986	923,597	–
1962	704,428	–	1987	928,356	–
1963	717,597	–	1988	882,079	–
1964	738,743	–	1989	831,159	82,454
1965	757,000	–	1990	823,610	117,577
1966	751,159	–	1991	862,638	147,137
1967	745,409	–	1992	866,110	134,067
1968	773,207	–	1993	866,248	149,933
1969	765,501	–	1994	876,097	150,565
1970	761,732	–	1995	862,232	152,097
1971	765,843	–	1996	878,572	114,168
1972	767,750	–	1997	913,770	92,167
1973	773,674	–	1998	937,485	87,777
1974	781,971	–	2000	979,645	94,478

Source: USDA–National Agricultural Statistics Service.

TABLE B.14   Atmospheric CO_2 Concentrations at Mauna Loa Observatory

Year	Average CO_2 Concentration (ppmv)	Year	Average CO_2 Concentration (ppmv)
1959	316.00	1982	341.09
1960	316.91	1983	342.75
1961	317.63	1984	344.44
1962	318.46	1985	345.86
1963	319.02	1986	347.14
1964	319.52	1987	348.99
1965	320.09	1988	351.44
1966	321.34	1989	352.94
1967	322.13	1990	354.19
1968	323.11	1991	355.62
1969	324.60	1992	356.36
1970	325.65	1993	357.10
1971	326.32	1994	358.86
1972	327.52	1995	360.90
1973	329.61	1996	362.58
1974	330.29	1997	363.84
1975	331.16	1998	366.58
1976	332.18	1999	368.30
1977	333.88	2000	369.47
1978	335.52	2001	371.03
1979	336.89	2002	373.07
1980	338.67	2003	375.61
1981	339.95		

Source: Adapted from C. D. Keeling, T. P. Whorf, and the Carbon Dioxide Research Group (2004); Scripps Institution of Oceanography (SIO), University of California, La Jolla, California USA 92093-0444, with permission of the publisher.

TABLE B.15   US National Violent Crime Rate

Year	Violent Crime Rate (per 100,000 Inhabitants)
1984	539.9
1985	558.1
1986	620.1
1987	612.5
1988	640.6
1989	666.9
1990	729.6
1991	758.2
1992	757.7
1993	747.1
1994	713.6
1995	684.5
1996	636.6
1997	611.0
1998	567.6
1999	523.0
2000	506.5
2001[a]	504.5
2002	494.4
2003	475.8
2004	463.2
2005	469.2

Source: http://www.census.gov/compendia/statab/hist_stats.html.

[a]The murder and nonnegligent homicides that occurred as a result of the events of September 11, 2001 are not included in the rate for the year 2001.

TABLE B.16   US Gross Domestic Product

Year	GDP, Current Dollars (Billions)	GDP, Real (1996) Dollars (Billions)
1976	1823.9	4311.7
1977	2031.4	4511.8
1978	2295.9	4760.6
1979	2566.4	4912.1
1980	2795.6	4900.9
1981	3131.3	5021.0
1982	3259.2	4919.3
1983	3534.9	5132.3
1984	3932.7	5505.2
1985	4213.0	5717.1
1986	4452.9	5912.4
1987	4742.5	6113.3
1988	5108.3	6368.4
1989	5489.1	6591.8
1990	5803.2	6707.9
1991	5986.2	6676.4
1992	6318.9	6880.0
1993	6642.3	7062.6
1994	7054.3	7347.7
1995	7400.5	7543.8
1996	7813.2	7813.2
1997	8318.4	8159.5
1998	8781.5	8508.9
1999	9274.3	8859.0
2000	9824.6	9191.4
2001	10,082.2	9214.5
2002	10,446.2	9439.9

Source: http://www.census.gov/compendia/statab/hist_stats.html.

TABLE B.17   Total Annual US Energy Consumption

Year	BTUs (Billions)	Year	BTUs (Billions)
1949	31,981,503	1978	79,986,371
1950	34,615,768	1979	80,903,214
1951	36,974,030	1980	78,280,238
1952	36,747,825	1981	76,342,955
1953	37,664,468	1982	73,286,151
1954	36,639,382	1983	73,145,527
1955	40,207,971	1984	76,792,960
1956	41,754,252	1985	76,579,965
1957	41,787,186	1986	76,825,812
1958	41,645,028	1987	79,223,446
1959	43,465,722	1988	82,869,321
1960	45,086,870	1989	84,999,308
1961	45,739,017	1990	84,729,945
1962	47,827,707	1991	84,667,227
1963	49,646,160	1992	86,014,860
1964	51,817,177	1993	87,652,195
1965	54,017,221	1994	89,291,713
1966	57,016,544	1995	91,199,841
1967	58,908,107	1996	94,225,791
1968	62,419,392	1997	94,800,047
1969	65,620,879	1998	95,200,433
1970	67,844,161	1999	96,836,647
1971	69,288,965	2000	98,976,371
1972	72,704,267	2001	96,497,865
1973	75,708,364	2002	97,966,872
1974	73,990,880	2003	98,273,323
1975	71,999,191	2004	100,414,461
1976	76,012,373	2005	99,894,296
1977	77,999,554		

Source: Annual Energy Review—Energy Overview 1949–2005, US Department of Energy–Energy information Center, http://www.eia.doe.gov/aer/overview.html.

TABLE B.18   Annual US Coal Production

Year	Coal Production (10^3 Short Tons)	Year	Coal Production (10^3 Short Tons)
1949	480,570	1978	670,164
1950	560,386	1979	781,134
1951	576,335	1980	829,700
1952	507,424	1981	823,775
1953	488,239	1982	838,112
1954	420,789	1983	782,091
1955	490,838	1984	895,921
1956	529,774	1985	883,638
1957	518,042	1986	890,315
1958	431,617	1987	918,762
1959	432,677	1988	950,265
1960	434,329	1969	980,729
1961	420,423	1990	1,029,076
1962	439,043	1991	995,984
1963	477,195	1992	997,545
1964	504,182	1993	945,424
1965	526,954	1994	1,033,504
1966	546,822	1995	1,032,974
1967	564,882	1996	1,063,856
1968	556,706	1997	1,089,932
1969	570,978	1998	1,117,535
1970	612,661	1999	1,100,431
1971	560,919	2000	1,073,612
1972	602,492	2001	1,127,689
1973	598,568	2002	1,094,283
1974	610,023	2003	1,071,753
1975	654,641	2004	1,112,099
1976	684,913	2005	1,133,253
1977	697,205		

Source: Annual Energy Review—Coal Overview 1949–2005, US Department of Energy–Energy Information Center.

TABLE B.19   Arizona Drowning Rate, Children 1–4 Years Old

Year	Drowning Rate per 100,000 Children 1–4 Years Old	Year	Drowning Rate per 100,000 Children 1–4 Years Old
1970	19.9	1988	9.2
1971	16.1	1989	11.9
1972	19.5	1990	5.8
1973	19.8	1991	8.5
1974	21.3	1992	7.1
1975	15.0	1993	7.9
1976	15.5	1994	8.0
1977	16.4	1995	9.9
1978	18.2	1996	8.5
1979	15.3	1997	9.1
1980	15.6	1998	9.7
1981	19.5	1999	6.2
1982	14.0	2000	7.2
1983	13.1	2001	8.7
1984	10.5	2002	5.8
1985	11.5	2003	5.7
1986	12.9	2004	5.2
1987	8.4		

Source: http://www.azdhs.gov/plan/report/im/dd/drown96/01dro96.htm.

TABLE B.20   US Internal Revenue Tax Refunds

Fiscal Year	Amount Refunded (Millions Dollars)	National Population (Thousands)
1987	96,969	242,289
1988	94,480	244,499
1989	93,613	246,819
1990	99,656	249,464
1991	104,380	252,153
1992	113,108	255,030
1993	93,580	257,783
1994	96,980	260,327
1995	108,035	262,803
1996	132,710	265,229
1997	142,599	267,784
1998	153,828	270,248
1999	185,282	272,691
2000	195,751	282,193
2001	252,787	285,108
2002	257,644	287,985
2003	296,064	290,850
2004	270,893	293,657
2005	255,439	296,410
2006	263,501	299,103

Source: US Department of Energy–Internal Revenue Service, SOI Tax Stats–Individual Time Series Statistical Tables, http://www.irs.gov/taxstats/indtaxstats/article/O,,id=96679,00.html.

TABLE B.21 **Arizona Average Retail Price of Residential Electricity (Cents per kWh)**

Year	Jan	Feb	Mar	Apr	May	Jun	Jul	Aug	Sep	Oct	Nov	Dec
2001	6.99	7.13	7.4	8.09	9.41	9.04	8.84	8.84	8.81	8.95	7.17	7.26
2002	7.01	7.17	7.46	7.69	9.37	8.97	8.65	8.78	8.79	8.99	7.37	7.46
2003	7.06	7.57	7.59	7.82	9.52	9.09	8.78	8.74	8.7	8.83	7.21	7.55
2004	7.27	7.49	7.61	8.05	9.26	9.1	8.88	8.87	8.96	8.79	8.05	7.86
2005	7.75	7.99	8.19	8.67	9.6	9.41	9.3	9.28	9.3	9.23	8.12	7.88
2006	8.05	8.21	8.38	8.92	10.19	10.05	9.9	9.88	9.89	9.88	8.74	8.56
2007	8.33	8.46	8.8	9.19	10.2	9.96	10.37	10.33	10.17	10.16	9.08	8.89
2008	8.85	9.02	9.38	10.02	11.03	11.06	10.95	10.86	10.63	10.46	9.55	9.61
2009	9.51	9.82	9.93	10.65	11.33	11.27	11.3	11.29	11.17	10.97	9.86	9.7
2010	9.57	9.84	9.98	10.24	11.75	11.74	11.78	11.59	11.52	10.96	10.14	10
2011	9.84	9.93	10.25	10.97	11.77	11.77	11.85	11.67	11.53	11.08	10.31	9.98
2012	10.01	10.26	10.44	11.17	11.88	11.9	11.86	11.83	11.66	11.36	10.73	10.41
2013	10.25	10.7	10.87	11.74	12.17	12.18	12.51	12.33	12.22	12.02	11.06	11.01
2014	10.92	11.23	11.32	11.97	–	–	–	–	–	–	–	–

Source: http://www.eia.gov/electricity/data.cfm#sales.

TABLE B.22 **Denmark Crude Oil Production (In Thousands of Tons)**

Year	Jan	Feb	Mar	Apr	May	Jun	Jul	Aug	Sep	Oct	Nov	Dec
2001	1372	1439	1499	1399	1340	1018	1121	1411	1560	1492	1578	1709
2002	1627	1457	1536	1560	1575	1431	1567	1267	1421	1619	1531	1592
2003	1617	1445	1598	1464	1482	1514	1406	1520	1560	1578	1574	1550
2004	1560	1335	1626	1645	1685	1617	1715	1471	1607	1726	1543	1731
2005	1577	1536	1632	1605	1568	1541	1518	1591	1459	1536	1485	1470
2006	1459	1351	1471	1330	1518	1377	1547	1364	1086	1456	1429	1450
2007	1266	1194	1290	1256	1290	1258	1240	1340	1159	1382	1264	1231
2008	1255	1024	1242	1101	1275	1138	1268	1141	1085	1196	1155	1156
2009	1201	1067	1140	1110	1081	1066	1112	1061	1129	1051	925	959
2010	1095	937	1014	1116	1061	906	1110	710	1014	1080	1009	1106
2011	987	791	964	925	1090	872	937	906	861	859	930	818
2012	826	830	854	867	866	860	853	820	724	824	819	838
2013	787	752	808	764	756	682	741	679	635	720	687	671
2014	675	637	691	659	–	–	–	–	–	–	–	–

Source: http://www.ens.dk/en/info/facts-figures/energy-statistics-indicators-energy-efficiency/monthly-statistics.

TABLE B.23   US Influenza Positive Tests (Percentage)

Week	1997	1998	1999	2000	2001	2002	2003	2004	2005	2006	2007	2008	2009	2010	2011	2012	2013	2014
1	–	19.30	9.39	24.97	13.37	9.74	5.62	12.43	18.57	11.72	6.97	8.55	8.44	4.01	25.17	3.22	35.42	29.18
2	–	20.92	14.76	21.30	17.80	13.85	8.80	9.33	16.94	13.22	9.24	13.59	10.86	4.76	29.28	4.40	34.26	27.33
3	–	25.61	20.30	16.73	22.11	19.74	11.49	6.01	21.34	14.12	11.58	16.80	16.05	5.63	33.52	5.97	30.65	26.39
4	–	28.07	22.07	12.68	23.22	21.37	16.32	3.98	23.30	15.73	18.19	22.71	20.76	5.11	34.34	8.10	29.80	23.90
5	–	26.72	27.32	10.66	20.93	22.80	20.50	2.92	26.87	16.69	22.38	28.46	23.93	5.16	35.4	10.88	26.65	20.18
6	–	24.37	28.29	10.15	18.99	22.79	26.36	1.84	26.55	17.95	28.18	31.75	25.27	4.56	34.66	12.62	22.83	18.73
7	–	20.13	25.51	7.47	15.43	23.86	25.32	1.79	25.55	19.57	25.80	31.82	24.81	4.78	34.85	14.84	20.68	16.47
8	–	14.24	25.23	6.37	13.94	24.91	22.55	1.48	23.46	20.45	23.80	30.40	24.54	5.42	30.98	19.00	19.76	13.53
9	–	9.69	23.38	3.71	11.46	22.92	22.22	0.68	19.95	22.59	22.57	27.65	24.35	6.14	27.51	23.82	18.81	11.94
10	–	7.60	20.59	3.42	7.20	21.18	18.69	0.59	16.04	22.43	20.25	23.68	22.38	6.61	23.79	28.18	18.01	10.84
11	–	5.42	15.62	2.70	5.91	18.52	17.12	0.58	16.94	19.74	19.53	22.01	20.55	5.71	20.28	30.72	16.47	11.11
12	–	4.10	14.31	1.76	5.78	15.12	14.64	1.33	14.66	18.47	16.87	20.39	16.74	4.42	16.12	27.53	15.63	12.53
13	–	2.79	8.26	1.23	4.38	13.75	9.96	0.92	11.04	17.98	13.76	17.78	13.83	4.35	12.8	21.56	14.48	13.22
14	–	2.22	7.92	1.26	3.27	12.41	8.55	0.51	9.12	13.57	11.03	14.54	10.66	2.76	9.81	22.08	13.91	14.39
15	–	2.87	6.45	1.70	2.23	10.86	5.16	0.96	5.27	11.36	12.14	11.77	7.19	2.05	7.12	20.45	10.79	15.00
16	–	1.48	2.63	1.90	2.58	13.09	5.01	1.66	4.19	9.89	9.67	8.51	7.68	1.26	4.42	20.64	8.62	13.72
17	–	1.29	2.53	1.44	1.28	8.57	4.13	0.80	2.68	7.48	9.75	5.37	12.46	0.71	3.35	17.27	8.11	13.08
18	–	1.40	1.73	0.94	0.59	10.23	5.56	0.69	2.27	7.75	6.75	4.63	15.02	0.80	1.91	15.99	6.81	12.62

19	—	0.87	1.12	0.21	0.52	8.76	2.50	0.16	1.73	7.22	4.98	2.56	18.09	0.62	1.10	15.17	5.82	10.90
20	—	0.96	1.13	0.23	0.68	6.00	2.04	0.36	2.06	5.63	3.93	1.57	26.79	0.52	0.84	13.33	4.87	9.81
21	—	NR	NR	NR	NR	NR	1.83	0.57	1.71	6.32	4.32	1.37	33.49	0.19	0.67	11.16	4.49	8.39
22	—	NR	NR	NR	NR	NR	0.85	0.73	1.37	4.69	1.77	0.99	39.48	0.28	0.53	10.91	4.58	8.77
23	—	NR	NR	NR	NR	NR	1.80	0.38	0.76	3.81	1.20	0.92	41.61	0.28	0.27	9.59	4.17	6.65
24	—	NR	NR	NR	NR	NR	1.02	0.24	0.99	1.97	1.70	0.80	43.05	0.44	0.43	9.97	4.28	6.62
25	—	NR	NR	NR	NR	NR	0.73	0.40	0.90	1.19	0.62	0.48	36.18	0.40	0.25	6.76	4.34	5.12
26	—	NR	NR	NR	NR	NR	0.46	0.41	0.58	1.18	1.17	0.45	31.78	0.22	0.26	6.74	3.56	4.17
27	—	NR	NR	NR	NR	NR	0.16	0.17	0.47	0.88	1.61	0.33	32.04	0.31	0.33	4.72	3.53	—
28	—	NR	NR	NR	NR	NR	0.47	0.27	0.20	0.77	1.92	0.47	27.54	0.68	0.60	4.58	4.13	—
29	—	NR	NR	NR	NR	NR	0.57	0.36	0.31	1.20	1.39	0.33	27.11	0.50	0.67	3.67	3.16	—
30	—	NR	NR	NR	NR	NR	0.63	0.00	0.18	1.97	3.18	0.86	27.56	1.36	0.31	5.13	3.18	—
31	—	NR	NR	NR	NR	NR	1.27	0.32	0.00	2.12	1.97	0.62	23.67	1.05	0.77	9.49	3.81	—
32	—	NR	NR	NR	NR	NR	0.32	0.16	0.41	1.33	2.94	0.64	22.55	1.73	0.91	7.54	3.39	—
33	—	NR	NR	NR	NR	NR	0.39	0.15	0.61	1.39	1.70	0.09	24.36	1.06	1.06	6.72	3.15	—
34	—	NR	NR	NR	NR	NR	0.00	0.49	0.39	1.67	1.97	0.55	22.03	2.03	1.00	4.18	2.69	—
35	—	NR	NR	NR	NR	NR	0.00	0.48	0.57	1.87	2.49	0.39	22.29	1.27	0.78	3.06	4.45	—
36	—	NR	NR	NR	NR	NR	0.39	0.36	0.1	1.48	2.2	0.45	24.01	1.54	0.63	3.05	4.03	—
37	—	NR	NR	NR	NR	NR	0.17	0.77	0.45	2.91	2.44	0.70	25.59	1.85	0.90	2.23	3.82	—
38	—	NR	NR	NR	NR	NR	0.37	0.45	0.16	3.51	3.14	1.06	26.99	1.62	0.80	3.08	3.52	—

(*continued*)

TABLE B.23 *(Continued)*

Week	1997	1998	1999	2000	2001	2002	2003	2004	2005	2006	2007	2008	2009	2010	2011	2012	2013	2014
39	–	NR	NR	NR	NR	NR	0.00	1.52	0.49	2.99	1.90	0.60	31.98	1.34	1.20	3.59	4.61	–
40	0.00	0.14	1.07	0.07	0.69	0.00	1.54	0.65	0.58	2.43	1.85	0.62	33.54	2.14	1.03	3.71	4.02	–
41	0.73	0.34	1.06	0.27	0.35	0.20	4.58	0.65	0.55	2.68	1.74	0.8	37.58	3.2	0.57	4.12	3.99	–
42	1.10	0.18	2.24	0.24	0.54	0.00	11.63	1.62	1.06	2.76	1.63	1.26	39.38	3.38	0.82	5.95	3.22	–
43	0.42	0.33	2.64	0.83	1.07	0.34	17.83	1.22	0.50	2.59	1.57	1.78	35.58	3.92	1.17	6.90	3.56	–
44	0.53	0.88	3.91	0.81	0.76	0.37	19.43	1.62	0.80	3.80	2.07	1.21	30.93	5.36	0.96	7.95	4.72	–
45	0.28	0.63	5.75	1.99	0.71	0.25	24.26	2.02	1.30	3.64	2.24	1.71	27.61	6.51	1.13	10.03	5.81	–
46	0.36	0.39	8.07	1.97	2.19	0.47	28.10	2.78	1.22	2.94	2.11	1.40	19.62	8.73	0.96	13.32	7.3	–
47	0.91	0.94	10.56	3.50	1.61	1.05	33.12	1.96	1.59	3.55	3.47	1.77	14.56	10.61	1.83	19.15	9.18	–
48	1.65	1.17	13.71	3.91	3.08	0.79	34.74	3.58	2.86	3.70	3.32	2.25	10.74	9.99	1.54	22.89	12.07	–
49	1.53	1.85	19.11	5.67	3.56	1.66	34.06	5.02	5.54	4.55	3.67	2.27	7.64	13.25	2.29	29.86	18.29	–
50	3.18	1.73	23.33	6.90	4.6	3.89	32.93	6.56	8.64	8.09	4.17	3.16	6.44	18.85	2.71	33.04	20.77	–
51	7.13	3.76	30.96	11.69	5.80	6.56	30.80	11.78	12.75	10.29	5.34	4.33	4.95	25.04	3.82	37.87	28.12	–
52	12.60	5.91	29.37	13.87	8.39	6.76	24.73	16.76	13.48	10.39	7.15	4.73	4.87	26.27	3.33	38.43	31.03	–
53	17.95	–	–	–	–	–	20.10	–	–	–	–	5.54	–	–	–	–	–	–

NR: Not Reported

Source: http://gis.cdc.gov/grasp/fluview/fluportaldashboard.html.

TABLE B.24 **Mean Daily Solar Radiation in Zion Canyon, Utah (Langleys)**

Year	Jan	Feb	Mar	Apr	May	Jun	Jul	Aug	Sep	Oct	Nov	Dec
2003	–	–	–	–	–	–	–	–	–	–	184	178
2004	212	229	453	503	619	615	573	535	464	262	208	175
2005	166	216	385	508	529	549	579	474	443	302	224	170
2006	184	272	310	477	572	583	508	509	431	291	211	177
2007	220	263	396	466	590	634	542	511	432	316	233	179
2008	176	270	415	569	542	647	569	499	459	333	208	157
2009	214	240	423	487	593	586	638	617	523	367	283	196
2010	213	282	440	546	672	703	665	595	570	322	244	153
2011	242	294	389	533	584	703	597	625	488	371	248	214
2012	246	321	425	560	713	733	550	517	485	367	241	158
2013	232	333	457	506	541	645	494	463	412	320	207	180
2014	205	265	401	479	549	642	–	–	–	–	–	–

Source: http://www.raws.dri.edu/cgi-bin/rawMAIN.pl?utZIOC.

TABLE B.25 US Motor Vehicle Traffic Fatalities

Year	Fatalities	Resident Population (Thousands)	Licensed Drivers (Thousands)	Registered Motor Vehicles (Thousands)	Vehicle Miles Traveled (Billions)	Annual Unemployment Rate (%)
1966	50,894	196,560	100,998	95,703	926	3.8
1967	50,724	198,712	103,172	98,859	964	3.8
1968	52,725	200,706	105,410	102,987	1016	3.6
1969	53,543	202,677	108,306	107,412	1062	3.5
1970	52,627	205,052	111,543	111,242	1110	4.9
1971	52,542	207,661	114,426	116,330	1179	5.9
1972	54,589	209,896	118,414	122,557	1260	5.6
1973	54,052	211,909	121,546	130,025	1313	4.9
1974	45,196	213,854	125,427	134,900	1281	5.6
1975	44,525	215,973	129,791	126,153	1328	8.5
1976	45,523	218,035	134,036	130,793	1402	7.7
1977	47,878	220,239	138,121	134,514	1467	7.1
1978	50,331	222,585	140,844	140,374	1545	6.1
1979	51,093	225,055	143,284	144,317	1529	5.8
1980	51,091	227,225	145,295	146,845	1527	7.1
1981	49,301	229,466	147,075	149,330	1555	7.6
1982	43,945	231,664	150,234	151,148	1595	9.7
1983	42,589	233,792	154,389	153,830	1653	9.6
1984	44,257	235,825	155,424	158,900	1720	7.5
1985	43,825	237,924	156,868	166,047	1775	7.2
1986	46,087	240,133	159,486	168,545	1835	7
1987	46,390	242,289	161,816	172,750	1921	6.2
1988	47,087	244,499	162,854	177,455	2026	5.5
1989	45,582	246,819	165,554	181,165	2096	5.3
1990	44,599	249,464	167,015	184,275	2144	5.6
1991	41,508	252,153	168,995	186,370	2172	6.8
1992	39,250	255,030	173,125	184,938	2247	7.5
1993	40,150	257,783	173,149	188,350	2296	6.9
1994	40,716	260,327	175,403	192,497	2358	6.1
1995	41,817	262,803	176,628	197,065	2423	5.6
1996	42,065	265,229	179,539	201,631	2484	5.4
1997	42,013	267,784	182,709	203,568	2552	4.9
1998	41,501	270,248	184,861	208,076	2628	4.5
1999	41,717	272,691	187,170	212,685	2690	4.2
2000	41,945	282,162	190,625	217,028	2747	4
2001	42,196	284,969	191,276	221,230	2796	4.7
2002	43,005	287,625	194,602	225,685	2856	5.8
2003	42,884	290,108	196,166	230,633	2890	6
2004	42,836	292,805	198,889	237,949	2965	5.5

TABLE B.25 (*Contionued*)

Year	Fatalities	Resident Population (Thousands)	Licensed Drivers (Thousands)	Registered Motor Vehicles (Thousands)	Vehicle Miles Traveled (Billions)	Annual Unemployment Rate (%)
2005	43,510	295,517	200,549	245,628	2989	5.1
2006	42,708	298,380	202,810	251,415	3014	4.6
2007	41,259	301,231	205,742	257,472	3031	4.6
2008	37,423	304,094	208,321	259,360	2977	5.8
2009	33,883	306,772	209,618	258,958	2957	9.3
2010	32,999	309,326	210,115	257,312	2967	9.6
2011	32,479	311,588	211,875	265,043	2950	8.9
2012	33,561	313,914	211,815	265,647	2969	8.1

Sources: http://www-fars.nhtsa.dot.gov/Main/index.aspx, http://www.bls.gov/data/.

TABLE B.26 Single-Family Residential New Home Sales and Building Permits (In Thousands of Units)

Year	Jan Sales	Jan Permits	Feb Sales	Feb Permits	Mar Sales	Mar Permits	Apr Sales	Apr Permits	May Sales	May Permits	Jun Sales	Jun Permits
1963	42	43.3	35	43.4	44	63.8	52	82.1	58	78.8	48	71
1964	39	42.7	46	48.5	53	67.6	49	73.1	52	70.9	53	72.3
1965	38	39.2	44	41.4	53	65.6	49	73.2	54	69.3	57	71.3
1966	42	39.1	43	38.1	53	69.6	49	66.2	49	60.5	40	58.8
1967	29	31.5	32	34.2	41	56	44	59.2	49	67.6	47	68.5
1968	35	38.9	43	44.7	46	61.4	46	72	43	70.7	41	60.2
1969	34	39.8	40	42.5	43	58.5	42	69.7	43	64.5	44	61.7
1970	34	27.7	29	34.4	36	50.8	42	63.5	43	58.3	44	62.7
1971	45	44.1	49	48.1	62	78.8	62	89.6	58	90.2	59	93.2
1972	51	61.8	56	64.8	60	94.2	65	95	64	103.2	63	104.1
1973	55	64.9	60	66.2	68	91.6	63	94.9	65	102.2	61	92
1974	37	38.7	44	44.8	55	66.5	53	76.8	58	73.3	50	63.7
1975	29	29.8	34	32.5	44	45.5	54	65.2	57	68.2	51	68.4
1976	41	46.3	53	53.6	55	81.4	62	87.3	55	81.3	56	89.1
1977	57	52.6	68	67.2	84	110.5	81	108.2	78	112.4	74	117.2
1978	57	62.9	63	67.9	75	110.1	85	119.7	80	124.3	77	130.4
1979	53	54.9	58	56.4	73	101.5	72	102.2	68	110	63	100.8
1980	43	45.2	44	47	44	49.1	36	48.4	44	49.6	50	61.2
1981	37	39.6	40	42.9	49	61.2	44	69.3	45	61	38	56.3
1982	28	25.4	29	27.9	36	44	32	46.6	36	45.8	34	50.6

1983	44	48.5	46	50.9	57	79.2	59	81.6	64	93	59	101.2
1984	52	60.2	58	72.6	63	88.6	61	92.3	59	98.2	58	90.3
1985	48	55.8	55	58.1	67	85.9	60	93.9	65	96.8	65	89
1986	55	65.3	59	61.1	89	89.3	84	114.8	75	109.3	66	110.8
1987	53	62.4	59	69.5	73	103.5	72	107.6	62	96.6	58	107.7
1988	43	50.5	55	63	68	98.2	68	93.1	64	98.7	65	105.9
1989	52	60.9	51	58.9	58	84.1	60	87.5	61	93.6	58	92
1990	45	60.7	50	60.7	58	82.8	52	79.2	50	83.1	50	79.7
1991	30	37.6	40	43.5	51	61.2	50	75.7	47	78.1	47	73.9
1992	48	55.2	55	61.1	56	82.4	53	88	52	82.7	53	91.6
1993	44	55.1	50	61.3	60	84.2	66	91.5	58	85.2	59	97
1994	46	63.4	58	69.2	74	104	65	102	65	107.7	55	109.2
1995	47	58.2	47	59.8	60	85.1	58	83.1	63	95.9	64	97.4
1996	54	66	68	74.4	70	95.7	70	109.9	69	109.2	65	100.7
1997	61	65.8	69	70.3	81	88.7	70	104.4	71	101.3	71	100.9
1998	64	70.1	75	78.1	81	105.1	82	113.6	82	107.3	83	115.8
1999	67	74.2	76	86.6	84	118.9	86	119.9	80	115.9	82	128
2000	67	78.3	78	89.1	88	119	78	107.6	77	119.3	71	114.5
2001	72	85.6	85	85.1	94	112.7	84	116.5	80	124.4	79	119.2
2002	66	88.7	84	95.5	90	111	86	125.4	88	127.1	84	118.9
2003	76	98	82	93.9	98	117.7	91	134.2	101	132.1	107	138.3
2004	89	103.4	102	108.4	123	154.8	109	155	115	150.2	105	159.3
2005	92	106.9	109	114.8	127	150.6	116	152.7	120	156	115	166.2

(continued)

TABLE B.26 (*Continued*)

Year	Jan Sales	Jan Permits	Feb Sales	Feb Permits	Mar Sales	Mar Permits	Apr Sales	Apr Permits	May Sales	May Permits	Jun Sales	Jun Permits
2006	89	114.3	88	115.6	108	146.7	100	130.8	102	144.5	98	139.3
2007	66	80.3	68	79.4	80	103.4	83	98.6	79	106.6	73	97.2
2008	44	48	48	48	49	54.1	49	63.4	49	61.9	45	59.4
2009	24	22.1	29	26.3	31	32.7	32	37.8	34	39.5	37	47
2010	24	31.4	27	35.4	36	50.9	41	46.1	26	39.9	28	42.9
2011	21	26.2	22	26.5	28	37.8	30	37.2	28	39.7	28	41.5
2012	23	30.5	30	35.7	34	42.9	34	44.5	35	50.3	34	48.1
2013	32	40.7	36	42.3	41	51.7	43	60.2	40	62.8	43	57.4
2014	33	41.1	35	41.2	39	51.4	40	57.6	49	59.1	-	-
1963	62	72.8	56	68.1	49	65.4	44	69.7	39	51	31	40.8
1964	54	68.2	56	61.1	48	61.9	45	60.8	37	50.8	33	42.2
1965	51	64.9	58	64.6	48	60.3	44	61.5	42	54	37	44.5
1966	40	47.3	36	46.9	29	40.2	31	36.9	26	32.7	23	27.1
1967	46	58.1	47	64.7	43	57.6	45	61.2	34	51.6	31	40.2
1968	44	64.3	47	62.2	41	60.8	40	65	32	51.9	32	42.5
1969	39	55.2	40	50.3	33	51.3	32	53.5	31	40.1	28	38.8
1970	44	59.6	48	58.3	45	60.9	44	62.2	40	51.3	37	57.1
1971	64	86	62	82.7	50	78.1	52	76	50	73.3	44	65.9
1972	63	89.1	72	101.1	61	84.6	65	98.1	51	76.5	47	60.8
1973	54	82.6	52	78	46	61.7	42	60.4	37	49.9	30	37.8
1974	48	61.7	45	56.6	41	46.9	34	48.2	30	36.4	24	30.3

1975	51	69.5	53	63.8	46	65.5	46	68	46	51.5	39	47.6
1976	57	82.5	59	80.6	58	78.5	55	77.4	49	73.8	47	61.8
1977	64	99.7	74	110.1	71	97.2	63	94.9	55	87.6	51	68.4
1978	68	100.9	72	107.3	68	96.6	70	104	53	87.5	50	71
1979	64	93	68	97.1	60	79.4	54	83.3	41	57.7	35	45.1
1980	55	74.6	61	75.3	50	80.2	46	76.3	39	55.3	33	48.2
1981	36	52.6	34	45.2	28	41.8	29	35.6	27	29.2	29	29.6
1982	31	46.8	36	47.6	39	52.4	40	55.5	39	54.8	33	48.9
1983	51	82.3	50	85.9	48	77.8	51	76.5	45	68.3	48	56.5
1984	52	79.1	48	80.7	53	70.8	55	75.8	42	62.7	38	51.2
1985	63	92.1	61	90	54	81.3	52	89.7	51	64.9	47	59.1
1986	57	106.2	52	91.3	60	94.8	54	93.5	48	66.4	49	74.7
1987	55	95.8	56	87.6	52	86.6	52	81.7	43	65.5	37	59.7
1988	57	85.3	59	95.9	54	85.1	57	82.9	43	71.8	42	63.4
1989	62	79	61	89.2	49	78.6	51	81.2	47	69.2	40	57.4
1990	46	70.9	46	72.5	38	57.7	37	62.5	34	47	29	37.1
1991	43	74.9	46	69.9	37	64.2	41	70.4	39	52.6	36	51.7
1992	52	83.3	56	76.6	51	80.1	48	80.3	42	63.8	42	65.8
1993	55	88.2	57	91	57	89.8	56	87.9	53	80.5	51	74.8
1994	52	90.9	59	100.9	54	91.5	57	85.9	45	74.8	40	68.9
1995	64	88.3	63	101.4	54	90.1	54	90.8	46	78.4	45	68.8

(continued)

TABLE B.26 (*Continued*)

Year	Jan Sales	Jan Permits	Feb Sales	Feb Permits	Mar Sales	Mar Permits	Apr Sales	Apr Permits	May Sales	May Permits	Jun Sales	Jun Permits
1996	66	101.9	73	97.6	62	85.9	56	90.8	54	71.5	51	66
1997	69	99.8	72	91.8	67	95.6	62	97.5	61	72.5	51	73.9
1998	75	111.2	75	104.4	68	102.5	69	103.8	70	86.6	61	89
1999	78	114.6	78	112.6	65	103.1	67	97.6	61	90.3	57	84.8
2000	76	98.8	73	111.6	70	95.8	71	102.8	63	87.7	65	73.7
2001	76	110.2	74	116.2	66	92.4	66	104.4	67	89.2	66	79.7
2002	82	122.4	90	119.8	82	110.1	77	123	73	96	70	94.6
2003	99	138.6	105	131	90	130.5	88	138.1	76	99.2	75	109.6
2004	96	145.3	102	145.6	94	134.5	101	128.5	84	114.6	83	113.8
2005	117	145.9	110	161.9	99	151.3	105	139.1	86	124	87	112.5
2006	83	111.6	88	121.5	80	97.7	74	98	71	82.2	71	76
2007	68	89.8	60	87.5	53	66.6	57	70.7	45	54.7	44	45.2
2008	43	55.7	38	48	35	45.9	32	40.4	27	26.2	26	24.6
2009	38	46.9	36	42.9	30	40.7	33	38.6	26	31.9	24	34.7
2010	26	37.5	23	37.2	25	34.3	23	31.5	20	29.6	23	30.6
2011	27	35.9	25	41.6	24	36.3	25	34.4	23	31.6	24	29.8
2012	33	47.3	31	49.8	30	43.3	29	49.6	28	40.4	28	36.3
2013	33	58.7	31	58	31	50.5	36	54.4	32	43.8	31	40.2
2014	–	–	–	–	–	–	–	–	–	–	–	–

Source: http://www.census.gov/housing/hvs/.

TABLE B.27   Best Airline On-Time Arrival Performance (Percentage)

Year	Jan	Feb	Mar	Apr	May	Jun	Jul	Aug	Sep	Oct	Nov	Dec
1995	73.83	78.91	79.3	81.3	80.68	75.94	80.08	79.85	85.58	82.09	77.85	67.66
1996	62.67	71.9	75.94	80.18	78.92	74.67	75.29	74.69	78.69	77.19	77.91	66.63
1997	68.41	75.21	78.1	79.85	83.16	76.14	77.5	78.56	85.03	81.55	78.22	73.54
1998	75.05	75.39	75.85	79.12	77.48	70.42	78.86	77.03	78.94	81.75	83.29	73.22
1999	67.66	78.91	78.09	75.73	76.19	70.88	71.07	76.11	79.35	80.06	81.41	78
2000	73.75	74.76	76.99	75.37	74.25	66.34	70.31	69.96	78.1	76.07	72.82	62.75
2001	75.42	72.73	75.22	79.29	81.47	75.18	78.12	76.18	67.66	84.78	84.71	80.22
2002	81.02	84.69	78.59	82.58	82.76	78.64	79.66	82.62	87.95	84.18	85.21	78.34
2003	83.32	76.54	82.58	86.85	84.93	82.36	79.66	79	85.63	86.39	80.2	76.04
2004	74.85	77.48	81.29	83.04	77.62	72.95	75.95	78.29	83.91	81	79.12	71.56
2005	71.39	77.58	76.94	83.44	83.67	75.2	70.92	75.16	82.66	81.26	80.03	71.01
2006	78.76	75.3	76.12	78.42	78.27	72.83	73.7	75.8	76.22	72.91	76.52	70.8
2007	73.11	67.26	73.27	75.71	77.9	68.07	69.78	71.64	81.7	78.21	80.03	64.34
2008	72.36	68.63	71.58	77.67	79.02	70.84	75.7	78.44	84.88	86.04	83.33	65.34
2009	77.02	82.6	78.4	79.14	80.49	76.12	77.6	79.68	86.17	77.27	88.59	71.99
2010	78.69	74.66	79.96	85.31	79.94	76.42	76.69	81.65	85.07	83.77	83.16	72.04
2011	76.3	74.54	79.24	75.5	77.06	76.92	77.84	79.34	83.88	85.54	85.3	84.37
2012	83.75	86.16	82.19	86.26	83.38	80.66	76.01	79.15	83.3	80.21	85.73	76.56
2013	80.98	79.62	–	–	–	–	–	–	–	–	–	–

Source: http://www.rita.dot.gov/bts/subject_areas/airline_information/airline_ontime_tables/2013_02/table_02.

TABLE B.28 US Automobile Manufacturing Shipments (Dollar in Millions)

Year	Jan	Feb	Mar	Apr	May	Jun	Jul	Aug	Sep	Oct	Nov	Dec
1992	5618	7205	7446	7420	8076	7692	4395	7045	7058	8219	7432	6427
1993	7091	8339	8231	7772	8012	8015	3831	6246	7399	8546	8186	7267
1994	7986	9482	9215	9092	9346	9653	4748	8832	9194	9583	9573	8560
1995	8541	10,048	9796	8781	8851	8877	4207	7898	8999	9167	8333	7236
1996	7379	8852	7817	8688	8993	8521	4900	7837	8895	8739	8213	6,425
1997	7698	8898	8228	8121	7804	8120	5151	7943	8179	9354	8490	7,380
1998	7248	9013	9280	8622	8706	7312	3874	8162	9006	9658	8396	7431
1999	7594	9144	9429	8550	8635	8738	4657	8797	8895	9280	8607	7477
2000	8231	9117	9817	8467	8919	9536	4437	7941	9267	9133	7720	6175
2001	6485	7142	8541	6986	8431	7999	4411	8182	6827	8372	7312	6358
2002	7227	7239	7541	7886	8382	7726	4982	8089	7465	8565	7149	5876
2003	7383	7308	7923	6959	7638	7374	4903	6937	8036	8520	6680	6575
2004	6670	7778	9103	7323	7306	7947	4212	7709	7875	7453	6837	6634
2005	6520	7302	7692	6966	7403	7906	4680	8289	8402	8472	7525	7209
2006	7454	8009	9829	7571	9143	9247	4424	8655	7621	8126	7278	6872
2007	6788	6897	8121	6468	7424	7482	4945	8010	6969	7962	7250	6412
2008	7328	7862	7229	6818	7092	7376	5156	7302	7386	7352	5673	4930
2009	3869	3995	4221	4039	3519	3994	3277	4635	4911	5554	5106	5287
2010	5373	5021	5969	5716	5788	6765	5397	7262	7422	6957	6389	6171
2011	5974	6324	8197	6448	6752	7521	5739	7761	7508	8362	7288	7253
2012	7308	8290	8919	8659	9594	9678	7141	9437	8886	10,081	9,898	9085
2013	9789	10,591	11,387	10,666	10,920	10,992	8076	11,166	10,989	12,195	10,996	8365
2014	8683	9705	9577	8356	9686	–	–	–	–	–	–	–

Source: http://www.census.gov/econ/currentdata/.

APPENDIX C

INTRODUCTION TO R

Throughout the book, we often refer to commercial statistical software packages such as JMP and Minitab when discussing the examples. These software packages indeed provide an effective option particularly for the undergraduate level students and novice statisticians with their pull-down menus and various built-in statistical functions and routines. However there is also a growing community of practitioners and academicians who prefer to use R, an extremely powerful and freely available statistical software package that can be downloaded from http://www.r-project.org/. According to this webpage,

R is an integrated suite of software facilities for data manipulation, calculation and graphical display. It includes

- *an effective data handling and storage facility,*
- *a suite of operators for calculations on arrays, in particular matrices,*
- *a large, coherent, integrated collection of intermediate tools for data analysis,*
- *graphical facilities for data analysis and display either on-screen or on hardcopy, and*

Introduction to Time Series Analysis and Forecasting, Second Edition.
Douglas C. Montgomery, Cheryl L. Jennings and Murat Kulahci.
© 2015 John Wiley & Sons, Inc. Published 2015 by John Wiley & Sons, Inc.

- *a well-developed, simple and effective programming language which includes conditionals, loops, user-defined recursive functions and input and output facilities.*

The term "environment" is intended to characterize it as a fully planned and coherent system, rather than an incremental accretion of very specific and inflexible tools, as is frequently the case with other data analysis software.

R, like S, is designed around a true computer language, and it allows users to add additional functionality by defining new functions. Much of the system is itself written in the R dialect of S, which makes it easy for users to follow the algorithmic choices made. For computationally-intensive tasks, C, C++ and Fortran code can be linked and called at run time. Advanced users can write C code to manipulate R objects directly.

Many users think of R as a statistics system. We prefer to think of it of an environment within which statistical techniques are implemented. R can be extended (easily) via packages. There are about eight packages supplied with the R distribution and many more are available through the CRAN family of Internet sites covering a very wide range of modern statistics.

In this second edition of our book, we decided to provide the R-code for most of the examples at the end of the chapters. The codes are generated with the novice R user in mind and we therefore tried to keep them simple and easy to understand, sometimes without taking full advantage of more sophisticated options available in R. We nonetheless believe that they offer readers the possibility to immediately apply the techniques covered in the chapters with the data provided at the end of the book or with their own data. This after all we believe is the best way to learn time series analysis and forecasting.

BASIC CONCEPTS IN R

R can be downloaded from the R project webpage mentioned above. Although there are some generic built-in functions such as mean() to calculate the sample mean or lm() to fit a linear model, R provides the flexibility of writing your own functions as in C++ or Matlab. In fact one of the main advantages of R is its ever-growing user community, who openly shares the new functions they wrote in terms of "packages." Each new package has to be installed and loaded from "Packages" option in order to be able to use its contents. We provide the basic commands in R below.

Data entry can be done manually using c() function such as

```
> temp<-c(75.5,76.3,72.4,75.7,78.6)
```

Now the vector temp contains 5 elements that can displayed using

```
> temp
[1] 75.5 76.3 72.4 75.7 78.6
```

or

```
> print(temp)
[1] 75.5 76.3 72.4 75.7 78.6
```

However for large data sets, importing the data from an ASCII file, for example, a .txt file, is preferred. If, for example, each entry of temp represents the average temperature on a weekday and is stored in a file named temperature.txt, the data can then be imported to R using read.table() function as

```
> temp<-read.table("temperature.txt",header=TRUE,Sep="")
```

This command will assign the contents of temperature.txt file into the data frame "temp." It assumes that the first row of the file contains the names of the individual variables in the file, for example, in this case "Day" and "Temperature" and the data are space delimited. Also note that the command further assumes that the file is in the working directory, which can be changed using the File option. Otherwise the full directory has to be specified, for example, if the file is in C:/Rcoding directory,

```
read.table("C:/Rcoding/temperature.txt",header=T,sep=",")
```

Now we have

```
> temp
  Day Temperature
1   1        75.5
2   2        76.3
3   3        72.4
4   4        75.7
5   5        78.6
```

We can access each column of the temp matrix by one of the two commands

```
> temp$Temperature
[1]  75.5 76.3 72.4 75.7 78.6
> temp[,2]
[1]  75.5 76.3 72.4 75.7 78.6
```

Now that the data are imported, we can start using built-in function such as the sample mean and the log transform of the temperature by

```
> mean(temp$Temperature)
[1]  75.7
> log(temp$Temperature)
[1]  4.324133 4.334673 4.282206 4.326778 4.364372
```

One can also write user-defined functions to analyze the data. As we mentioned earlier, for most basic statistical functions there already exists packages containing the functions that would serve the desired purpose. Some basic examples of functions are provided in the R-code of the examples.

As indicated in the R project's webpage: "One of R's strengths is the ease with which well-designed publication-quality plots can be produced, including mathematical symbols and formulae where needed. Great care has been taken over the defaults for the minor design choices in graphics, but the user retains full control." In order to show the flexibility of plotting options in R, in the examples we provide the code for different plots for time series data and residual analysis with various options to make the plots look very similar to the ones generated by the commercial software packages used in the chapters.

Exporting the output or new data can be done through write.table() function. In order to create, for example, a new data frame by appending to the original data frame the log transform of the temperature and export the new data frame into a .txt file, the following commands can be used

```
> temp.new<-cbind(temp,log(temp$Temperature))
> temp.new
  Day Temperature log(temp$Temperature)
1   1        75.5              4.324133
2   2        76.3              4.334673
3   3        72.4              4.282206
4   4        75.7              4.326778
5   5        78.6              4.364372
> write.table(temp.new," C:/Rcoding/Temperaturenew.txt")
```

BIBLIOGRAPHY

Abraham, B. and Ledolter, J. (1983). *Statistical Methods for Forecasting*. John Wiley & Sons, Hoboken, NJ.

Abraham, B. and Ledolter, J. (1984). A note on inverse autocorrelations. *Biometrika* **71**, 609–614.

Akaike, H. (1974). A new look at the statistical model identification. *IEEE Trans. Autom. Control* **19**, 716–723.

Atkins, S. M. (1979). Case study on the use of intervention analysis applied to traffic accidents. *J. Oper. Res. Soc.* **30**, 651–659.

Attanasio, O. P. (1991). Risk, time varying second moments and market efficiency. *Rev. Econ. Studies* **58**, 479–494.

Bartlett, M. S. (1946). On the theoretical specification and sampling properties of autocorrelated time series. *J. R. Stat. Soc. Ser. B* **8**, 27–41.

Bates, J. M. and Granger, C. W. J. (1969). The combination of forecasts. *Oper. Res. Q.* **20**, 451–468.

Berger, J. O. (1985). *Statistical Decision Theory and Bayesian Analysis*. Springer-Verlag, New York.

Bisgaard, S. and Kulahci, M. (2005). Interpretation of time series models. *Qual. Eng.* **17**(4), 653–658.

Bisgaard, S. and Kulahci, M. (2006a). Studying input–output relationships, part I. *Qual. Eng.* **18**(2), 273–281.

Introduction to Time Series Analysis and Forecasting, Second Edition.
Douglas C. Montgomery, Cheryl L. Jennings and Murat Kulahci.
© 2015 John Wiley & Sons, Inc. Published 2015 by John Wiley & Sons, Inc.

Bisgaard, S. and Kulahci, M. (2006b). Studying input–output relationships, part II. *Qual. Eng.* **18**(3), 405–410.

Bisgaard, S. and Kulahci, M. (2011). *Time Series Analysis and Forecasting by Example*. John Wiley & Sons, Hoboken, NJ.

Bodurtha, J. N. and Mark, N. C. (1991). Testing the CAPM with time varying risks and returns. *J. Finance* **46**, 1485–1505.

Bollerslev, T. (1986). Generalized autoregressive conditional heteroskedasticity. *J. Econometrics* **31**, 307–327.

Bollerslev, T., Chou, R. Y., and Kroner, K. F. (1992). ARCH modeling in finance: a review of the theory and empirical evidence. *J. Econometrics* **52**, 5–59.

Box, G. E. P. and Luceño, A. (1997). *Statistical Control by Monitoring and Feedback Adjustment*. John Wiley & Sons, Hoboken, NJ.

Box, G. E. P. and Pierce, D. A. (1970). Distributions of residual autocorrelations in autoregressive-integrated moving average time series models. *J. Am. Stat. Assoc.* **65**, 1509–1526.

Box, G. E. P. and Tiao, G. C. (1975). Intervention analysis with applications to economic and environmental problems. *J. Am. Stat. Assoc.* **70**, 70–79.

Box, G. E. P., Jenkins, G. M., and Reinsel, G. (2008). *Time Series Analysis, Forecasting and Control*, 4th edition. Prentice-Hall, Englewood Cliffs, NJ.

Brockwell, P. J. and Davis, R. A. (1991). *Time Series: Theory and Methods*, 2nd edition. Springer-Verlag, New York.

Brockwell, P. J. and Davis, R. A. (2002). *Introduction to Time Series and Forecasting*, 2nd edition. Springer-Verlag, New York.

Brown, R. G. (1963). *Smoothing, Forecasting and Prediction of Discrete Time Series*. Prentice-Hall, Englewood Cliffs, NJ.

Brown, R. G. and Meyer, R. F. (1961). The fundamental theorem of exponential smoothing. *Oper. Res.* **9**, 673–685.

Chatfield, C. (1996). *The Analysis of Time Series: An Introduction*, 5th edition. Chapman and Hall, London.

Chatfield, C. and Yar, M. (1988). Holt–Winters forecasting: some practical issues. *The Statistician* **37**, 129–140.

Chatfield, C. and Yar, M. (1991). Prediction intervals for multiplicative Holt–Winters. *Int. J. Forecasting* **7**, 31–37.

Chow, W. M. (1965). Adaptive control of the exponential smoothing constant. *J. Ind. Eng.* **16**, 314–317.

Clemen, R. (1989). Combining forecasts: a review and annotated bibliography. *Int. J. Forecasting* **5**, 559–584.

Cochrane, D. and Orcutt, G. H. (1949). Application of least squares regression to relationships containing autocorrelated error terms. *J. Am. Stat. Assoc.* **44**, 32–61.

Cogger, K. O. (1974). The optimality of general-order exponential smoothing. *Oper. Res.* **22**, 858–867.

Cook, R. D. (1977). Detection of influential observation in linear regression. *Technometrics* **19**, 15–18.

Cook, R. D. (1979). Influential observations in linear regression. *J. Am. Stat. Assoc.* **74**, 169–174.

Cox, D. R. (1961). Prediction by exponentially weighted moving averages and related methods. *J. R. Stat. Soc. Ser. B* **23**, 414–422.

Dagum, E. B. (1980). *The X-11-ARIMA Seasonal Adjustment Method.* Statistics Canada, Ottawa.

Dagum, E. B. (1983). *The X-11-ARIMA Seasonal Adjustment Method.* Technical Report 12-564E, Statistics Canada, Ottawa.

Dagum, E. B. (1988). *The X-11-ARIMA/88 Seasonal Adjustment Method: Foundations and User's Manual.* Statistics Canada, Ottawa.

Dalkey, N. C. (1967). *Delphi.* P-3704, RAND Corporation, Santa Monica, CA.

Daniels, H. E. (1956). The approximate distribution of serial correlation coefficients. *Biometrika* **43**, 169–185.

Degiannakis, S. and Xekalaki, E. (2004). Autoregressive conditional heteroscedasticity (ARCH) models: a review. *Qual. Technol. Quant. Management* **1**, 271–324.

Dickey, D. A. and Fuller, W. A. (1979). Distribution of the estimates for autoregressive time series with a unit root. *J. Am. Stat. Assoc.* **74**, 427–431.

Durbin, J. and Watson, G. S. (1950). Testing for serial correlation in least squares regression I. *Biometrika* **37**, 409–438.

Durbin, J. and Watson, G. S. (1951). Testing for serial correlation in least squares regression II. *Biometrika* **38**, 159–178.

Durbin, J. and Watson, G. S. (1971). Testing for serial correlation in least squares regression III. *Biometrika* **58**, 1–19.

Engle, R. F. (1982). Autoregressive conditional heteroscedasticity with estimates of the variance of United Kingdom Inflation. *Econometrica* **50**, 987–1007.

Engle, R. F. and Bollerslev, T. (1986). Modelling the persistence of conditional variances. *Econometric Rev.* **5**, 1–50.

Engle, R. F. and Kroner, K. F. (1995). Multivariate simultaneous generalized ARCH. *Econometric Theory* **11**, 122–150.

French, K. R. G., Schwert, G. W., and Stambaugh, R. F. (1987). Expected stock returns and volatility. *J. Financial Econ.* **19**, 3–30.

Fricker, R. D. Jr. (2013). *Introduction to Statistical Methods for Biosurveillance with an Emphasis on Syndromic Surveillance.* Cambridge University Press, New York, NY.

Fuller, W. A. (1995). *Introduction to Statistical Time Series*, 2nd edition. John Wiley & Sons, Hoboken, NJ.

Gardner, E. S. Jr. (1985). Exponential smoothing: the state of the art. *J. Forecasting* **4**, 1–28.

Gardner, E. S. Jr. (1988). A sample of computing prediction intervals for time-series forecasts. *Manage. Sci.* **34**, 541–546.

Gardner, E. S. Jr. and Dannenbring, D. G. (1980). Forecasting with exponential smoothing: some guidelines for model selection. *Decision Sci.* **11**, 370–383.

Glosten, L., Jagannathan, R., and Runkle, D. (1993). On the relation between the expected value and the volatility of the nominal excess return on stocks. *J. Finance* **48**(5), 1779–1801.

Goodman, M. L. (1974). A new look at higher-order exponential smoothing for forecasting. *Oper. Res.* **2**, 880–888.

Granger, C. W. J. and Newbold, P. (1986). *Forecasting Economic Time Series*. 2nd edition. Academic Press, New York.

Greene, W. H. (2011). *Econometric Analysis*, 7th edition. Prentice-Hall, Upper Saddle River, NJ.

Hamilton, J. D. (1994). *Time Series Analysis*. Princeton University Press, Princeton, NJ.

Haslett, J. (1997). On the sample variogram and the sample autocovariance for nonstationary time series. *The Statistician.* **46**, 475–486.

Hill, T., Marquez, L., O'Conner, M., and Remus, W. (1994). Artificial neural network models for forecasting and decision making. *Int. J. Forecasting* **10**, 5–15.

Holt, C. C. (1957). *Forecasting Seasonals and Trends by Exponentially Weighted Moving Averages*. Office of Naval Research Memorandum No. 52, Carnegie Institute of Technology.

Izenman, A. J. and Zabell, S. A. (1981). Babies and the blackout: the genesis of a misconception. *Soc. Sci. Res.* **10**, 282–299.

Jenkins, G. M. (1954). Tests of hypotheses in the linear autoregressive model, I. *Biometrika* **41**, 405–419.

Jenkins, G. M. (1956). Tests of hypotheses in the linear autoregressive model, II. *Biometrika* **43**, 186–199.

Jenkins, G. M. (1979). *Practical Experiences with Modelling and Forecasting Time Series*. Gwilym Jenkins & Partners Ltd., Jersey, Channel Islands.

Jenkins, G. M. and Watts, D. G. (1969). *Spectral Analysis and its Applications*. Holden-Day, San Francisco.

Jones-Farmer, A. L., Ezell, J. D., and Hazen, B. T. (2014). Applying control chart methods to enhance data quality. *Technometrics* **56**(1), 29–41.

Ledolter, J. and Abraham, B. (1984). Some comments on the initialization of exponential smoothing. *J. Forecasting* **3**, 79–84.

Ledolter, J. and Chan, K. S. (1996). Evaluating the impact of the 65 mph maximum speed limit on Iowa rural interstates. *Am. Stat.* **50**, 79–85.

Lee, L. (2000). *Bad Predictions*. Elsewhere Press, Rochester, MI.

Ljung, G. M. and Box, G. E. P. (1978). On a measure of lack of fit in time series models. *Biometrika* **65**, 297–303.

Lütkepohl, H. (2005). *New Introduction to Multiple Time Series Analysis*. Springer-Verlag, New York.

Makridakis, S., Chatfield, C., Hibon, M., Lawrence, M. J., Mills, T., Ord, K., and Simmons, L. F. (1993). The M2 competition: a real time judgmentally based forecasting study (with comments). *Int. J. Forecasting* **9**, 5–30.

Marris, S. (1961). The treatment of moving seasonality in Census Method II. *Seasonal Adjustment on Electronic Computers*. Organisation for Economic Co-operation and Development, Paris, pp. 257–309.

McKenzie, E. (1984). General exponential smoothing and the equivalent ARIMA process. *J. Forecasting* **3**, 333–344.

McKenzie, E. (1986). Error analysis for Winters' additive seasonal forecasting system. *Int. J. Forecasting* **2**, 373–382.

Montgomery, D. C. (1970). Adaptive control of exponential smoothing parameters by evolutionary operation. *AIIE Trans.* **2**, 268–269.

Montgomery, D. C. (2013). *Introduction to Statistical Quality Control*, 7th edition. John Wiley & Sons, Hoboken, NJ.

Montgomery, D. C. and Weatherby, G. (1980). Modeling and forecasting time series using transfer function and intervention methods. *AIIE Trans.* **12**, 289–307.

Montgomery, D. C., Johnson, L. A., and Gardiner, J. S. (1990). *Forecasting and Time Series Analysis*, 2nd edition. McGraw-Hill, New York.

Montgomery, D. C., Peck, E. A., and Vining, G. G. (2012). *Introduction to Linear Regression Analysis*, 5th edition. John Wiley & Sons, Hoboken, NJ.

Muth, J. F. (1960). Optimal properties of exponentially weighted forecasts. *J. Am. Stat. Assoc.* **55**, 299–306.

Myers, R. H. (1990). *Classical and Modern Regression with Applications*, 2nd edition. PWS-Kent Publishers, Boston.

Nelson, B. (1991). Conditional heteroskedasticity in asset returns: a new approach. *Econometrica* **59**, 347–370.

Newbold, P. and Granger, C. W. J. (1974). Experience with forecasting univariate time series and the combination of forecasts. *J. R. Stat. Soc. Ser. A* **137**, 131–146.

Pandit, S. M. and Wu, S. M. (1974). Exponential smoothing as a special case of a linear stochastic system. *Oper. Res.* **22**, 868–869.

Percival, D. B. and Walden, A. T. (1992). *Spectral Analysis for Physical Applications*. Cambridge University Press, New York, NY.

Priestley, M. B. (1991). *Spectral Analysis and Time Series*. Academic Press.

Quenouille, M. H. (1949). Approximate tests of correlation in time-series. *J. R. Stat. Soc. Ser. B* **11**, 68–84.

Raiffa, H. and Schlaifer, R. (1961). *Applied Statistical Decision Theory*. Harvard University Press, Cambridge, MA.

Reinsel, G. C. (1997). *Elements of Multivariate Time Series Analysis*, 2nd edition. Springer-Verlag, New York.

Roberts, S. D. and Reed, R. (1969). The development of a self–adaptive forecasting technique. *AIIE Trans.* **1**, 314–322.

Shiskin, J., Young, A. H., and Musgrave, J. C. (1967). *The X-11 Variant of the Census Method II Seasonal Adjustment Program*. Technical Report 15, U.S. Department of Commerce, Bureau of the Census, Washington, DC.

Shiskin, J. (1958). Decompostion of economic time series. *Science* **128**, 3338.

Schwarz, G. (1978). Estimating the dimension of a model. *Ann. Stat.* **6**, 461–464.

Solo, V. (1984). The order of differencing in ARIMA models. *J. Am. Stat. Assoc.* **79**, 916–921.

Sweet, A. L. (1985). Computing the variance of the forecast error for the Holt–Winters seasonal models. *J. Forecasting* **4**, 235–243.

Tiao, G. C. and Box, G. E. P. (1981). Modeling multiple time series with applications. *J. Am. Stat. Assoc.* **76**, 802–816.

Tiao, G. C. and Tsay, R. S. (1989). Model specification in multivariate time series (with discussion). *J. R. Stat. Soc. Ser. B* **51**, 157–213.

Tjostheim, D. and Paulsen, J. (1982). Empirical identification of multiple time series. *J. Time Series Anal.* **3**, 265–282.

Trigg, D. W. and Leach, A. G. (1967). Exponential smoothing with an adaptive response rate. *Oper. Res. Q.* **18**, 53–59.

Tsay, R. S. (1989). Identifying multivariate time series models. *J. Time Series Anal.* **10**, 357–372.

Tsay, R. S. and Tiao, G. C. (1984). Consistent estimates of autoregressive parameters and extended sample autocorrelation function for stationary and nonstationary ARIMA models. *J. Am. Stat. Assoc.* **79**, 84–96.

Tukey, J. W. (1979). *Exploratory Data Analysis*. Addison-Wesley, Reading, MA.

U.S. Bureau of the Census. (1969). *X-11 Information for the User*. U.S. Department of Commerce, Government Printing Office, Washington, DC.

Wei, W. W. S. (2006). *Time Series Analysis: Univariate and Multivariate, Methods*. Addison Wesley, New York.

West, M. and Harrison, J. (1997). *Bayesian Forecasting and Dynamic Models*, 2nd edition. Springer-Verlag, New York.

Wichern, D. W. and Jones, R. H. (1977). Assessing the impact of market disturbances using intervention analysis. *Manage. Sci.* **24**, 329–337.

Winkler, R. L. (2003). *Introduction to Bayesian Inference and Decision*, 2nd edition. Probabilistic Publishing, Sugar Land, Texas.

Winters, P. R. (1960). Forecasting sales by exponentially weighted moving averages. *Manage. Sci.* **6**, 235–239.

Wold, H. O. (1938). *A Study in the Analysis of Stationary Time Series*. Almqvist & Wiksell, Uppsala, Sweden. (Second edition 1954.)

Woodridge, J. M. (2011). *Introductory Econometrics: A Modern Approach*, 5th edition. Cengage South-Western, Independence, KY.

Yar, M. and Chatfield, C. (1990). Prediction intervals for the Holt–Winters' forecasting procedure. *Int. J. Forecasting* **6**, 127–137.

Yule, G. U. (1927). On a method of investigating periodicities in disturbed series, with reference to Wolfer's sunspot numbers. *Philos. Trans. R. Soc. London Ser. A* **226**, 267–298.

INDEX

Introduction to Time Series Analysis and Forecasting, Second Edition.
Douglas C. Montgomery, Cheryl L. Jennings and Murat Kulahci.
© 2015 John Wiley & Sons, Inc. Published 2015 by John Wiley & Sons, Inc.

WILEY SERIES IN PROBABILITY AND STATISTICS

ESTABLISHED BY WALTER A. SHEWHART AND SAMUEL S. WILKS

Editors: *David J. Balding, Noel A. C. Cressie, Garrett M. Fitzmaurice, Geof H. Givens, Harvey Goldstein, Geert Molenberghs, David W. Scott, Adrian F. M. Smith, Ruey S. Tsay, Sanford Weisberg*
Editors Emeriti: *J. Stuart Hunter, Iain M. Johnstone, Joseph B. Kadane, Jozef L. Teugels*

The *Wiley Series in Probability and Statistics* is well established and authoritative. It covers many topics of current research interest in both pure and applied statistics and probability theory. Written by leading statisticians and institutions, the titles span both state-of-the-art developments in the field and classical methods.

Reflecting the wide range of current research in statistics, the series encompasses applied, methodological and theoretical statistics, ranging from applications and new techniques made possible by advances in computerized practice to rigorous treatment of theoretical approaches.

This series provides essential and invaluable reading for all statisticians, whether in academia, industry, government, or research.

† ABRAHAM and LEDOLTER · Statistical Methods for Forecasting
  AGRESTI · Analysis of Ordinal Categorical Data, *Second Edition*
  AGRESTI · An Introduction to Categorical Data Analysis, *Second Edition*
  AGRESTI · Categorical Data Analysis, *Third Edition*
  AGRESTI · *Foundations of Linear and Generalized Linear Models*
  ALSTON, MENGERSEN and PETTITT (editors) · Case Studies in Bayesian
    Statistical Modelling and Analysis
  ALTMAN, GILL, and McDONALD · Numerical Issues in Statistical Computing for
    the Social Scientist
  AMARATUNGA and CABRERA · Exploration and Analysis of DNA Microarray and
    Protein Array Data
  AMARATUNGA, CABRERA, and SHKEDY · Exploration and Analysis of DNA
    Microarray and Other High-Dimensional Data, *Second Edition*
  ANDĚL · Mathematics of Chance
  ANDERSON · An Introduction to Multivariate Statistical Analysis, *Third Edition*
* ANDERSON · The Statistical Analysis of Time Series
  ANDERSON, AUQUIER, HAUCK, OAKES, VANDAELE, and
    WEISBERG · Statistical Methods for Comparative Studies
  ANDERSON and LOYNES · The Teaching of Practical Statistics
  ARMITAGE and DAVID (editors) · Advances in Biometry
  ARNOLD, BALAKRISHNAN, and NAGARAJA · Records
* ARTHANARI and DODGE · Mathematical Programming in Statistics
  AUGUSTIN, COOLEN, DE COOMAN and TROFFAES (editors) · Introduction to
    Imprecise Probabilities
* BAILEY · The Elements of Stochastic Processes with Applications to the Natural
    Sciences

*Now available in a lower priced paperback edition in the Wiley Classics Library.
†Now available in a lower priced paperback edition in the Wiley–Interscience Paperback Series.

*Now available in a lower priced paperback edition in the Wiley Classics Library.

†Now available in a lower priced paperback edition in the Wiley–Interscience Paperback
Series.

BOX and DRAPER · Response Surfaces, Mixtures, and Ridge Analyses, *Second Edition*

BOX, HUNTER, and HUNTER · Statistics for Experimenters: Design, Innovation, and Discovery, *Second Editon*

BOX, JENKINS, and REINSEL · Time Series Analysis: Forecasting and Control, *Fourth Edition*

BOX, LUCEÑO, and PANIAGUA-QUIÑONES · Statistical Control by Monitoring and Adjustment, *Second Edition*

* BROWN and HOLLANDER · Statistics: A Biomedical Introduction

CAIROLI and DALANG · Sequential Stochastic Optimization

CASTILLO, HADI, BALAKRISHNAN, and SARABIA · Extreme Value and Related Models with Applications in Engineering and Science

CHAN · Time Series: Applications to Finance with R and S-Plus®, *Second Edition*

CHARALAMBIDES · Combinatorial Methods in Discrete Distributions

CHATTERJEE and HADI · Regression Analysis by Example, *Fourth Edition*

CHATTERJEE and HADI · Sensitivity Analysis in Linear Regression

CHEN · The Fitness of Information: Quantitative Assessments of Critical Evidence

CHERNICK · Bootstrap Methods: A Guide for Practitioners and Researchers, *Second Edition*

CHERNICK and FRIIS · Introductory Biostatistics for the Health Sciences

CHILES and DELFINER · Geostatistics: Modeling Spatial Uncertainty, *Second Edition*

CHIU, STOYAN, KENDALL and MECKE · Stochastic Geometry and Its Applications, *Third Edition*

CHOW and LIU · Design and Analysis of Clinical Trials: Concepts and Methodologies, *Third Edition*

CLARKE · Linear Models: The Theory and Application of Analysis of Variance

CLARKE and DISNEY · Probability and Random Processes: A First Course with Applications, *Second Edition*

* COCHRAN and COX · Experimental Designs, *Second Edition*

COLLINS and LANZA · Latent Class and Latent Transition Analysis: With Applications in the Social, Behavioral, and Health Sciences

CONGDON · Applied Bayesian Modelling, *Second Edition*

CONGDON · Bayesian Models for Categorical Data

CONGDON · Bayesian Statistical Modelling, *Second Edition*

CONOVER · Practical Nonparametric Statistics, *Third Edition*

COOK · Regression Graphics

COOK and WEISBERG · An Introduction to Regression Graphics

COOK and WEISBERG · Applied Regression Including Computing and Graphics

CORNELL · A Primer on Experiments with Mixtures

CORNELL · Experiments with Mixtures, Designs, Models, and the Analysis of Mixture Data, *Third Edition*

COX · A Handbook of Introductory Statistical Methods

CRESSIE · Statistics for Spatial Data, *Revised Edition*

CRESSIE and WIKLE · Statistics for Spatio-Temporal Data

CSÖRGŐ and HORVÁTH · Limit Theorems in Change Point Analysis

DAGPUNAR · Simulation and Monte Carlo: With Applications in Finance and MCMC

DANIEL · Applications of Statistics to Industrial Experimentation

DANIEL · Biostatistics: A Foundation for Analysis in the Health Sciences, *Eighth Edition*

*Now available in a lower priced paperback edition in the Wiley Classics Library.

*Now available in a lower priced paperback edition in the Wiley Classics Library.
†Now available in a lower priced paperback edition in the Wiley–Interscience Paperback Series.

GEWEKE · Contemporary Bayesian Econometrics and Statistics
GHOSH, MUKHOPADHYAY, and SEN · Sequential Estimation
GIESBRECHT and GUMPERTZ · Planning, Construction, and Statistical Analysis of Comparative Experiments
GIFI · Nonlinear Multivariate Analysis
GIVENS and HOETING · Computational Statistics
GLASSERMAN and YAO · Monotone Structure in Discrete-Event Systems
GNANADESIKAN · Methods for Statistical Data Analysis of Multivariate Observations, *Second Edition*
GOLDSTEIN · Multilevel Statistical Models, *Fourth Edition*
GOLDSTEIN and LEWIS · Assessment: Problems, Development, and Statistical Issues
GOLDSTEIN and WOOFF · Bayes Linear Statistics
GRAHAM · Markov Chains: Analytic and Monte Carlo Computations
GREENWOOD and NIKULIN · A Guide to Chi-Squared Testing
GROSS, SHORTLE, THOMPSON, and HARRIS · Fundamentals of Queueing Theory, *Fourth Edition*
GROSS, SHORTLE, THOMPSON, and HARRIS · Solutions Manual to Accompany Fundamentals of Queueing Theory, *Fourth Edition*
* HAHN and SHAPIRO · Statistical Models in Engineering
HAHN and MEEKER · Statistical Intervals: A Guide for Practitioners
HALD · A History of Probability and Statistics and their Applications Before 1750
† HAMPEL · Robust Statistics: The Approach Based on Influence Functions
HARTUNG, KNAPP, and SINHA · Statistical Meta-Analysis with Applications
HEIBERGER · Computation for the Analysis of Designed Experiments
HEDAYAT and SINHA · Design and Inference in Finite Population Sampling
HEDEKER and GIBBONS · Longitudinal Data Analysis
HELLER · MACSYMA for Statisticians
HERITIER, CANTONI, COPT, and VICTORIA-FESER · Robust Methods in Biostatistics
HINKELMANN and KEMPTHORNE · Design and Analysis of Experiments, Volume 1: Introduction to Experimental Design, *Second Edition*
HINKELMANN and KEMPTHORNE · Design and Analysis of Experiments, Volume 2: Advanced Experimental Design
HINKELMANN (editor) · Design and Analysis of Experiments, Volume 3: Special Designs and Applications
HOAGLIN, MOSTELLER, and TUKEY · Fundamentals of Exploratory Analysis of Variance
* HOAGLIN, MOSTELLER, and TUKEY · Exploring Data Tables, Trends and Shapes
* HOAGLIN, MOSTELLER, and TUKEY · Understanding Robust and Exploratory Data Analysis
HOCHBERG and TAMHANE · Multiple Comparison Procedures
HOCKING · Methods and Applications of Linear Models: Regression and the Analysis of Variance, *Third Edition*
HOEL · Introduction to Mathematical Statistics, *Fifth Edition*
HOGG and KLUGMAN · Loss Distributions
HOLLANDER, WOLFE, and CHICKEN · Nonparametric Statistical Methods, *Third Edition*
HOSMER and LEMESHOW · Applied Logistic Regression, *Second Edition*

*Now available in a lower priced paperback edition in the Wiley Classics Library.
†Now available in a lower priced paperback edition in the Wiley–Interscience Paperback Series.

*Now available in a lower priced paperback edition in the Wiley Classics Library.
†Now available in a lower priced paperback edition in the Wiley–Interscience Paperback Series.

KLEIBER and KOTZ · Statistical Size Distributions in Economics and Actuarial
  Sciences
KLEMELÄ · Smoothing of Multivariate Data: Density Estimation and Visualization
KLUGMAN, PANJER, and WILLMOT · Loss Models: From Data to Decisions,
  Third Edition
KLUGMAN, PANJER, and WILLMOT · Loss Models: Further Topics
KLUGMAN, PANJER, and WILLMOT · Solutions Manual to Accompany Loss
  Models: From Data to Decisions, *Third Edition*
KOSKI and NOBLE · Bayesian Networks: An Introduction
KOTZ, BALAKRISHNAN, and JOHNSON · Continuous Multivariate Distributions,
  Volume 1, *Second Edition*
KOTZ and JOHNSON (editors) · Encyclopedia of Statistical Sciences: Volumes 1 to 9
  with Index
KOTZ and JOHNSON (editors) · Encyclopedia of Statistical Sciences: Supplement
  Volume
KOTZ, READ, and BANKS (editors) · Encyclopedia of Statistical Sciences: Update
  Volume 1
KOTZ, READ, and BANKS (editors) · Encyclopedia of Statistical Sciences: Update
  Volume 2
KOWALSKI and TU · Modern Applied U-Statistics
KRISHNAMOORTHY and MATHEW · Statistical Tolerance Regions: Theory,
  Applications, and Computation
KROESE, TAIMRE, and BOTEV · Handbook of Monte Carlo Methods
KROONENBERG · Applied Multiway Data Analysis
KULINSKAYA, MORGENTHALER, and STAUDTE · Meta Analysis: A Guide to
  Calibrating and Combining Statistical Evidence
KULKARNI and HARMAN · An Elementary Introduction to Statistical Learning
  Theory
KUROWICKA and COOKE · Uncertainty Analysis with High Dimensional
  Dependence Modelling
KVAM and VIDAKOVIC · Nonparametric Statistics with Applications to Science and
  Engineering
LACHIN · Biostatistical Methods: The Assessment of Relative Risks, *Second Edition*
LAD · Operational Subjective Statistical Methods: A Mathematical, Philosophical,
  and Historical Introduction
LAMPERTI · Probability: A Survey of the Mathematical Theory, *Second Edition*
LAWLESS · Statistical Models and Methods for Lifetime Data, *Second Edition*
LAWSON · Statistical Methods in Spatial Epidemiology, *Second Edition*
LE · Applied Categorical Data Analysis, *Second Edition*
LE · Applied Survival Analysis
LEE · Structural Equation Modeling: A Bayesian Approach
LEE and WANG · Statistical Methods for Survival Data Analysis, *Fourth Edition*
LEPAGE and BILLARD · Exploring the Limits of Bootstrap
LESSLER and KALSBEEK · Nonsampling Errors in Surveys
LEYLAND and GOLDSTEIN (editors) · Multilevel Modelling of Health Statistics
LIAO · Statistical Group Comparison
LIN · Introductory Stochastic Analysis for Finance and Insurance
LINDLEY · Understanding Uncertainty, *Revised Edition*
LITTLE and RUBIN · Statistical Analysis with Missing Data, *Second Edition*
LLOYD · The Statistical Analysis of Categorical Data
LOWEN and TEICH · Fractal-Based Point Processes
MAGNUS and NEUDECKER · Matrix Differential Calculus with Applications in
  Statistics and Econometrics, *Revised Edition*

*Now available in a lower priced paperback edition in the Wiley Classics Library.
†Now available in a lower priced paperback edition in the Wiley–Interscience Paperback Series.

PANKRATZ · Forecasting with Dynamic Regression Models

PANKRATZ · Forecasting with Univariate Box-Jenkins Models: Concepts and Cases

PARDOUX · Markov Processes and Applications: Algorithms, Networks, Genome and Finance

PARMIGIANI and INOUE · Decision Theory: Principles and Approaches

* PARZEN · Modern Probability Theory and Its Applications

PEÑA, TIAO, and TSAY · A Course in Time Series Analysis

PESARIN and SALMASO · Permutation Tests for Complex Data: Applications and Software

PIANTADOSI · Clinical Trials: A Methodologic Perspective, *Second Edition*

POURAHMADI · Foundations of Time Series Analysis and Prediction Theory

POURAHMADI · High-Dimensional Covariance Estimation

POWELL · Approximate Dynamic Programming: Solving the Curses of Dimensionality, *Second Edition*

POWELL and RYZHOV · Optimal Learning

PRESS · Subjective and Objective Bayesian Statistics, *Second Edition*

PRESS and TANUR · The Subjectivity of Scientists and the Bayesian Approach

PURI, VILAPLANA, and WERTZ · New Perspectives in Theoretical and Applied Statistics

† PUTERMAN · Markov Decision Processes: Discrete Stochastic Dynamic Programming

QIU · Image Processing and Jump Regression Analysis

* RAO · Linear Statistical Inference and Its Applications, *Second Edition*

RAO · Statistical Inference for Fractional Diffusion Processes

RAUSAND and HØYLAND · System Reliability Theory: Models, Statistical Methods, and Applications, *Second Edition*

RAYNER, THAS, and BEST · Smooth Tests of Goodnes of Fit: Using R, *Second Edition*

RENCHER and SCHAALJE · Linear Models in Statistics, *Second Edition*

RENCHER and CHRISTENSEN · Methods of Multivariate Analysis, *Third Edition*

RENCHER · Multivariate Statistical Inference with Applications

RIGDON and BASU · Statistical Methods for the Reliability of Repairable Systems

* RIPLEY · Spatial Statistics

* RIPLEY · Stochastic Simulation

ROHATGI and SALEH · An Introduction to Probability and Statistics, *Second Edition*

ROLSKI, SCHMIDLI, SCHMIDT, and TEUGELS · Stochastic Processes for Insurance and Finance

ROSENBERGER and LACHIN · Randomization in Clinical Trials: Theory and Practice

ROSSI, ALLENBY, and McCULLOCH · Bayesian Statistics and Marketing

† ROUSSEEUW and LEROY · Robust Regression and Outlier Detection

ROYSTON and SAUERBREI · Multivariate Model Building: A Pragmatic Approach to Regression Analysis Based on Fractional Polynomials for Modeling Continuous Variables

* RUBIN · Multiple Imputation for Nonresponse in Surveys

RUBINSTEIN and KROESE · Simulation and the Monte Carlo Method, *Second Edition*

RUBINSTEIN and MELAMED · Modern Simulation and Modeling

RUBINSTEIN, RIDDER, and VAISMAN · Fast Sequential Monte Carlo Methods for Counting and Optimization

*Now available in a lower priced paperback edition in the Wiley Classics Library.
†Now available in a lower priced paperback edition in the Wiley–Interscience Paperback Series.

RYAN · Modern Engineering Statistics

RYAN · Modern Experimental Design

RYAN · Modern Regression Methods, *Second Edition*

RYAN · Sample Size Determination and Power

RYAN · Statistical Methods for Quality Improvement, *Third Edition*

SALEH · Theory of Preliminary Test and Stein-Type Estimation with Applications

SALTELLI, CHAN, and SCOTT (editors) · Sensitivity Analysis

SCHERER · Batch Effects and Noise in Microarray Experiments: Sources and Solutions

* SCHEFFE · The Analysis of Variance

SCHIMEK · Smoothing and Regression: Approaches, Computation, and Application

SCHOTT · Matrix Analysis for Statistics, *Second Edition*

SCHOUTENS · Levy Processes in Finance: Pricing Financial Derivatives

SCOTT · Multivariate Density Estimation: Theory, Practice, and Visualization

* SEARLE · Linear Models

† SEARLE · Linear Models for Unbalanced Data

† SEARLE · Matrix Algebra Useful for Statistics

† SEARLE, CASELLA, and McCULLOCH · Variance Components

SEARLE and WILLETT · Matrix Algebra for Applied Economics

SEBER · A Matrix Handbook For Statisticians

† SEBER · Multivariate Observations

SEBER and LEE · Linear Regression Analysis, *Second Edition*

† SEBER and WILD · Nonlinear Regression

SENNOTT · Stochastic Dynamic Programming and the Control of Queueing Systems

* SERFLING · Approximation Theorems of Mathematical Statistics

SHAFER and VOVK · Probability and Finance: It's Only a Game!

SHERMAN · Spatial Statistics and Spatio-Temporal Data: Covariance Functions and Directional Properties

SILVAPULLE and SEN · Constrained Statistical Inference: Inequality, Order, and Shape Restrictions

SINGPURWALLA · Reliability and Risk: A Bayesian Perspective

SMALL and MCLEISH · Hilbert Space Methods in Probability and Statistical Inference

SRIVASTAVA · Methods of Multivariate Statistics

STAPLETON · Linear Statistical Models, *Second Edition*

STAPLETON · Models for Probability and Statistical Inference: Theory and Applications

STAUDTE and SHEATHER · Robust Estimation and Testing

STOYAN · Counterexamples in Probability, *Second Edition*

STOYAN and STOYAN · Fractals, Random Shapes and Point Fields: Methods of Geometrical Statistics

STREET and BURGESS · The Construction of Optimal Stated Choice Experiments: Theory and Methods

STYAN · The Collected Papers of T. W. Anderson: 1943-1985

SUTTON, ABRAMS, JONES, SHELDON, and SONG · Methods for Meta- Analysis in Medical Research

TAKEZAWA · Introduction to Nonparametric Regression

TAMHANE · Statistical Analysis of Designed Experiments: Theory and Applications

*Now available in a lower priced paperback edition in the Wiley Classics Library.

†Now available in a lower priced paperback edition in the Wiley–Interscience Paperback Series.

TANAKA · Time Series Analysis: Nonstationary and Noninvertible Distribution
   Theory
THOMPSON · Empirical Model Building: Data, Models, and Reality, *Second Edition*
THOMPSON · Sampling, *Third Edition*
THOMPSON · Simulation: A Modeler's Approach
THOMPSON and SEBER · Adaptive Sampling
THOMPSON, WILLIAMS, and FINDLAY · Models for Investors in Real World
   Markets
TIERNEY · LISP-STAT: An Object-Oriented Environment for Statistical Computing
   and Dynamic Graphics
TROFFAES and DE COOMAN · Lower Previsions
TSAY · Analysis of Financial Time Series, *Third Edition*
TSAY · An Introduction to Analysis of Financial Data with R
TSAY · Multivariate Time Series Analysis: With R and Financial Applications
UPTON and FINGLETON · Spatial Data Analysis by Example, Volume II:
   Categorical and Directional Data
† VAN BELLE · Statistical Rules of Thumb, *Second Edition*
VAN BELLE, FISHER, HEAGERTY, and LUMLEY · Biostatistics: A Methodology
   for the Health Sciences, *Second Edition*
VESTRUP · The Theory of Measures and Integration
VIDAKOVIC · Statistical Modeling by Wavelets
VIERTL · Statistical Methods for Fuzzy Data
VINOD and REAGLE · Preparing for the Worst: Incorporating Downside Risk in
   Stock Market Investments
WALLER and GOTWAY · Applied Spatial Statistics for Public Health Data
WEISBERG · Applied Linear Regression, *Fourth Edition*
WEISBERG · Bias and Causation: Models and Judgment for Valid Comparisons
WELSH · Aspects of Statistical Inference
WESTFALL and YOUNG · Resampling-Based Multiple Testing: Examples and
   Methods for p-Value Adjustment
* WHITTAKER · Graphical Models in Applied Multivariate Statistics
WINKER · Optimization Heuristics in Economics: Applications of Threshold
   Accepting
WOODWORTH · Biostatistics: A Bayesian Introduction
WOOLSON and CLARKE · Statistical Methods for the Analysis of Biomedical Data,
   Second Edition
WU and HAMADA · Experiments: Planning, Analysis, and Parameter Design
   Optimization, *Second Edition*
WU and ZHANG · Nonparametric Regression Methods for Longitudinal Data Analysis
YAKIR · Extremes in Random Fields
YIN · Clinical Trial Design: Bayesian and Frequentist Adaptive Methods
YOUNG, VALERO-MORA, and FRIENDLY · Visual Statistics: Seeing Data with
   Dynamic Interactive Graphics
ZACKS · Examples and Problems in Mathematical Statistics
ZACKS · Stage-Wise Adaptive Designs
* ZELLNER · An Introduction to Bayesian Inference in Econometrics
ZELTERMAN · Discrete Distributions—Applications in the Health Sciences
ZHOU, OBUCHOWSKI, and MCCLISH · Statistical Methods in Diagnostic
   Medicine, *Second Edition*

*Now available in a lower priced paperback edition in the Wiley Classics Library.
†Now available in a lower priced paperback edition in the Wiley–Interscience Paperback
Series.